Soil factors in crop production in a semi-arid environment

Soil factors in crop production in a semi-arid environment

Edited by
J.S. Russell and E.L. Greacen

University of Queensland Press

in association with the Australian Society of Soil Science Incorporated

Typeset, printed and bound by Academy Press Pty. Ltd., Brisbane

Distributed in the United Kingdom, Europe, the Middle East, Africa, and the Caribbean by Prentice-Hall International, International Book Distributors Ltd., 66 Wood Lane End, Hemel Hempstead, Herts., England

National Library of Australia
Cataloguing-in-publication data

Soil factors in crop production in a semi-arid environment.

Index.
ISBN 0 7022 1303 9.

1. Desert soils — Addresses, essays, lectures.
I. Russell, John Samuel, ed. II. Greacen, Emmet Lewis, joint ed. III. Australian Society of Soil Science.

631.4

Contents

Contributors

Numbers in parenthesis indicate the pages on which the authors' contributions begin

BOWEN, G.D. (197) Division of Soils, C.S.I.R.O., Adelaide, South Australia.
BUTLER, B.E. (9) Division of Soils, C.S.I.R.O., Canberra, A.C.T.
CARTWRIGHT, B. (197) Division of Soils, C.S.I.R.O., Adelaide, South Australia.
CLARKE, A.L. (279) Wheat Research Institute, Toowoomba, Queensland.
COLWELL, J.D. (105) Division of Soils, C.S.I.R.O., Canberra, A.C.T.
EMERSON, W.W. (78) Division of Soils, C.S.I.R.O., Adelaide, South Australia.
GREACEN, E.L. (163) Division of Soils, C.S.I.R.O., Adelaide, South Australia.
HUBBLE, G.D. (9) Division of Soils, C.S.I.R.O., St. Lucia, Queensland.
LADD, J.N. (127) Division of Soils, C.S.I.R.O., Adelaide, South Australia.
MILLINGTON, R.J. (224) Division of Land Use Research, C.S.I.R.O., Canberra, A.C.T.
NORRISH, K. (33) Division of Soils, C.S.I.R.O., Adelaide, South Australia.
OADES, J.M. (127) Waite Agricultural Research Institute, University of Adelaide, South Australia.
PECK, A.J. (301) Division of Land Resources Management, C.S.I.R.O., Nedlands, Western Australia.
PICKERING, J.G. (33) Division of Soils, C.S.I.R.O., Adelaide, South Australia.
ROSE, C.W. (224) Department of Environmental Science, Griffith University, Nathan, Queensland.
RUSSELL, J.S. (3,279) Division of Tropical Crops and Pastures, C.S.I.R.O., St. Lucia, Queensland.
SIMS, H.J. (243) Victorian Department of Agriculture, Melbourne, Victoria.
SMILES, D.E. (54) Division of Environmental Mechanics, C.S.I.R.O., Canberra, A.C.T.
SMITH, A.N. (262) Agricultural Research Institute, New South Wales Department of Agriculture, Wagga Wagga, New South Wales.
WILLIAMS, C.H. (105) Division of Plant Industry, C.S.I.R.O., Canberra, A.C.T.

Foreword

Much of Australian agriculture is practised in a semi-arid environment characterized not only by low levels of annual rainfall but also by marked dry seasons and considerable variability from year to year. While water is considered a scarce resource, and much of our soil management is aimed at its conservation, nevertheless the agricultural systems which have evolved in Australia tend to be based on plants able to complete their cycle of growth, flowering and seed setting before the exhaustion of soil moisture reserves. This is not always commensurate with full utilization of the water available, particularly in the wetter periods of the year, and indeed usage by agricultural systems may be less than that by the native vegetation which it replaces.

Thus soil management must also take account of the environmental effects induced by changes in patterns of water use brought about by agricultural development. This topic is considered in the last chapter of this monograph, but it is likely to be one to which Australian soil scientists may well need to give greater attention in the future.

On behalf of the membership of the Society I should like to express our great appreciation of the actions taken by Federal Council in 1969 in order to initiate the production of this publication, and of the work of the Monograph Committee under Mr. G.D. Hubble as convener in managing the project. The editors, Dr. J.S. Russell and Dr. E.L. Greacen, as well as the authors, are to be congratulated on the production of a volume which will bring credit to the Society and to Australian soil science.

M.J. Mulcahy
Federal President,
Australian Society of Soil Science Incorporated
April 1975

Preface

Since its beginning, the Australian Society of Soil Science Incorporated has been interested in developing publications to promote both the advancement of soil science and the scientific standing of the Society. Largely on its initiative, the *Australian Journal of Soil Research* was established in 1963, but the Society now has only a minimal role in its publication. The present book is its first venture into substantial publishing, integrating the work of a number of authors over a wide field.

The first moves in this project were made in 1969 when the then Federal Council of the Society appointed a monograph committee. This body, comprising Dr. J.K. Leslie, Dr. J.S. Russell, and myself was asked to report on feasibility, to suggest a subject and general outline, and to submit proposals for publication. While its enquiries were proceeding, the Organizing Committee for the 9th International Congress of Soil Science, through Dr. Hallsworth, president of the Congress, transferred $3,000 of surplus Congress funds to the Society for the project. This gift made publication financially possible. The Monograph Committee reported to Council in May 1970, its recommendations were formally adopted, and it was made responsible for managing the project. Drs. J.S. Russell and E.L. Greacen were subsequently appointed editors and secured contributors.

The project has depended, throughout, on the generous help of many individuals. Mr. G.H. Burvill, Professor C.M. Donald, Dr. E.G. Hallsworth, Mr. J.K.M. Skene, and Professor R.O. Slatyer served as consultant members of the committee, commenting and offering suggestions on proposals. The authors have given much time and effort to the preparation of their chapters; and Drs. A.D. Rovira, A.J. Peck, D.S. McIntyre, J.W. Holmes, A.E. Martin, R.J. Gilkes, J.W. Parsons, T.J. Marshall and K. Norrish, and Messrs. G.B. Stirk, P.J. Ross, N.J. Barrow, K.P. Barley, G. Blackburn, H.A. Nix, and G.D. Hubble painstakingly refereed the manuscripts. The Committee gratefully acknowledges and thanks all of these persons for their contributions and excellent cooperation.

Finally, the Society is greatly indebted to the editors Dr. J.S. Russell and Dr. E.L. Greacen for their persistent effort and good management in bringing the project to fruition.

G.D. Hubble
Convener, Monograph Committee

Part A

Introduction

CHAPTER 1

Introduction

J.S. Russell

The soil is a significant component of arid ecosystems. While climatic factors and their variation are important in the definition of aridity, plant growth in arid areas involves a climate-soil-plant interaction. Permanent crop or pasture production and ecosystem stability are difficult to obtain in most environments disturbed by man. But in semi-arid areas special problems arise due to the low and variable rainfall, the limited options available for increasing yields and the lack of diversity of species that can be grown.

The significant role of the soil in the climate-soil-plant interaction emphasizes the need for a fuller scientific understanding of the soil and its various interactions. Though it is true that the potential both for manipulation of the soil and for inputs to the soil is usually low in dry areas there is nevertheless the possibility of selecting the most suitable soils for given land uses and of devising practices which maximize plant yield and which are consistent with the maintenance of ecosystem stability.

Large areas of the world can be classified as arid. The arid and extremely arid areas—the deserts—occupy significant areas of all the continents except Europe. Between these deserts and the more humid areas lie the semi-arid lands. In Africa and Australia in particular these areas are huge crescents fringing the deserts. In Africa the main desert is the Sahara. The Australian deserts have no single name but are a grouping of many deserts such as the Great Sandy Desert, Gibson Desert, Great Victoria Desert, Nullarbor Plain and the Simpson Desert.

The definition of degrees of aridity such as arid and semi-arid usually depends upon climatic criteria and the areas so classified will vary with the limiting criteria used. Meigs (1953) presented a world classification in which he recognized three main groupings, extremely arid, arid and semi-arid. Areas were classified as extremely arid if rainfall records showed at least one complete year without rain. The boundary between arid and semi-arid areas was based on the rainfall-evaporation index of Thornthwaite (1948). However an important distinction between arid and semi-arid areas was that, in the latter, rainfall was considered sufficient for certain crops and cultivation was possible. This has also been considered as a major distinction between arid and semi-arid in this book rather than any formal climatic criteria.

The arid areas not only cover a large part of the earth's surface, they also support a significant number of the world's population. It has been estimated (Amiran 1966) that the arid zone comprises 36 per cent of the world's land surface. Of this slightly less than half (16 per cent) can be classified as semi-arid. Some 380 million people live in the arid zone and of this 72 per cent or 280 million live in the semi-arid areas. Although some of these people live on irrigated lands which derive their water from more humid areas most would be supported from water originating from within their environment.

Land use in semi-arid areas poses special problems. The rainfall is not only low, it is also usually quite variable. Moreover an additional phenomenon, which can be described as medium-term climatic variation, occurs. With this phenomenon above-average wet periods and below-average dry periods occur. These fluctuations from the long-term average, which may have only minor effects in humid areas, can have quite major effects in semi-arid areas. They are particularly important when new areas or crops are in a pioneering phase. There is frequently an extension of more intensive land use in periods of above-average rainfall years into drier areas leading to a subsequent retreat in periods of average or below-average rainfall years. Meinig (1963) has described this process for the northern limits of cultivation in South Australia in his book *On the Margins of the Good Earth* and the human and societal effects of such changes brought about essentially by medium-term climatic variation and man's optimistic desire to extend cultivation as much as possible.

One of the consequences of medium-term variation in rainfall is that the boundary between areas where cultivation of the soil and economic crop growth is possible and areas where it is not possible is continually changing. In broad terms the boundary is between the areas where agronomic and ecologic practices are feasible. Thus in periods of above-average rainfall the use of agronomic practices such as tilling the soil can advance into the drier areas. But in extended periods of below-average rainfall years continuation of such practices may not only be uneconomic they may result in soil deterioration and loss.

During these long dry periods the importance of soil may be brought to the notice of those living beyond the semi-arid areas. The major losses of soil through wind erosion in the Great Plains of North America in the 1930s created an awareness of the importance of sound soil management practices and led to scientific research into factors affecting soil erosion and fertility. Similar spectacular effects were also observed in Australia during the late 1930s and 1940s. At one stage the increase in wind erosion and the apparent threat posed by large areas of drifting sands led to investigation as to the possible effects on the better watered areas (Ratcliffe 1947).

Generally speaking the drier limit of the semi-arid lands discussed in this book has been taken to be those areas in southern Australia where wheat can be grown and in northern Australia where grain sorghum culture is possible. Also the areas where these crops are grown or could be grown on suitable soils approximate the semi-arid areas as a whole. In climatic terms the limiting climatic criteria of the wheat-growing areas in southern Australia are similar to those used by Meigs, that is a Thornthwaite index of more than 20 and less than 40. However in eastern and northern Australia these criteria are not satisfactory to define the limits of crop growth. In northern Australia it is likely that the sorghum-growing areas are defined by values of the Thornthwaite index closer to 15 to 35 than those used by Meigs. Expressed in annual rainfall terms the drier limit of cultivation in southern Australia is as low as 250 mm. With decreasing latitude this value increases and in northern Australia the limiting value approaches 750 mm.

Cropping of the semi-arid areas of Australia has been practised for more than 120 years. Although early agriculture began in the Sydney area of New South Wales and in Tasmania the first large-scale attempts at growing crops in a semi-arid environment began in the northward spread of settlement from Adelaide in South Australia in the 1840s and 1850s with cultivation of the red-brown earths for winter cereals, mainly wheat. These initial attempts were frequently painful for the farmer and damaging to the soil and the environment. However there has been a gradual evolution of practice with time. This evolution is most advanced in the winter rainfall areas of southern Australia. For these areas it is possible to state that practices which can maintain soil productivity and stability are available for most environments even if they are not always used. Also the drier limits of the cropping boundary have become stabilized. But in the northern semi-arid subtropical areas where cultivation of sorghum is relatively new, land use is still evolving and practices which can maintain ecosystem stability as well as crop yields have not yet been achieved. In addition well-defined cropping boundaries have yet to be established.

Much of the soil-plant research that has taken place in Australia has occurred in a semi-arid environment. There has been a feedback between the environment and the type of research undertaken. It is significant for example that some of the most satisfactory water balance models at this point in time are Australian models. There is in fact much less interest in water balance in Europe and North America. Thus a unique feature of much Australian soil-plant research is that it operates in an environment where water is limiting and where the soils are old and frequently of low fertility. For most semi-arid areas irrigation is not an available option and research must be aware that improved land use has to operate within the prevailing climatic environment.

Scientific research of semi-arid soils is frequently more difficult than on soils in humid areas. This is especially so with field studies involving soil-plant interactions. Plant yields are usually low and the annual variation is high. In comparison with wetter areas work over a longer period of time is usually necessary to sample the variable climatic environment and to achieve equivalent results. Also the use of large inputs is not economically possible in practice and this acts as a constraint on the type of research undertaken.

This book was seen as a means of gathering together a large amount of information about semi-arid soils in general but particularly covering Australian soils. It was seen as presenting current concepts and accumulated knowledge to the present with the integration of past work and pointing in the future to areas of work where present techniques are likely to make a contribution. It attempts to emphasize the importance of soil science in contributing to our understanding of ecosystems in semi-arid areas.

The first part of the book is concerned with the properties of semi-arid soils. These are considered from their morphological, mineralogical, physical and chemical aspects. This section is more than an inventory and attempts to achieve a quantitative basis on which the climate and plant interactions can be understood.

The second part of the book looks at first order interactions, that is between soils and both climate and plants. This section examines mechanisms and processes. It looks at the possibilities of systems analysis, simulation and modelling in understanding these systems.

The third part of the book looks at second order interactions, that is between soils, climate, plants and man. This assesses the significant practices that man can use in growing plants on semi-arid soils such as cultivation, fertilization and crop sequential practices and examines their effects on ecosystem stability in particular and on the long-term yield and soil trends. The evolutionary change of fallowing practices with time is also examined in an historical and scientific context.

It is hoped that the contributions in this book will add to a further understanding of the climate-soil-plant-man interaction in semi-arid areas. The limited water resources in these areas and the delicate and fragile balance that is maintained in nature emphasize the need for careful soil management. Throughout most of these areas man is not likely to be able to alter crop yields spectacularly as has been achieved in some arid areas through irrigation or in humid areas through the use of large amounts of nutrients. Such improvements as are made are likely to be achieved slowly through an improved understanding of the various components of the ecosystem, including the soil, and how they interact together.

REFERENCES

Amiran, D.H.K. (1966). In "Arid Lands: A Geographical Appraisal". (Ed. E.S. Hills). pp. 219-31. Methuen, London.

Meigs, P. (1953). *Reviews of Research on Arid Zone Hydrology*. Vol. I, 203-10. UNESCO, Paris.

Meinig, R.W. (1963). "On the Margins of the Good Earth". J. Murray, London.

Ratcliffe, F. (1947). "Flying Fox and Drifting Sand". Angus and Robertson, Sydney.

Thornthwaite, C.W. (1948). *Geog. Rev.* **38**, 55-94.

Part B

Properties of semi-arid soils in Australia

CHAPTER 2

Morphologic properties

B.E. Butler and G.D. Hubble

INTRODUCTION

The main concern of this chapter is soil morphology and its role as a factor in crop production in semi-arid regions. It is necessary at an early stage to define the problem in environmental terms and, because of interactions between soil and climatic factors, to subdivide the climatic range. The role of several morphological categories in moderating conditions for growth is examined, and five broad morphologic classes of soil are used for discussion of the data in the literature on crop production and soil morphology relationships. Only rain-fed cropping is considered; irrigation agriculture, important in many areas, is beyond the scope of this examination.

THEORY OF SOIL MORPHOLOGY IN RELATION TO SOIL USE

Though morphologic concepts and practices are widely used in soil science and technology there is little by way of general theory of the place of morphology in soil science. It seems necessary as a prelude to the specific task of this chapter to state this theory.

Morphology is the study or science of form. In soil science it is the study of those properties that can be sensed by sight or handling. These are categorized as colour, texture, structure, consistence, thickness of horizons and of the soil as a whole, the nature of the boundaries between horizons, and the presence of concretions and other accessory materials.

The morphological properties of soils are unique in the ease with which they are determined. It is because of this easy accessibility that morphology has become the basis for the most common identification of soils as well as much of the language for their discussion. Nevertheless, other properties are of more importance in determining their use or in expressing their essential character. Instances of important soil properties for use could be cited as water storage capacity and available phosphorus for semi-arid cropping, or the nature of the exchange complex and the associated cations for expressing the important intrinsic character of a soil. These properties, however, require comprehensive laboratory measurement and as a consequence such data are always in short supply, and there is a strong tendency to use the morphological properties as substitutes or extenders in identifying soils with required use or intrinsic properties.

This use of morphology as substitute or surrogate data can be made where a correlation has been established, but it would not be sound to assume a correlation, or to assume that a correlation established in one locality would apply in another locality. However, the question of correlation among soil properties is not very well explored, and generalizations must be tentative. Webster and Butler (1976) examined the association among individual soil properties of 110 profiles from 400 ha near Canberra and found very poor associations between different soil properties. Norris (1971) in a multivariate study of 146 profiles from 190,000 ha near Griffith found that though a taxonomy based on all morphology showed considerable correspondence with a taxonomy based on all laboratory data from the same soils, the allocation of soils to classes on the basis of individual properties was often poor. Yet it is on the status of individual properties that success in land use may depend.

In spite of difficulties with validation it is essential to make the fullest possible use of morphological data; the alterntive, of operating solely on analytical properties, is unacceptable.

The problems of the validity of morphology-land use relationships may be grouped according to their nature:

(i) Relationships involving morphological properties themselves, as for example surface crusts preventing seedling emergence. These cases of *direct involvement* present no problem, though calibration in the cut-off range is needed.

(ii) Relationships where there is a *logical connection* between the morphological properties and the use characteristic, but the particular relationship has not been worked out. An example would be the putative connection between soil texture, pore size distribution, and the moisture characteristic. This limits the usefulness of morphology to our understanding of it, and here not much progress has been made.

(iii) Relationships where there is an *empirical* but proven *correlation* between a morphological and a use value in a certain region. An example would be the association between soil blackness and available phosphorus. Where such correlations have been established they are very useful, but they could not be presumed to apply beyond the region.

(iv) Relationships where land use value has been established, by means of field plots, for a certain *morphological soil type.* Where such relationships have been soundly established with sufficient replication to cover soil type variability, they are very valuable, but they are empirical; like (iii) above they could not be presumed to apply beyond their region. They cannot be generalized because the relationship is not with any particular soil property.

In examining the relation between morphology and crop production in semi-arid environments it appears that those relationships based on *direct involvement* (i), and on *logical connection* (ii), will provide the best prospects for examination. Relationships based on *empirical correlations* (iii) or *morphological soil* type (iv) will be little mentioned here because, from their nature, they cannot be generalized. They could, nevertheless, have a most valuable role in regions where a groundwork of study and correlation of use data and morphological data has been made. Then morphology would be used as an extension vehicle, a role for which it is particularly suited.

But few studies are available in the literature in which individual morphological properties are correlated or evaluated against use values or analytical properties. The authors are left therefore to make use of field plot data in which morphology was not part of the experimental design, and to draw what generalization about morphology may seem reasonable, considering the importance of the role that soil morphology is expected to play in providing a link between research criteria and their application to varying fields situations. This lack of correlation studies is a serious omission, and indicates an unfortunate state of isolation on the part of morphology.

THE SEMI-ARID ZONES AND THEIR SOIL REQUIREMENTS FOR CROPPING

The demands which cropping makes on soils depend in part on the environment, and this for our purpose is semi-arid; however, there are other environmental and management variables which add to the range of demands on the soil.

The semi-arid lands (Meigs 1953) are found, generally, in zones extending as haloes around the desert belts of the world. These, as illustrated in figure 1, extend across the land masses at about 20°N and 20°S of the equator. Continental shape and relief have strong modifying effects on this generalization, but it is true enough to serve as a basis for discussion. The deserts tend to reach the coasts on the western sides of the land masses, but on the east there is a belt of well-watered country between them and the ocean. Semi-arid zones occur as E—W belts along the temperate side of each desert, bordering each on the east, and then swinging westward parallel to the tropical margin of each desert.

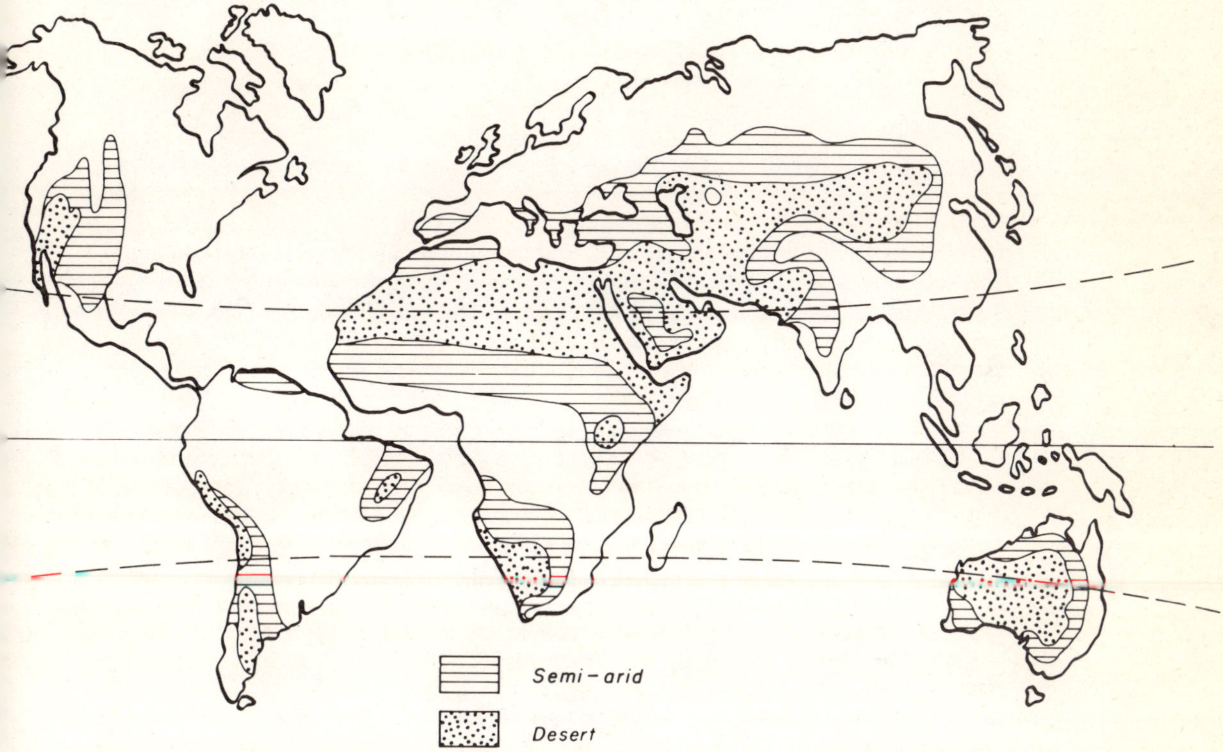

Fig. 1. The main semi-arid zones of the world are shown as haloes around the desert zones.

The environments of these regions offer great contrasts in growing season conditions. Taylor (1960) recognized six types of dry-land climates based on the distribution of rainfall between growing season and the rest of the year, the dessicating power of climate during the crop season, and seasonal temperatures. However, many semi-arid lands around the world do not fit this grouping.

In Australia, three climatic subtypes of the semi-arid zone are recognized (see fig. 2). These provide a convenient basis for a brief outline of important differences in semi-arid climates in relation to cropping. Two of them are similar to types recognized by Taylor (1960), the third is transitional in character and has a much wider climatic range. The limits of the Australian zone are taken from cropping experience with interpolation based mainly on rainfall.

(i) The *temperate semi-arid zone* on the cooler sides of the deserts receives most of its rain in the cool season which is also the main growing season. The climate is often called mediterranean; the main crops are wheat and barley. Rainfall and temperature data for Roseworthy and Northam in figure 3 are representative. Soil

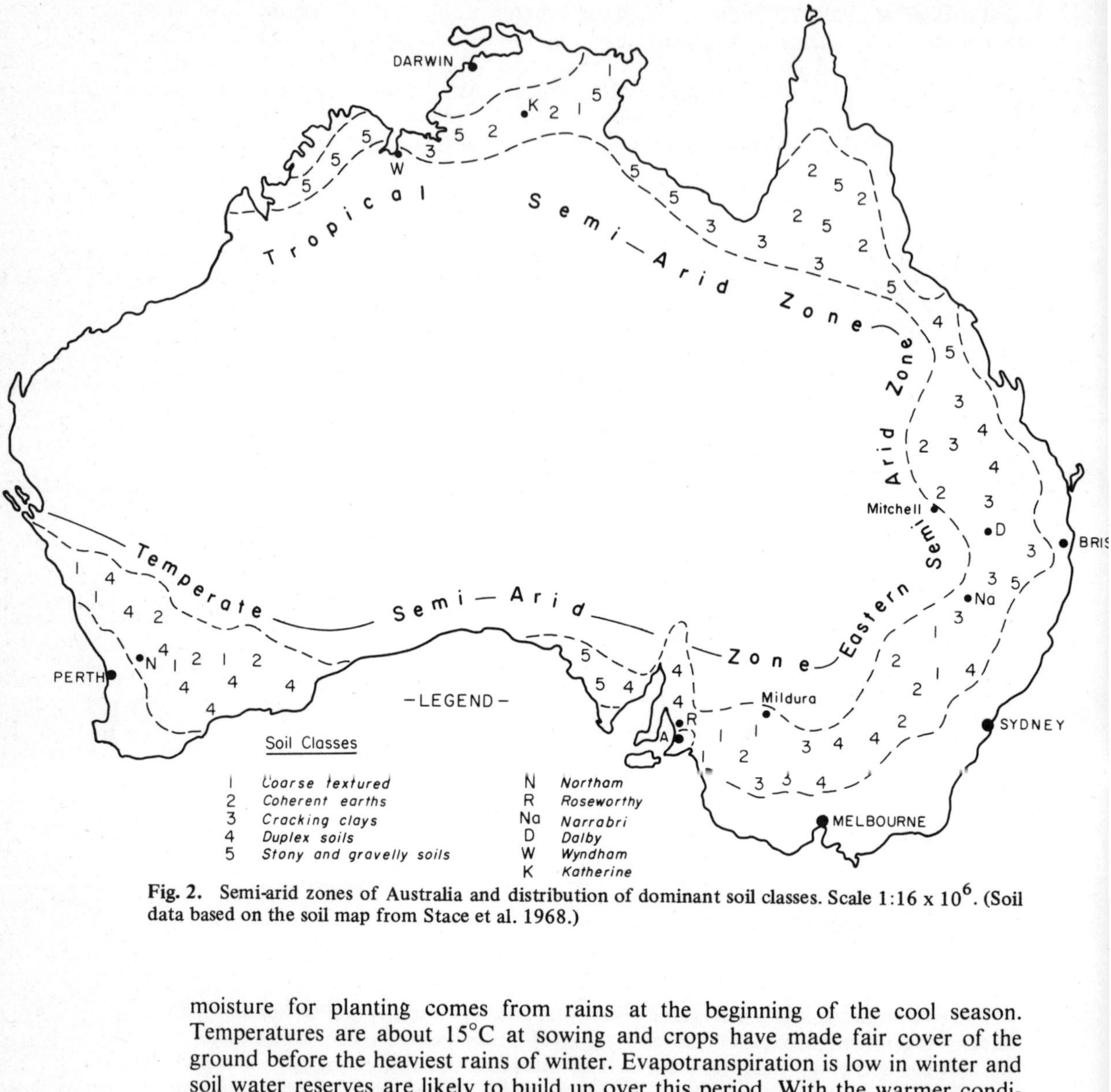

Fig. 2. Semi-arid zones of Australia and distribution of dominant soil classes. Scale 1:16 x 10^6. (Soil data based on the soil map from Stace et al. 1968.)

moisture for planting comes from rains at the beginning of the cool season. Temperatures are about 15°C at sowing and crops have made fair cover of the ground before the heaviest rains of winter. Evapotranspiration is low in winter and soil water reserves are likely to build up over this period. With the warmer conditions of spring, the crop develops rapidly to flowering and harvest drawing on stored soil water and the declining rainfall.

(ii) The *tropical semi-arid zone* along the warmer sides of the deserts receives its rain in the hotter season. This is the growing season and the main crops are sorghum, millet, and grain legumes. Data for Katherine and Wyndham in figure 3 illustrate the character of rainfall and seasonal temperatures. Cropping conditions are in marked contrast to those of the temperate semi-arid zone owing to higher rainfall intensities, temperatures, and evapotranspiration. After a long dry season with temperatures in the 25° to 30°C range, the rains come in against a background of rising temperatures. This is a critical time when cultivation and sowing must take advantage of the earliest falls before the heavy rains which may occur at the break of

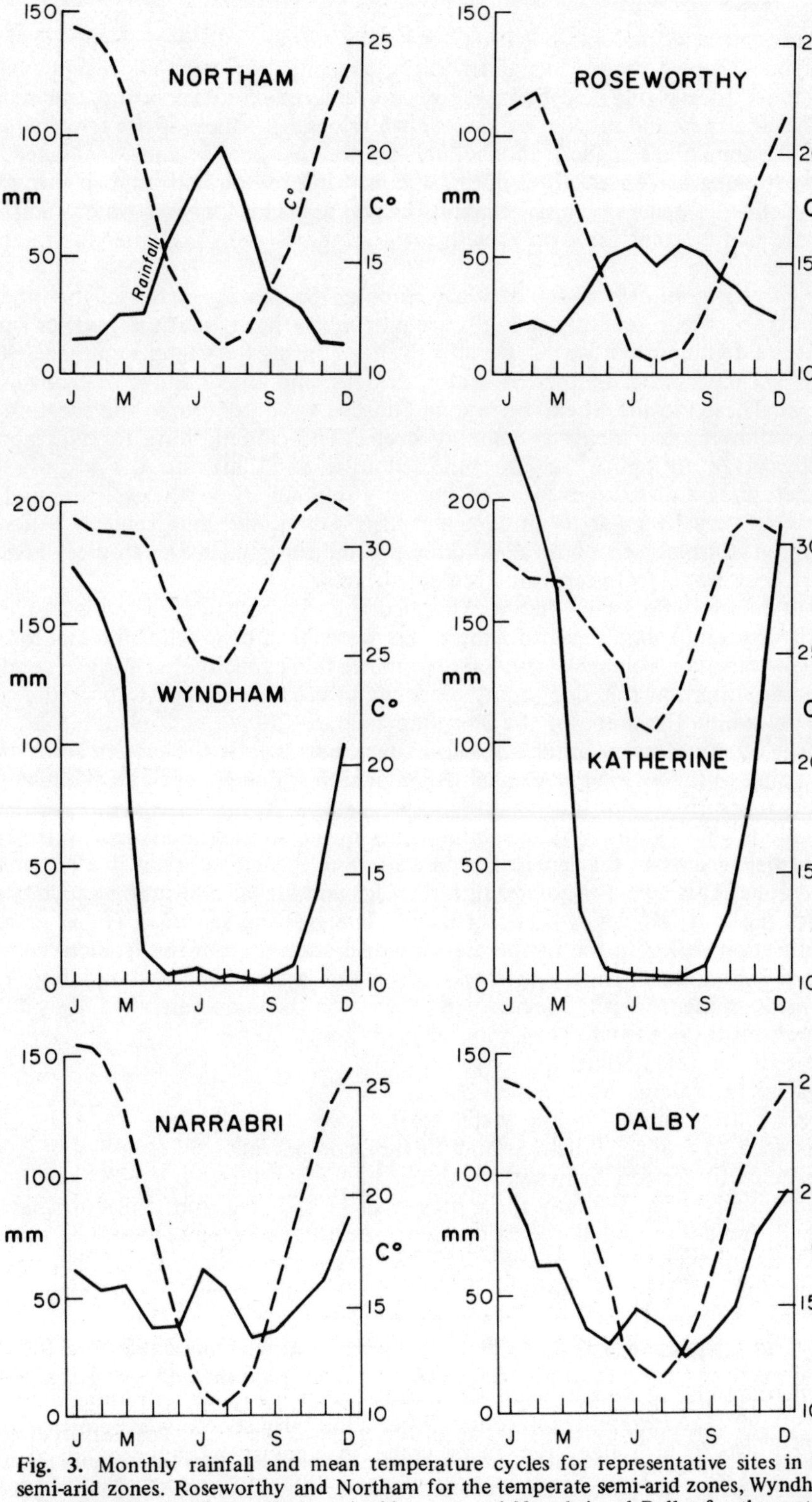

Fig. 3. Monthly rainfall and mean temperature cycles for representative sites in the semi-arid zones. Roseworthy and Northam for the temperate semi-arid zones, Wyndham and Katherine for the tropical semi-arid zones, and Narrabri and Dalby for the eastern semi-arid zones.

season and leach soluble nutrients, especially nitrate, into deeper subsoil thereby delaying crop growth. Sustained high capacity for rainfall acceptance and resistance to crusting and compaction are required to avoid serious reduction of water efficiency and erosion or surface waterlogging. Much of the growing-season rainfall—sometimes called monsoonal—comes as heavy and prolonged falls. Temperatures are about 30°C during the growing season and atmospheric saturation deficit is generally high, making heavy demands on soil water. Soil water storage is important for crop growth between major rains and after the rains have finished.

(iii) *The eastern semi-arid zone* combines some of the characteristics of the other two main types. Mean monthly rainfalls are more even throughout the year, or summer rains are dominant but with some effective falls during the winter months. Cropping may be either in the summer or winter, or both, and a wide range of crops may be grown. These include wheat, barley, and linseed as winter crops, and sorghum, millet, sunflower, and maize as summer crops. The climatic data for Narrabri and Dalby given in figure 3*a* are representative and illustrate the slightly lower temperatures and more even mean monthly rainfalls. Nevertheless, seasonal rains are often very irregular in amount and distribution, and long rainless periods are common. Summer cropping conditions are milder than in the tropical zone, but short heat waves between rains remain a hazard.

In all three semi-arid zones the most important characteristic of a soil is its capacity to absorb and hold rainfall and to supply this water to a crop with minimal loss. The character of the main volume of the soil determines this capacity, but surface conditions also play an important role due to any tendency to crust or seal, or to promote run-off and erosion which the slope of the site may favour.

Fallow is often necessary in the semi-arid zones, particularly the eastern zone, and the soil's response to fallow may be critical. A stable tilth will mean both an effective fallow and a resistance to erosion during the fallow.

Interactions with climate on the requirements of the soil are indicated by the fact of higher intensity rains in the tropical semi-arid zone, generally, than in the temperate semi-arid zone. This calls for greater tilth stability and higher infiltration capacity in the former for the same efficiency of water usage. The growing season is cooler and has a lower saturation deficit in the temperate semi-arid zone than in the tropical zone: this calls for lower moisture-supplying rate than in the latter. Climatic conditions favour water erosion in the tropical zone and wind erosion in the temperate zone; thus different soil properties are needed for cropping in each zone.

These may be summarized as:

(1) Capacity for rainfall acceptance
(2) Capacity to store and supply water to the crop
(3) Resistance to sealing, crusting, and surface compaction
(4) Resistance to erosion by water or wind.

The question is: how well may these properties be assessed from a morphological examination? Some, like resistance to crusting, sealing, and compaction, are subject to direct observation under the heading of soil consistence. But most of the other soil properties are not directly observable as morphology, though something may be deduced about them from the morphological features, texture, structure, and consistence. The association with these morphological features is inferred from the logical connection between particle size grading and aggregation on the one hand, and pore size distribution and the moisture characteristic on the other.

Thus, in the terminology used above in the general theory of soil morphology, we should look for direct involvement and logical connections between texture, structure, consistence, and gross physical dimensions, and the four soil properties listed above. The literature contains few studies of their relationship to morphological attributes, though, from general experience, many workers believe that good relationships exist. Taylor (1960) and Kohnke (1968) describe relationships between texture and infiltration. Structure is also involved in infiltration insofar as decrease in infiltration is proposed as a measure of structural stability (see Middleton 1930). Jamison (1956), Taylor (1960), and Salter and Williams (1965) have shown relationships between soil texture and available water-holding capacity (WHC). Salter and Williams found that structure and organic matter content were also significant in the relationship. Resistance to sealing and crusting is a direct expression of structural stability but is also related to texture. Lutz (1934) showed the importance of soil structure in giving resistance to soil erosion.

But nutrient supply is also a factor in crop production, and, though one would not count on nutrient level being expressed in morphology, except as a local empirical correlation, one should look for possible relationships between morphology and nutrient supply through soil water factors.

MORPHOLOGICAL CATEGORIES AND THEIR RELATION TO CROP PRODUCTION FACTORS

From the preceding discussion it is appropriate to examine texture, structure, consistence, and gross physical dimensions for possible direct or associated connection with crop production factors.

(i) *Texture* (field texture) is a measure of the overall physical character of a soil sample based on "feel" after kneading moist in the hand for a few minutes. It largely reflects particle size grading; classes like sand, loam, clay, and intergrades are recognized. High sensitivity can be developed as a personal skill and large numbers of textural classes are proposed—fourteen basic classes in the United States system (Soil Survey Staff 1951) and nineteen in the Australian system (Northcote 1971), each with further qualification based on grain size of the sand fraction where this is dominant. To standardize these personal assessments, correlation with laboratory mechanical analyses has been effected through a texture triangle (Marshall 1947; Soil Survey Staff 1951). Unfortunately the size fractions recognized differ, and a universal standard has not been established. Nor can field textures assessed by feel be precisely related to laboratory particle size data since differences in clay mineral type, associated organic matter, exchangeable cations, sesquioxides, and structure may also be involved. Thus field texture classes are crude units each embodying an appreciable range of properties. Although experience has shown that many important soil properties vary with texture none of the relationships applies under all circumstances (Black 1968). In fact, precise relationships are unlikely.

Black notes that the only direct effect of particle size on plant growth is the mechanical resistance to root growth and shoot emergence offered by large particles. Texture is, however, associated with many aspects of soil water behaviour that affect plant growth and these become more critical in crop production in semi-arid regions. Salter and Williams (1965) record a considerable range in WHC of soils from 2 cm to 8.4 cm of water per 30 cm depth. Variation is associated largely with soil texture, values being greatest in the fine sandy loam to clay loam range, and decreasing towards both coarse sand and clay ends of the spectrum. But the relationship involves organic matter content and structure too. Thus the effect of texture on WHC is complicated. Clay soils may have either a relatively low or a high WHC; indeed, soils most famous for this characteristic are the black clays of the tropic zones. This difference in soils of the same texture must be mainly due to structural differences. Some sands, which have a low WHC according to the general texture relationship, may have a high WHC if organic matter is obvious in the hand sample.

Table 1 illustrates for some Australian soils the general relationship between field tex-

Table 1. Estimated average water-holding capacities (WHC) for subsurface soils of different textural classes from semi-arid regions of Australia

Field texture grading	Clay + silt mean content wt. %, oven dry fine earth	Water-holding capacity mm/cm depth		Available WHC (A) / Total WHC (NA+A) %
		Non-available * (NA)	Available † (A)	
Sands – loamy sands (8)‡	10	0.3	1.4	81
Sandy loam – fine sandy loams (3)	37	1.0	1.9	64
Loams (4)	42	1.8	2.0	55
Sandy clay loams – clay loams (6)	51	1.3	1.9	59
Medium-heavy clays				
(*a*) grey, brown and red clays (7)	73	2.7	1.1	29
(*b*) black earths ‖ (5)	86	3.1	2.2	42

SOURCE. Data from Stace et al. 1968 except for the black earths.

* Water retained at 15 bar } oven-dry basis.
† Water retained at 0.1 bar minus water retained at 15 bar } oven-dry basis.
‡ Number of samples.
‖ Berndt (unpublished data) and Stirk 1972. The water-holding capacity values for these soils are from direct determination on field samples taken at wilting point and field capacity.

ture grading, laboratory-determined values of soil water content when crop plants wilt permanently, and the WHC of the soil. Owing to the different basis of determination, the WHC data for the black earths are not directly comparable with those for the other soils included in the table. However, they represent real field conditions and are in line with the acknowledged superior cropping performance of these soils as compared with the other cracking clays.

The effect of soil texture on efficiency of rainfall usage is well known. It expresses itself as the greater production on sandy soils from frequent light rains compared with greater production from loam and clay soils from few heavy rains. This situation arises, firstly, from the much greater amount of water retained by the clay soil against uptake by plants as the soil dries out after wetting, and, secondly, the much deeper penetration of a light rain in a sandy soil than in a clay. Thus, using the values given in table 1 and allowing for the difference between sandy and clay soils in the amount of water retained at air-dryness, a 25 mm rain would wet to field capacity approximately 16 cm depth of the average air-dry sand and about 22 mm of the fall, less evaporation loss, would be available to the growing crop. For the average air-dry clay, comparable figures would be approximately 9 cm depth of wetting and only 11 mm of the rainfall, less evaporation loss, available to the crop. Under similar drying conditions, evaporation losses following light rain tend to be higher from clay soils because the water is held closer to the surface and upward capillary conductivity is higher than for sands.

Differences in the ease with which soil water is released at increasing suction are also important in semi-arid cropping. Water release characteristics for soils of a range of textures (Taylor 1960) show, for example, that a coarse sand gives up most of its water at low suctions while some medium-textured soils and clays continue to release water as the suction increases to near wilting point. For this reason, sands are generally unsatisfactory soils for cropping in tropical semi-arid regions where evapotranspiration is high and rains infrequent. Crops grow rapidly while water is available but quickly exhaust the limited supply and fail if further rain is delayed. On clays, stored water is released more slowly as suction increases, the initial storage is much greater if the soil was wet to depth, and a crop can continue to grow for a long period without rain.

From glasshouse experiments with fertilized lucerne (*Medicago sativa*) on two clays and two sand soils of different character and under two watering regimes (simulating infrequent-heavy, and frequent-light rains), Fergus and Stirk (personal communication) concluded that both sands and clays exhibit some undesirable features for cropping in a semi-arid environment. Simple separation of the soils on the basis of recognized textural classes did not completely explain the results which they considered were influenced by other physical properties that affect depth of wetting and evaporation losses, e.g. the moisture characteristic and the hydraulic conductivity.

Soil texture has a major influence on rainfall acceptance; Kohnke (1968) shows a tenfold water entry rate for loamy sand compared with a clay loam. But here, too, the effect of structure is large. Generally speaking, a coarse sand has very high rainfall acceptance when dry or wet. Clay soils generally contrast in having high rainfall acceptance when dry, but swelling clays quickly close their cracks and infiltration into the wet soil is very slow. However, a well-structured clay soil, i.e. one with many coarse pores and high stability, continues to absorb rainfall at a moderate rate even when wet. Medium-textured soils, loams, often have a moderate rainfall acceptance when dry or wet.

For many soils, the rate of rainfall acceptance is controlled by the tendency to form surface crusts or seals. These effectively limit water entry to very low final figures, besides causing a potentially drastic barrier to seedling emergence. Crusting, on drying, results from a combined effect of structural weakness and heavy pounding rain on an unprotected, bare surface of soils having unfavourable size grading. One of the main purposes of cultivation and management is to mitigate the deleterious effects of crusting and compaction on rainfall acceptance by producing a rough porous surface which has the capacity to hold a large amount of rainfall for the time it needs to soak in. This porous surface will tend to slake down to its previous, uncultivated state. If the soil is of low structural grade or stability, and if rainfall is heavy, this may happen quickly, with deleterious effects to seedling emergence and to rainfall acceptance. Serious collapse of the surface tilth and subsequent hard setting may also occur under sustained light rain where the underlying soil at 5-10 cm depth has low hydraulic conductivity. Such weakly structured surface soils are prone to erosion from subsequent rain showers. Under these circumstances, the cultivator needs to look to various devices such as contour banking or bunding, the success of which will depend on the character of the rain and the structural stability of the soil.

(ii) *Structure*. From the foregoing it is clear that texture needs to be associated with structure when assessing the water relations of a soil. Here it is necessary to distinguish *structure* as used by the soil physicist and *structure* as used by the soil morphologist. The soil physicist refers to aggregation of primary particles and the stability of aggregates in water when subject to standardized attrition, determined by standardized laboratory techniques. The morphologist's descriptions are based on different criteria: "the very fragments or clods into which the soil breaks up" (Zakharov 1927): or in more detail, the shape, size, and degree of particulateness of the soil mass and the pieces into which it may break. There is no established correlation between the two concepts of structure, though relationships are plausible.

The categories of structure (Soil Survey Staff 1951) are the grade, class, and type described by Zakharov (1927). It is not known if each of these categories is equally important in moderating the role of texture in the soil-water relations discussed above. A study by Ayres, Button, and de Jong (1972) shows disappointing consistency in differences in aeration between the main structural types. It would seem that structural *grade*, i.e. the degree to which aggregation is manifest in the soil material, is important in soil-water relations, and so too the *class* of structure, i.e. the size of the aggregates; but the *type*, i.e. the shape of aggregates, is of unproved significance in appraising water relationships. The more direct statement of class as a size in millimetres (Butler 1955) is more serviceable than the usual custom (e.g. Soil Survey Staff 1951) of having "very fine" mean different sizes for different types of structure.

The grade and size of structure are together the best expression of the degree of structural development, for soil water considerations, in a soil material. Structure often consists of peds in a series of sizes, and a small amount of hand working will break the larger peds into smaller ones. The smaller the final ped size, down to about 2 mm, and the smaller the amount of work needed to produce it, the higher the degree of structural development. Description in these terms permits a relative evaluation of soils, and will be used here.

The stability of structure–its resistance to forces tending to disrupt and disperse the aggregates into their constituent primary particles–is important in relation to crusting and the continued high permeability of a soil when wet. It seems to relate better to consistence than to structure, and is therefore considered below.

(iii) *Consistence* is a statement of the strength and kind of cohesion of a hand sample of soil material as it occurs in the field. Like texture it is determined by manipulation, but it refers to the soil's physical form *before* it is homogenized by kneading, and to the particular water content. The contrast between the initial cohesion and strength of the soil material, and those same characteristics after kneading can be small or great, and is a measure of the amount of fine structural organization in the original soil sample. Consistence varies also according to the water content of the sample; the contrast between wet and dry can be profound. Usually both consistence descriptions are given.

In contrast to the lengthy description of consistence given by the Soil Survey Staff (1951) a description in terms of coalescence and pulverescence has been proposed (Butler 1955). Both statements refer to the amount of change in physical state occurring when the soil is manipulated between thumb and fingers with just sufficient force to cause a brittle or plastic deformation. Coalescence refers to the degree of plastic deformation in more clayey soils when moist. Pulverescence refers to the degree of brittle deformation in less clayey soils when dry. Experience has shown that a high coalescence is characteristic of soils with weak structure which collapse when wet to give a low rate of infiltration. Such soils crust severely on drying. On the other hand low coalescence is characteristic of soils with well-developed structure, with continuing high or moderate permeability when wet and with minor tendency to crusting.

Experience has associated high pulverescence with medium- and coarse-textured soils which are massive or crusted in the natural state or tend to crust seriously after rain on a newly cultivated surface.

(iv) *Gross physical dimensions*. Limitations to the rooting volume of a soil are set by the soil's depth to rock, or by the amount of stones, concretions, cemented horizons or hardpan in the soil. These morphological features impose a direct limitation on the capacity of the soil to supply water and nutrients. Clearly these limitations only apply if they lie within both the potential depth of rooting of the crop, and the depth of soil wetted by rain.

The potential root zone varies with crop species modified by climatic and soil factors–notably soil texture, structure, and fabric. Under favourable conditions, cereals and grain legumes do not root much below 120-180 cm (E.W. Russell 1973). At Katherine, in the tropical semi-arid zone of Australia, on a very deep, permeable red earth of clay texture but low WHC, bullrush millet had reduced the water content almost to wilting point to a depth of 150 cm (Begg et al. 1964) when it failed owing to drought, eight weeks after emergence. Subsequent sampling showed that some roots penetrated to 360 cm depth in moist soil but were apparently unable to supply water from below 150 cm fast enough to prevent wilting. Of three other crops grown on the same soil, sorghum appeared to have the best-developed root system and cotton the worst with peanuts intermediate (Slatyer 1955). Cotton roots were well developed in the top 60 cm but sparse below. Shallow effective rooting of cotton is also suggested by the accumulation of nitrate between 60 and 120 cm depth in a nitrogen-fertilized clay soil; on the other hand sorghum depleted nitrate to 120 cm depth (Hipp and Gerard 1973). For wheat grown in deep black earths of the Darling Downs, Queensland, data indicate that little water is normally used from below 120 cm depth. In a year of above average stored water at planting and low growing-season rainfall, only 10 per cent of available water stored to 120 cm depth used by the crop came from the 90-120 cm layer (Waring, Fox, and Teakle 1958*b*).

A shallower, 75 cm-depth limit used for the textural groups (see below) has been chosen arbitrarily to give an available water capacity of about 7.5 cm or more in all but poor sands. If this depth of soil is wet to field capacity, it should be able to meet the water needs of a crop transpiring at moderate rates for up to three weeks without rain in the temperate and eastern zones. Soil depth becomes more important as cropping shifts into the hot summer months and onto soils of lower WHC which are also likely to wet to greater depths in this environment. In the tropical semi-arid zone, a minimum depth of 150 cm for the coherent earths and 200 cm for fine sands seems desirable, but even deep coarse sands are likely to be unsuitable for cropping.

Stones, gravel, or hard nodules below tillage depth reduce the WHC of a soil in proportion to the volume of the potential root zone they occupy. Flaggy stone, oriented more or less parallel to the soil surface, may have some additional effect in delaying root penetration into deeper moist soil. Weathered mineral material below the solum and even extending into crevices in hard rock may be effective to varying degrees in meeting crop needs and must be considered in relation to soil depth. The effect of pans within the soil and hard rock at its base depends on their porosity, continuity, and depth. The lower and finer their porosity, the greater their continuity, and the shallower they occur within the potential rooting and wetting depth, the less suitable the soil is for cropping. Roots of wheat, sorghum, and several other crops including lucerne failed to penetrate stratified layers of a deep sandy soil at depths as shallow as 30 cm even when adequate water and nutrients were supplied (Worker et al. 1973). This soil contained 96 per cent sand but consisted of a succession of layers of fine sand with a little silt + clay and thicker layers of coarse sand. The bulk density of individual cores ranged from 1.73 to 1.85.

In Australia, the availability and productivity of semi-arid croplands is little affected by these features. Most of the soils exceed 1 m depth: black earths overlying basalt in the uplands of the Darling Downs, Queensland, are an exception. Some in the Mallee (*Eucalyptus* spp.) scrub lands of the south overlie hard limestone at shallow depth and contain considerable broken stone. These soils are quite unsatisfactory in drought years and produce good yields only when the growing season rainfall is plentiful and evenly distributed. Many of the solonized brown soils (Stace et al. 1968) of this temperate semi-arid zone contain much limestone rubble and discontinuous carbonate-cemented pans well within rooting depth of cereal crops but there is no published evidence of their effect on crop production.

GENERALIZED MORPHOLOGIC PROFILES OF AUSTRALIA IN RELATION TO CROP PRODUCTION

Morphology cannot be applied to field assessments except in the context of soil profiles, that is to say a particular assembly of possibly diverse morphological characteristics in a set of horizons. The possible variety of profiles is great and discussion of them is only possible here if they are grouped into a few classes. The classes proposed for use in this discussion are generalized on the bases of physical morphological features, texture, structure, consistence, and gross physical dimensions as discussed above.

For purposes of discussion five soil groups are proposed:

(1) Coarse-textured (sandy) soils ⩾ 75 cm deep
(2) Coherent earth (loamy) soils with gradual or no horizon changes with depth, ⩾ 75 cm deep
(3) Cracking clay soils, ⩾ 75 cm deep
(4) Duplex soils having a marked texture increase between 5 and 75 cm from the surface, ⩾ 75 cm deep
(5) Shallow and stony or gravelly soils, < 75 cm deep.

The soil groups are illustrated in figure 4 and a rough indication of their distribution in Australia is given in figure 2. They are broad classes only, covering a range of character in physical form, and unlimited by considerations of chemical nature or pedological criteria. Table 2 relates these soil groups to the more generally recognized soil classes in current use; it is apparent that the soil groups represent a high level of generalization.

Cropping experience, predominantly Australian, on each of these soil groups in turn, in semi-arid environments, will now be examined.

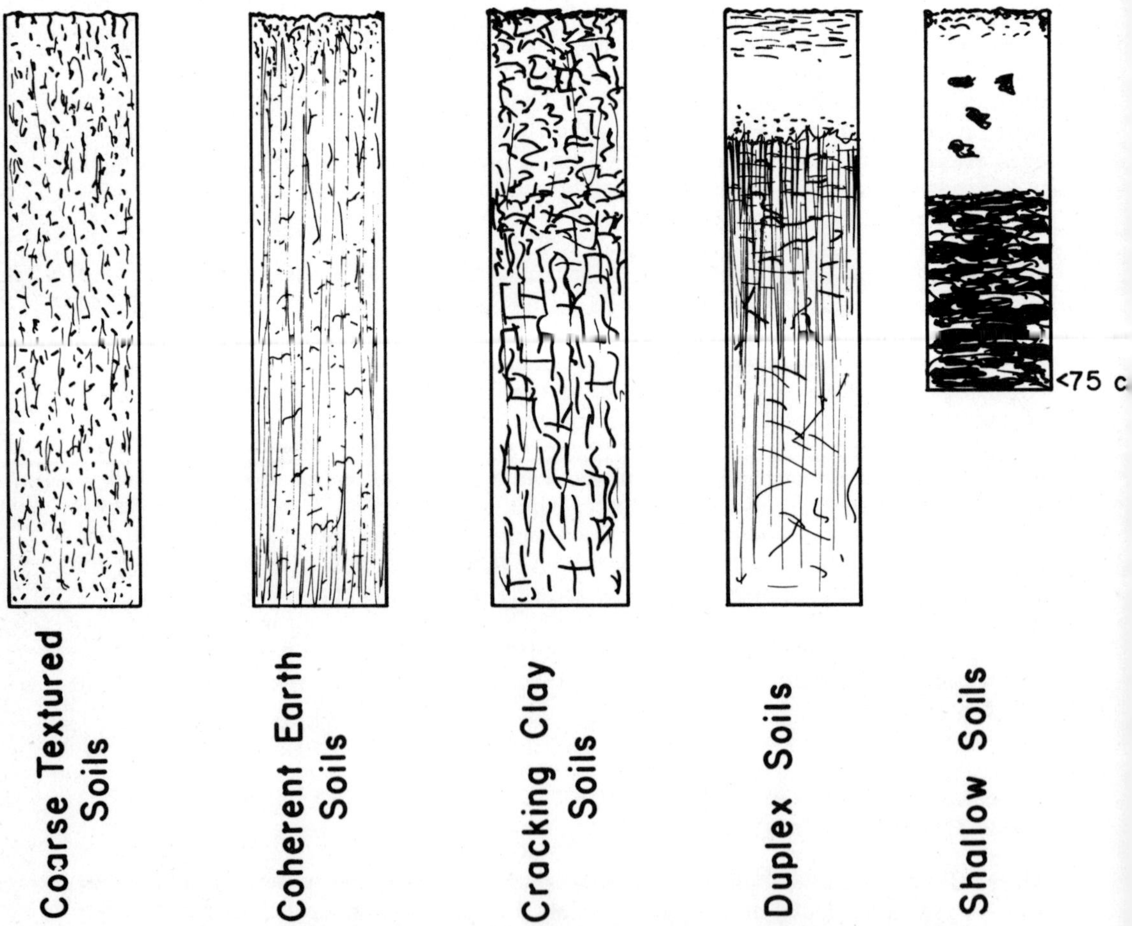

Fig. 4. Diagrammatic expression of the five classes of soil discussed in the text; from left to right, coarse textured soils; coherent earth soils; cracking clay soils; duplex soils; shallow, stony, and gravelly soils.

Table 2. Soils of the Australian semi-arid croplands related to the texture profile groups and to other soil classifications

Texture profile group	Australian soil groups*	Approximate equivalents in other classification systems		
		Factual Key sections†	World Soil Map units ‡	U.S. Soil Taxonomy orders ‖
1. Coarse-textured (sandy) soils	Earthy sands	Uc5	Arenosols	Entisols
	Siliceous sands	Ucl	Regosols	
2. Coherent earth	SOLONIZED BROWN SOILS	Gcl, Gc2, Gn1	Xerosols	Alfisols
	Red earths	Gn2	Ferralsols	Oxisols and Ultisols
	Yellow earths	Gn2	Ferralsols	
	Krasnozems	Gn3, Gn4	Ferralsols and Nitosols	Oxisols
	Euchrozems	Gn3	Ferralsols and Nitosols	Alfisols or Inceptisols
	Prairie soils	Gn3, Gn4, Uf6	Phaeozems	Mollisols
	Chernozems	Um6, Gn3		
3. Cracking clays	BLACK EARTHS	Ug5	Vertisols	Vertisols (pellusterts)
	Rendzinas	Uf6	Rendzinas	Mollisols
	GREY, BROWN and RED CLAYS	Ug5	Vertisols	Vertisols (chromusterts)
4. Duplex soils	RED-BROWN EARTHS	Dr2 . Db1	Luvisols	Alfisols
	Non-calcic brown soils	Dr2 . Db1		
	SOLODIZED SOLONETZ and SOLODIC SOILS	Dr2 . Db1 . Db2 Dy2 . Dy3 . Dy5 Dd1 . Dd2	Solonetz Planosols	Alfisols (natrustalfs and natrudalfs)
5. Shallow and stony or gravelly soils	Lithosols – and shallow and stony phases of several of the above groups.	Ucl . Uc4 . Uc6 Um1 . Um2 . Um4 Um5 . Um6.	Lithosols and stony soils of several other units	Entisols and stony soils of several other orders

* Stace et al. 1968 (groups of major importance in capitals).
† Northcote 1971.
‡ F.A.O. 1968, 1970.
‖ Soil Survey Staff 1975.

(i) *Coarse-textured soils* have a high rainfall acceptance rate but a low WHC. The latter disadvantage can be compensated by greater depth of soil, so, to a depth of 3 m, greater depth is marginally advantageous. The coarsest sands have little available water and can be of little value for cropping; but finer sands, and especially silty and loamy sands, have a moderate capacity and are ideal for meeting crop needs under semi-arid conditions when most of the crop growth occurs in the cooler season and evapotranspiration demand is low. However, as already noted, sands are generally unsatisfactory for cropping in the hot growing season of tropical semi-arid lands.

Experience in Victorian at Walpeup ("Field Day" 1970), Western Australia (Wild 1959), and South Australia (French 1963*a*) is in accord on the finding that bare fallow on sands is not effective for moisture conservation. French (1963*a*) records that lighter soils under fallow at Wanbi and Minnipa stored an average of only 12 mm of usable water. However, he indicates that a sand soil may have some water storage potential if it has a clay subsoil. The beneficial effect of fallow on these soils in earlier years is now understood to have been due largely to greater nitrogen mineralization and accumulation of nitrate compared to continuously cropped soils. The nitrogen needs of crops on these soils are now met by higher soil fertility resulting from the inclusion of legume leys in the rotation, or by nitrogenous fertilizers.

Sands have the special advantage of making the best use of small amounts of rain; Willatt (1967) in Rhodesia found that planting can be started on sands after only 25-38 mm of rain, whilst for clay soils about 100 mm of rain would be required.

Sand soils are prone to wind erosion under the seasonally dry conditions of the semi-arid zones; and relict sand dunes are common in some regions. Note of this proneness is given by "Field Day" (1970), Wild (1959), and French (1963*a*). The erosion hazard depends on the season, surface cover, and management, but the intrinsic liability of the soil to erode is expressed by its coherence when dry. Incoherent sands are most prone to erosion, and increasing coherence means decreasing erodibility.

Water erosion hazard is low in sand soils, but may become serious when a less permeable horizon occurs at shallow depth as in the Blain soil at Katherine (Arndt, Phillips, and Norman 1963).

There is little information available on the use of sand soils in tropical semi-arid zones, but the work of Arndt, Phillips and Norman (1963) showed that the sandier soils were at a disadvantage compared with the Tippera clay loam; crops would wilt a few days after rain on the Blain sand. On the other hand rainfall acceptance of Blain sand was higher than that of the Tippera soil, with the result that crops on the Blain sand kept growing longer into the dry season.

(ii) *Coherent earth soils* cover a range in textures from sandy loam to clay at the surface, and there may be a gradual increase in clay content with depth. Typically these soils do not crack on drying, and have a low development of plastic properties. Nevertheless, some of them may be texturally clays from the surface downward. These soils include as an important subgroup, especially common in the tropical semi-arid zones, the clays of low plasticity (low coalescence when moist); these possess a high resistance to erosion, but in many cases a low WHC and a low tendency to crusting. The low WHC is noted by Slatyer (1954) in reference to the Tippera soil at Katherine, and associated by him with the clay mineral type (kaolinite).

At Katherine, the Tippera soil with a mean clay content of 54 per cent to 114 cm depth, held, between 10 and 180 cm depth, only 140 mm of available water (0.8 mm/cm) 3 days after the cessation of rains which had wet the profile beyond this depth. Sixty per cent of this was lost over the following dry-season bare fallow of 203 days and the remaining 46 mm of available water was held below 40 cm depth (Arndt, personal communication). Both wet and dry season fallows on this soil had no observed effects on the subsequent peanut and cotton crops (Katherine Research Station 1959). Pidgeon

(1972) has examined a similar association between low WHC and kaolinitic or ferralitic soils in Uganda.

The effect of low WHC means a loss of water efficiency in these soils when heavy rains occur. The water entry rate may be only moderate, but much of the water entering the soil is lost as drainage after the storage capacity of the root zone is filled. This, with the heavier rainfalls and high evapotranspiration of crops between rains, makes a minimum depth of 150 cm desirable for these soils in the tropical semi-arid zone. Another disadvantage is that this tendency renders the soils relatively ineffective for fallowing in regard to soil moisture storage. However, Pereira et al. (1958) report increased water storage of 63-76 mm, to 180 cm depth and crop yield increases of 30-100 per cent, as a result of fallowing of red earths in semi-arid, tropical East Africa. This storage is of the same order as that reported by French (1963*a*) for South Australian soils, though much lower than often achieved for cracking clays. Miles (1957), on Queensland experience, states that red loam soils on basalt have a poorer capacity for storing water than the black clays on the same parent material. This tendency may explain the finding of Kohn, Storrier and Cuthbertson (1966) that the soils at Wagga show little moisture storage from fallow. The soils are red earths (Northcote 1966) and would fall within the category discussed here.

Soils which fall into the earth group include those tending to plastic clays with depth, such as some red-brown earths, and red and yellow podzolics (Stace et al. 1968). These soils merge across from the duplex group to the earths, according to the contrast between A and B horizons; those with the lesser contrast are included in the earth group. These soils with the more plastic clay subsoil—having moderate coalescence when moist—have a higher WHC and consequently store more water under fallow. A number of instances for this group of soils may be cited, though some of those listed include cracking clay soils too. Wild (1959) reports valuable storage of moisture in fallows on heavier soils in lower rainfall areas. French (1963*a*) also found that moisture storage from fallow was significant on the heavier soils. Mowatt (1970) states that the heavier soils respond well when fallowed. However, he records deleterious effects of fallow on the structure of some red-brown earth soils, leading to sheet erosion. Tuohey, Robson, and Rooney (1972) report on the advantageous effects of fallow on moisture storage in the heavier soils in the Victorian Wimmera district; some of these soils would be red-brown earths, others would be cracking clays. Reports from the mallee region of northwestern Victoria ("Field Day" 1970) indicate water storage by fallow in their "normal land" in contrast to its lack of effectiveness on their "light soils".

Other soils in the earth group as used here are normal, medium-textured soils—loam to fine sandy clay loams. These, from the relationship of Salter and Williams (1965), one would expect to have higher WHC. This seems to be borne out by Miles's (1957) statement that, at Biloela, the "immature alluvial soils" have a WHC equal to the black clays in the district.

A weakness of the earth soils is their tendency to crust or seal under pounding rain. The problem is probably worse in tropic regions, soil for soil, because of the heavier rains and the higher saturation deficits during crop emergence. The best indication of proneness to crusting is the consistence characteristic of pulverescence when dry. Soils with a high pulverescence are liable to crusting. A degree of coarse (5 cm) structure may be associated with crusting, but more often the surface soil is massive.

(iii) *Cracking clay soils* crack deeply when dry. Though their texture does not vary beyond the clay range they cover the widest structural range, and vary from being attractive agronomically because of their tilth and water availability, to being unworkable because of their coarse, hard structure and unfavourable water relations. All cracking clay soils tend to be coarser in structure and more coalescent with depth. No defined line can be drawn for the limit of suitability, but surface structure and consistence are a good guide to relative suitability for cropping. The better-structured soils are good because of their favourable tilth and high WHC. When dry these soils have a structure which breaks down with small force to peds of 3 mm and less; when moist their coalescence is low. Soils which are unsuitable for use because of bad tilth and low WHC have a coarse structure of more than 10 cm peds when dry and these can be broken in the hand only with the greatest difficulty to 25 to 12 mm peds. When moist, these soils have a high coalescence.

The unfavourably structured soils of this group are generally avoided for cropping and consequently little information is available on them. The favourably structured soils are extensively used and studied; they have a reputation as heavy crop producers. In a semi-arid region, bare fallow is generally needed to provide more water than current accession. Swartz (1972) cites a fallowed soil which conserved 230 mm of available water.

Initially, rainfall acceptance is high, but infiltration gradually decreases as the depth of wetting increases, presumably owing to the clogging or closing of the larger pores as the clay swells. The better soils of this type have no tendency to surface crusting although the clogging of the larger pores by fine aggregates, < 0.5 mm, alone greatly reduces infiltration from heavy rains. In fact, in subtropical semi-arid areas where heavy storm rains in early summer may make a significant part of the annual rainfall, the most effective deep wetting of the clays is the result of rapid and deep penetration down cracks after the onset of surface flow and before the cracks close by swelling of the clay. The conductivity of the subsoil is much lower once the cracks have closed, and further downward movement and infiltration are very slow, resulting in run-off losses from heavy rains. Since the surface soil has low coherence, serious erosion ensues at this stage if rain continues, the topography favours run-off, and adequate controls have not been provided.

In a three-year study of these clays in the semi-arid zone of eastern Australia, Waring, Fox and Teakle (1958*a*) found that available water in 120 cm depth of soil at the planting of a winter cereal crop following a short (six-seven months) summer bare fallow, averaged 122 mm. Of this only 56 mm, representing 17 per cent of the rainfall, was gain during the fallow period, the other 66 mm being available water remaining in the soil at the previous harvest. The latter was closely correlated with rainfall during the two months preceding harvest and the authors suggested that one of the factors contributing to this correlation was better water penetration owing to soil cracking which normally occurs under the crop at this time.

In the eastern semi-arid zone the farmer has the opportunity of cropping in either the hotter or the cooler season; he can use bare fallow to carry soil water across from either season to the other according to the best prospects. However, cracking clays occur in other climatic zones too, and their prospects vary accordingly. In the clay soils of the Wimmera, Victoria, significant available water is stored under bare fallow (Tuohey, Robson and Rooney 1972) but this needs to be a fallow ploughed in the previous winter (Hore and Sims 1955). In South Australia, French (1963*a*) found that black, self-mulching clays were the most effective in storing water under conditions of fallow.

Gilgai microrelief is common in this group of soils and affects their cropping performance as discussed later.

(iv) *Duplex soils* have two horizons of strongly contrasting texture and physical character. A considerable amount of variation occurs in the degree of contrast between the horizons, the sharpness of change, and the depth in the soil at which it occurs. But in all cases the upper horizon is much less clayey than the lower. The term duplex soil is used by Northcote (1971) in a sense similar to here, but the contrast and sharpness of the boundary between A and B horizons is required to be more marked here than are Northcote's limits.

This group of soils merges, on the one hand, toward the earth group (as the contrast between horizons becomes less marked and more gradual) and on the other hand, toward the poorly-structured cracking clays as the upper horizon becomes thinner or is lost by erosion or by mixing with the subsoil by cultivation. Intergrades to adjacent types occur but the overall characteristics are as follows. There would be about 20 cm of sandy loam passing abruptly to a dense sandy clay. The upper horizon generally has a weak structural development; it is hard and brittle when dry with high pulverescence and tends to crust or set hard, as a mass, after rain. In the sandier soils this hardness and crusting tendency is weaker, but in soils with a sandy clay loam texture it is strongly developed. The lower horizon is a dense sandy clay or clay, extremely hard when dry, very firm and of high coalescence when moist, with either coarse structural development or none.

Cropping experience on duplex soils has not been isolated in the literature. On the one hand, it seems that farming has tended to avoid the more extreme of these soils; on the other hand, they have not been separated from other soil types in the experimental designs of agronomic trials. However, much crop research and experimentation in the Australian temperate semi-arid zone has been carried out on some of the better of these soils.

These soils have a number of special traits in regard to water. Rainfall accession of any moderate amount causes partial waterlogging of the A horizon, and this condition may persist for weeks. The soil under these conditions loses its physical strength and becomes prone to puddling; and depending on topography erosion can be a hazard.

The efficiency of water use on these soils varies greatly with the texture profile and the nature of the main horizon boundary; the WHC in sandy- and silty-textured A horizons tends to be low; where the A horizon is shallow, waterlogging above the heavier subsoil would probably inhibit root development in most crops. The clayey subsoil would provide some storage of available water to carry growth through to the dry season though the material does not in many instances favour root proliferation. The latter is especially true of some strongly solonized soils with subsoil bulk densities exceeding about 1.8 g/cc; generally the soil pores are too fine for roots to enter readily, and commonly there is little development of structure to facilitate root penetration.

There is little information on the effectiveness of fallowing for water storage on these duplex soils. However, in the temperate semi-arid zone French (1963*a*) records the importance in this regard of a clay layer of more than 25 per cent clay content at 15-30 cm depth; the water storage under fallow, in these heavier soils, of 25-65 mm, contrasts with an average of less than 12 mm in light soils. Most of the soils dry out appreciably during the summer months, but whereas little water loss occurs below 30 cm depth on the soils with clay subsoils, the lighter soils without them continue drying to greater depth.

Within limits, suitability of the duplex soils for crop production increases with depth to the texture change and decreases with increasing texture contrast. The effect of a strong texture contrast depends on crop and rainfall factors; the hazard is greatest when a soil with a high infiltration rate and low water capacity changes abruptly to a clay subsoil of low permeability below common tillage depth but well within both the root zone and the depth that would otherwise be wet by the heavier rains. These conditions occur, for example, when a loamy sand surface soil overlies a poorly-structured sandy clay or finer subsoil between 15 and 40 cm depth as in some solodized solonetz and solodic soils of the eastern semi-arid zone in southern Queensland. In some instances, the effect of an abrupt texture change at shallower depth can be ameliorated and the WHC of the overlying soil increased by mixing some of the clay subsoil into the lighter-textured surface soil by deep cultivation. Generally, the poor physical properties of the subsoils do not favour this unless further amelioration by gypsum is possible (Sims and Rooney 1965; Matheson 1969).

Where the upper zone provides a reasonable depth of drained soil for rooting—40-60 cm—the underlying clay subsoil increases WHC in wet years in much the same way as do polythene films or asphalt barriers; Sumner and Gilfillan (1971) have found that layers of this sort in sandy soil doubled the amount of water held in the 0-60 cm layer and gave yield increases of up to 50 per cent. These deeper soils have the advantage too of waterlogging only after the heavier rains. Generally, water stored in the clay is available for carrying the crop through to maturity.

At the other end of the range, texture-contrast soils posing the greatest physical hazards are those with thin, < 10 cm, silty-loam surface soils of low stability. These usually overlie very slowly permeable clays. The surface soils may be saturated by rains of no more than 30-40 mm, resulting in complete collapse of the tilth which on drying sets throughout its depth to form a hard mass. Complete or near failure of the crop results when such conditions occur shortly after sowing, and the infiltration of subsequent rains is much reduced. In standard simulated rainfall tests on a wide range of soils from Australian cereal-producing lands, Arndt (unpublished data) has recorded some of the lowest values of rainfall acceptance and highest dry-soil strengths, for soils of this kind.

Millington (1961), in a study of soils near Adelaide, found that even when waterlogging was not obvious, crop growth was restricted by wet surface conditions.

Surface crusting is associated particularly with the shallower duplex soils. The morphological indications for it are in terms of texture, structure, and consistence. Silty clay loam textures, very weak grade or apedal structures, and high pulverescence when dry indicate strong crusting tendencies; sand texture, a moderate grade of structure, and low pulverescence when dry indicate little or no tendency to crusting.

In tropical semi-arid zones, rainfall during heavy storms may have a marked effect in breaking down a cultivated soil surface. Arndt, Phillips and Norman (1963) report "sealing" of soils at Katherine under these circumstances even when soil textures are clay loam and loamy sand.

Soil erosion by water is a serious hazard in the case of shallow duplex soils under cultivation. The soil characteristics favouring erosion are an incoherent or water unstable surface soil resting at 10 or 20 cm on a relatively impermeable layer. Where topography concentrates run-on, soils of this description are very prone to sheet erosion, especially in regions of tropic storms. Arndt, Phillips and Norman (1963) on earth-type soils at Katherine provide an interesting comparison of the susceptibility of soils to erosion and sealing. Blain soil, with a sandy, water-unstable surface and less permeable subsoil, is much affected by surface erosion; Tippera soil, with a water-stable clay loam surface, sheds much water but does not erode. These data agree with the general indication of susceptibility to erosion of sands and silts in contrast to the more stable clays. Among the clay loams and clays the indication of resistance to erosion is low coalescence when moist.

(v) *Shallow, stony, and gravelly soils.* Though these soils may vary greatly in character their common trait is a significant limitation in the amount of soil, available rooting volume, and hence WHC. A limit of 75 cm depth has been drawn but obviously the shallower the soil the less it serves crop needs.

The stone or gravel in these soils may be limestone, laterite, basalt, or various kinds of country rock in different districts, and the fine earth will vary in character with the stone. Stone and gravel are undesirable because they exclude their volume of soil.

The soils of this class may have the character of any of those listed above, i.e. they are stony phases or variants of them; consequently they cover quite a wide range in character of tilth, water entry rates, and WHC. Some may be suitable for special crops, or even for general cropping in favourable circumstances, but, in most cases, serious impairment of crop production will be caused by the restricted soil volume.

INTERACTION WITH MORPHOLOGIC FACTORS

Nutrient level and nutrient cycling

There is a broad relationship between soil texture and content of the major nutrients Ca, Mg, K, P, and N in the surface soil, very sandy soils usually having the poorest and cracking clays the highest nutrient status; this is illustrated in table 3 for some

Table 3. Relationship of nutrient status to textural class for surface 10 cm of a range of soils of Australian semi-arid lands

Textural class	Clay %	% (Oven-dry weight)			Exchangeable cations (me/100g soil)		
		N	P	K	Ca	Mg	K
Sand to sandy loam (6) *	*7* 3–14	*0.04* 0.01–0.08	*0.01* 0.01–0.03	*0.76* 0.12–2.2	*2.1* 0.3–4.8	*1.0* 0.2–1.5	*0.33* 0.1–0.9
Loam to clay loam (6)	*24* 22–27	*0.13* 0.09–0.23	*0.04* 0.02–0.06	*1.1* 0.07–1.9	*6.8* 3.0–14	*3.9* 2.2–5.9	*1.3* 0.4–2.1
Clays (7)	*52* 40–73	*0.16* 0.09–0.23	*0.05* 0.02–0.14	*1.1* 0.41–1.5	*22* 6.5–45	*12* 5.3–29	*1.6* 0.9–2.8

SOURCE. Data from Stace et al. 1968. Means and ranges of values.
* Number of samples.

Australian soils. Data of Russell (1967) show similar relationships for surface soils to 15 cm depth in the temperate semi-arid zone of South Australia. Mean values for sands and clay loams respectively are 0.04 and 0.14 per cent for total nitrogen, 70 and 290 ppm for total phosphorus, and 0.35 and 2.50 me/100 g soil for exchangeable potassium contents; the 15 atmosphere water contents were 2.6 and 15.7 per cent respectively.

There is evidence for a relationship between the soil texture profile and nutrient cycling based on the difference in soil water response to rainfall; this is associated with the mineralization of soil organic matter, with leaching, and with the effect of water supply on nutrient uptake. The practice of bare fallowing for water conservation can greatly increase mineralization and the accumulation of nitrate for the succeeding crop. In the temperate, semi-arid region of South Australia, with essentially dry summers, French (1963*a*) has recorded increases in nitrate accumulation of more than 50 per cent (five-year average) at planting as a result of fallowing. He found that sands and clay loams were more effective than other soils in producing available nitrogen. Some of the highest accumulations of mineral nitrogen have been found in cracking clay soils of moderate organic matter content, especially when the fallow spans a summer period with sporadic rains. Thus, Waring and Teakle (1960) record, for the black earths of eastern Australia, mineral nitrogen to 120 cm depth at planting averaging 140 kg/ha over a three-year period. This is double the mean value for nitrogen recovered in grain and straw.

The fate of the nitrate accumulated depends very much on seasonal rainfall and soil permeability, the latter controlled mainly by the texture-structure profile. In years of average rainfall or less, and especially in the medium- to fine-textured soils, the nitrate is usually retained in the upper 30-60 cm of soil where it is readily available to the crop during the early weeks of growth. But when rainfall is above average, and especially when heavy rains occur near seeding time, much of the nitrate may be leached into the deep subsoil. This occurs even in the slowly-permeable clay soils. Both Martin and Cox (1956) and Waring and Teakle (1960) record that rises in nitrate nitrogen in the deeper subsoils of the black earths on the Darling Downs, Queensland, were correlated with increases in water content and could be explained on the basis of leaching. In very permeable soils, nitrate is easily leached by heavy rains and is often washed beyond the root zone where it is unavailable to the crop unless brought closer to the surface by lateral seepage. Russell (1967) found that responses to nitrogen fertilizers in the semi-arid cereal areas of South Australia are largely determined by seasonal rainfall; there was little response when seasonal rainfall was below 100-150 mm; above this level, up to 400 mm, there were increased responses but with a marked interaction with site related to soil factors. Referring to these results, Reuter (1967) states that the most consistent responses were obtained on sandy soils which are low in organic matter, easily leached, and often not fallowed. Heavy rains before or after seeding result in severe nitrogen deficiency on these soils.

For a long time fallowing has been the means of providing available nitrogen for cereal crops in many countries, but its exploitation in a fallow-crop rotation has produced a steady decline in soil nitrogen contents and crop yields. Data for north American soils cultivated for thirty years or more (Haas and Evans 1957) show that under small grains, soil nitrogen losses for an alternate crop-fallow rotation were substantially higher than for continuous cropping, ranging from 13 to 50 per cent with an average of 29 per cent. Losses of similar order have been recorded in Australia; on the black earths of the Darling Downs, 20 per cent of the total nitrogen in the surface 15 cm was lost over twenty-five years of cultivation (Martin and Cox 1956).

Very sandy soils, which have low inherent nitrogen contents, which mineralize organic nitrogen readily (French 1963*b*), and which are more vulnerable to leaching, can be expected to become nitrogen deficient earlier in their cropping history than finer-textured soils. In the temperate semi-arid zone of southern Australia, the use of legume leys, notably barrel medic (*Medicago truncatula*) in the drier areas and subterranean clover (*Trifolium subterranean*) where rainfall is adequate, in rotation with cereal crops has reversed the fall in soil nitrogen content and crop yield, and soil fertility is increasing substantially. In this, also, there is an interaction with soil texture. Russell (1968) comments that the low nitrogen status of the coarse- as compared with the fine-textured soils is due both to the relationship already noted between inherent fertility and soil texture, and to the failure of annual pasture legumes such as *Medicago* spp. to grow on many coarse-textured soils in contrast to their prolific growth on fine textured soils.

Another aspect of the climate-soil-plant interaction in semi-arid cropping is the effect of extended dry periods in reducing nutrient uptake from the 10-20 cm of surface soil, normally richest in nutrients. This is likely to be most significant in the early weeks of crop growth, on medium- to fine-textured soils, when fertilizers have been drilled at seeding; it applies especially to phosphorus uptake (Russell 1968; Simpson and Lipsett 1973). Light infrequent showers may not wet these soils to the depth of fertilizer placement. In these circumstances, and when inherently fertile surface soils are dry for long periods, the nutritional characteristics of the moist subsoil become especially important. The opposite effect, that of the osmotic suction of fertilizer salts in reducing water availability in sands of low available water capacity, has been noted also (Fergus and Stirk, unpublished data).

Site characteristics

Considerable variation in the performance of soils of given morphology arises from different site characteristics: slope, microrelief, and stoniness. Stoniness has already been considered above, but it may be mentioned further that stoniness may also be effective in reducing erosion.

The main significance of slope is in relation to loss of rainfall by run-off, and consequent soil erosion. The hazards increase with increasing gradient and length of slope (Zingg 1940; Free 1946). Zingg reports that total soil loss in run-off was increased 2.6 and 3 times respectively by doubling the degree and the horizontal length of slope. When rainfall intensity exceeds the infiltration rate and water begins to flow over the soil, these two factors determine the down-slope build up in volume and depth of flow, its velocity, and hence its erosive power. However, it is difficult to set critical values in relation to water erosion because they depend so much on interaction with other factors including the texture, structure, stability, and infiltration rate of the surface soil, the hydraulic conductivity of the subsurface soil, the cropping system, and management. Erosion risks certainly become more serious for slopes greater than 4 per cent. Bare fallowing for water conservation in semi-arid lands greatly increases the exposure of the surface soil to erosive forces. It is therefore particularly important that conservative limits for slopes be adopted in developing these lands for cropping.

Gilgai microrelief can have pronounced effects on crop production in semi-arid lands. These patterns of small mounds and depressions, association with swelling clays and with duplex soils that have thick subsoil horizons of these materials, are strongly developed in eastern Australia. They occur also in parts of Africa and north America and have probably been obliterated in long-cultivated lands, as in parts of India.

In Australia, the forms, patterns, and vertical interval or amplitude vary greatly (Hallsworth, Robertson and Gibbons 1955), the latter ranging from less than 10 cm to more than 1.5 m. They affect cropping in three ways. Firstly, striking contrasts in crop growth result from soil differences associated with the different parts of the microrelief, and may continue long after the microrelief has been levelled by cultivation. On the black earths of the Darling Downs, Queensland, Thompson and Beckman (1959) found wheat plants on the puffs (mounds) were yellowish and spindly, poorly tillered, and had small ears often incompletely filled with grain, compared with normal growth in the depression. Despite apparently more favourable soil physical properties on the puff, and the possibility of higher available soil water in the depressions, the differences were considered more likely to be due to different nutrient levels in soils of these sites. Differences of this kind reach extreme proportions as shown in figure 5, where the soil micropattern includes both duplex soils and cracking clays.

Fig. 5. Photograph of a wheat crop growing on the contrasting soils of a gilgai complex. Contrast the poorer growth on the hard-setting solodic (duplex) soil with heavier growth (marked by the spade) on a mound of cracking clay. (CSIRO photograph.)

Secondly, microrelief can affect water distribution leading to lateral variation in water storage in the soil, to flooding of the larger depressions and damage to the crop in wet years, and, in the case of linear gilgai on slopes, to concentration of run-off resulting in rill and gully erosion. Thirdly, the larger "melon hole" gilgai of the cleared brigalow (*Acacia harpophylla*) forest lands seriously interfere with mechanized farming and, where deep and closely spaced, preclude arable farming over considerable areas. Levelling of these areas can expose the extremely acid and saline subsoils of lower nutrient status that lie beneath the gilgai banks (J.S. Russell 1973).

DISCUSSION

The reader may be daunted by the presentation of such complex interactions of a few morphological features, and at the same time surprised that so many morphological features and combinations of features common in pedological discussions are apparently not considered. This complexity is due to the fact that the whole ambit of continental range has been considered; the number of combinations of factors is great even though limited to semi-arid conditions.

The generalization aimed at in this chapter is, paradoxically, the reason why other morphological associations and entities like morphological soil types have not been discussed; they can only be generalized within their own region, and tend to disappear when relationships of three or four adjacent regions are pooled to provide the broader generalization. The empirical correlations and morphological soil type associations, mentioned in earlier discussion, dissipate when the ambit of study is broadened to cover a continent or more.

This study also suffers from entailed imperfections in its "experimental design". It would have been more fruitful to use data and experience for each kind of soil in each kind of semi-arid zone; this would have revealed the interactions of different environments and morphological properties in crop production and prospects. Unfortunately this is not possible as, on the one hand, there is a degree of confounding between soils and climates and, on the other hand, years of agricultural experience had led, in each district, to some soils being avoided for cropping whilst others more suitable provide the only quotable experience.

CONCLUSIONS

The foregoing material and discussions bring the conclusion that the importance of morphological properties (in the strict sense) in semi-arid cropping is dominated by their effects on the soil water regime and the consequences of treatments, mainly cultivation and fallowing, to improve it. The soil texture profile is the outstanding property in these relationships; soil depth for its part in determining potential water storage capacity, and slope for its influence on possible loss of water and soil in run-off, are probably next in importance. The significance of structure and consistence is largely in modifying the effects of texture, especially in the surface soil in relation to tilth, water entry, and seedling emergence when climatic conditions are unfavourable, but also in the subsoil where they affect water movement and root penetration. Apart from broad correlations between texture and nutrient status of surface soils, the influence of morphological properties on the nutrient regime is through their control of water availability and movement.

Most of the documented cropping experience in the tropic and eastern semi-arid zones is based on cracking clays and, to a lesser degree, the coherent earth soils. Experience on duplex soils in these zones is more restricted, and it seems that the poorer of these are largely avoided by farmers on the basis of past experience. Coarse-textured soils are rarely used in the tropical semi-arid zone, and one is left to reason that they are of little value owing to their water relations; they are certainly of less value than those in the temperate semi-arid zones.

The limited experience of fallowing the coherent earths in the tropical semi-arid zone of East Africa (Pereira et al. 1958) shows useful water storage between the 30-60 cm and 150-180 cm depths. These authors conclude, nevertheless, that bare fallow is dangerous and inappropriate for this region owing to erosion hazards during the long soil exposure involved. However, they found fallows protected by sowing a quick-growing, short-season annual grass were satisfactory from the important aspects of both soil conservation and improved crop yield.

We have no experience to offer on the effect of fallow on cracking clays in the tropical semi-arid zone (e.g. Katherine). In the absence of winter season rains these fallows would need to be carried through from one wet season to the next, increasing the chance of water loss as compared with fallows on cracking clays in the temperate and eastern semi-arid zones. Experience in the Ord region of northwestern Australia (Beech, Norman and McIntyre 1968) does not clearly indicate how much available water is stored by fallow on cracking clays nor how much would remain at the end of the dry season. If the clays are wet deeply through cracks, fallows should be no less effective on them in storing water than on the coherent earths of East Africa. Owing to their lower infiltration rates, erosion hazards on the clays would be even greater than on the coherent earths. Serious erosion has occurred on sloping sites under summer bare fallow in the subtropical part of the eastern semi-arid zone and clays in the tropical semi-arid zone would be subject to the same hazard.

The eastern semi-arid zones generally offer better prospects than the temperate or tropical zones: the eastern semi-arid zone has wider cropping options; either summer or winter crops can be sown according to the seasonal opportunities. Fallows are more effective and may be short in duration. In contrast, both the temperate and the tropic semi-arid zones have essentially one-season options, and of these the latter carries the most severe risk. Fallows in both zones, to be effective, must be maintained from one wet season through the year to the next.

Modification of both the soil water regime and the tendency to surface sealing and crusting, and the prevention of soil erosion, remain the major soil management problems associated with morphological properties in the successful cropping of semi-arid lands.

REFERENCES

Arndt, W., Phillips, L.J., and Norman, M.J.T. (1963). CSIRO Aust. Div. Land Res. Reg. Surv. Tech. Paper No. 23.

Ayres. K.W., Button, R.G., and de Jong, E. (1972). *Can. J. Soil Sci.* 52, 311-21.

Beech, D.F., Norman, M.J.T., and McIntyre, G.A. (1968). *Aust. J. Exp. Agric. Anim. Husb.* **8**, 72-80.

Begg, J.E., Bierheuzen, J.F., Lemon, E.R., Misra, D.K., Slatyer, R.O., and Stern, W.R. (1964). *Agric. Meteorol.* 1, 294-312.

Black, C.A. (1968). "Soil-plant Relationships". 2nd Edn. John Wiley, New York.

Butler, B.E. (1955). *J. Aust. Inst. Agric. Sci.* 21, 239-49.

F.A.O. (1968). World Soil Resources Reports No. 33.

———. (1970). Key to soil units for the soil map of the world. (AGL: SM/70/2.)

"Field Day" (1970). Mallee Research Station Walpeup, Victoria.

Free, G.R. (1946). *J. Amer. Soc. Agron.* **38**, 207-17.

French, R.J. (1963*a*). *J. Dep. Agric. S. Aust.* 67, 42-8.

———. (1963*b*). *J. Dep. Agric. S. Aust.* 67, 76-9.

Haas, H.J., and Evans, C.E. (1957). U.S. Dep. Agric. Tech. Bull. No. 1164.

Hallsworth, E.G., Robertson, Gwen K., and Gibbons, F.R. (1955). *J. Soil Sci.* 6, 1-31.

Hipp, B.W., and Gerard, C.J. (1973). *Agron. J.* 65, 712-4.

Hore, H.L., and Sims, H.J. (1955). *J. Dep. Agric. Vict.* 53, 263-4.

Jamison, V.C. (1956). *Soil Sci.* **81**, 459-69.

Katherine Research Station Progress Report 1946-56. (1959), CSIRO, Melbourne.
Kohn, G.D., Storrier, R.R., and Cuthbertson, E.G. (1966). *Aust. J. Exp. Agric. Anim. Husb.* 6, 233-41.
Kohnke, H. (1968). "Soil Physics". McGraw-Hill, New York.
Lutz, J.F. (1934). Mo. Agr. Exp. Stn. Res. Bull. No. 212.
Marshall, T.J. (1947). Coun. Sci. Ind. Res. Aust. Bull. No. 224.
Martin, A.E., and Cox, J.E. (1956). *Aust. J. Agric. Res.* 7, 169-83.
Matheson, W.E. (1969). *J. Dep. Agric. S. Aust.* 72, 402-6.
Meigs, P. (1953). *Reviews of Research on Arid Zone Hydrology*. Vol. I, 203-10. UNESCO, Paris.
Middleton, H.E. (1930). U.S. Dep. Agric. Tech. Bull. No. 178.
Miles, L.G. (1957). *J. Aust. Inst. Agric. Sci.* 23, 190-5.
Millington, R.J. (1961). *Aust. J. Agric. Res.* 12, 397-408.
Mowatt, P.J. (1970). S. Aust. Dep. Agric. Ext. Bull. No. 12.
Norris, J.M. (1971). *J. Soil Sci.* 22, 69-80.
Northcote, K.H. (1966). "Atlas of Australian Soils". Explanatory data for sheet 3. Sydney-Canberra-Bourke-Armidale area. Melbourne Univ. Press. Melbourne.
———. (1971). "A Factual Key for the Recognition of Australian Soils". Rellim, Adelaide.
Pereira, H.C., Wood, R.A., Brzostowski, H.W., and Hosegood, P.H. (1958). *Emp. J. Expt. Agric.* 26, 213-28.
Pidgeon, J.D. (1972). *J. Soil Sci.* 23, 431-41.
Reuter, D.J. (1967). *J. Dep. Agric. S. Aust.* 70, 188-98.
Russell, E.W. (1973). "Soil Conditions and Plant Growth". 10th Edn. Longmans, London.
Russell, J.S. (1967). *Aust. J. Exp. Agric. Anim. Husb.* 7, 453-62.
———. (1968). *Aust. J. Exp. Agric. Anim. Husb.* 8, 736-48.
———. (1973). *Trop. Grassl.* 7, 209-18.
Salter, P.J., and Williams, J.B. (1965). *J. Soil Sci.* 16, 310-7.
———. (1967). *J. Soil Sci.* 18, 174-81.
Simpson, J.R., and Lipsett, J. (1973). *Aust. J. Agric. Res.* 24, 199-209.
Sims, H.J., and Rooney, D.R. (1965). *J. Dep. Agric. Vict.* 63, 401-9.
Slatyer, R.O. (1954). *J. Aust. Inst. Agric. Sci.* 20, 46-7.
———. (1955). *Aust. J. Agric. Res.* 6, 365-77.
Soil Survey Staff (1951). Soil Survey Manual. U.S. Dep. Agric. Handbk. No. 18, Washington.
———. (1975). Soil Taxonomy. U.S. Dep. Agric. Handbk. No. 436, Washington.
Stace, H.C.T., Hubble, G.D., Brewer, R., Northcote, K.H., Sleeman, J.R., Mulcahy, M.J., and Hallsworth, E.G. (1968). "A Handbook of Australian Soils". Rellim, Adelaide.
Stephens, C.G. (Ed.) (1953). "Soil Surveys for Land Development". FAO, Rome.
Stirk, G.B. (1972). Symposium on physical aspects of swelling clay soils. Univ. of New England, Armidale, pp.53-60.
Sumner, M.E., and Gilfillan, E.G. (1971). Proc. South Afr. Sugar Tech. Assoc. 45th Congress. pp.165-8. Soil Sci. Dept. Univ. Pietermaritzburg, Natal.
Swartz. G.L. (1972). *Qld. J. Agric. Anim. Sci.* 29, 407-22.
Taylor, S.A. (1960). *In* "Plant Water Relations in Arid and Semi-arid Conditions": Reviews of Research UNESCO Paris, 191-203.
Thompson, C.H., and Beckmann, G.G. (1959). CSIRO Aust. Div. Soils, Soils and Land Use Ser. No. 28.
Tuohey, C.L., Robson, A.D., and Rooney, D.R. (1972). *Aust. J. Exp. Agric. Anim. Husb.* 12, 414-9.
Waring, S.A., Fox, W.E., and Teakle, L.J.H. (1958*a*). *Aust. J. Agric. Res.* 9, 205-16.
———. (1958*b*). *Aust. J. Agric. Res.* 9, 717-29.
Waring, S.A., and Teakle, L.J.H. (1960). *Aust. J. Agric. Res.* 11, 27-41.
Webster, R., and Butler, B.E. (1976). *Aust. J. Soil Res.* 14, 1-24.
Wild, A.S. (1959). *J. Dep. Agric. West. Aust.* 8, 249-50.
Willatt, S.T. (1967). *Rhodesia Agric. J.* 64, 4-11.
Worker, G.F., Ayers, R.S., Karah, M.A., and Thomason, R.E. (1973). *J. Agric. Sci.* 81, 513-6.
Zakharov, S.A. (1927). Achievements of Russian science in the morphology of soils. Acad. Sci. USSR (Leningrad) Russ. Ped. Invest. No. 2.
Zingg, A.W. (1940). *Agric. Engng.* 21, 59-64.

CHAPTER 3

Clay mineralogic properties

K. Norrish and J.G. Pickering

INTRODUCTION

This chapter summarizes the information that is available on the clay mineralogy of soils in the semi-arid areas (250-750 mm rainfall) of Australia. There are not a lot of published papers that deal specifically with the clay mineralogy of these soils. There are some by early workers in the field (Shearer and Cole 1939*a*, 1939*b*; Cole 1943; Hosking 1940); some are concerned with weathering and particle size (Hosking, Neilson and Carthew 1957; Hosking 1957), and a few are on the mineralogy of specific great soil groups (Norrish and Rogers 1956; Radoslovich 1958; Turton, Marsh and McKenzie 1961; Turton et al. 1962). In the laboratories of the C.S.I.R.O. Division of Soils, clay mineral analyses have been carried out in Adelaide since 1948, and in Brisbane since 1961. For the soils under discussion the C.S.I.R.O. files contain results on more than 400 profiles, and this represents most of the data available. Some of these are recorded in *A Handbook of Australian Soils* (Stace et al. 1968); many results are scattered in publications where mineralogy is only an incidental part of the paper; some of the results are contained only in internal reports.

The soils have been considered within the framework of Great Soil Groups as set out in Stace et al. 1968. Only those groups used extensively for agricultural or horticultural purposes have been included.

Analyses were made on the $<2\mu m$ fractions of the soil by X-ray diffraction methods based on techniques and diagnostic tests fully described by Brown (1961).

However for palygorskite special procedures were required. It occurs in several of the profiles under discussion and is usually accompanied by a high illite content. The recognition and estimation, by X-R-D methods, of palygorskite in this situation presents certain difficulties (Heller 1961). Both illite and palygorskite are characterized by having their strongest diffraction peak at 10Å. Under the electron microscope palygorskite is easily recognized because of its fibrous habit, but it is difficult to make even a semi-quantitative estimate of amount from a micrograph. To overcome these difficulties the (200) line of palygorskite at 6.44Å is used. This is not a strong line, but when the clay is heated its intensity increases with temperature up to approximately 80°C, and remains at a constant level in the temperature range 80-180°C. In this range, the intensity is about double that of the unheated sample. The effect is reversible.

As a routine procedure a diffraction trace is obtained with the sample heated at 120°C and then if a slow scanning speed is used, somewhat less than 5 per cent of palygorskite can be detected, even in the presence of illite.

In making semi-quantitative estimates of the components, the potassium content and exchange capacity of the clay were considered together with the diffraction data in the following manner. The kaolin/illite ratio was established from diffractometer traces, using the peak height of basal reflections, weighted in favour of illite by a factor of two. Allowing for the amounts of other non-clay minerals present, the percentage ranges of kaolin and illite were then calculated, using this ratio. From these amounts of kaolin and

Table 1. Typical examples of clay mineralogy from each Great Soil Group

All data relate to <2μm fraction

Anal-ysis no.	Sample no.	Great Soil Group	Profile location	*Factual Key	Depth	Per cent <2μm	Cation Exchange Capacity me %	Per cent range						
								K	I	R	M	Q	H/Go	V or Vi
1	VW/71/C13	Grey, brown and red clays	Warracknabeal, Vic.	Ug5.24	30–40 cm	46	30	20–30	50–65	10–20	–	–	–	–
2	B555/3	Grey, brown and red clays	Chinchilla, Qld.	Ug5.24	12–30 cm	50	44	50–65	–	–	30–40	10–20	–	–
3	B561/2	Black earths	Darling Downs, Qld.	Ug5.16	10–20 cm	48	82	20–30	5–10	–	50–65	5–10	1–5	–
4	A310/3	Black earths	County Grey, S.A.	–	34–38 cm	55	52	5–10	50–65	20–30	–	<1	–	–
5	A655/2	Rendzinas	County Robe, S.A.	–	8–15 cm	–	43	10–20	65–80	10–20	–	–	–	–
6	VW/70/C5	Solodized solonetz and solodic soils	Glenorchy, Vic.	Db2.43	20–30 cm	45	33	20–30	50–65	10–20	–	1–5	–	–
7	B558/4a	Solodized solonetz and solodic soils	Chinchilla, Qld.	Dy5.43	30–38 cm	30	30	50–65	–	20–30	–	10–20	1–5	–
8	SA/69/C8	Soloths	Wanilla, S.A.	Dy3.41	20–30 cm	49	35	40–50	10–20	20–30	–	–	5–10	–
9	482/4b	Soloths	Monivae, Vic.	Dr2.42	40–50 cm	47	31	65–80	–	10–20	–	<1	1–5	1–5(Vi)
10	B559/1b	Soloths	Chinchilla, Qld.	Dd2.41	10–13 cm	20	26	20–30	–	20–30	–	50–65	–	–
11	SA/69/C28	Solonized brown soils	Bute, S.A.	Gcl.12	20–30 cm	18	47	20–30	50–65	20–30	–	<1	–	–
12	P550/1	Solonized brown soils	Merredin, W.A.	Gcl.12	0–10 cm	16	36	50–65	30–40	10–20	–	<1	–	–
13	A1004/6	Red-brown earths	Urrbrae, S.A.	Dr2.23	26–30 cm	25	30	30–40	40–50	10–20	–	1–5	5–10	–
14	B633A/3	Red-brown earths	Quilpie, Qld	Dr2.32	20–30 cm	45	26	65–80	10–20	–	10–20	5–10	–	–
15	B637/3	Red earths	Quilpie, Qld.	Um5.2	20–30 cm	34	18	50–65	10–20	5–10	–	10–20	–	–
16	617/4a	Red earths	Harden, N.S.W.	Gn2.14	30–40 cm	41	24	40–50	30–40	–	–	<1	10–20	–
17	A970/4a	Terra rossas	Kybybolite, S.A.	Uc6.13	50–60 cm	15	34	40–50	20–30	10–20	–	<1	10–20	–
18	A1011/2	Euchrozems	Inverell, N.S.W.	Dr4.11	10–20 cm	51	28	65–80	–	10–20	–	<1	10–20	–
19	A989/3b,4a	Desert loams	Whyalla, S.A.	Dr1.33	30–40 cm	–	48	20–30	20–30	20–30	–	<1	1–5	10–20P
20	P554/2	Lateritic podzolic	Northam, W.A.	Dy5.51	10–20 cm	47	10	>80	–	–	–	–	–	–
21	VIC/71/P18	Unspecified podzolic	Ballarat, Vic.	Gn3.11	20–30 cm	27	32	50–65	10–20	10–20†	–	–	–	5–10 (V)
22	SA/72/P1	Grey-brown podzolic	Eden Valley, S.A.	–	0–10 cm	9	21	20–30	(triocta) 65–80	–	–	–	–	–
23	5471	Podzols	Mt. Burr, S.A.	–	0–10 cm	5	–	50–65	(10–20)		–	1–5	–	10–20 (V)
24	15849A	Solonized brown soil	Willalooka, S.A.	–	15–20 cm	57	55	10–20	(>80)		–	–	–	–

* Northcote 1971.
† Regularly interstratified material.

Abbreviations:
K = Kaolin
I = Illite
R = Randomly interstratified material
M = Montmorillonite
Q = Quartz
H/Go = Hematite and/or Goethite
V = Vermiculite
Vi = Inhibited Vermiculite
P = Palygorskite

illite, a theoretical value of the cation exchange capacity (CEC) is determined using average values for these minerals (derived from Grim 1953).

The excess of the measured CEC over the "theoretical" value was used to express the amount of randomly interstratified material, and where applicable, montmorillonite. The amounts of kaolin and illite are then reduced proportionately.

The presence of this randomly interstratified material is evident on diffraction patterns as central scatter, and with somewhat broader, but rational, basal reflections.

In the summaries, the term randomly interstratified material is included with illite.

Kaolin, illite and montmorillonite are the most commonly occurring minerals and the names are used throughout in their generic sense; see Brown 1961, Grim 1953, Loughnan 1969.

SOILS

The mineralogy of the different Great Soil Groups is briefly summarized below using all the information available. Table 1 shows mineral analyses of typical soil clays; table 2 shows average potash contents and exchange capacities of clays, and table 3 gives silicate analyses of some clays.

Figures 1 and 2 show the locations of profiles studied. One dot on the map may represent more than one profile.

Grey, brown and red clays. Fig. 1*a* (76 profiles)

The soils of this extensive Great Soil Group have been studied in considerable numbers. Profiles from the Lower North and Barossa Valley in South Australia, northwestern Victoria and the Riverina of New South Wales show a consistent mineralogical pattern.

Table 2. Average K_2O % and CEC (me %) for all the samples < 2 μm of each Great Soil Group

		Mean K_2O%	Mean cation exchange capacity me %
Solonized brown soils		2.57	39.6
Red-brown earths		2.60	38.1
Solodized solonetz and solodic	South	1.36	32.1
	North	0.33	36.1
Grey, brown and red clays	South	3.35	46.4
	North	0.92	60.7
Soloths	South	1.19	32.1
	North	0.42	38.1
Rendzinas		1.84	54.0
Black earths	South	2.71	49.7
	North	0.82	61.3
Grey-brown podzolic		2.09	31.9
Terra rossas		1.74	40.8
Red earths		1.02	22.9
Red podzolic		2.73	30.5
Brown podzolic		0.34	21.0
Grey-brown and red calcareous soils		2.47	48.6
Yellow podzolic		1.68	30.4
Lateritic podzolic		1.26	27.4
Euchrozems		0.35	47.5
Desert loams	South	2.23	43.8
	North	1.16	47.1

NOTE. The distinction between north and south is an arbitrary one, but the wide separation between locations should be obvious from the maps of figure 1 and figure 2.

Table 3. Silicate analyses* of calcium saturated $<2\mu m$ fractions of soils (oven-dry) of table 1

Analysis no.	Sample no.	Fe_2O_3	MnO	TiO_2	CaO	K_2O	P_2O_5	SiO_2	Al_2O_3	MgO	H_2O	Total
1	VW/71/C13	8.67	0.05	0.95	1.54	2.35	0.08	52.99	20.95	1.97	10.65	100.20
2	B555/3	5.92	0.07	1.17	1.09	0.60	0.08	56.09	22.34	0.76	12.40	100.52
3	B561/2	7.18	0.09	1.18	2.12	0.87	0.14	51.51	19.90	1.02	15.12	100.03
4	A310/3	7.67	0.04	0.87	1.66	2.99	0.16	49.18	18.35	2.56	16.55	100.03
5	A655/2	7.27	0.03	0.85	1.95	2.95	0.06	49.13	17.66	2.97	16.85	99.72
6	VW/70/C5	10.41	0.02	0.89	0.75	3.04	0.08	47.19	26.49	1.38	11.43	101.68
8	SA/69/C8	11.11	0.02	1.07	0.79	1.68	0.06	44.09	27.13	0.91	13.37	100.23
9	482/4b	21.41	0.03	2.84	0.45	0.29	0.13	34.34	23.62	0.16	16.47	99.74
10	B559/1b	5.86	0.03	1.90	0.67	0.16	0.11	61.56	14.50	0.14	14.62	99.55
11	SA/69/C28	8.05	0.14	0.85	1.60	2.84	0.10	48.67	20.53	3.16	14.33	100.27
12	P550/1	4.76	0.14	0.62	1.25	2.56	0.12	44.63	26.41	1.99	17.13	99.61
16	617/4a	10.29	0.05	1.18	0.42	2.01	0.13	42.24	27.91	0.79	14.05	99.07
17	A970/4a	12.74	0.15	1.19	0.59	1.53	0.24	42.00	20.41	0.63	15.71	95.19
18	A1011/2	13.95	0.05	1.02	0.49	0.35	0.15	38.44	30.38	0.23	16.30	101.36
20	P554/2	3.76	0.01	0.93	0.20	0.20	0.03	42.02	36.19	0.09	15.80	99.23
21	VIC/71/P18	8.19	0.02	0.81	0.69	2.27	0.06	43.60	30.64	0.86	14.23	101.37
22	SA/72/P1	10.76	0.04	1.03	0.62	1.36	0.11	41.00	29.76	1.32	14.27	100.27
24	15849A	9.79	0.04	0.81	0.98	3.86	0.06	49.21	22.47	2.49	11.83	101.54

* For details of analytical methods see Norrish and Hutton 1969.

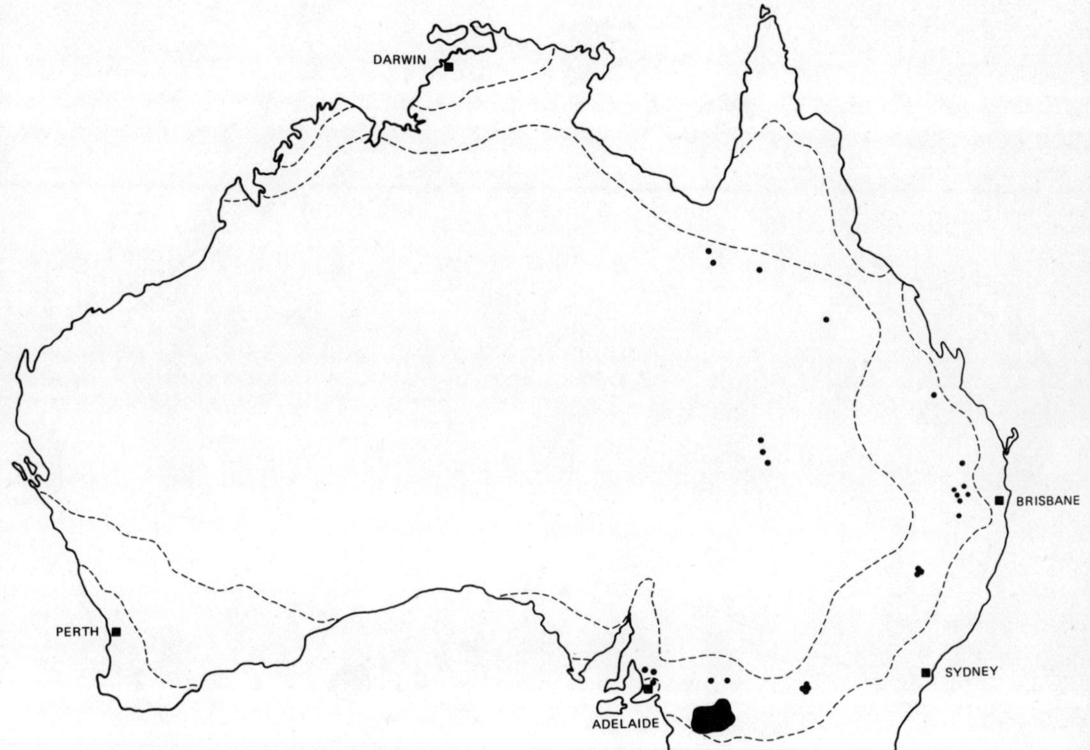

Fig. 1a. Grey, brown and red clays.

Such soils are dominated by illite, with kaolin always present, usually to the extent of 20-30 per cent, but occasionally higher. Montmorillonite is found in only 3 of the 54 profiles from the above areas. Iron oxides have not been detected in the clay fraction of most of these soils.

In those areas of northern New South Wales and Queensland under review the mineralogy of this group shows a marked change. Montmorillonite is nearly always present and is frequently the dominant clay mineral, often increasing with depth in the profile. Kaolin is also more abundant than in southern areas, but illite is only a minor constituent and frequently absent. Iron oxides do not occur except in the odd profile, with felspar occurring in profiles near Narrabri.

The higher illite in the southern soils and the high montmorillionite in the northern soils are evident from the potash and CEC figures given in table 2.

Solonized brown soils. Fig. 1*b* (51 profiles)

Norrish (1949) studied this group of soils and found the clay mineralogy particularly consistent. As with several other Great Soil Groups with widespread occurrence in

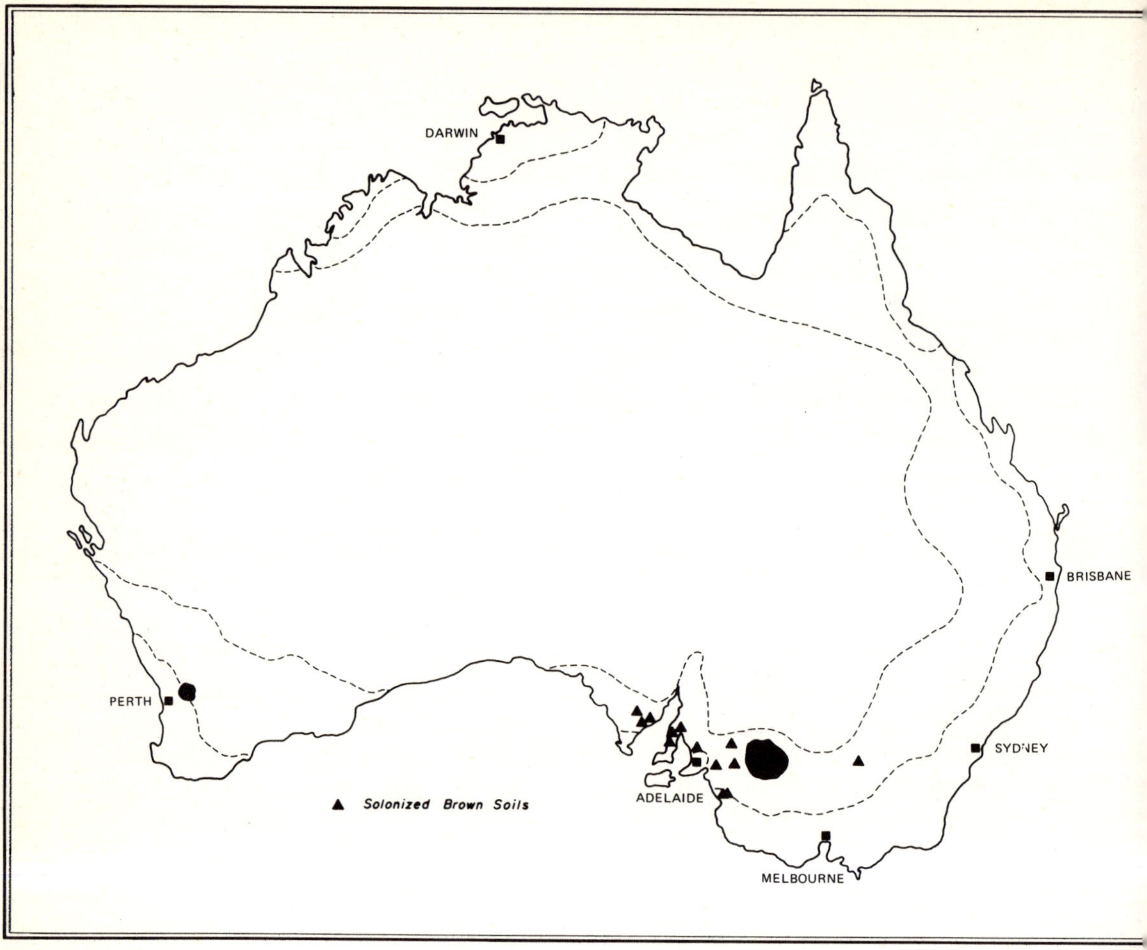

Fig. 1b. Solonized brown soils.

southern Australia, illite is the dominant mineral, with kaolin always present. In the few profiles examined from the Merredin district, Western Australia, the kaolin content usually exceeds the illite content, but elsewhere (South Australia, Victoria and New South Wales), kaolin is generally limited to 30-40 per cent. Goethite and/or hematite are detected in about 20 per cent of the profiles examined, most of them from Eyre Peninsula, South Australia.

Red brown earths. Fig. 1*c* (95 profiles)

Profiles of this agriculturally important group have been examined from all of the mainland states of Australia (Radoslovich 1958). In South Australia, soils from Eyre Peninsula, Barossa Valley, the Lower North, and the Adelaide plains have been analyzed. Other areas studied include the Wimmera in Victoria and from Bendigo northeast through much of central New South Wales and into southern Queensland. A few profiles from the Barkly Tablelands (Northern Territory-Queensland) and Merredin (Western Australia) were also studied.

In the southern part of the continent, the clay mineralogy is consistently dominated by illite, with kaolin always present, usually varying between 20 and 40 per cent and

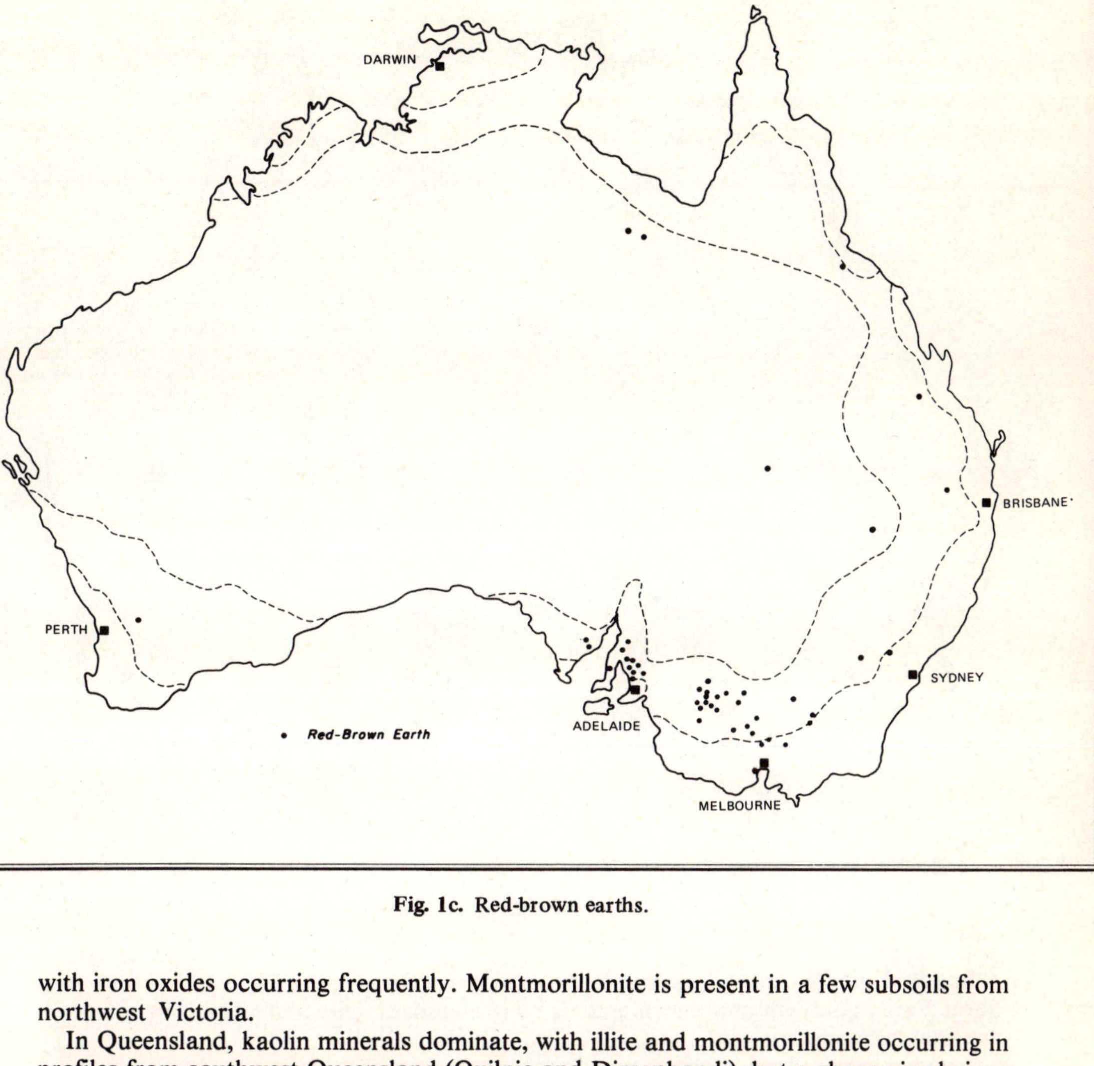

Fig. 1c. Red-brown earths.

with iron oxides occurring frequently. Montmorillonite is present in a few subsoils from northwest Victoria.

In Queensland, kaolin minerals dominate, with illite and montmorillonite occurring in profiles from southwest Queensland (Quilpie and Dirranbandi), but a change is obvious in the wetter areas further to the north and east. Profiles from Oakey, Rolleston and Charters Towers are dominated by halloysite, with a complete absence of illite and with montmorillonite sometimes present as a minor constituent.

Profiles from the Barkly Tablelands show the same illite/kaolin pattern as found in southern areas with varying amounts of montmorillonite usually present. The few profiles from Merredin, Western Australia are dominated by kaolin with lesser amounts of illite.

Black earths. Fig. 1*d* (17 profiles)

Only a small number of profiles of this somewhat scattered group have been studied, with most samples coming from the Darling Downs, Queensland. In this region, montmorillonite is the most abundant clay mineral, with kaolin present to 30 per cent or less,

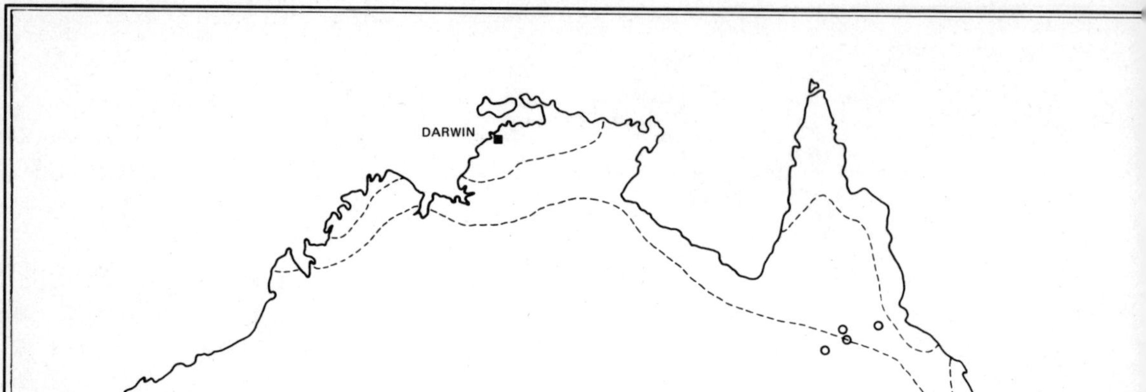

Fig. 1d. Black earths and euchrozems.

although 2 profiles from Tara are dominated by kaolin. Further west in the more arid Surat Basin, although montmorillonite is quite abundant, illite and felspar are present in considerable amounts.

Isolated profiles from Victoria and South Australia are dominated by illite, with lesser amounts of kaolin normally present, although one profile from the Barossa Valley contains a moderate amount of montmorillonite, in addition.

The general mineralogical difference between the southern and northern members of this group are shown by the higher K_2O in the former and the increased CEC for the latter.

Euchrozems. Fig. 1*d* (8 profiles)

A small number of profiles of this Great Soil Group have been studied, mostly from Queensland. Kaolin is present in all cases and is usually the only layer lattice silicate present. Montmorillonite occurs in only 2 profiles, in one as the dominant clay mineral and in the other at approximately the same concentration as kaolin. Iron oxides are always present, varying from about 10 to 30 per cent. Probably both hematite and goethite are present in each profile.

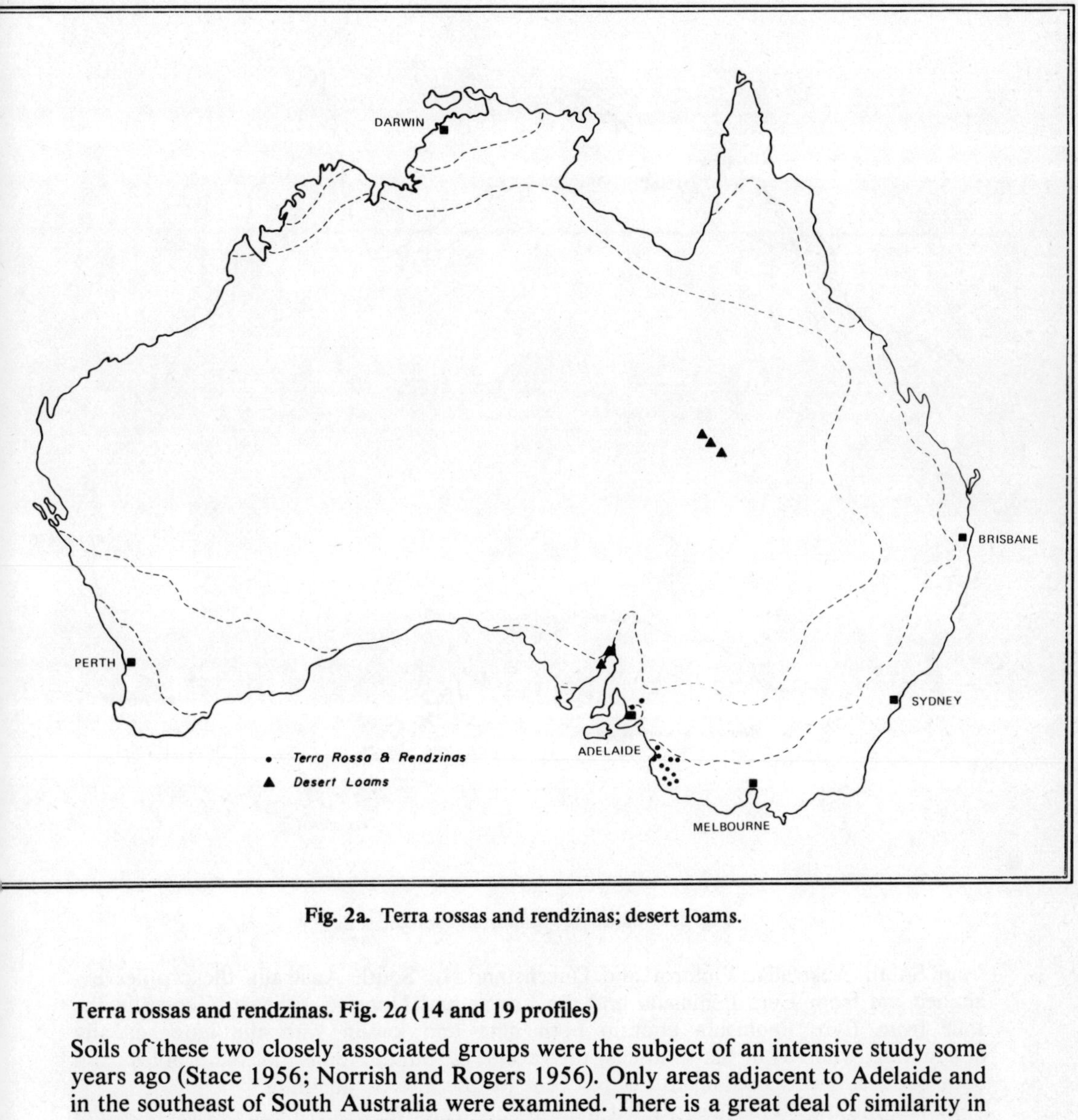

Fig. 2a. Terra rossas and rendzinas; desert loams.

Terra rossas and rendzinas. Fig. 2*a* (14 and 19 profiles)

Soils of these two closely associated groups were the subject of an intensive study some years ago (Stace 1956; Norrish and Rogers 1956). Only areas adjacent to Adelaide and in the southeast of South Australia were examined. There is a great deal of similarity in the mineralogy of these two Great Soil Groups. Illite and kaolin are the dominant clay minerals, with minor amounts of chlorite (a residue from the parent material) frequently present in subsoils from near Adelaide. Except for a few profiles, the iron oxide content of these soils is generally lower than the limit of detection.

More recent studies on additional profiles of terra rossas and rendzinas from the same general areas show mineralogy consistent with the above. However, one rendzina profile from the southeast contains 40-50 per cent montmorillonite, and another rendzina profile from the same district is dominated by palygorskite that is residual from the parent material.

Solodized solonetz and solodic soils. Fig. 2*b* (86 profiles)

Although this group of soils occurs extensively in all states most profiles analyzed are

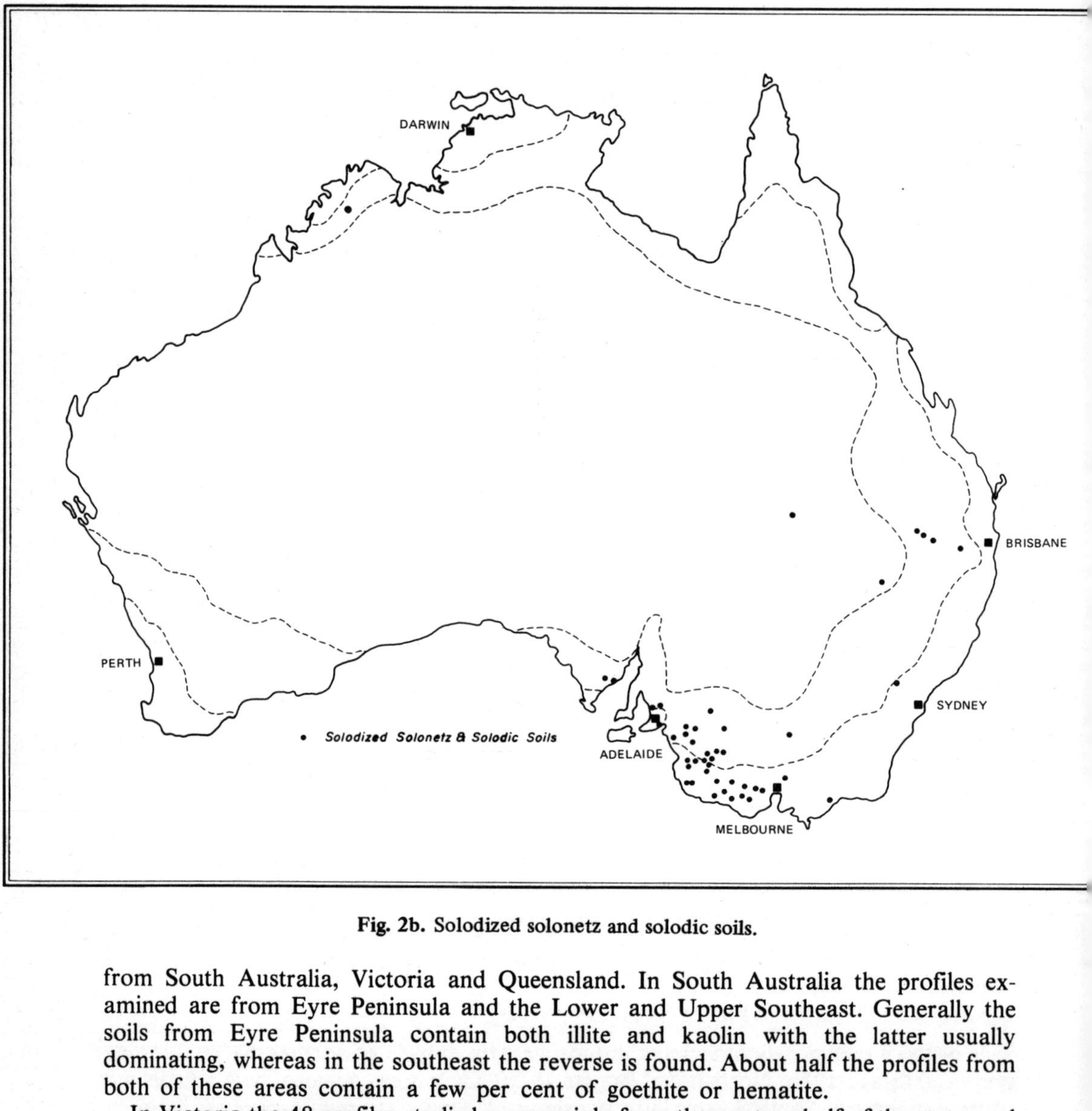

Fig. 2b. Solodized solonetz and solodic soils.

from South Australia, Victoria and Queensland. In South Australia the profiles examined are from Eyre Peninsula and the Lower and Upper Southeast. Generally the soils from Eyre Peninsula contain both illite and kaolin with the latter usually dominating, whereas in the southeast the reverse is found. About half the profiles from both of these areas contain a few per cent of goethite or hematite.

In Victoria the 48 profiles studied were mainly from the western half of the state, and kaolin is again the most abundant clay mineral. Illite is a major component in about half of these soils, and is a minor component or absent from the remainder which are formed from basalt. Montmorillonite occurs in 9 profiles, but varies markedly in distribution and concentration. Vermiculite is a very minor component in some profiles, whereas about half the profiles contain detectable amounts of goethite or hematite. Three profiles from the Colac region, Victoria, contain hydrated halloysite.

Compared to solonized brown soils, red brown earths and grey, brown and red clays, the southern soils of this group have a lower average illite content that is reflected in the K_2O content of table 2.

In southern Queensland, this group of soils is also kaolinitic, with lesser amounts of montmorillonite and occasionally illite. Felspar occurs in the clay fraction in 2 profiles, and hematite/goethite are present in only 2 of the Queensland profiles.

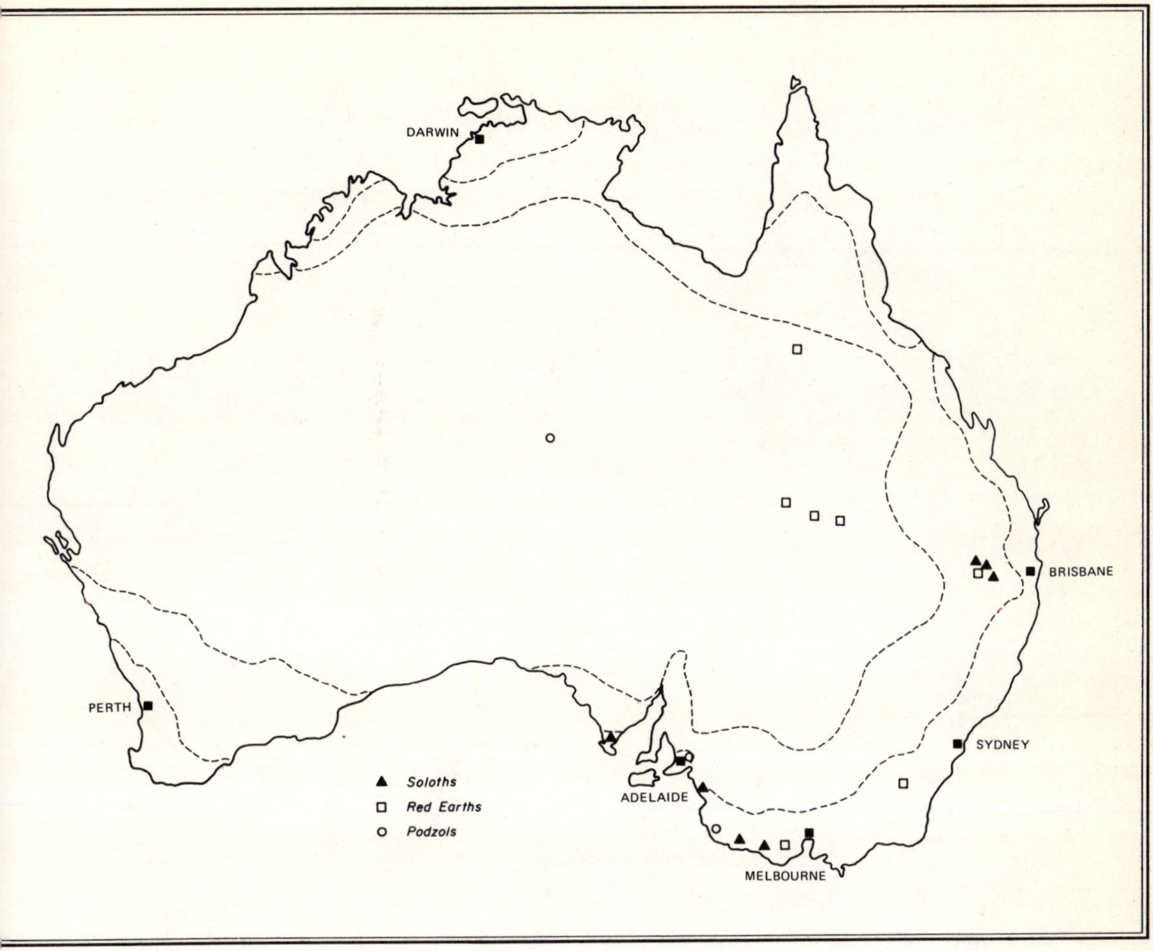

Fig. 2c. Soloths, red earths, podzols.

Soloths. Fig. 2*c* (12 profiles)

Of the 12 soloth profiles analyzed, all contained kaolin, which in South Australia is associated with quite variable amounts of illite, and small amounts of goethite. The Victorian soils are formed on basalt, have relatively low potash contents, and are virtually devoid of illite; small amounts of iron oxides are present. One profile contains 10-20 per cent felspar in the clay fraction, and another profile has quite minor amounts of inhibited vermiculite.

In Queensland the kaolin is always accompanied by quite variable amounts of montmorillonite which may increase with depth in the profile. Iron oxides are just detectable in a few profiles.

Red earths. Fig. 2*c* (10 profiles)

The largest single area of this Great Soil Group occurs in southern Queensland, and the mineralogy of profiles from this region shows a kaolin content of 70 per cent or higher in most cases. Iron oxides are not always detectable, and where present are generally less than 10 per cent. Illite is present in a majority of profiles but in only minor amounts.

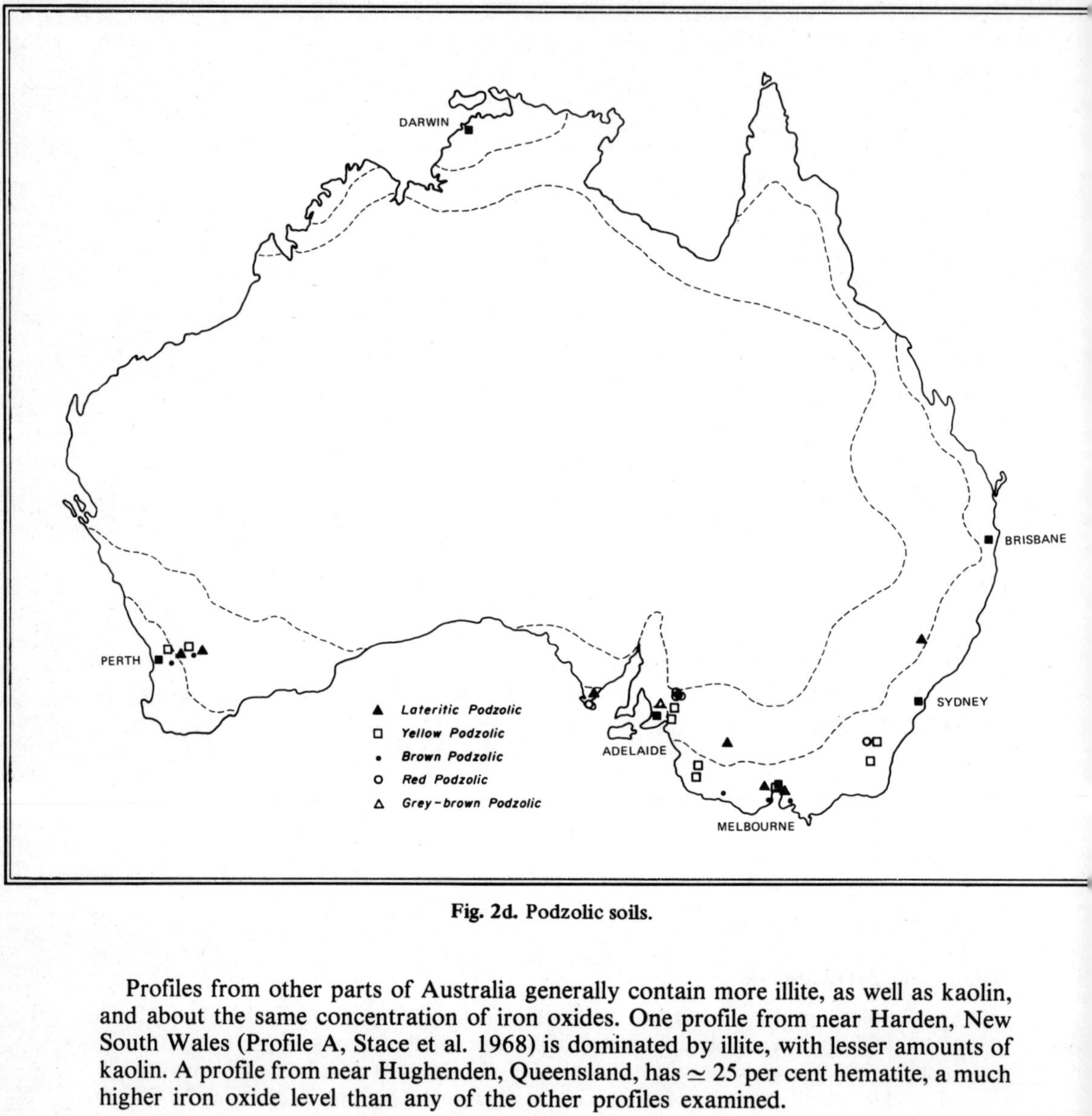

Fig. 2d. Podzolic soils.

Profiles from other parts of Australia generally contain more illite, as well as kaolin, and about the same concentration of iron oxides. One profile from near Harden, New South Wales (Profile A, Stace et al. 1968) is dominated by illite, with lesser amounts of kaolin. A profile from near Hughenden, Queensland, has $\simeq 25$ per cent hematite, a much higher iron oxide level than any of the other profiles examined.

The podzolic group of soils. Fig. 2*d* (45 profiles)

Profiles of the various podzolic groups have been analyzed from the York-Merredin area (Western Australia) and Eyre Peninsula, Barossa Valley and the southeast (South Australia). In addition isolated profiles from various localities in Victoria and New South Wales have been examined.

In the Western Australian lateritic podzolic soils there is a consistent mineralogical pattern. These soils are dominated by kaolin, with quite minor amounts of illite, inhibited vermiculite or montmorillonite commonly present, usually in the deeper parts of the horizon. Gibbsite is often present in the upper parts of the profile and can be present to the extent of 30-40 per cent of the clay fraction. Iron oxides are frequently present. It should be noted that the mica in some of these highly weathered profiles has been shown to be a relic of the parent material, and is relatively coarse-sized particles of muscovite, rather than the fine-grained illite that normally occurs in soils. Excluding the above mentioned lateritic podzolic soils from Western Australia, all the podzolic groups show quite a variable mineral composition. In these soils although kaolin is always present and is usually the dominant mineral, illite is present in 75 per cent of profiles examined in considerable proportions. In half of these profiles containing illite, the concentration of this mineral is fairly uniform in all horizons. In the other half of the samples examined the illite content decreases with depth, often very rapidly. In several profiles where this occurs the illite is trioctahedral.

Montmorillonite is present as a quite minor constituent in several profiles but assumes major proportions in a grey brown podzolic profile from the Barossa Valley.

Inhibited vermiculite is present in small amounts in several profiles, and normal vermiculite is commonly present in about 25 per cent of the profiles. Talc is seen as a quite minor constituent in 2 profiles from the Barossa Valley. The montmorillonite, inhibited vermiculite, vermiculite and talc are usually in the deeper parts of the profile. Felspar was detected in the clay fractions of a few profiles.

Small amounts of gorceixite (barium aluminium phosphate) and related phosphate minerals have been detected in grey brown and yellow podzolic soils in South Australia and lateritic podzolic soils in Western Australia. These minerals occur in the deeper part of the profile.

Podzols. Fig. 2*c* (6 profiles)

Podzol profiles from the southeast of South Australia and the Barossa Valley are all dominated by kaolin minerals, with minor amounts of illite present. Montmorillonite is a minor constituent at depth in some profiles, and vermiculite can be present also in small amounts. Iron oxides are present in a minority of profiles.

Desert loams. Fig. 2*a* (13 profiles)

Although the majority of desert loams occur where the annual rainfall is less than 250 mm, there are some areas of this Great Soil Group occurring where the rainfall exceeds 250 mm p.a. Profiles from the Port Augusta-Whyalla region of South Australia and from southwest Queensland have been studied.

The South Australian profiles are all dominated by illite, with lesser amounts of kaolin, and nearly always contain minor amounts of palygorskite. The palygorskite level may be as high as 10-20 per cent, but is usually lower. In some cases the palygorskite is only detectable by electron microscopy. In addition, iron oxides are always present in a minor amount. This mineralogical pattern is particularly consistent for desert loams in South Australia regardless of the rainfall zone.

In southwestern Queensland a significantly different mineralogy is evident. Montmorillonite and kaolin are the major minerals found with illite occurring in only minor amounts. Iron oxides are not detected.

Palygorskite is not reported in any of these Queensland profiles, but because of the difficulties (discussed elsewhere) in detecting this mineral, it could be present in small amounts.

MINERALS

Illite

Except for the highly weathered soils, e.g. podzolic, illite is usually the dominant mineral of the southern soils. Although the mica in the clay fraction of soils varies from well-crystallized mica, muscovite, to very fine-grained illite, it is the latter that is most widespread. This illite is very fine grained, often $< 0.1\mu m$, see figure 3; it has a low potash content < 6 per cent K_2O—see analysis 24, table 3, and the cation exchange capacity is higher than that usually associated with illites. Norrish (1949) showed that for the illite of the solodized solonetz soils there was an inverse relationship between K_2O content and CEC. The sum of the CEC and K_2O (expressed in me per cent) was constant approximately at 150 me per cent. Radoslovich (1958) has demonstrated the same relationship for the red brown earths. This chemical behaviour suggests that the illite is randomly interstratified (degraded). However the X-ray diffraction pattern does not show evidence of interstratification. The mica basal spacing is always at 10 Å, though the reflection may be broadened, and its higher orders are rational.

The high exchange capacity and small particle size of these minerals gives them physical properties that resemble montmorillonite. Clay soils can show considerable swelling, even when no montmorillonite is present in the clay, so that the soils may have undesirable engineering properties (Aitchison, Sprigg and Cochrane 1954; Aitchison and Holmes 1953*a*, *b*).

Prescott and Piper (1932) showed that acid extractable K_2O was correlated with the clay content of Mallee (solonized brown) soils; 3 per cent K_2O corresponding to 100 per cent clay. Piper (1938) showed that the red brown earth soils of South Australia contained appreciable amounts of potash in their clay and silt fractions; 1.5-2.9 per cent K_2O for 100 per cent silt + clay. These high potash contents reflect the high illite in the

Fig. 3. Micrograph of a typical soil illite. (CSIRO photograph.)

clay, and all those soils high in illite have high total potash contents. Although the potassium of micas was once regarded as completely non-exchangeable, and unavailable for plant growth, a considerable amount of research (see review by Norrish, 1973) has been carried out on this subject in the last decade, and the results show that when the potassium in solution is low in concentration the structural potassium of mica can be exchanged with other ions. Micas such as phlogopites and biotites (rare in the soils under discussion) are capable of supplying potassium in solution at concentrations in the range 2-100 ppm (Raussell-Colom et al. 1965; Scott and Smith 1966; Newman 1969; Rousseaux et al. 1973). This concentration is adequate for plant growth. Muscovite and illite will only supply potassium to solutions that are approximately 0.1 ppm or less in potassium (Scott, Hunziker and Hanway 1960; Chute and Quirk 1967; Scott and Smith 1966). Concentrations in this range are probably sufficient for plant growth as Asher and Ozanne (1967) have shown that if the potassium in solution is > 0.4 ppm then growth is not limited; from 0.4-0.04 ppm growth was limited with the lowest concentrations giving severe potassium deficiency. It seems then that soils high in illite are unlikely to show serious potassium deficiency except where the rate of demand is high.

It is probably not coincidence that the level of potassium required for plant growth matches that which can be supplied by illites.

A further important result from the work on micas is the surprising discovery that it is more difficult to exchange the lattice potassium of fine-grained micas than of coarse-grained micas (Scott 1968; Scott and Smith 1967). This behaviour undoubtedly explains the stability of illites in soil.

Too little is known of the geology and history of most of the soils to speculate in detail on the origin of the illite. In certain cases (Radoslovich 1958; Norrish and Rogers 1956) the illite appears to have its origin in the fine-grained micas of the parent rocks, but its overall distribution is not consistent with this as a general explanation. Except for those soils that have had a history of severe leaching (lateritic podzolics), and those that have basalt as a parent material, the soils of southern Australia are fairly consistently high in illite. Illite tends to be absent or low in the northern soils. The difference is obviously a climatic one. Many of the southern soils are exceedingly old and this indicates that under a Mediterranean-type climate illites may be stable indefinitely. It seems likely that illites may even form in arid soils providing there is a source of potassium.

Kaolin

Some kaolin is present in almost all the soils considered here. It tends to be the dominant clay mineral of the northern soils and those southern soils that have had a history of severe weathering (lateritic podzolics) or have formed on parent materials with little or no potassium. Many of the Victorian soils are high in kaolin and this is undoubtedly due to basalt being the parent material of many of them.

Where kaolin is present with other clay minerals, it is often difficult to identify it more specifically, but halloysite is found in some soils while well-crystallized kaolinite is often the dominant mineral of the highly weathered soils; see figure 4. Kaolin certainly forms in soils.

Montmorillonite

The only soil group in which montmorillonite is usually dominant is the black earths, and this is for the northern soils that usually have basalt as their parent material. The black earths of southern Australia are often very high in illite. Unusual soils high in montmorillonite occur erratically in a variety of the other Great Soil Groups and its dominance generally reflects local geology or drainage.

In some of the more highly weathered soils the montmorillonite and vermiculite have

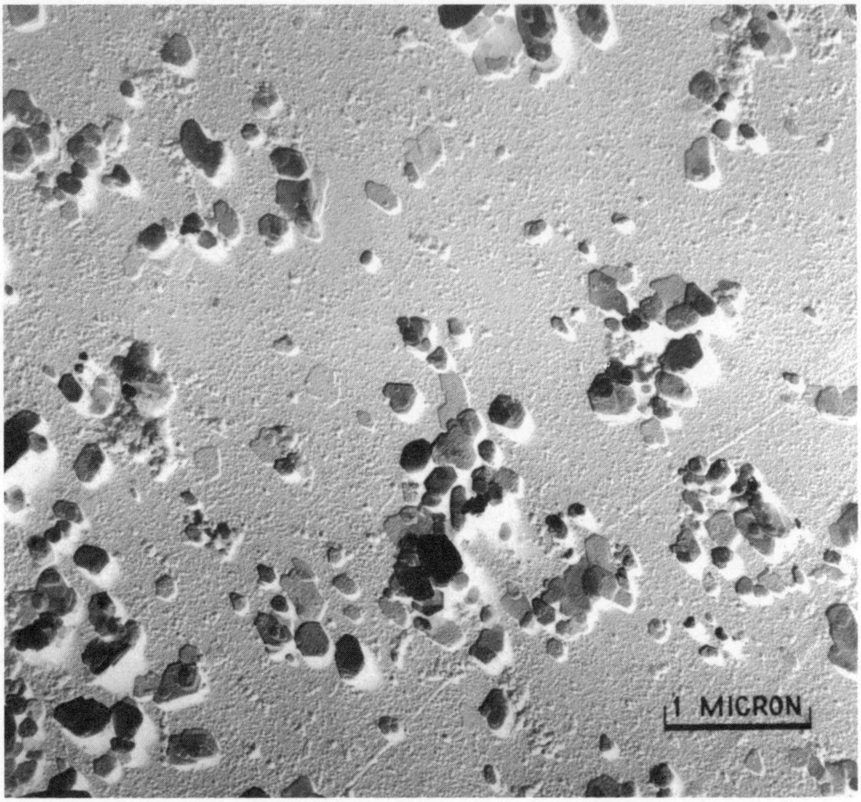

Fig. 4. Micrograph of a soil kaolin. (CSIRO photograph.)

polymerized hydroxy-aluminium ions in the inter-layer region. These "chloritized" or "inhibited" minerals exhibit reduced swelling and exchange capacities.

Montmorillonite may form directly in the soil or from illites that have lost potassium.

Palygorskite

Apart from several other isolated occurrences, palygorskite occurs in the more arid soils. In the soils having a rainfall of less than 250 mm p.a., the mineral is comparatively common, and present in sufficient amount to be easily identified. Often the mineral is present as a very fine coating, see figure 5, on the surfaces of soil aggregates and minerals (Singer and Norrish 1974) and in these circumstances the mineral may not be identified by the normal methods of analysis. Singer and Norrish (1974) have studied the chemistry of the palygorskite and of these arid soils and have concluded that palygorskite forms in such soils where the soil solution is high in Si and Mg and has a high pH. As mentioned earlier, it is difficult to identify small quantities of palygorskite in the presence of illite, and it seems likely that this mineral may be very common in low concentrations in the more arid soils.

In those instances where palygorskite is reported in the higher rainfall areas its presence is as a residue of parent material (Rogers, Martin and Norrish 1954) or the result of some other special sequence of events (Beattie 1970).

Fig. 5. Micrograph of a soil palygorskite. (CSIRO photograph.)

Primary rock minerals

Only rarely are primary rock minerals observed in the clay fractions of soils. Felspar is found in particular localities, but generally such minerals as felspars, amphiboles, pyroxenes, chlorites, biotites and phlogopites are absent from the clay. Many of the soils are old and the above results indicate considerable weathering. Similar soils of the northern hemisphere are often reported to have moderate concentrations of primary minerals in the clay fraction. The presence of these minerals is indicative of less weathering, younger soils, and often the primary minerals in such soils were reduced to clay size by glacial processes. None of the semi-arid Australian soils have been formed from recent glacial till.

Accessory minerals

The clay minerals usually comprise the major portion of the fine fraction of soils, but various accessory minerals are present in small amounts. The oxides and hydroxides, etc. can be of importance in influencing the physical or chemical properties of a soil even when their concentration is low.

No systematic study of these minerals has been made for the group of soils under consideration and very often pretreatments are designed to remove the minerals before analyses. The following discussion is intended to draw attention to the investigations of these minerals in Australian soils.

Silica

Small amounts of quartz are present in almost all clays; the amount ranges from a fraction of a per cent up to about 20 per cent for the $< 2\mu$ fraction of sandy soils. The other crystalline polymorphs of silica are rare in soils.

Amorphous silica in the form of plant phytoliths is undoubtedly present in most surface soils.

Silica can act as a cement and in soils more arid than those considered here hardpans due to silica cementation are common.

Iron oxides

Goethite, α FeOOH, hematite, α Fe_2O_3 and maghemite, γ Fe_2O_3, are comparatively common minerals in the semi-arid soils. The red colour of soils is due to hematite, but despite this, goethite (yellow in colour) is the most abundant of these oxides. It is rare for a soil to contain only hematite, but goethite may be the only iron oxide. Mostly mixtures occur and then even a very small amount of hematite imparts a red colour. Maghemite occurs in small amounts in the clay fractions of soils, and is often present in high concentrations in the ironstone gravel, as are hematite and goethite (Taylor and Schwertmann 1974). Maghemite is a ferro-magnetic mineral and is usually responsible for the magnetism of soils.

Norrish and Taylor (1961) have studied soil goethites and show that they contain considerable amounts of aluminium in solid solution, up to 30 mole per cent diaspore. They also found that soil goethites are exceedingly fine grained, ca 0.02 μm. The high specific surface area of soil goethites is of importance since a small concentration of the mineral may affect soil properties.

It is very likely that free iron oxides fix phosphorus (see review by Wild, 1950) and perhaps other nutrients, and they may affect soil structure (Lutz 1936; McIntyre 1956).

Aluminium oxides

Gibbsite, $Al(OH)_3$, is the only aluminium oxide or hydroxide found in these soils, and then it is probably a relic from a past, more severe weathering. There have been no studies of the chemistry of soil gibbsite. It has been suggested that free aluminium oxides are important in fixing phosphorus (Muljadi, Posner and Quirk 1966; see also review by Wild, 1950). This is unlikely in the soils under consideration as, apart from the rarity of gibbsite, where it occurs its particle size is comparatively large. Generally in soils the particle size of gibbsite is in the micron range so that, even given a surface activity equal to that of iron oxides, its lower surface area will reduce its effects compared to iron oxides.

Manganese oxides

Taylor, McKenzie and Norrish (1964) have examined the mineralogy and chemistry of manganese oxides, and some of their samples were from the soils under consideration. Their results show that although lithiophorite, birnessite and hollandite were common mineralogical forms, in all cases the chemistry of the oxides was complex and it did not vary much with mineral form. The manganese oxides had high concentrations, compared to the soils, of cobalt, nickel and barium. Taylor, McKenzie and Norrish (1964) found that, like the iron oxides, the manganese oxides have a very small particle size of between 0.02 and 0.1μm. Taylor and McKenzie (1966) showed that for a wide variety of soils most of the soil cobalt (79 per cent) was associated with the manganese oxides. In subsequent laboratory studies McKenzie (1967, 1970) has shown that manganese oxides fix cobalt irreversibly, and Adams et al. (1969) found that in the field the uptake by pastures of applied cobalt could be related to the manganese content and pH of the soil. McKenzie (1972) has reviewed this work on these minerals.

Titanium oxides

Both rutile and anatase occur in small amounts in almost all soil clays. The anatase concentration is usually greater than that of rutile.

Carbonates

The presence of carbonate is one of the important morphological features of the semi-arid soils of Australia. Of the soils discussed here, all may contain carbonate except red earths, the various podzolic soils and podzols. Generally carbonates are removed by acid washing prior to separating the clay from a soil so that for most soils there is no mineralogical information on these minerals. In some cases the acid wash was omitted and in other soils, where the carbonate occurred in recognizable concentrations, it was sampled by hand picking.

Calcite is the usual soil carbonate, but in many soil profiles dolomite has also been found. Invariably where dolomite occurs it underlies calcite; the upper part of the calcareous horizon is calcite; both dolomite and calcite occur below this, the dolomite increasing in relative concentration with depth. In the lower parts of some calcareous horizons only dolomite is present.

This differentiation is undoubtedly a soil chromatographic process and is the result of dolomite being more soluble than calcite in carbonic acid.

Phosphates

Norrish (1968) has studied the mineralogy and chemistry of some of the phosphate minerals of these soils. The minerals identified were all members of the plumbogummite group of minerals, with gorceixite (Ba member), florencite (rare earths) and crandallite (Ca), being the common minerals. Usually the minerals had a very complex chemistry and contained Ba, Ca, Sr and rare earths. These minerals occurred in the coarse clay fraction of the soil and are considered to have formed in the soil. In some soils these minerals account for all the phosphorus in the soil, but generally they contained only a fraction of the total phosphorus.

Gorceixite and the related minerals occur in a variety of Great Soil Groups and over a variety of parent materials. The existence of the minerals in lateritic podzolic soils indicates a high resistance to weathering. Where the mineral accounts for most of the phosphorus of the soil, the evidence is that the phosphorus was not available for plant growth.

Although these minerals are responsible for low phosphorus availability in some Australian soils, they were not identified in many other phosphorus deficient soils.

Sulphates

Gypsum, barite, alunite and jarosite all have been identified in these soils.

Gypsum is moderately common in the more arid areas.

Barite has not been identified in the clay fraction of the soil, but occasionally occurs as nodules in the deeper horizons.

Alunite has been found in the clay fractions of several solonized brown soils. It is more common, and is present in higher amounts, in very arid soils. Alunite is common in the salt lakes of the arid and semi-arid areas (King 1953).

Very close to the ocean, alunite is often observed in clay soils as soft white clayey lumps. Its formation would seem to be the result of an interaction of clay with sea water.

Jarosite has been found as pale yellow, fine-grained lumps in swampy soils from various areas. Like alunite, jarosite has also been found in soils very near the ocean. In this case the jarosite was present in the iron-rich horizon of a relic lateritic podzolic soil, and again it would seem that it has resulted from a reaction of the soil (iron oxides) with sea water.

GEOGRAPHICAL DISTRIBUTION

Insofar as we can generalize the broad-scale distribution of clay minerals is as follows. In the south where the rainfall is highest in the winter and there is a hot dry summer, illite is usually the dominant clay mineral irrespective of Great Soil Groups. There are general and local exceptions to this. Illite will be low in those cases where the parent material of the soil lacks the necessary potash, as is the case for many of the Victorian soils formed from basalt. Likewise soils formed on relic laterites contain little or no mica, the kaolin persisting as the main mineral. The podzols that come from the higher rainfall areas generally have a reduced illite content and a higher kaolin content. In an area of soils high in illite, impeded drainage caused by local topography can give rise to soils in which the illite is largely depleted in potash, i.e. montmorillonite or a randomly interstratified mixture of the two minerals becomes dominant.

In the more arid southern soils, palygorskite is often present in small amounts.

The northern soils have a much lower average illite content than the southern soils, and this is clearly shown by the potash averages of table 2. Montmorillonite or kaolin dominates these soils. For some soils there is a tendency for montmorillonite to be highest in the drier areas of western Queensland with kaolin becoming higher in the eastern soils. However the pattern is erratic and obviously influenced by parent material.

REFERENCES

Adams, S.N., Honeysett, J.L., Tiller, K.G., and Norrish, K. (1969). *Aust. J. Soil Res.* **7**, 29-42.

Aitchison, G.D. and Holmes, J.W. (1953*a*). *Aust. J. Appl. Sci.* **4**, 244-59.

———. (1953*b*). *Aust. J. Appl. Sci.* **4**, 260-73.

Aitchison, G.D., Sprigg, R.C., and Cochrane, G.W. (1954). Geol. Surv. S. Aust. Bull. No. 32.

Asher, C.J., and Ozanne, P.G. (1967). *Soil Sci.* **103**, 155-61.

Beattie, J.A. (1970). *Aust. J. Soil Res.* **8**, 145-56.

Brown, G. (1961). "The X-ray Identification and Crystal Structures of Clay Minerals". Mineralogical Society, London.

Chute, J.H., and Quirk, J.P. (1967). *Nature* (*Lond.*) 213, 1156-7.

Cole, W.F. (1943). *Soil Sci.* **56**, 153-71.

Grim, R.E. (1953). "Clay Mineralogy". McGraw-Hill, New York.

Heller, L. (1961). *Acta Universitatis Carolinae. Geologica Supplementum* **1**, 173-80.

Hosking, J.S. (1940). *J. Coun. Sci. Ind. Res. Aust.* **13**, 206-16.

———. (1957). Clay mineralogy and parent materials. 2nd Aust. Conf. Soil Sci. Melbourne Paper No. 7.

Hosking, J.S., Neilson, M.E., and Carthew, A.R. (1957). *Aust. J. Agric. Res.* **8**, 45-74.

King, D. (1953). *Econ. Geol.* **48**, 689-703.

Loughnan, F.C. (1969). "Chemical Weathering of the Silicate Minerals". American Elsevier, New York.

Lutz, J.F. (1936). *Soil Sci. Soc. Amer. Proc.* **1**, 43-5.

McIntyre, D.S. (1956). *J. Soil Sci.* **7**, 302-6.

McKenzie, R.M. (1967). *Aust. J. Soil Res.* **5**, 235-46.

———. (1970). *Aust. J. Soil Res.* **8**, 97-106.

———. (1972). *Z. Pflanzenernahrung Bodenk.* **131**, 221-42.

Muljadi, D., Posner, A.M., and Quirk, J.P. (1966). *J. Soil Sci.* **17**, 212-47.

Newman, A.C.D. (1969). *J. Soil Sci.* **20**, 357-73.

Norrish, K. (1949). Soil colloids of the Solonized Brown Soils. Paper to ANZAAS, Hobart, 1949.

———. (1968). *Trans. 9th Int. Congr. Soil Sci.* 2, 713-23.

———. (1973). Proc. 1972 Int. Clay Conf. (Madrid). pp. 417-32.

Norrish, K., and Hutton, J.T. (1969). *Geochim. Gosmochim. Acta.* **33**, 431-53.

Norrish, K., and Rogers, L.E.R. (1956). *J. Soil Sci.* **7**, 294-301.

Norrish, K., and Taylor, R.M. (1961). *J. Soil Sci.* **12**, 294-306.

Northcote, K.H. (1971). "A Factual Key for the Recognition of Australian Soils". Rellim, Adelaide.

Piper, C.S. (1938). *Trans. Roy. Soc. S. Aust.* **62**, 53-100.

Prescott, J.P., and Piper, C.S. (1932). *Trans. Roy. Soc. S. Aust.* **56**, 118-47.

Radoslovich, E.W. (1958). *J. Soil Sci.* **9**, 242-51.

Raussell-Colom, J.A., Sweatman, T.R., Wells, C.B., and Norrish, K. (1965). "Experimental Pedology". (Eds. E.G. Hallsworth and D.V. Crawford.) pp. 40-72. Butterworths, London.

Rogers, L.E.R., Martin, A.E., and Norrish, K. (1954). *Min. Mag.* 30, 534-40.
Rousseaux, J.M., Nathan, Y., Vielvoge, L.A., and Herbillon, A. (1973). Proc. 1972 Int. Clay Conf. (Madrid). pp. 449-56.
Scott, A.D. (1968). *Trans. 9th Int. Congr. Soil Sci.* 2, 649-60.
Scott, A.D., Hunziker, R.R., and Hanway, J.J. (1960). *Soil Sci. Soc. Amer. Proc.* 24, 191-4.
Scott, A.D., and Smith, S.J. (1966). Proc. 14th Nat. Conf. Clays and Clay Minerals. pp. 69-81.
———. (1967). Proc. 15th Nat. Conf. Clays and Clay Minerals. pp. 357-73.
Shearer, J., and Cole, W.F. (1939*a*). *J. Roy. Soc. West. Aust.* 26, 121-31.
———. (1939*b*). *J. Roy. Soc. West. Aust.* 26, 133-7.
Singer, A., and Norrish, K. (1974). *Amer. Min.* 59, 508-17.
Stace, H.C.T. (1956). *J. Soil Sci.* 7, 280-93.
Stace, H.C.T., Hubble, G.D., Brewer, R., Northcote, K.H., Sleeman, J.R., Mulcahy, M.J., and Hallsworth, E.G. (1968). "A Handbook of Australian Soils". Rellim, Adelaide.
Taylor, R.M., and McKenzie, R.M. (1966). *Aust. J. Soil Res.* 4, 29-39.
Taylor, R.M., McKenzie, R.M., and Norrish, K. (1964). *Aust. J. Soil Res.* 2, 235-48.
Taylor, R.M., and Schwertmann, U. (1974). *Clay Minerals* 10, 289-98.
Turton, A.G., Marsh, N.L., and McKenzie, R.M. (1961). CSIRO Aust. Div. Soils, Div. Report 8/61.
Turton, A.G., Marsh, N.L., McKenzie, R.M., and Mulcahy, M.J. (1962). CSIRO Aust. Soil Publ. No. 20.
Wild, A. (1950). *J. Soil Sci.* 1, 221-38.

CHAPTER 4

Air-water-heat relationships

D.E. Smiles

INTRODUCTION

It is the purpose of this review to offer a general survey of the transport of air, heat and water in soils in a semi-arid environment.

Within this framework some aspects of soil physics have been consciously neglected as being of marginal relevance to the chapter. Furthermore, the chapter does not aim to provide a definitive account of the subject but seeks only to indicate the present level of knowledge and to provide a fairly basic bibliography to assist the reader to pursue particular topics in greater depth, should he so desire.

We begin by noting that the interactions of air, water and heat with soils, while apparently dissimilar, yet maintain some unity in the similarity of their flux and continuity equations, and also because their transfer coefficients depend strongly on soil structure. This does not imply that the transfer coefficients can be quantitatively predicted if the structure is defined although quantitative judgments are possible, and biological response to particular structural changes is predictable. The monograph edited by Shaw (1952) surveys information of this type. Having made this statement it should be noted that gas flow through soils is generally described in terms of the amount of air-filled pore space, while considerable effort is applied to the problem of predicting hydraulic conductivity-moisture content relations from the moisture characteristic (and by implication the pore size distribution) using methods introduced by Childs and Collis-George (1950).

It should also be noted that the movements of heat, gases and liquid water are not independent aspects of the physical relations of soils. For example, the transfer of water vapour is essentially a gaseous transport phenomenon that has associated with it the transfer of latent heat. As the chapter develops, problems of this nature will be identified.

SOIL WATER

Rigid soils

In rigid soils it is conventional to model water movement mathematically from bases provided by:

(1) The soil moisture characteristic
(2) Equations describing the volume flux of water (and water vapour) in response to the gradient of a scalar potential, and
(3) The law of continuity of matter.

Soil moisture characteristic

The hysteretic relationships between the volumetric moisture content θ of a soil, and the soil moisture potential Ψ are called moisture characteristics. They are assumed to be characteristic properties of stable non-swelling soils and arise because of the capillary and adsorptive interaction of the water and solid. The physical bases of the forces involved are well known (see for example, Childs 1969) although a knowledge of these is not essential to the measurement of Ψ. The value of $\Psi(\theta_1)$ is conveniently defined, for example, as the work per unit weight of water necessary to remove an infinitesimal element of water from a pool of water at chemical equilibrium with the soil water, and at the same height in the gravitational field to the soil at moisture content θ_1. When $\Psi = 0$, a non-swelling soil is saturated with water.

It should be noted that if the potential of the soil water is defined relative to a pool of pure water, then it will include an osmotic component due to the presence of solutes in the water. The importance of the osmotic potential in liquid water movement is at present under examination, but it has become conventional to assume, in dealing with non-saline soils, that its contribution is insignificant. In very dry soils, however, and where vapour transfer is important, the osmotic potential differences cannot be neglected. For example, the principal difference between the pressure membrane and vacuum desiccator methods for determining $\Psi(\theta)$ of soils arises because equilibration in the former equipment occurs through a membrane permitting passage of solution, but in the latter, the "membrane" is an air gap and permits passage only of solvent. It should be noted that ionic distributions associated with the net negative charge of clay mineral surfaces in the soil are not considered to contribute to the osmotic potential.

Hysteresis in the soil moisture characteristic was examined systematically by Haines (1930) while Poulovassilis (1962) applied the independent domain theory of Everett (1955) to permit quantitative prediction of the phenomenon.

In essence, the independent domain theory permits the construction of a diagram within which each elemental domain is identifiable in terms of a volume, a Ψ value at which it wets, and one at which it drains. The diagram permits the prediction of θ, or Ψ, of the material following a defined sequence of wetting and draining cycles. Topp and Miller (1966) and Topp (1969) have questioned this theory, and Topp (1971) has suggested modifications to permit the consideration of limited dependence.

As Philip (1972*a*) points out, there is indeed no *a priori* reason why the theory should be exact. Its merit, however, lies in the fact that for most practical situations (Talsma 1970; Poulovassilis 1970) it is apparently adequate and computationally more convenient than the amended version of Topp (1971) which, in its turn, is subject to most of the conceptual weaknesses of the original theory based on the independent domain concept.

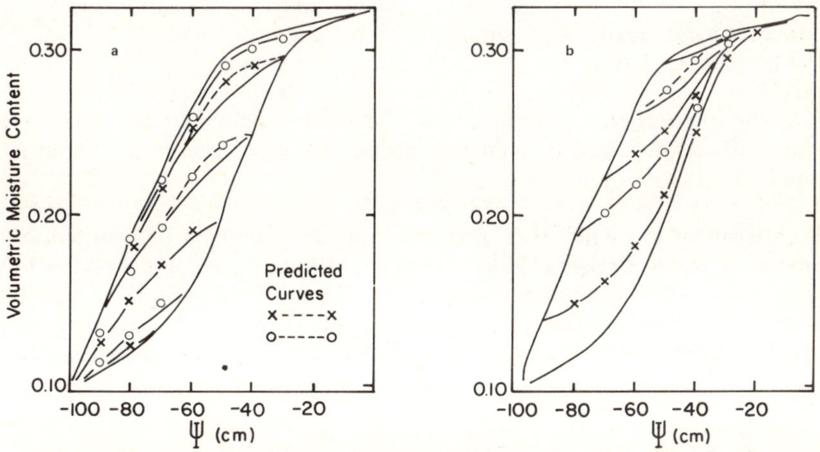

Fig. 1. Moisture characteristics showing hysteresis, for a non-swelling fine sandy soil. In these diagrams the continuous lines represent experimental data while the broken lines and points, x, o represent points predicted using the independent domain theory of Poulovassilis (1962). Figure 1*A* refers to draining situations while figure 1*B* refers to wetting. The results are due to Talsma (1970).

Typical hysteretic behaviour in $\Psi(\theta)$ in a fine sandy soil is shown in figure 1. These data are reported by Talsma (1970) and permit comparison of draining (*a*) and wetting (*b*) scanning curves. The predicted curves were based on the theory of Poulovassilis (1962), and demonstrate that while this approach is not perfect, yet its predictive value is considerable.

The flux equation

The isothermal flow of liquid water in soils is described by Darcy's Law in the form

$$\bar{v} = -K \text{ grad } \Phi \tag{1}$$

In this equation $\bar{v}$ is the macroscopic volume flux density relative to the solid matrix, K is the hydraulic conductivity and Φ is the total potential of the soil water, defined for a non-swelling soil in equations 2 and 3.

$\bar{v}$ is defined formally in terms of the rest frame of the soil particles since this definition is crucial to the extension of the theory of water movement to swelling soils.

K, as it was originally defined in Darcy's experiments, applied to saturated sands. Buckingham (1907), however, recognized that it was likely to depend on the water content and in particular to be dependent on the "thickness and extent" of the water films. Later Richards (1931) discussed the flow of water about air-filled voids as if they were solid, and Childs and Collis-George (1950) demonstrated experimentally that Darcy's Law is indeed appropriate to water flow in unsaturated soils. The majority of experiments indicate K to be effectively a single valued function of the water content, θ, and through the moisture characteristic, K may be considered a function of Ψ also. Since $\Psi(\theta)$ is hysteretic, so also is $K(\Psi)$.

It should be noted that the uniqueness of $K(\theta)$ has also been the subject of some controversy (see for example, Poulovassilis 1969). The magnitude of any effect appears, however, to be so small relative to the errors involved in the measurements, and to the magnitude of the variation of K with θ, that the phenomenon appears practically not to be a problem.

Figure 2 shows characteristic $\Psi(\theta)$ and $K(\theta)$ functions for wetting of the light clay soil of Moore (1939).

Richards' (1931) contention that the liquid water flows about the air-filled void space as if it were part of the solid matrix has implicitly received theoretical confirmation by Philip (1972*b*) who has shown that significant momentum transfer from the water to the gas phase through interfaces in the soil is unlikely, and by implication the geometry of the interfaces during steady flow should be constant.

In a saturated, rigid soil

$$\Phi = Z + P \tag{2}$$

where Z is the gravitational component of the total potential of the water measured positive upwards relative to a convenient datum, and P is the pressure potential. In an unsaturated soil

$$\Phi = Z + \Psi. \tag{3}$$

Some criticism of the validity of Darcy's Law when applied to field situations has been reported. Swartzendruber (1968) and Philip (1969), for example, review in general the conditions for which it should, in theory, be valid. It now appears that much of the criticism arose from examination of spurious results of poorly executed or interpreted experiments. Smiles (1969) discusses the problem as it applies to saturated swelling materials.

The continuity equation

The general flow equation for water in a rigid soil in which sources or sinks (for example, plant roots) occur is formulated in terms of continuity of matter. It is written

$$\frac{\partial \theta}{\partial t} + q = -\text{div } \bar{v} \tag{4}$$

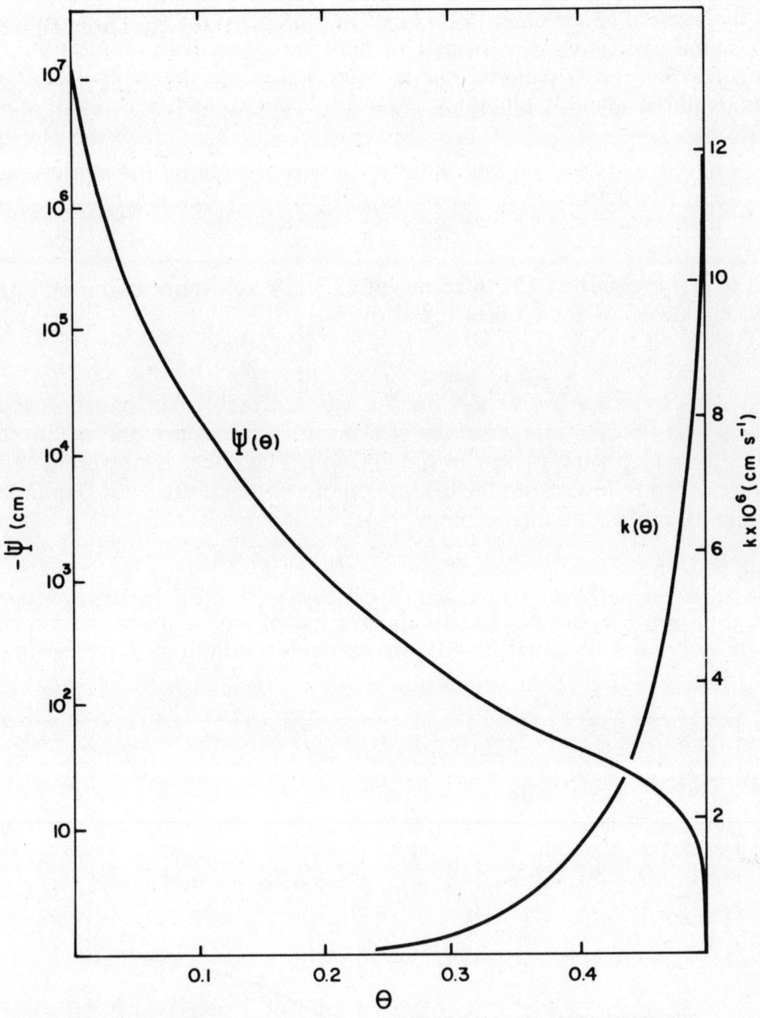

Fig. 2. Graphs showing $K(\theta)$ and a wetting $\Psi(\theta)$ relation for the light clay soil of Moore (1939).

where t is time, and q is the local rate of removal of water per unit volume by sinks.

Combined with equation (1), equation (4) becomes

$$\frac{\partial \theta}{\partial t} + q = \text{div } (K \text{ grad } \Phi) \tag{5}$$

Little work is reported on non-steady flow in terms of this equation however because it is argued that suitable approximations to most field flow situations are either one-dimensional or exhibit cylindrical or radial symmetry.

For one-dimensional vertical flow, and where sources or sinks can be neglected, equation (5) is written

$$\frac{\partial \theta}{\partial t} = \frac{\partial}{\partial z}\left(K\frac{\partial \Psi}{\partial z}\right) - \frac{\partial K}{\partial z} \tag{6}$$

where z is the vertical coordinate. Boundary conditions (9) for equation (6) make it convenient to define z positive downwards.

In this equation, the first term on the right-hand side describes flow due to the moisture potential gradient, while the second term represents the gravitational contribution.

If flow is non-hysteretic, the soil uniform, and $\Psi(\theta)$ unique for static and dynamic situations, the introduction of a soil water diffusivity D defined by

$$D = K\frac{d\Psi}{d\theta} \tag{7}$$

permits us to write equation (6) in terms of either Ψ or (more commonly) θ only. In terms of θ, equation (6) then takes the form

$$\frac{\partial\theta}{\partial t} = \frac{\partial}{\partial z}(D\frac{\partial\theta}{\partial z}) - \frac{dK}{d\theta}\frac{\partial\theta}{\partial z} \tag{8}$$

Figure 3 shows the $D(\theta)$ function for the soil represented in figure 2. It should be noted that for this function, the increase in D for $\theta < 0.06$ represents an increased contribution due to water vapour flow in the dry soil. The effect is not predicted from the data of figure 2, but follows from the discussion of soil air. It does not significantly affect the behaviour described in this section.

Infiltration

The study of infiltration, drainage and capillary rise of water in soils is to a large extent the study of solutions of equation (8). In particular, solutions have been sought for

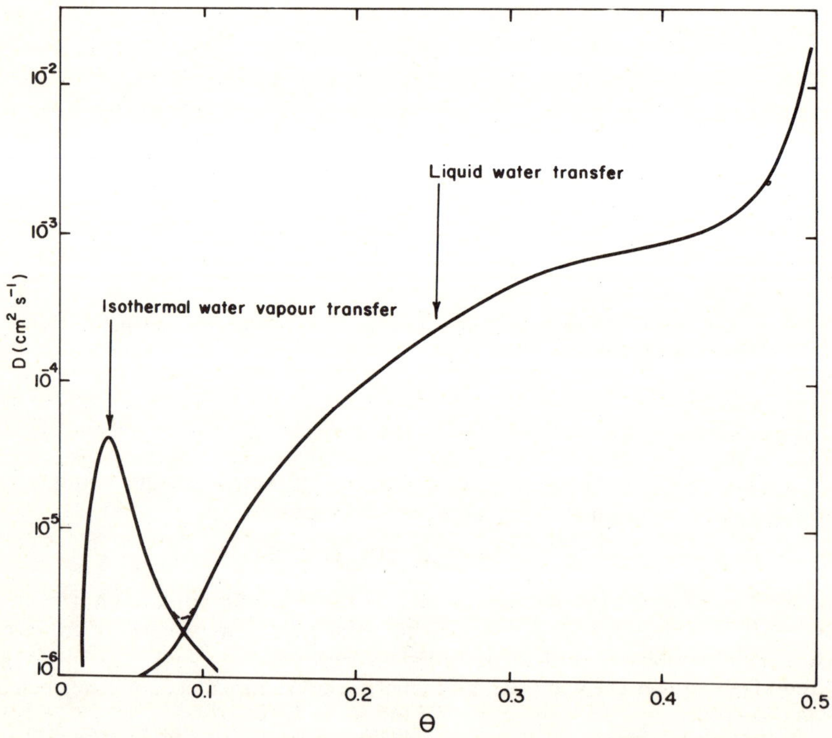

Fig. 3. The relationship between the soil water diffusivity and the moisture content θ. The liquid water curve is derived from figure 2 using the equation $D = K\,d\Psi/d\theta$. The vapour diffusivity is calculated using $\Psi(\theta)$ and the method of Philip and de Vries (1957).

situations where water is applied and maintained at the surface of a very long column of uniform soil within which there is initially a uniform moisture profile, i.e. for conditions described by

$$\begin{aligned} \theta = \theta_n \quad & z \geqslant 0 \quad t = 0 \\ \theta = \theta_o \quad & z = 0 \quad t > 0 \end{aligned} \tag{9}$$

The solution of equation (8) subject to conditions (9) may be written in the form (Philip 1969),

$$z(\theta, t) = \chi_1 t^{1/2} + \chi_2 t + \chi_3 t^{3/2} + \ldots \tag{10}$$

where the χ's are solutions of ordinary differential equations and in particular, χ_1 is the solution of the sorption (gravity free) problem where the second term on the right-hand side of equation (8) is zero.

When the signs of the even-numbered coefficients on the right-hand side of equation (10) are reversed, the equation describes the phenomenon known as capillary rise (Philip 1966*b*).

If θ_n is not zero, K_n will be finite, and the cumulative infiltration, i(t), in the column, is given by

$$i(t) = \int_{\theta_n}^{\theta_o} z(\theta, t) d\theta + K_n t \tag{11}$$

The substitution of equation (10) in (11) gives

$$i(t) = St^{1/2} + (A + K_n)t + Bt^{3/2} \ldots \tag{12}$$

in which $S = \int_{\theta_n}^{\theta_o} \chi_1 d\theta$ is the sorptivity (Philip 1957) and A, B, ... are the corresponding integrals of χ_2, χ_3, ... In practice, it is found that $1 \gg (A + K_n)t/St^{1/2} \gg Bt^{3/2}/(A + K_n)t \gg \ldots$, so that for short times infiltration can often be described by the simple two-parameter equation

$$i = St^{1/2} + A't, \tag{13}$$

where A′ is given approximately by $A' \simeq (K_o - K_n)/3$ (Youngs 1968; Philip 1969; Talsma 1969*a*; Collis-George 1974; Smiles and Knight 1976) and where K_o, K_n are values of hydraulic conductivity at θ_o and θ_n, respectively. On the other hand, the long-term cumulative infiltration is described by

$$i = G + K_o t. \tag{14}$$

The correspondence between equations (13) and (14) is discussed by Philip (1966*a*). It should be noted, however, that

$$St^{1/2} \neq G.$$

Figure 4 shows the principal characteristics of one-dimensional infiltration. This figure shows the way the moisture profile develops with time. Thus, early in the process the (well-defined) wetting front advances at a rate proportional to $(\text{time})^{1/2}$. This is interpreted physically as the period when the process is dominated by the moisture potential gradient and flow is described exactly by the solution of equation (8) but with the second term on the right-hand side negligibly small. The pattern of flow then gradually changes until the "shape" of the wetting front becomes constant and it proceeds at a constant velocity

$$V = \frac{K_o - K_n}{\theta_o - \theta_n}$$

and a constant surface flux $v_o = K_o$. This continuing phase is controlled by the gravitational potential gradient and is referred to as the "profile at infinity".

Figure 5 shows the i(t) function for the situation described in figure 4. The early stages of this graph plot as a straight line on a $t^{1/2}$ scale; the slope of this straight line is S. The final stages plot as a straight line with respect to t. The slope of this part is K_o.

The transition from the early stages of infiltration to the profile at infinity presents conceptual problems that are resolved in part by methods described by Philip (1974). In particular, Philip and Knight (1974) developed a rapidly converging integral method of solution of the infiltration equation. This method depends on the accurate estimation of

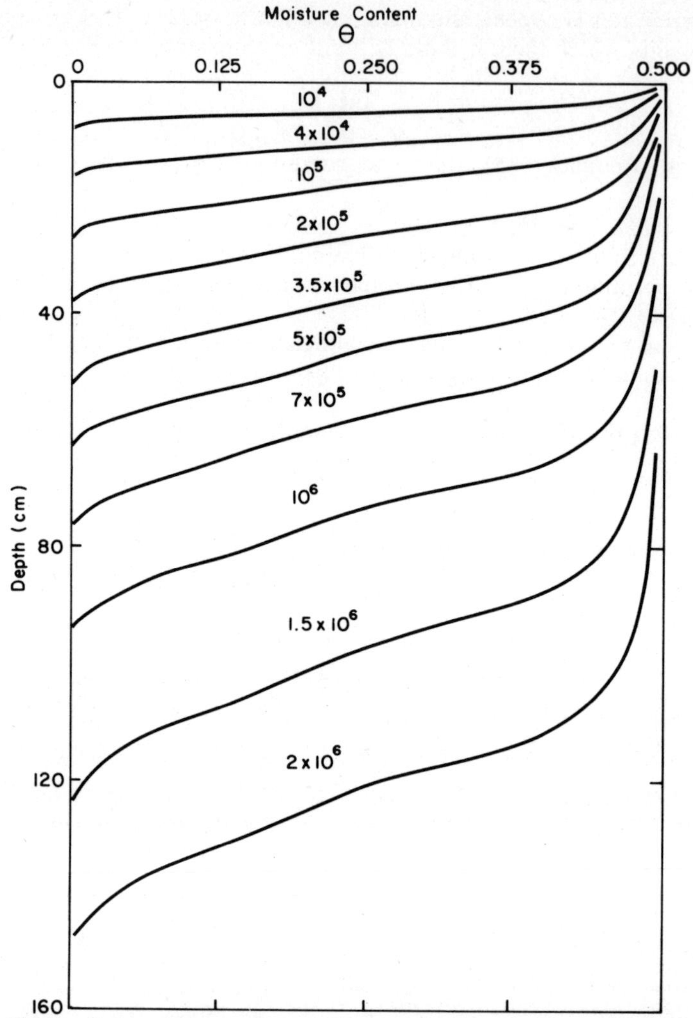

Fig. 4. Soil moisture profiles computed using the solution of equation (6) subject to conditions (9) for the soil with properties shown in figures 2 and 3. The number on each profile represent the infiltration time in seconds at which the particular moisture content profile is developed.

the "shape" of the moisture content profile, and then on the rapid convergence of a simple iterative scheme. The profile shape is determined by referring to the flux/water content relation (Philip 1973), the forms of which are quite restricted.

The extension of infiltration theory to more complicated systems is in its infancy. Even so, a number of situations have been examined:

(i) Infiltration into layered soils. Childs and Bybordi (1969) consider the movement of water into a layered profile in which the conductivity of each layer is less than that of the soil above. They use the infiltration model of Green and Ampt (1911) assuming that the hydraulic conductivity of the soil may be considered effectively to be zero at all θ except θ_{max} (saturation) where the conductivity is a constant.

(ii) Infiltration from cylindrical and spherical cavities. These two cases are related to

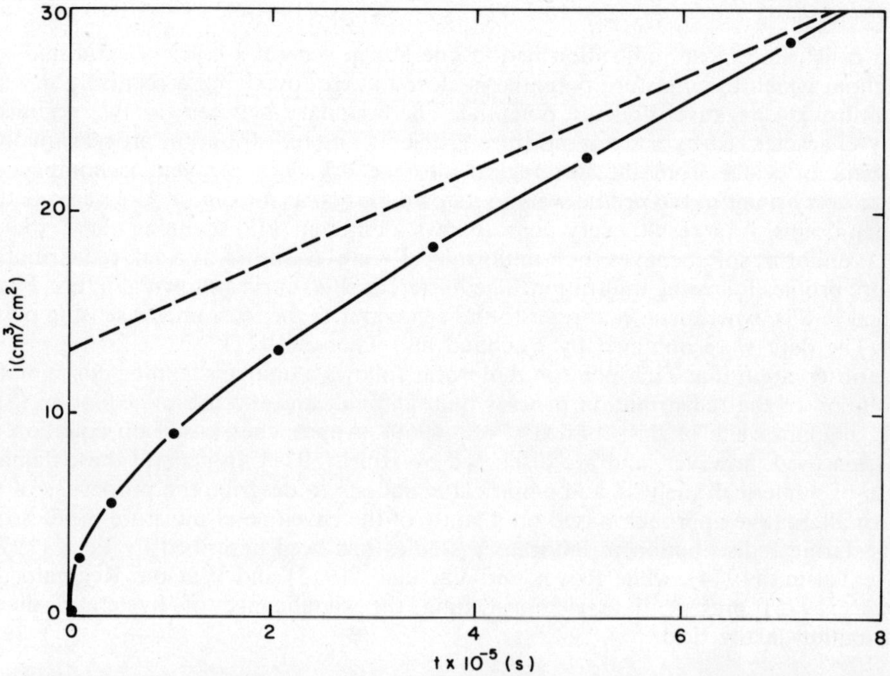

Fig. 5. Cumulative infiltration, i(t) for the infiltration process shown in figure 4. The points on the continuous curve represent sequentially the first 7 profiles in figure 4. The broken line is an extrapolation towards t = 0 of the final 3 profiles of figure 4 when i (t) is linear.

the field situations of furrow and trickle irrigation respectively. They have been considered in terms of the early stages of infiltration for which diffusion-type equations are appropriate; and large time steady state situations where the gravitational potential dominates. The early stages are examined theoretically by Philip (1969) and practically by Talsma (1969*b*) and Talsma, Parlange and Groenevelt (1972) and the continuing steady state by Philip (1968), Youngs (1970), and Raats (1970). The situation is summarized by Philip (1968) who notes that one-dimensional infiltration will result in an infinitely deep wetted region and the soil is completely wetted. On the other hand, furrow and trickle irrigation ultimately produce a wetted region that is effectively finite horizontally although the depth of penetration is infinite. This prediction is discussed by Philip (1968) and by Youngs (1970). It should be noted that trickle irrigation is essentially a constant flux situation, but the arguments above are apposite.

The infiltration problem has traditionally been considered in terms of the step function boundary conditions (9). Recently, however, attention has been given to the constant flux boundary condition. Few experimental data are available although Philip (1969, 1974), Parlange (1972) and Youngs (1972) discuss approaches to the problem.

For the particular case of sustained steady rainfall, v_r, for example, two situations may be observed:

(i) $v_r < K_{max}$ (where K_{max} is the saturated hydraulic conductivity): then all rain infiltrates and ultimately a uniform profile comparable with the long-term one in figure 4 develops. This profile has a zone of uniform moisture content θ_r such that $K(\theta_r) = v_r$.

(ii) $v_r > K_{max}$, initially all rain may infiltrate, if the sorptivity of the soil is great enough. Ultimately, however, the gravitational control of infiltration dominates, the soil surface becomes saturated and run-off of water may occur.

Detailed examination of these situations is presented for example by Rubin and Steinhardt (1963, 1964*a*, 1964*b*) and by Mein and Larson (1971).

Redistribution

At the conclusion of an infiltration period one has in general a layer of saturated soil, throughout which the moisture potential is close to zero, overlying a relatively dry soil with a strongly negative moisture potential. The boundary between the two regions is usually characterized by a steep moisture gradient. The redistribution process involves movement of water from the wet region into the dry. The problem is complicated because every point in the profile wets to its own particular maximum moisture content and then drains. As a result, every point follows a different $\Psi(\theta)$ scanning curve; that is, even a uniform soil behaves non-uniformly. Figure 6 shows typical redistribution moisture profiles following infiltration. The hysteretic $\Psi(\theta)$ curves shown in figure 7 were obtained in this experiment, and refer to the behaviour at the tensiometer sensing points Cl-10. The data were obtained by Vachaud and Thony (1971).

The observation that each point in the profile follows a unique scanning curve makes the solution of the redistribution process quite difficult and the use of equation (8) is strictly inappropriate in this context (Peck 1969). Approaches based on equation (8) have been used, however, and are discussed by Hillel (1971). In general these demand the use of numerical analysis and empirical equations to describe the properties of the soil. An alternative approach based on a study of the envelope of moisture profiles that occurs during redistribution in laboratory studies has been described by Peck (1971), and by Talsma (1974), while Royer and Vachaud (1975) and Watson, Reginato and Jackson (1975) present data demonstrating the significance of hysteresis during redistribution in the field.

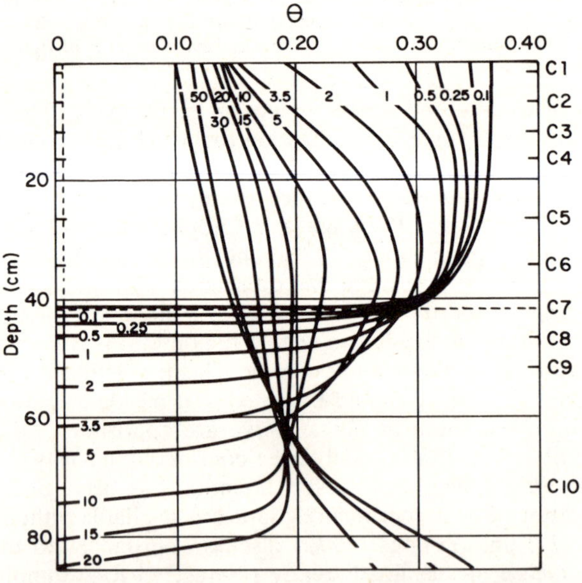

Fig. 6. Redistribution moisture profiles obtained by Vachaud and Thony (1971). Moisture contents were determined non-destructively using gamma beam attenuation methods. The numbers on the graphs represent the redistribution times (hrs) at which the profiles were observed. C1–10 are tensiometer sensing points used to produce figure 7. The process by which each point in the profile wets to a unique maximum moisture content, and then drains is clearly seen in this diagram. Note that the 30 hrs and subsequent profiles are influenced by the impermeable floor of the soil column.

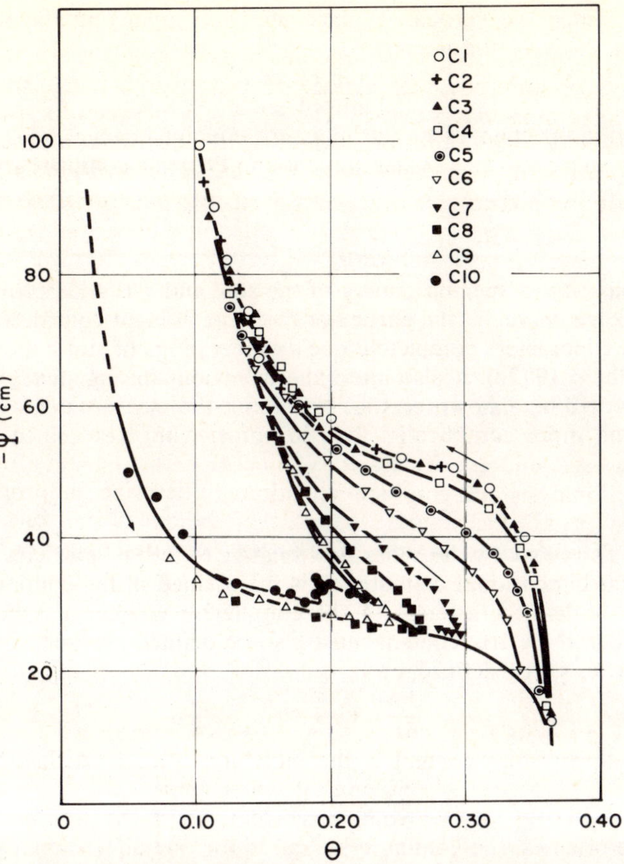

Fig. 7. Moisture characteristic curves corresponding to the redistribution phenomenon shown in figure 6. Note that the soil at levels C1–6 drained continuously, at levels C7–10 wetted and then drained, and at C10 then commenced to wet again as water accumulated at the bottom of the containing vessel.

Swelling soils

The theory of the interaction of water with rigid soils has been extended in principle to swelling soils. Two major difficulties arise with these materials in addition to the problem of interpretation of Darcy's Law.

Overburden

Firstly it is necessary to recognize that the total potential of the soil water now contains an additional term, the overburden potential, Ω, and care must be taken in the definition of Ψ. Thus, if z, the space coordinate, is defined positively downwards and the potential components defined per unit weight of water then

$$\Phi = \Psi - Z + \Omega. \quad (15)$$

Ψ is, as before, the potential arising from the interaction of the water and the solid but now defined with respect to the unloaded soil sample (Philip 1970; Groenevelt and Bolt 1972). As before $Z = -z$, and Ω arises because of the forces resisting the one-dimensional increase in bulk volume that follows an increase in water content. These forces arise because of the weight of the column above the point in question plus any surface load P(0). Since the vertical displacement accompanying the water content change depends on the way the void ratio

$$e = \frac{\text{volume of void}}{\text{volume of solid}}$$

(as a volume description) changes as the moisture content changes, it is necessary to describe the relationship between e, water content and P(0) in a complete specification of the properties of the material.

In general

$$\Omega = \alpha[P(0) + \int_o^z \gamma \, dz] \qquad (16)$$

where γ is the apparent wet specific gravity of the soil, and α is an appropriate average slope of the "shrinkage curve" at the particular moisture content considered. $0 \leqslant \alpha \leqslant 1$.

The overburden effect alters completely one's expectations of static moisture profiles in swelling soils (Philip 1972*c*). It also alters the behaviour during steady and transient flow. It should be noted here, however, that in practice the measurements of potential in swelling soils are no more complicated than in non-swelling ones since a manometer measures $(\Psi + \Omega)$.

Continuity

Secondly, because Darcy's Law describes flow relative to the soil particles, care must be taken to ensure that dimensional consistency is maintained in the continuity equation. Thus, if the space is defined in terms of the cumulative volume of solid (Smiles and Rosenthal 1968) then the water content must also be defined per unit volume of solid, i.e. if we introduce m-space defined by

$$\frac{dm}{dz} = (1 + e)^{-1} \qquad (17)$$

then the appropriate moisture content is the moisture ratio, ϑ, defined by

$$\vartheta = \frac{\text{volume of water}}{\text{volume of solid}}$$

and in the absence of sinks, the continuity equation for one-dimensional flow becomes

$$-\frac{\partial \vartheta}{\partial t} = \frac{\partial v}{\partial m} \qquad (18)$$

After incorporation of Darcy's Law and equation (17), equation (18) becomes

$$\frac{\partial \vartheta}{\partial t} = \frac{\partial}{\partial m}\left[\frac{K}{1+e}\frac{\partial \Phi}{\partial m}\right] \qquad (19)$$

The expansion of this equation in terms of the components of Φ is involved but the basic properties of the flow process can be identified if we consider only the case of a saturated soil. For this situation $\alpha = 1$*, $\vartheta = e$ and substitution of equations (15) and (16) in (19) results in the equation

$$\frac{\partial e}{\partial t} = \frac{\partial \vartheta}{\partial t} = \frac{\partial}{\partial m}\left[D_m \frac{\partial e}{\partial m}\right] - \frac{\partial}{\partial m}[(1-\gamma)K] \qquad (20)$$

in which $$D_m = \frac{K}{(1+e)}\frac{d\Psi}{de}. \qquad (21)$$

Infiltration and gravity-affected flow

Equation (20) is of the same form as equation (8) and if considered in terms of conditions analogous to equation (9), viz.

$$\begin{array}{lll} e = e_o & m = 0 & t > 0 \\ e = e_n & m > 0 & t = 0 \end{array} \qquad (22)$$

can be treated mathematically in the same way.

* Note that, strictly speaking, α is not equal to 1 when the slope of the unloaded shrinkage curve is less than 1. The error is not significant in the example discussed here.

The basic difference between the formulations for rigid and swelling soils is that equation (20) contains $(1 - \gamma)$ in the second term on the right-hand side. This term is always negative and its effect is to modify the apparent effect of gravity. Thus, if $\gamma = 2$, as it will in a saturated soil where $e \simeq 0.7$ (and where the solid specific gravity is 2.7), then the process of infiltration will be exactly analogous to that of capillary rise in a rigid soil (Philip 1966*b*). As e increases, the analogy persists but with the gravitational effect progressively reduced. The physical consequence of an increasing e will thus be to increase the period during which the infiltration phenomenon, and gravity-affected flow in general, are described by $\sqrt{t}$ behaviour.

This prediction, which is supported by other properties of the material as well (Smiles 1974), is supported by limited field data including information on the settlement of an embankment in Tasmania (Cooper and Meyer 1972) where $\sqrt{t}$ behaviour is still observed after 540 days. By comparison $\sqrt{t}$ behaviour in infiltration in non-swelling field soils rarely persists for more than fifteen minutes (Talsma 1969*a*).

The process can also be considered in terms of the changes of gravitational potential energy of the solid and liquid phases of the system. In a rigid soil, infiltration results in a net decrease of potential energy of the water relative to some convenient datum, while the potential energy of the solid remains unchanged. During infiltration in a swelling soil, however, the potential energy of the solid increases while that of the water decreases, although at a rate less than would occur in a rigid soil.

Figure 8 shows the situation for a two-phase system. In this diagram the volumetric water and solid contents $[\theta, (1 - \theta)$ respectively$]$ are plotted as (mutually exclusive) functions of $zt^{-1/2}$. This form of presentation is possible because an alternative set of for-

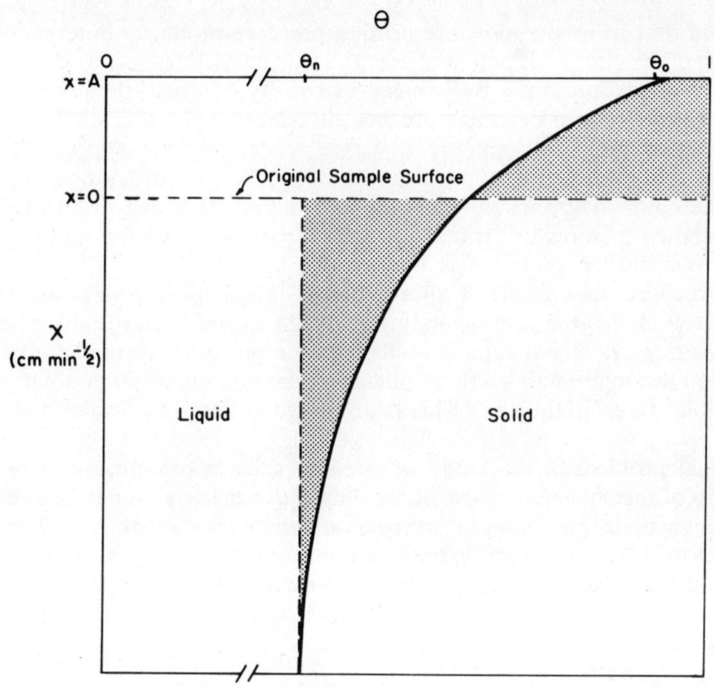

Fig. 8. Diagram showing the spatial distribution of solid and water, in a two-phase swelling system, during the sorption phase of infiltration. The stippled areas represent the spatial distributions of liquid and solid that are "interchanged" during the infiltration process. The effect is accompanied by a net increase in the gravitational potential energy of the system.

mulations for equations (19) and (22) is possible in z-space, leading to moisture profiles that retain their similarity in terms of $zt^{-1/2}$ for extended periods of time (Smiles 1974).

In figure 8 $zt^{-1/2} = \chi = 0$ gives the height of the clay surface at $t = 0$ while $\chi = A$ represents the surface at all later times.

Continuity of matter demands that

$$\int_{O}^{A}(1-\theta)d\chi = \int_{\infty}^{O}(\theta-\theta_n)d\chi \qquad (23)$$

where θ_n corresponds to ϑ_n. This diagram describes the spatial distribution of the solid and the water as long as similarity in terms of $mt^{-1/2}$ is preserved in the infiltration profile.

The change of the gravitational potential energy $\triangle PE_w$ of the water distribution in the time interval 0 to t is given by

$$\triangle PE_w = PE_w(t) - PE_w(t=0) \qquad (24)$$

where $PE_w(t)$ is the potential energy of the spatial water distribution at time t in $A \leqslant \chi \leqslant \infty$ relative to some convenient datum, and $PE_w(t=0)$ is the potential energy of the water distribution in the same space (z) at $t = 0$. Choosing (for convenience) $\chi = 0$ as datum we then find that

$$\triangle PE_w = tg\rho_w[\int_{A}^{O}(\theta-1)\chi\, d\chi + \int_{O}^{\infty}(\theta_n-\theta)\chi\, d\chi] \qquad (25)$$

while the potential energy change of the solid distribution $\triangle PE_s$, calculated analogously is given by

$$\triangle PE_s = tg\rho_s[\int_{A}^{O}(1-\theta)\chi\, d\chi + \int_{O}^{\infty}(\theta-\theta_n)\chi\, d\chi] \qquad (26)$$

In these two equations g is the acceleration due to gravity while ρ_w and ρ_s are densities of water and solid respectively.

The net change of gravitational potential energy $\triangle PE$ of the system is then

$$\triangle PE = tg(\rho_s-\rho_w)[\int_{A}^{O}(1-\theta)\chi\, d\chi + \int_{O}^{\infty}(\theta-\theta_n)\chi\, d\chi] \qquad (27)$$

It will be noted that while the moisture profiles preserve similarity in terms of $zt^{-1/2}$, the rate of change of gravitational potential energy is linear with respect to t′, and always increases during infiltration into a two-phase system. By contrast, the potential energy of the water in a rigid system decreases during infiltration while that of the solid remains unchanged.

Finally it should be noted that the importance of the overburden potential during infiltration resides not in its absolute magnitude, but in its space gradient. Thus overburden contributes to moisture transfer from the beginning of the infiltration process although the overburden potential is then a minimum.

It will be recalled that figure 8 shows the situation for a two-phase (clay-water) system. Figure 9 shows the corresponding diagram for infiltration into a bed of stable swelling soil aggregates when water is applied to the surface at a small constant suction. In this situation the soil remains a three-phase system and the water content profile more closely resembles those in figure 4. This is discussed in detail by Smiles and Colombera (1975).

The principal problem in the study of swelling soils at present is that while a considerable body of theory exists, most of the illustrative discussion has been based on arbitrary, although plausible, material properties, rather than reliable experimental data. There is therefore a need to examine real materials to develop insights into the properties of these soils at the same time taking care to ensure that the restraints that these soils experience during an experiment are relevant to and consistent with the premises on which the theory is based.

SOIL TEMPERATURE

Heat may be transferred in soil as sensible heat by conduction, as latent heat, or by radiation. It is conventional however to consider that conduction is the principal method

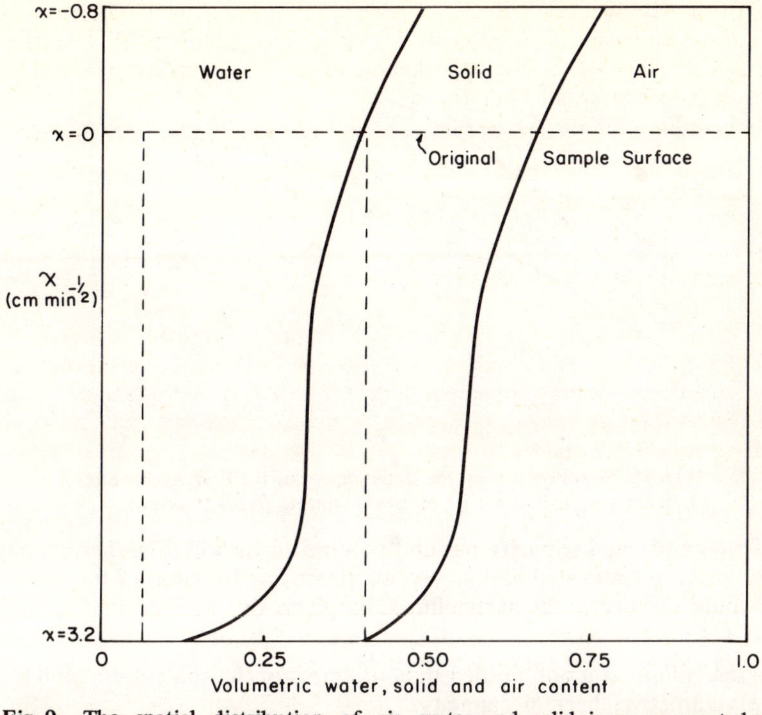

Fig. 9. The spatial distribution of air, water and solid is an aggregated swelling soil, at the surface of which water is applied at a small constant suction.

and this forms a basis for the discussions by Keen (1931), van Wijk (1963), Rose (1966) and Kirkham and Powers (1972). Modern authors rely heavily on heat conduction theory (Carslaw and Jaeger 1959).

The flux equation

The flux of heat $\bar{v}_h$ by conduction is described in a manner analogous to those of water and gas by the equation

$$\bar{v}_h = -\lambda \text{ grad } T \tag{28}$$

in which T is the temperature and λ the thermal conductivity. λ may vary with position but to a good approximation in field soils, is independent of the temperature.

λ varies principally with the particle size distribution of a soil, the water content θ, the bulk density ρ, and the organic matter content in organic soils. In general, the greater the bulk density and the higher the water content the greater will be the thermal conductivity since both these situations increase the effective interparticle thermal contact within the soil. Most semi-arid soils contain little organic matter except as a mulch. Typical relations between λ and θ for different soils are given, for example, by van Wijk (1963). Some representative results are shown in figure 10.

Continuity

In the absence of sources or sinks for heat the continuity equation, in terms of temperature, is written

$$C\frac{\partial T}{\partial t} = -\text{div } \bar{v} \tag{29}$$

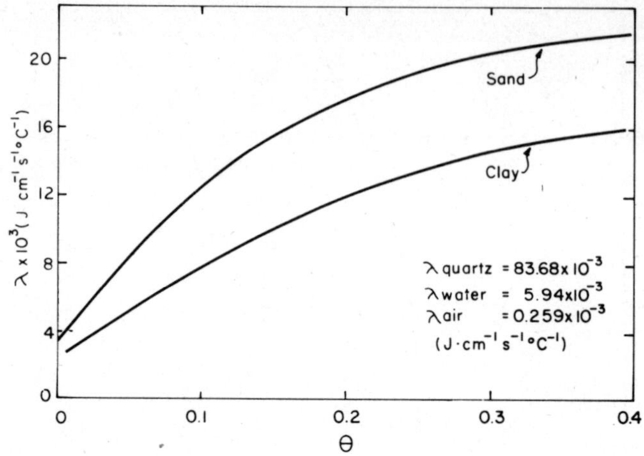

Fig. 10. Graphs showing the dependence of the thermal conductivity λ, on the volumetric moisture content θ, for two soils.

in which C is the thermal capacity per unit volume of the soil. The thermal capacity per unit volume of an unsaturated soil is also an increasing function of the water content θ and of the bulk density ρ. In particular, C is given by

$$C = \sum_i n_i c_i \tag{30}$$

where n_i is the volume fraction of the i-th component of the soil system, and c_i is the corresponding volumetric thermal capacity.

Values of c are tabulated, for example, in van Wijk (1963), or in Geiger (1965).

Equation 29, for the most important field case, of one-dimensional flow in a uniform soil, becomes

$$\frac{\partial T}{\partial t} = \frac{\lambda}{C}\frac{\partial^2 T}{\partial x^2} \tag{31}$$

in which x is distance, and t the time. Since both λ and C are functions of θ, so also is λ/C. Some typical results of van Wijk (1963) are shown graphically in figure 11.

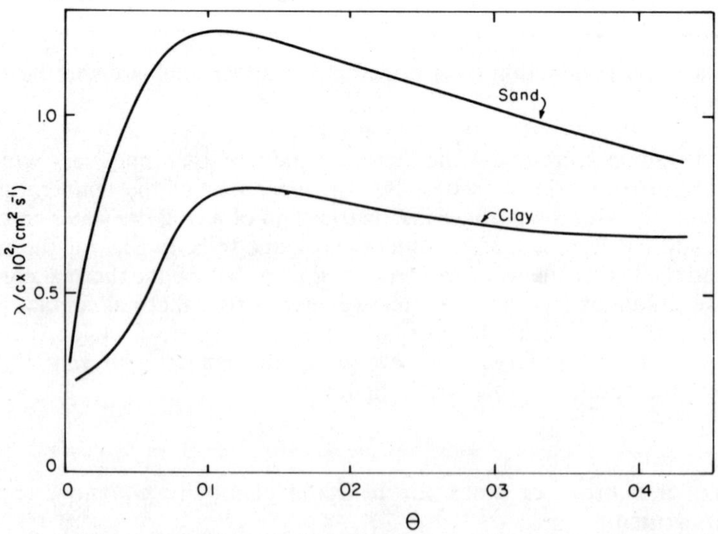

Fig. 11. The variation of the thermal diffusivity D = λ/C with the volumetric moisture content θ, for two soils.

Consideration of heat flow in soils has been largely an investigation of solutions of this equation or of its cylindrical or spherical forms when subject to appropriate boundary conditions.

Surface temperatures

The varying temperature regime within the soil effectively arises as a consequence of soil surface temperature variation. The basic pattern of this variation is imposed by insolation, and the way in which the net radiation, R_n, at the soil surface is partitioned. The simplest energy balance equation takes the form

$$R_n = LE + G + H. \tag{32}$$

In this equation, E is the evaporation rate of water from the surface, L is the latent heat of evaporation at the soil surface temperature, G is the flux of heat away from the surface into the soil, and H represents the energy loss from the surface in convective exchange with the air. Detailed discussions of the surface energy balance of (bare) soils are presented, for example, in van Wijk (1963), Geiger (1965), and Rose (1966). Well-documented reviews of the problem for vegetated surfaces are presented, for example, by Lemon (1963) and more recently by Denmead (1972).

A characteristic response of the soil surface temperature, T_s, to the attenuated net radiation at the bottom of a wheat crop is shown in figure 12 [from Denmead (1972)]. G

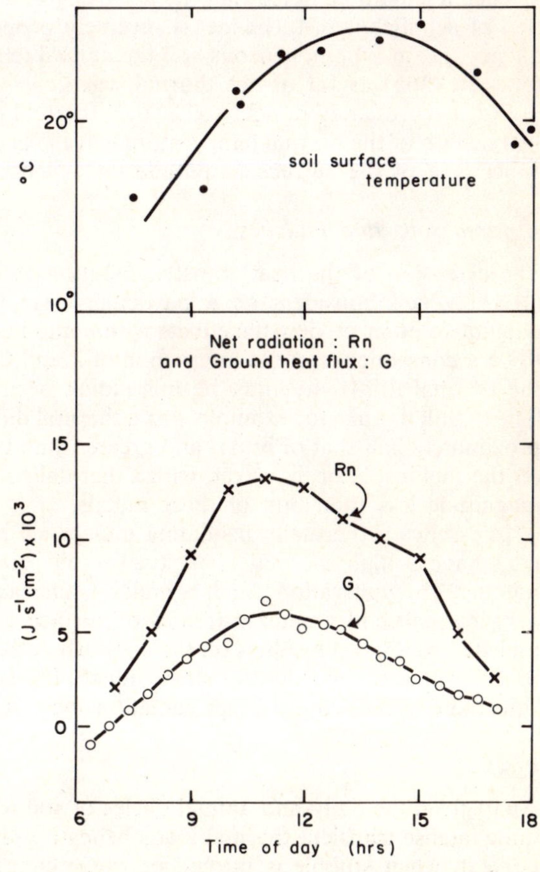

Fig. 12. Variation of soil surface temperature (fig. 12*a*), net radiation, R_n, and ground heat flux, G, (fig. 12*b*) for a wheat crop at Deniliquin, New South Wales, during twelve daylight hours.

is also shown in this diagram which refers to a crop at Deniliquin in New South Wales. Solutions of the heat flow equation for the essentially harmonic soil surface temperature variation shown in figure 12 are discussed in great detail by van Wijk (1963) and by Kirkham and Powers (1972).

Temperature profiles

The basic patterns of the resultant profiles are the same as those observed with a "simple" sinusoidal surface temperature with a period $2\pi/\omega$, viz.

(i) The wavelength, ℓ, of the temperature wave is given by

$$\ell = 2\pi\left(\frac{2\lambda}{C\omega}\right)^{1/2} \qquad (33)$$

(ii) The amplitude of the wave diminishes as

$$\exp\left(-x[C\omega/2\lambda]^{1/2}\right) \qquad (34)$$

(iii) The fluctuations are propagated into the soil with a velocity

$$\left(\frac{2\lambda\omega}{C}\right)^{1/2} \qquad (35)$$

It is of general interest to note that combination of equations (33) and (34), together with the criterion that the amplitude at depth x^* should be 1 per cent (say) of that at the surface indicates that

$$x^* \approx 0.73\ell$$

i.e. the disturbance is virtually undetectable beyond about 0.7 wavelengths.

Then it should be noted that the wavelength, and by implication, the depth of penetration of significant disturbance, is inversely proportional to the square root of the frequency, i.e. in a homogeneous soil the annual temperature wave will penetrate about 19 times ($\approx\sqrt{365}$) as far as the diurnal wave.

For a representative field soil with a thermal diffusivity λ/C of 1×10^{-7} m^2 s^{-1}, the wavelength of the diurnal temperature wave is about 0.38 m, and its amplitude decays to 1 per cent of the surface amplitude, at a depth of about 0.28 m.

Layered soils and mulches

The extension of the heat transfer equation to layered soils is discussed in van Wijk (1963) which considers as a particular case the thermal mulch problem. From a qualitative point of view the effects within and beneath such a mulch may be considered to be a consequence of the interaction of λ and C while noting that arguments based on the thermal diffusivity may be misleading because it is essentially a composite coefficient. Still dry air, for example, has a thermal diffusivity of about 1.9×10^{-5} m^2 s^{-1}, approximately half that of brass, and greater than that of mild steel. The air is a much better thermal insulator, however, with a thermal conductivity which is about four orders of magnitude less than that of these metals.

In essence a thermally insulating mulch will have a low thermal conductivity and it may have a high thermal capacity. In soils however a mulch is often achieved (fortuitously) by cultivation. Such a mulch is characterized by lower λ and C than the underlying soil, and greater extremes of surface temperature than those of adjacent unmulched soil. Crop residues on the soil surface behave similarly but with the added complication that the albedo of a straw mulch, for example, may be greater than that of the soil, thereby reducing the net radiation load at the surface.

Fires

Apart from the daily and annual cycles of soil temperature the soil surface is subject to quite intense transient thermal loads beneath grass and bushfires, and in an agricultural context, when stubble is burned or sugar cane is fired before harvest.

Until recently this problem has not been considered in great detail although a good deal of empirical data have been collected, in particular, by people seeking to define the effects of burning on soil organisms. Nevertheless the physical processes involved appear to be well understood. In essence, the burning process yields radiant energy and sensible heat. The latter component will not affect the soil surface temperature except insofar as the cooler air entrained by the strong convection over the fire may marginally diminish the temperature rise, due to radiant energy, of the soil surface at the periphery of the fire. The problem then is to solve the energy balance equation in the presence of the ephemeral but intense net radiation flux and advection due to the fire.

Scotter (1970) has attempted, with some success, to predict soil temperature profiles under these conditions. He simplified the problem, by treating not the surface heat flux but the surface temperature as an appropriate boundary condition, and further assumes that this boundary temperature may be taken as a step function from ambient to 100°C and then back to ambient. The duration of the step is taken to be equal to the duration of the fire temperature in excess of 100°C at the site. The constant maximum of 100°C is justified (for moist soils) by Vines (1968) who considers it to be a consequence of the vapourization process of the soil water. These assumptions appear to be valid for Scotter's conditions but may be inappropriate for stubble firing over relatively dry soils and would certainly be invalid for windrow burning of timber following land clearing.

The second major assumption implicit in Scotter's approach was that transfer of latent heat and liquid water could be ignored. Aston and Gill (1976) recognize this in an approach that numerically solves simultaneously the heat, water vapour, and liquid water flow equations. The approach fits their data, as well as data published by Scotter (1970), very well. The general problem of "coupled" heat and water flow is considered in more detail below.

Finally it should be noted that evaporation of water may result in significant cooling of the soil and that considerable heat may be released as a very dry soil is wetted. The former situation is in principle analyzable in terms of the energy balance equation, while the latter may be predicted by integrating beneath the moisture characteristic and making a suitable adjustment for latent heat transfer.

SOIL AIR

The nature of the soil atmosphere and its behaviour is not as well understood in detail as are the properties of the liquid phase.

The problem arises in part because of the difficulty of accurately measuring point concentrations and fluxes of gases in the soil. The difficulty carries over to the biological problems of transfer of water vapour, oxygen and carbon dioxide to and from organisms. Despite these reservations, there is no argument that the existence of living organisms distributed through the soil implies that soil ventilation is important in maintaining gas concentration levels acceptable for biological function, and some data are available to indicate rates of transfer across the soil surface.

Keen (1931) appears to be the first to have systematically reviewed the possible processes contributing to soil aeration. He demonstrated that gaseous diffusion is probably adequate to account for the rates of transfer of oxygen and carbon dioxide observed in some field situations. This work appears to have been interpreted to mean that mass flow of gases resulting from barometric pressure change, mass flow of water or atmospheric turbulence, is unimportant. More recently, however, Fukuda (1955), Farrell, Greacen and Gurr (1966) and Scotter and Raats (1963, 1969) have examined the effect of periodic pressure changes at the soil surface on gaseous transfer. The principal effect is considered to arise because of dispersive enhancement of the diffusion coefficient. The conclusions reached are that within the top 10 cm of the soil oscillatory pressure change may produce considerable vertical and horizontal transfer of gas. Kimball and Lemon (1971) confirm these conclusions although they evidently believe that in field soils the effect is of second order when compared with molecular diffusion.

In what follows it will be convenient to treat processes firstly in terms of diffusion theory and then to cite, where possible, relevant information on other ventilation processes. In addition, it is convenient to consider the transfer of water vapour separately from the transfer of other important gases since physical aspects of the processes differ.

Gas transfer

As with liquid water transfer, this problem may be introduced as if it were isothermal. Furthermore, it is convenient to consider that the movement of the gas is restricted to the continuous air-filled pore space of the soil because the transfer coefficients of gases dissolved in water are approximately four orders less than those of the gases in air.

The flux equation

If, then, it is assumed that the transfer process is essentially one of diffusion, the flux density (g/cm^2sec) of the gas may be written

$$\bar{v}_g = -D^* \text{ grad } \rho \tag{36}$$

in which ρ is the density and D^* the effective diffusion coefficient of the gas in the soil.

D^* depends on the air-filled porosity ϵ of the soil, on the geometry of the system, α, and on D_o, the diffusion coefficient of the gas in air.

This dependency is often written $D^* = \alpha\epsilon D_o$. For many situations, $\alpha = 2/3$. A full discussion is presented by Currie (1970).

As with the case of water vapour flow (which is discussed in detail later), the gas laws permit the flux equation to be recast in terms of partial pressure gradients. Then by considering ρ to be a function of both the temperature T, and the pressure, the consequences of variation in the gradients of each of these factors may be explored.

Substitution of the flux equation into the appropriate continuity equation, which may contain source and sink terms, results in a transient flow equation, many solutions of which are examined by Kirkham and Powers (1972).

As Kirkham and Powers point out there is still considerable difficulty in defining the effective diffusivity and also problems of measurement of gas concentrations. In the case of oxygen, which is biologically the most important gas, the use of the platinum electrode technique promises some hope, although early claims for the bare electrode have not been substantiated. In particular, the measurement of the flux of oxygen to the bare electrode as an indication of the availability of oxygen in the soil has been shown to depend on many properties of the soil, and the electrode geometry; see for example McIntyre (1970) for a full examination of the bare platinum electrode method. On the other hand, the membrane-covered electrode does provide a reasonably reliable measurement of oxygen concentration.

Despite these difficulties, the problems of predicting fluxes and concentrations in inert materials with well-defined boundary conditions are not great. The major difficulties arise when an attempt is made to describe the geometry of biologically important situations, and the conditions that exist in the immediate environment of a respiring soil organism. Griffin (1972) reviews approaches to this problem from a microbiological point of view and concludes that the problem is insoluble at present because the techniques are not yet available for measuring cell surface concentrations or fluxes to individual cells *in situ*.

Water vapour transfer and evaporation

This phenomenon is essentially non-isothermal in soils, firstly because (varying) temperature gradients almost always exist, and then because water vapour transfer in general involves change of state at some point in the process. The appropriate flux and continuity equations must therefore take account of this simultaneous transfer of mass and energy. In the first instance, however, the problem is discussed isothermally and then, in a separate section, non-isothermally.

Flux equation and potentials

The transfer of water vapour in soil $\bar{v}_{vap}$ (cm sec^{-1}), is conventionally defined, as is that for gases, in terms of the equation

$$\bar{v}_{vap} = -\frac{D^*}{\rho_w} \text{ grad } \rho \tag{37}$$

in which ρ_w is the density of liquid water, ρ is the water vapour density of the soil air and D^* is the diffusion coefficient of water vapour in the soil.

The difference in flux units between equations (36) and (37) arises because it is convenient for equation (37) to be consistent with Darcy's Law (equation 1) for liquid water flow. As with gas, D^* depends on the diffusion coefficient of water vapour in air, the air-filled porosity of the soil, and its geometry, with the additional complication that enhanced transfer appears to occur as a result of spatially intermittent liquid phase transfer across isolated regions of liquid water (Philip and de Vries 1957; Rose 1963*b*).

This approach (equation 37) is not analogous to that giving rise to equation (1) for liquid water flow; however, an equation analogous to equation (1) may be derived by noting that the soil water potential, Ψ, is related to the vapour density ρ by the equation

$$\Psi = \frac{RT}{g} \log_e\left(\frac{\rho}{\rho_o}\right) \tag{38}$$

in which R is the gas constant, and ρ_o is the saturated water vapour density at absolute temperature T.

The flux equation then takes the form (similar to equation 1):

$$\bar{v}_{vap} = -K_{vap} \text{ grad } \Psi \tag{39}$$

in which

$$K_{vap} = \frac{g\rho}{\rho_w RT} D^*$$

or alternatively, recalling that Ψ is a function of the moisture content θ

$$\bar{v}_{vap} = -D_{vap} \text{ grad } \theta \tag{40}$$

where

$$D_{vap} = \frac{g\rho D^*}{\rho_w RT}\frac{d\Psi}{d\theta}.$$

In contrast with the isothermal liquid water diffusivity which is greatest at high water contents, D_{vap} is greatest in relatively dry soils. This results from the dependence of D^* on ϵ, and because $d\Psi/d\theta$ is large close to $\theta = 0$. Figure 3 shows $D_{vap}(\theta)$ superimposed on $D(\theta)$ for the Yolo light clay soil.

The process with which water vapour transfer is most closely associated is that of evaporation from soil to the atmosphere. The general problem is reviewed by Hillel (1971) who also includes an extensive bibliography. It is, however, useful to briefly examine this process here, both as a soil phenomenon, and as a consequence of meteorological conditions. It is usual to discuss the process for a bare soil in terms of the general isothermal water flow equation (8), but with a diffusivity which is the sum of the isothermal liquid and vapour diffusivities. The conditions

$$\begin{aligned} \theta &= \theta_n & z > 0 \quad & t = 0 \\ E &= D\frac{\partial \theta}{\partial z} - K & z = 0 \quad & t > 0. \end{aligned} \tag{41}$$

are considered to be appropriate.

In this equation E is the evaporation rate. E depends on the water vapour density gradient above the soil surface and a transfer coefficient that, in still air, is equal to the diffusion coefficient of water vapour in air. In turbulent conditions this transfer coefficient may be increased by several orders of magnitude.

In general, E is equal to either the evaporation rate, E_s, from a saturated soil exposed to the same atmospheric conditions, or some lesser value determined by the transmission properties of the soil. Because of the difficulties of relating E_s to the environmental properties controlling it, these properties are conveniently specified in terms of E_s. A full account is given by Philip (1957).

Two distinct situations have been examined in detail.

Steady evaporation from a water table

For this situation the flow equation reduces to

$$z_w = \int_{\theta_o}^{\theta_{sak}} \left(\frac{D}{E + K}\right) d\theta \qquad (42)$$

where θ_{sat} is the saturated moisture content at the water table, θ_o is the moisture content of the soil surface, and z_w is the water table depth. This equation may be integrated numerically to give moisture content profiles for different values of z_w and E, and the effect of z_w on E can be explored for both uniform and layered soils. It follows from this equation that for any water table depth there is a maximum steady evaporation rate, E_{max}, that the soil can support. If E_s is less than E_{max}, then the evaporation rate will be E_s. If E_s exceeds E_{max} however, water transport in the soil limits the rate of evaporation.

Evaporation from soil in the absence of a water table

The examination of this problem by Philip (1957) appears to be basic to all subsequent work. Philip's analysis is consistent with the previously observed "stages" of drying of a soil and anticipates a third:

(i) During stage 1 the soil is sufficiently wet that the evaporation rate is controlled by and equal to E_s. According to Rose (1968*a*, 1968*b*) this first stage corresponds to moisture content in excess of "field capacity". This is followed by a second stage in which

(ii) E is controlled not by the properties of the environment but by water transfer properties of the soil. During much of this period the cumulative loss of water is proportional to $\sqrt{t}$ (while the soil profile is effectively uniform and semi-infinite). It is interesting to note that this $\sqrt{t}$ behaviour need not imply a constant soil surface water content such as that imposed by Rose (1968), but may occur as the soil surface water content decreases because, as Philip (1967) points out, the sorptivity of the drying soil is virtually independent of the surface moisture content when this is sufficiently small.

(iii) During the final stage, when the soil profile is relatively dry, the direction and magnitude of the (vapour) flux is affected by soil temperature gradients. Water then tends to move from regions of higher to regions of lower temperature in accordance with laws described below.

HEAT AND WATER TRANSFER

The equation (39) describing water vapour flow in soils permits a consideration of transfer in response to gradients of both moisture content and temperature. For this situation Philip and de Vries (1957) showed that $\bar{v}_{vap}$ may be written

$$\bar{v}_{vap} = -D_{T_{vap}} \text{ grad } T - D_{\theta_{vap}} \text{ grad } \theta \qquad (43)$$

and furthermore, that liquid water transfer, with some simplifying assumptions, could be similarly described in terms of gradients of temperature and moisture content.

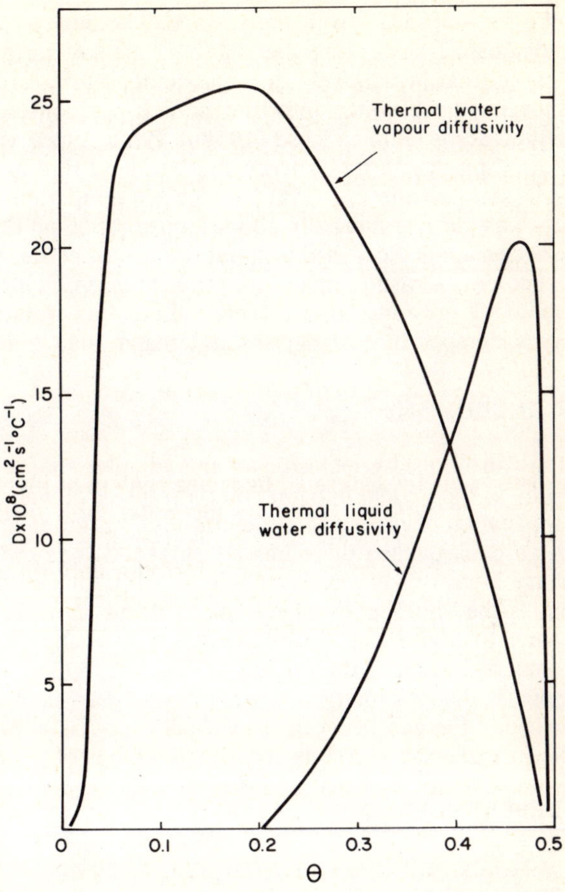

Fig. 13. The thermal water vapour diffusivity, and the thermal liquid water diffusivity calculated by the method of Philip and de Vries (1957) for the soil whose properties are shown in figures 2 and 3.

Figure 13 shows $D_{T_{vap}}$, the coefficient of transfer of water vapour, moving in response to a temperature gradient, and the corresponding coefficient for liquid water transfer. These curves were calculated by Philip and de Vries (1957), and refer to Yolo light clay. Comparison with figure 3 (showing the isothermal diffusivity curves) indicates that while liquid transfer in response to temperature gradients must be small in comparison with that due to moisture content gradients, the transfer of water vapour may be significant since temperature gradients an order greater than moisture content gradients at the surface of relatively dry soils, may not be unusual. The relative magnitudes of the contributions of temperature and moisture gradients to the total flux of water will depend on θ.

An alternative approach to the problem of heat and water transfer; based on the theory of thermodynamics of irreversible processes, was developed in particular in the school of Sterling Taylor. This approach is reviewed, and an extensive bibliography presented, by Nielsen et al. (1972), who also discuss comparative experimental tests of the two approaches. Neither theory was completely satisfactory in the tests described although it should be noted that the test of Philip and de Vries theory is probably improper. Philip and de Vries calculated diffusivity functions for illustrative purposes only. There appears to be no suggestion that the calculated values should replace ones determined in properly instrumented experiments.

It should be very profitable to examine the application of these theories to problems of heat and water transfer beneath the soil surface in quite dry environments. Various rationally based strategies may then become evident to ensure that water is transferred into the soil away from the surface. The burning of stubble for example may result in significant transfer into the soil from the soil surface. It is perhaps worth noting in this respect that in relatively dry soils temperature driven vapour transfer may result in net movement of water into the soil, as a consequence of the periodic soil surface temperature regime. Rose (1968*a*, 1968*b*), for example, examined such a situation at Alice Springs. In his experiments he found that in a loamy sand soil experiencing considerable variation (~ 35C°) in surface temperature the water content at a depth of 15 cm was varying diurnally about a mean of about 0.16. Rose estimated that temperature-driven vapour flow in this situation was of comparable magnitude to liquid water flow. Since the average temperature gradient was downwards the net vapour flux was into the soil. One presumes that the net vapour flux would be upward during winter, when the mean temperature at depth was greater than that at the soil surface.

CONCLUSIONS

It is an implied conclusion of this chapter that while there now exists a considerable body of theoretical information on transfer of water, gases and heat in soil, the practical application of theory to agriculturally important situations is not particularly advanced. As a consequence there has developed serious criticism of the direction of research in some areas of soil physics. Childs (1968), for example, when discussing drainage theory notes the "further investigations to throw more light on the basic principles of flow of water to drains tend to be matters of crossing t's and dotting i's". Philip (1972*a*) has similarly criticized the pursuit of detail beyond a point of diminishing returns. The basic point is that the major problems in many areas of soil physical research appear to lie in bridging the gap between theoretical expertise and biological practice. A first step must lie in exploring methods for simplification of existing theory.

REFERENCES

Aston, A.R., and Gill, A.M. (1976). *Aust. J. Soil Res.* **14**, 55-66.

Buckingham, E. (1907). U.S. Dep. Agric. Bur. Soils, Bull. No. 38.

Carslaw, H.S., and Jaeger, J.C. (1959). "Conduction of Heat in Solids". 2nd Edn. Oxford Univ. Press, London.

Childs, E.C. (1968). *Trans. 9th Int. Congr. Soil Sci.* **1**, 349-56.

———. (1969). "An Introduction to the Physical Basis of Soil Water Phenomena". Wiley-Interscience, London.

Childs, E.C., and Bybordi, M. (1969). *Water Resour. Res.* **5**, 446-59.

Childs, E.C., and Collis-George, N. (1950). *Proc. Roy. Soc.* **A201**, 392-405.

Collis-George, N. (1974). *Soil Sci.* **117**, 282-7.

Cooper, I.D., and Meyer, P.A. (1972). 6th Conf. Aust. Road Res. Board, Canberra. Paper No. 82.

Currie, J.A. (1970). In "Sorption and Transport Processes in Soils". pp.152-71. Soc. Chem. Ind. Monograph No. 37. London.

Denmead, O.T. (1972). In "The Biology and Utilization of Grasses". (Eds. V.B. Youngner and C.M. McKell). pp. 157-71. Academic Press, New York.

Everett, D.H. (1955). *Faraday Soc. Trans.* **51**, 1551-7.

Farrell, D.A., Greacen, E.L., and Gurr, C.G. (1966). *Soil Sci.* **102**, 305-13.

Fukuda, H. (1955). *Soil Sci.* **79**, 249-58.

Geiger, R. (1965). "The Climate near the Ground". 4th Edn. Harvard Univ. Press, Cambridge, Mass.

Green, W.H., and Ampt, G.A. (1911). *J. Agric. Sci.* **4**, 1-24.

Griffin, D.M. (1972). "Ecology of Soil Fungi". Chapman and Hall, London.

Groenevelt, P.J., and Bolt, G.H. (1972). *Soil Sci.* **113**, 238-45.

Haines, W.B. (1930). *J. Agric. Sci.* **20**, 97-116.

Hillel, D. (1971). "Soil and Water: Physical Principles and Processes". Academic Press, New York.

Keen, B.A. (1931). "The Physical Properties of Soils". Longmans Green, London.

Kimball, B.A., and Lemon, E.R. (1971). *Soil Sci. Soc. Amer. Proc.* **35**, 16-21.

Kirkham, D., and Powers, W.L. (1972). "Advanced Soil Physics". Wiley-Interscience, New York.

Lemon, E.R. (1963). In "Environmental Control of Plant Growth". (Ed. L.T. Evans). pp.55-78. Academic Press, New York.

McIntyre, D.S. (1970). *Adv. Agron.* 22, 235-81.
Mein, R.G., and Larson, C.L. (1971). Univ. Minn. Water Resour. Res. Center Bull. No. 43.
Moore, R.E. (1939). *Hilgardia.* 12, 383-426.
Nielsen, D.R., Jackson, R.D., Cary, J.W., and Evans, D.D. (Eds.) (1972). "Soil Water". Amer. Soc. Agron., Madison.
Parlange, J.-Y. (1972). *Soil Sci.* **114**, 1-4.
Peck, A.J. (1969). *Soil Sci. Soc. Amer. Proc.* 33, 980-1.
———. (1971). *Aust. J. Soil Res.* 9, 59-71.
Philip, J.R. (1957). *J. Met.* **14**, 354-66.
———. (1966*a*). In "Water in the Unsaturated Zone". UNESCO Symp. Wageningen, 1, 471-8.
———. (1966*b*). In "Water in the Unsaturated Zone". UNESCO Symp. Wageningen, 2, 559-64.
———. (1967). *J. Appl. Meteorol.* 6, 581-2.
———. (1968). *Trans. 9th Int. Congr. Soil Sci.* **1**, 1-9.
———. (1969). *Adv. Hydrosci.* 5, 215-305.
———. (1970). *Water Resour. Res.* **6**, 1248.
———. (1972*a*). *Soil Sci.* **113**, 294-300.
———. (1972*b*). *Z. Angew. Math. Phys.* 23, 353-72.
———. (1972*c*). 8th Congr. Irrig. Drain., Varna, Bulgaria.
———. (1973). *Soil Sci.* **116**, 328-35.
———. (1974). *Soil Sci.* **117**, 257-64.
Philip, J.R., and Knight, J.H. (1974). *Soil Sci.* **117**, 1-13.
Philip, J.R., and de Vries, D.A. (1957). *Trans. Amer. Geophys. Union.* **38**, 222-32.
Poulovassilis, A. (1962). *Soil Sci.* 93, 405-12.
———. (1969). *J. Soil Sci.* 20, 52-6.
———. (1970). *Soil Sci.* **109**, 5-12.
Raats, P.A.C. (1970). *Soil Sci. Soc. Amer. Proc.* 34, 709-14.
Richards, L.A. (1931). *Physics* **1**, 318-33.
Rose, C.W. (1966). "Agricultural Physics". Pergamon Press, London.
———. (1968*a*). *Aust. J. Soil Res.* 6, 31-44.
———. (1968*b*). *Aust. J. Soil Res.* 6, 45-57.
Rose, D.A. (1963*a*). *Br. J. Appl. Phys.* **14**, 256-62.
———. (1963*b*). *Br. J. Appl. Phys.* **14**, 491-6.
Royer, J.M., and Vachaud, G. (1975). *Soil Sci. Soc. Amer. Proc.* 39, 221-3.
Rubin, J., and Steinhardt, R. (1963). *Soil Sci. Soc. Amer. Proc.* 27, 246-51.
———. (1964*a*). *Soil Sci. Soc. Amer. Proc.* **28**, 1-5.
———. (1964*b*). *Soil Sci. Soc. Amer. Proc.* **28**, 614-9.
Scotter, D.R. (1970). *Aust. J. Soil Res.* **8**, 273-9.
Scotter, D.R., and Raats, P.A.C. (1968). *Water Resour. Res.* **4**, 1201-6.
———. (1969). *Soil Sci.* **108**, 170-6.
Shaw, B.T. (Ed.) (1952). "Soil Physical Conditions and Plant Growth". Amer. Soc. Agron., Academic Press, New York.
Smiles, D.E., (1969). *Aust. J. Soil Res.* 7, 91-8.
———. (1974). *Soil Sci.* **117**, 140-7.
Smiles, D.E., and Colombera, P.M. (1975). Proc. Int. Centre for Heat and Mass Transfer, 7th Int. Seminar, Dubrovnik, Yugoslavia 1974. pp.77-85. Scripta, Washington.
Smiles, D.E., and Knight, J.H. (1976). *Aust. J. Soil Res.* **14**, 103-8.
Smiles, D.E., and Rosenthal, M.J. (1968). *Aust. J. Soil Res.* 6, 237-48.
Swartzendruber, D. (1968). *Soil Sci. Soc. Amer. Proc.* 32, 11-8.
Talsma, T. (1969*a*). *Aust. J. Soil Res.* 7, 269-76.
———. (1969*b*). *Aust. J. Soil Res.* 7, 277-84.
———. (1970). *Water Resour. Res.* 6, 964-70.
———. (1974). *Aust. J. Soil Res.* 12, 15-26.
Talsma, T., Parlange, J.-Y., and Groenevelt, P.H. (1972). *Aust. J. Soil Res.* 10, 209-14.
Topp, G.C. (1969). *Soil Sci. Soc. Amer. Proc.* 33, 645-51.
———. (1971). *Soil Sci. Soc. Amer. Proc.* 35, 219-25.
Topp, G.C., and Miller, E.E. (1966). *Soil Sci. Soc. Amer. Proc.* 30, 156-62.
Vachaud, G., and Thony, J.L. (1971). *Water Resour. Res.* 7, 111-27.
Vines, R.G. (1968). *Aust. J. Bot.* 16, 499-514.
Watson, K.K., Reginato, R.J., and Jackson, R.D. (1975). *Soil Sci. Soc. Amer. Proc.* 39, 242-6.
van Wijk, W.R. (Ed.) (1963). "Physics of Plant Environment". North Holland, Amsterdam.
Youngs, E.G. (1968). *Soil Sci.* **106**, 157-63.
———. (1970). In "Sorption and Transport Processes in Soils". pp. 107-19. Soc. Chem. Ind. Monograph No. 37, London.
———. (1972). *J. Phys. D: Appl. Phys.* 5, 1592-5.

CHAPTER 5

Physical properties and structure

W.W. Emerson

INTRODUCTION

Soil structure is concerned with the organization and arrangement of soil particles into aggregates and the stability of such arrangements to stresses. The stresses may be external, caused by the mechanical action of raindrops or by cultivation. The stresses may also be internal, caused firstly when a dry soil is wetted, by differential swelling and entrapped air (Henin 1938). Secondly, the osmotic stress which arises between the clay particles, if the wetting solution has a low electrolyte concentration, may cause swelling, and in free solution, dispersion. In a semi-arid environment particle arrangements should be directed towards attaining the maximum use of water. For growing crops, a porous, stable structure is required, whereas the reverse is necessary for an area used for water harvesting.

The most important variable controlling the organization of soil aggregates is the clay fraction, because of its large surface area. Therefore a major theme of this chapter is how the properties of clays, such as the size of the clay particles and the forces between the particles, affect the stability of soil aggregates.

The first four sections of the chapter are devoted to basic mechanisms using simple systems. Firstly for slaking and dispersion, aggregates formed from particles of a single clay mineral are considered. Then for clay-organic matter bonding, the interaction between montmorillonite and polymers known to reduce slaking and dispersion (soil conditioners) is discussed. Similarly for the interaction between clays and oxides, the stabilization of montmorillonite by inorganic compounds is considered.

The next four sections are concerned with soil aggregates, beginning with the various methods used to characterize soil aggregates, such as their stability in water, and strength. The results of each method are related as far as possible to the mechanisms outlined earlier. Based on this discussion, a model of a soil aggregate stabilized by organic matter is evolved.

The last three sections are devoted to the practical importance of soil physical properties with special reference to Australian conditions. Aspects discussed in detail are crop growth and the construction of satisfactory earth structures, such as contour banks and dams. The main physical properties considered are aggregate swelling and dispersion. Other properties treated insofar as they affect water entry and movement are the soil-water contact angle and the cracking of clay soils.

PROPERTIES OF AGGREGATES OF CLAY MINERAL PARTICLES

Particle arrangement

Particles of the three most common clay minerals in soils, illite, kaolin and montmorillonite, differ markedly in average size, particularly thickness. The thickness of kaolinite crystals is about 50 nm compared with that of an individual sheet of montmorillonite, 1 nm, with illite intermediate.

One of the main causes of the macroscopic breakup (slaking) of dry clay aggregates on wetting is the "crystalline" swelling (Dettmann 1958) which is due to the hydration of the exchangeable cations. It might be anticipated that the thinner the particles, the greater the specific surface area and crystalline swelling, and hence the more severe the slaking. However, the stresses induced by swelling depend on the rate of swelling, which in turn is dependent on the permeability of the clay, as well as how far the stresses can be relieved by bending of the particles. To visualize these factors it is helpful to consider simple models of the particles.

For montmorillonite and kaolinite the suggested models are single sheets of octavo paper and dull books respectively, the latter each consisting of fifty pages of slightly thinner, quarto paper. The word dull is used deliberately to indicate that the books are only opened in exceptional circumstances, such as in the presence of 8N KAc (Wada 1961). The structure of orientated aggregates of kaolinite and montmorillonite can almost be simulated by dropping the books and sheets respectively at random onto the floor. The simulation is better for kaolinite since Taylor and Norrish (1966) found very little difference in the orientation, as measured by X-rays, of Na-kaolinite dried from a flocculated or a dispersed suspension. There is evidence that the montmorillonite sheets become linked edge to edge (McEwen and Pratt 1957) to form ribbons, so that a dry aggregate consists of interleaved ribbons or a tactoid (Emerson 1962). The orientation between the sheets is very good since the N_2 surface area of orientated aggregates is about 5 m^2/g compared with the total surface area of about 800 m^2/g (Greene-Kelly 1964).

Compared with montmorillonite, the kaolinite aggregate will be more porous, with the pores much larger, and the particles less well orientated. The model for fine-grained soil illite would be intermediate. There is some visual evidence of tactoid formation in a replica of the fracture surface of a compressed core of Willalooka illite (Aylmore and Quirk 1962). Diagrammatic representation of the suggested organization of kaolinite and illite aggregates, induced by drying from suspension, is shown in figure 2, inserts (*a*) and (*b*).

Slaking

The slaking of a dry aggregate of Ca-montmorillonite will now be compared with that of a kaolinite aggregate. The crystalline swelling of the former will be about 25 (50/2) times greater than the latter since there are the same number of exchangeable cations per unit surface area between two adjoining montmorillonite sheets as on the external surfaces of individual kaolinite particles. However, the montmorillonite will take water up only slowly because of the very small pores between the sheets. Further the stresses due to swelling can be relieved by unbending of the sheets, but not of the thicker kaolinite particles. Bending will also reduce the pressure set up in air entrapped within the aggregate, the second main cause of slaking. Bolt and Koenigs (1972) have calculated the average excess pressure which could be applied to the walls of a pore in rigid material by entrapped air and shown that it tends to a limit of 1 bar. If the particles separate sufficiently due to swelling, then the pressure in the air within a clay aggregate can fall below atmospheric (Stroosnyder and Koorevaar 1972) which would help to hold the aggregate together. Thus an initially dry montmorillonite aggregate, when immersed in water, very slowly swells from the outside, air bubbles remain entrapped within the matrix, and the aggregate stays intact. By contrast, the kaolinite aggregate crumbles rapidly to pieces as the air is released (Emerson 1964).

So far only initially dry aggregates have been considered. The wetter the aggregates initially, the greater the fraction of the crystalline swelling already completed and the larger the size of the pore filled with water. Both factors would tend to reduce slaking. Kaolinite aggregates with their larger pores still slake due to entrapped air after prior equilibration with a relative vapour pressure of 0.98, only two-thirds of the pore space being filled with water. In the case of orientated aggregates of Ca-illite (Willalooka), with its smaller particles and hence smaller pores, at a relative vapour pressure of 0.92, the crystalline swelling is complete, the pores are all filled with water and no slaking occurs. This is despite the fact that only one-third of the total swelling of the illite has been completed (Emerson 1964).

The dispersion of unsheared aggregates

When a clay aggregate is immersed in water, after any initial slaking, clay particles may slowly diffuse out into solution. Such dispersion is usually restricted to clays containing some exchangeable monovalent cations. The percentage of the isomorphous replacement required to be balanced by Na for example, before dispersion occurs, depends on the composition of the other cations, particularly exchangeable Al. A kaolinite with five-sixths of the charge balanced by Na ions, the rest by Al ions (Cashen 1959) does not disperse in water. The degree of dispersion also depends on the salt concentration in the water used, being reduced for example as the Ca-ion concentration in the water is increased.

As a corollary, any increase in the strength of the bonds between clay particles, due for example to adding to polymer, can be assessed by applying an osmotic stress between the particles. This is done by first replacing most of the cations on the exchange sites by Na-ions. Then the swelling and dispersion of the treated clay in dilute NaCl solutions are checked against that of the pure clay (Emerson 1956). A convenient test solution for illite and montmorillonite is N/100 NaCl adjusted to pH 7 (Clapp and Emerson 1965*a*), a solution in which the untreated Na-clays just disperse.

Dispersion of wet, sheared aggregates

Dry clay aggregates whose negative charge is balanced by divalent cations do not usually disperse when immersed in water. Such aggregates, if wetted slowly, attain a certain maximum water content at suctions near zero. In order for the clay to disperse in water, mechanical work has to be done first on the clay at a water content greater than this maximum water content. The water content at which the uniformly remoulded wet clay just shows perceptible dispersion when immersed in water is termed the water content for dispersion (Emerson and Smith 1970). This water content will be greater the greater the Ca-ion concentration in the water, until with a concentration equal to the flocculating concentration of a clay suspension dispersion no longer occurs.

In the case of Ca-illite or Ca-montmorillonite the water content for dispersion may not be greatly in excess of the maximum water uptake of the dry clay (Emerson 1967). The results for a preparation of Ca-Willalooka are shown in table 1. The presence of silt and fine sand sized quartz reduces the water content for dispersion proportionately. Thus silty loam soils containing a small amount of similar illite will be particularly liable to dispersion in the field. This condition will be compounded if the water content for dis-

Table 1. The water retention and water content for dispersion of Ca-illite preparations

% Ca-illite	% Silt (2–20μ)	% Fine sand (20–200μ)	Water content for dispersion	Water content at 100 cm suction
100	0	0	66	46
25	50	25	16	16
17	33	50	12	16

persion is reduced by the presence of exchangeable Mg rather than Ca and also by small percentages of Na.

In contrast to illite and montmorillonite, Ca-kaolinite cannot be dispersed by remoulding wet, probably because of exchangeable Al present. From a limited study of mixtures of illite and kaolinite it appears that when the amount of illite present is more than sufficient to cover the basal surfaces of the kaolinite particles, the mixtures disperse at the same water content as the pure illite (Emerson 1968).

Modifications for soil aggregates

Domains

In soil aggregates, the arrangements of clay particles in orientated aggregates already discussed will be modified by the presence of the silt and sand sized, mainly quartz, skeleton. In laboratory experiments on mixtures of sand and clay extracted from a red-brown earth, Brewer and Haldane (1957) showed that there was an upper limit to the clay content at which orientation of the clay around sand grains, caused by wetting and drying, was visible in the light microscope. The coatings were about 4μ thick. Quirk and Panabokke (1962*b*) measured the desorption isotherms for three clays using non-polar liquids. They showed that there was peak in the pore volume distribution at an equivalent water suction in the range 900-300 cm. If it is assumed that this volume represents pores between elements of randomly packed aggregates of orientated clay, then the equivalent diameter of these elements is in the range 9-28μ (Haines 1927). Such volume elements have independently been called domains by Emerson (1959*a*) and Aylmore and Quirk (1960). The concept of domains gives a more realistic way of thinking about the organization of clay particles in soils than, say, a card-house type structure. It is a less exact picture for kaolinitic soils because of the inherently poorer orientation with kaolinitic domains. The concept is particularly useful when considering the reactions of stabilizing agents with soils.

Slaking

In soil aggregates the importance of swelling as compared with entrapped air in causing slaking will be reduced for two reasons. Not only will there be less clay present, but there will be larger pores both between domains and between the large primary particles. In fact entrapped air has been shown to be the dominant cause of slaking in silt loam soils (Emerson and Grundy 1954; Panabokke and Quirk 1957).

Slaking in this case was found to increase rapidly with the rate of wetting of the dry aggregates (Emerson and Grundy, 1954). These authors also found that the water content of rapidly wetted aggregates brought to a low suction was greater than that of aggregates brought to the same suction by slow wetting. It was presumed to be due to water being held in cracks caused mainly by the pressure of entrapped air. This process was investigated in more detail by Quirk and Panabokke (1962*a*). These authors refer to the internal disruption caused by wetting aggregates rapidly at low suctions as "incipient failure". This distinguishes it from visual failure, i.e. slaking, when the suction helping to keep the aggregates together is reduced to zero.

BONDING OF MONTMORILLONITE BY ORGANIC POLYMERS

Synthetic polymers

The interaction of clays, usually montmorillonite, with linear ethylenic polymers has been studied most, since these were the first compounds to be proposed as soil conditioners. One of these, polyvinyl alcohol (PVA) (Geoghegan 1950) which is uncharged, links individual clay particles together. It is deduced from the increased (001) spacing that the links consist of overlapping double polymer chains (Emerson and Raupach 1964). Some of the hydroxyls along the chains probably form hydrogen bonds with the oxygen atoms of the clay surface, the exchangeable cations not being involved (Emerson and Raupach 1964). The other polymers such as "krilium" (Ruehrwein and Ward 1952) may be regarded as being derived from PVA by the substitution of adjacent pairs of OH groups with COOH and NH_2 groups. These enable the polymer to coordinate with exchangeable di- and tri-valent cations on the external surfaces of the clay domains, the negative charge on the polymer preventing penetration within domains.

Polysaccharides

The reaction of this group of polymers with clays has been studied in detail not only because such polymers are produced in soil but also because they can be obtained in a pure form from cultures of particular strains of bacteria. Qualitatively the polymers react in the same kind of way as the ethylenic polymers, depending on the presence or absence of uronic carboxyl groups (Clapp and Emerson 1972). The major difference is the molecular weight (M.W.) required to bond the clay particles together. For example, the non-ionic dextran produced by *Leuconostoc meserentoides* with a M.W. of the order of 10^8 (Jeanes 1966) is strongly adsorbed by montmorillonite and bonds the particles together giving an expanded (001) spacing (Emerson 1963; Clapp and Emerson 1972). However the hydrolysed products of this dextran with a M.W. of the order of 10^5 are only slightly adsorbed (Clapp, Olness and Hoffmann 1968), or not at all (Greenland and Parfitt 1970). On the other hand, PVA with a M.W. of 25,000 is strongly adsorbed by montmorillonite (Greenland 1963). The difference is thought to be due to the dextran molecules being more rigid and branched and also containing fewer hydroxyl groups. This prevents it from collapsing onto the surface of the clay to form a series of H-bonds.

Polysaccharides can be oxidized with periodate, the classical glycol splitting being followed in the presence of excess periodate by over-oxidation of the chain itself. The oxidation conditions required to break the polysaccharide montmorillonite clay bonds are more severe even than those which give over-oxidation in free solution (Clapp and Emerson 1972). On this basis it is suggested that the bonds between clay particles do not consist of single polymer chains which would be easily scissored, but intertwined strands of polymer in which the cis-hydroxy groups are screened.

BONDING OF MONTMORILLONITE BY INORGANIC COMPOUNDS

The stabilization of clay minerals by the addition of lime and cement has been extensively studied by Ingles and his co-workers (Ingles and Metcalf 1972). Insoluble gels of calcium silicates are formed, either directly by hydration in the case of cement, or indirectly by alkaline attack on the clay in the case of lime (Ingles 1970). Clay is stabilized when a continuous skeleton of material is formed (Herzog 1964), within which the crystalline swelling of the clay is unaffected (Emerson 1959*a*).

The interaction between clay minerals and synthetic oxides has been studied by Blackmore (1973) who precipitated up to 50 per cent by weight of iron (expressed as Fe_2O_3) in the presence of a suspension of Na-montmorillonite. With the maximum addition, none of the dried clay could then be re-dispersed even by vigorous stirring in a polyphosphate solution. El Rayah and Rowell (1973) precipitated the equivalent of 2.5 per cent by weight of Fe_2O_3 or Al_2O_3 onto Na-montmorillonite in suspension and showed that, after drying, the swelling of the clay in dilute NaCl solutions was much reduced compared with untreated clay, e.g. the clay did not disperse in 1.10^{-4}N NaCl. El-Swaify and Emerson (1975) using Na-illite, deduced that the Fe_2O_3 and Al_2O_3 were present on the surface of the clay as 4 nm spheres and thin layers respectively. These acted as polycations linking the clay particles together.

MEASUREMENT OF THE STABILITY OF SOIL AGGREGATES

Slaking

Estimates of structural stability have been based mainly on determining the breakup of aggregates in water. This has been measured usually either on a macroscale, by using a derivative of Tiulin's (1928) wet sieving technique, e.g. Downes and Leeper (1939), or on a microscale following on Middleton's (1930) end-over-end shaking technique, e.g. Quirk (1950). In the first method very little mechanical action is applied to the aggregates and it is mainly the immediate breakup on wetting (slaking) which is measured. The latter depends greatly on the rate of wetting, so that conditions such as the initial relative vapour pressure of the aggregates have to be standardized (Panabokke and Quirk 1957). A variation of this is to estimate the change in size of the wetted aggregates from the suction range over which the inter-aggregate pores drain (Childs 1940; Collis-George and Laryea 1972). In the micro-method, fracturing of the aggregates during the initial wetting is combined with mechanical breakdown due to abrasion during shaking.

Dispersion

In neither of the previous methods is subsequent dispersion of clay from the wetted aggregates usually distinguished from the initial slaking, although Tiulin (1928) originally specified that tap water should be used, presumably to avoid dispersion. Brewer and Blackmore (1956) using a similar technique and distilled water noted that breakup of one soil on the sieves was very slow and that the slaked fragments dispersed.

A method of characterizing aggregates based on the ease of dispersion of the aggregates in water has been developed (Emerson 1967). Six classes of aggregates are distinguished (table 2). In class 1, dispersion is caused by the osmotic pressure developed between the clay particles as the soluble salts diffuse out. At the other end of the scale in class 6, the forces between the clay particles are attractive, however far the particles are forced apart by mechanical action. Loveday and Pyle (1973) have combined the degree and rate of dispersion of dry, unsheared aggregates and of aggregates remoulded at the water content corresponding to 100 cm suction into a single value, the dispersion index.

The qualitative observations of dispersion on which the scheme in table 1 is based can easily be made quantitative (Emerson and Bakker 1973), but the force which is exerted during the prescribed half minute allowed for remoulding the soil uniformly is uncontrolled. This also means that highly subplastic soils (Butler 1955), which require several minutes of remoulding before a uniform sample is obtained, are outside the classification. Coughlan, Fox and Hughes (1973*b*) have used a vibrating table shaker to apply reproducible stresses to aggregates at different water contents.

Consistency

Butler (1955) has distinguished four classes of clay soils based mainly on the consistence of the soil in the hand. The subplastic soils already mentioned are those which increase in texture on prolonged kneading, e.g. from a gravel to a clay. Superplastic soils show the reverse effect changing from heavy to light clay, while the consistence of normal labile soils is unchanged. Self-mulching soils are similar to the last group, except that after puddling in the field, natural wetting and drying return the soil to a granular state. some measurements have been made of the mechanical strength of typical soils in Butler's classes (Brewer and Blackmore 1956; Greacen 1960). Aggregates of these soils, apart from the subplastic soil, have now been classified (Emerson, unpublished) according to their dispersion characteristics (cf. table 2) and the results are included in table 3.

Table 2. A classification of soil aggregates according to their dispersion in water

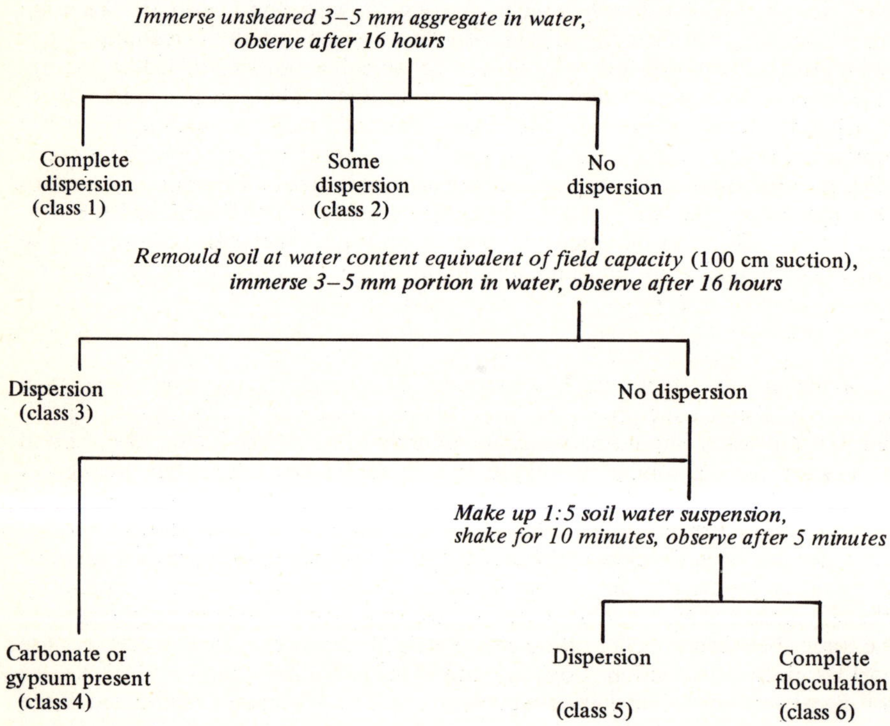

Table 3. Some properties of aggregates from four clay soils

Soil description *	Dispersion class †	Dry bulk density ‡	Equivalent strength ‖ (g/cm^2)
Superplastic	1	2.0	30
Normal labile	2	1.7	130
Self-mulching	4	1.6	200
(Strong structure)	5	–	1,000

* Butler 1955.
† Following Emerson 1967.
‡ Brewer and Blackmore 1956.
‖ Greacen 1960.

Superplastic soils are in class 1, that is the intrinsic attractive forces between the clay particles are very weak. It is suggested that continuous kneading improves the orientation between *individual* clay particles, leading to a decrease in shear strength. The absence of strong attractive forces between the particles also means that on drying a puddled, wet lump of soil, it will dry as a whole to yield a dense aggregate (cf. table 3). On the other hand, self-mulching clays are in class 4, dispersion being prevented by the presence of carbonate. If such a soil is puddled then, around the carbonate particles, the net attractive forces between the clay particles will increase as the concentration of divalent ions in solution increases. This would cause the adjoining clay particles to pack more closely, i.e. local syneresis to take place. Such a process would leave large pores, which on slight drying would be sites of crack initiation. The normal labile clays, which have intermediate consistence properties, also fit into an intermediate dispersion class, namely class 2, showing slight dispersion.

Strength, determined by mechanical stress

Greacen (1960) has suggested that the increased suction which develops when a soil at a given water content is continuously remoulded is a measure of the effective strength of the undisturbed soil at that water content. Physically it means that as the clay particles become disaggregated, the water is shared between the particles, the average size of the water-filled pores is decreased, and therefore the suction is increased. The values for aggregates of four soils, initially at a suction of 10 cm, are shown in table 3. These values were derived using two approximations, that there is a linear relationship for all continuously remoulded soils between the logarithm of the suction and the water content, and that the suctions developed at the liquid and plastic limits are the same for all soils. The results are in general agreement with the expected increase in attractive forces between the clay particles in going from dispersion classes 1 to 5. For the soil which dispersed completely in water, the effective strength was not much greater than the initial suction, whereas the effective strength of the strong aggregates (from the 0-10 cm of Beerik krasnozem) was 100 times greater. The aggregates of the last soil were completely stable to wet sieving, yet it will be noted that there was some dispersion on shaking a suspension of the soil in water.

The strength of dry soil aggregates has been measured by a drop-shatter test (Marshall and Quirk 1950). Farrell, Greacen and Larson (1967) investigated this method over a range of water contents using remoulded spheres of a loam soil. They measured the cumulative drop height (CDH) to cause a fixed degree of breakup. They also measured the tensile strain energy (TSE) needed to fail cores of the same soil, either by direct axial stress or indirectly by the Brazilian test (cf. Williams, Greenland and Quirk 1967*b*). The ratio of CDH to TSE was approximately constant, suggesting that breakup in the drop-shatter test is due to tensile fracture (Griffith 1924). In general much more quantitative information is required on the shear strength of soil aggregates, so that, for example, ranges of water content over which soils may be cultivated can be predicted.

Strength, determined by osmotic stress

Since it is difficult experimentally to apply known mechanical stresses to undisturbed aggregates, the Na-saturation method in which an osmotic stress is applied to the aggregates has been used. The method of measuring the swelling and dispersion of the aggregates with reduction in NaCl concentration originally was the hydraulic conductivity of a bed of aggregates (Emerson 1954; Greenland, Lindstrom and Quirk 1962). However it has been found to be more convenient to measure the fraction of the total clay and silt present which is dispersed at a particular salt concentration (Clapp and Emerson 1965*b*; Stefanson 1971).

BONDING OF AGGREGATES BY SOIL ORGANIC MATTER

Swelling of aggregates

It is well known that the longer a soil has been down to grass, the more resistant the surface soil aggregates are to slaking, e.g. Clarke and Marshall 1947. This resistance could be due to a reduction in the swelling of the clay particles in the aggregates. Dettmann and Emerson (1959) compared the swelling of surface aggregates of two clay soils from adjoining old grassland and old arable fields and found no difference, within experimental error. Also the percentage reduction in the water uptake of dry aggregates exposed to a relative vapour pressure of 0.98 was less than 5 per cent as between grassland and arable aggregates and between grassland aggregates and aggregates made from peroxidized grassland soil aggregates. A similar agreement in the water uptake at 30 cm suction of aggregates from grassland and arable on adjoining plots was found by Quirk and Panabokke (1962*a*).

However, in earlier comparisons Emerson (1954), percolating aggregate beds with decreasing concentrations of NaCl, found that distilled water could be percolated through old grassland aggregates. By contrast, the permeability of the corresponding old arable aggregates fell to zero when the concentration of NaCl was reduced to that required to flocculate a suspension of clay from the aggregates (fig. 1). Subsequently the swelling of the aggregates was measured in the NaCl solution of sufficient concentration just to prevent dispersion of the arable aggregates (Emerson 1959*a*). The linear swelling of the grassland aggregates was only one-fifth that of the arable aggregates. These comparisons showed that organic matter does not affect the hydration of the dry clay present, so that the "crystalline" swelling of the clay is not appreciably reduced, but that it provides sufficient restraint to inhibit slaking and also greatly reduce double layer swelling.

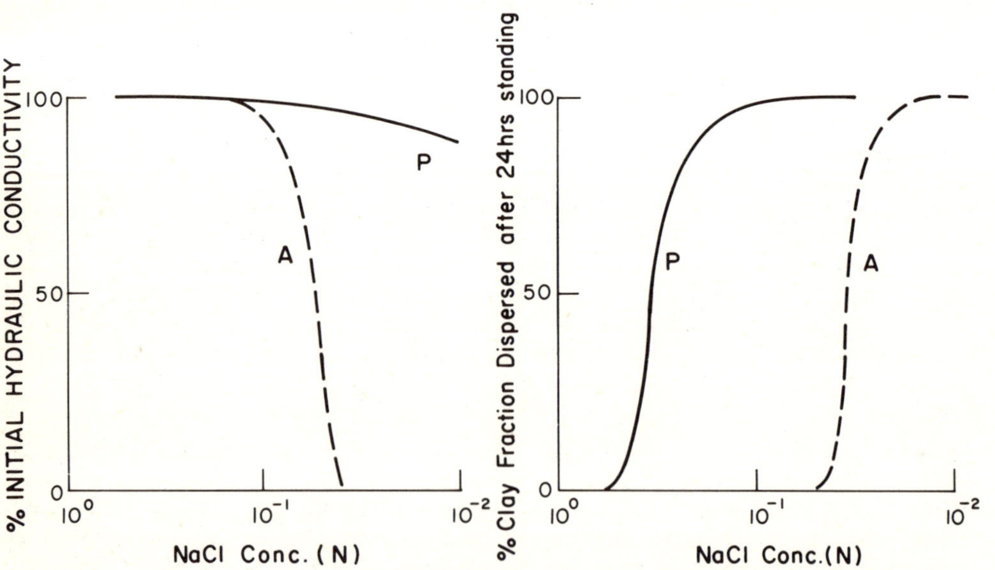

Fig. 1. Changes in the properties of aggregates of an illitic soil, taken from the 0–10 cm layer of two adjoining fields, (*P*) old pasture and (*A*) old arable, washed initially with N NaCl and then equilibrated with decreasing concentrations of NaCl (after Emerson 1954 and Dettmann and Emerson 1959).

a. The decrease in hydraulic conductivity of beds of aggregates.
b. The increase in dispersion of 1:5 suspensions of aggregates.

Dispersion of aggregates

Further insight into the clay-organic matter bonds may be derived from the stability of suspensions formed from surface soil aggregates by ultrasonics. Edwards and Bremner (1967) showed that it was possible to disperse effectively Ca-saturated surface soils in water only using ultrasonics, and that the suspensions formed were stable. Further that the suspensions usually contained as many $<2\mu$ particles as those derived by mechanically shaking peroxidized soil in a solution of calgon, e.g. after removing organic matter and substituting Na on the exchange sites. Subsequently it was found (Emerson 1971) that for acid, kaolinitic soils in particular, the effect was dramatic, suspensions of the subsoils rapidly and completely flocculating in water, whereas suspensions of the surface soil aggregates in water were stable. The cause of the flocculation of the kaolinite subsoil as mentioned already is the Al on the exchange sites. It follows that in dispersions of surface soil aggregates, the Al on clay particles is screened by organic matter, but that in the natural aggregates this organic matter can be cross-linked to hold all the particles together.

The same kind of results had been found earlier when comparing grassland and arable aggregates (Emerson 1954). The undisturbed grassland aggregates were much more resistant to applied osmotic stress, but the concentration of NaCl required to flocculate a suspension of the grassland aggregates was ten times as high (fig. 1). In other words, once the clay particles are sufficiently separated then the role of organic matter is changed from that of a bonding agent to a peptizing agent.

If the bonding by organic matter is very strong, then it may be necessary to add a dispersing agent, such as calgon, to separate all the $<2\mu$ particles, as shown by comparison with the International Soil Science Society method (Emerson 1971). If the bonding is weak, then remoulding the wet soil by hand with a spatula may be sufficient to achieve a fully dispersed, stable suspension, at least at the 2μ level (Emerson 1968). Thus surface soil aggregates can appear to be very stable as measured by wet sieving, but disperse completely after remoulding wet.

Aggregate organization

The suggested way in which polymeric organic matter links together clay and silt sized particles in soil aggregates is given diagrammatically in figure 2 (after Emerson 1959*a*). The silt-sized particles are designated quartz because this is a common constituent of many soils. The organic matter is shown bonded to the exterior surfaces of clay domains, the clay still being free to shrink and swell. However when the clay swells, the stresses are transmitted via the organic matter bonds, so that slaking is prevented. The bonds also restrict the double layer swelling of the Na-soil. In the case of kaolinitic soils, as mentioned already, the distinction between domains and coarse clay particles becomes blurred so that some organic matter may penetrate into the domains.

Details of the proposed organic matter-clay bonds are shown in the insert. Linkage to the clay surface is through the cations balancing the isomorphous replacement charge and by hydrogen bonding. The size of the individual polymers is not thought to be sufficient to bridge the gaps between clay domains. Therefore cross-linking of molecules through Al, Fe and hydrogen bonds is shown. Hydrogen bonding between molecules is invoked because it seems unlikely that shearing with a spatula could break covalently bonded organic compounds. One consequence of the model is that the amount of polymeric organic matter per unit area would decrease with decreasing soil particle size. This has been shown to be true for total carbon by Alexandrova, Yurlova and Lobitskaya (1964) who separated different-sized fractions from soil suspensions formed by trituration.

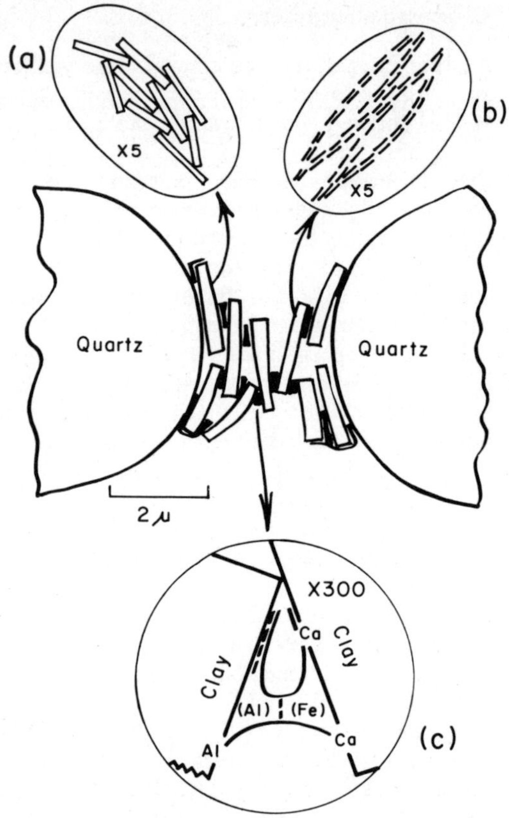

Fig. 2. Model of a soil aggregate stabilized by organic matter showing clay domains, organic matter and quartz.
Magnified inserts: (*a*) Kaolinite domain.
(*b*) Illite domain.
(*c*) Organic matter-clay interaction.
Dotted lines indicate H-bonding.

Biological factors

In this chapter so far the soil has been treated as an inert material and no mention has been made of the effect of biological factors on structure. At least in soils of low clay content some bonding mechanism other than that shown in figure 2*a* is required to link together the coarse and fine sand particles if slaking is to be prevented. Such links could be fungal hyphae (Hubbell and Chapman 1946; Bond and Harris 1964), the actual bonding of the hyphae to the particles being via bacterial gums. These links are shown diagrammatically in figure 3, and their scale should be compared with that of figure 2*a*. This second model also allows the clay to fully hydrate, and provides for some links between the clay domains.

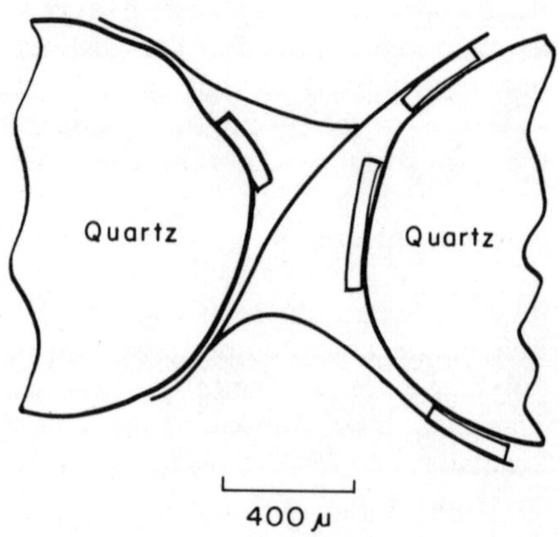

Fig. 3. Model for interaction of mineral particles and fungal hyphae. The smaller particles represent clay aggregates.

The organic matter-mineral bonds illustrated in figures 2*a* and 2*b* will all be susceptible to microbial attack. The polymeric organic matter could be protected due to its location in very small pores between clay domains (Rovira and Greacen 1957). Some of the fungal hyphae in soil also appear to be very resistant to decomposition (Bond and Harris 1964). However, if aggregates are sheared wet, so that these bonds are broken, then the organic matter would become more accessible to attack both because of physical accessibility (Rovira and Greacen 1957) and because reactive chemical groups would be exposed.

The composition of bonding organic matter

It has not yet been possible to stabilize clay or soil synthetically by adding polymeric compounds such as polysaccharides or humic acids extracted from soil organic matter. The presence of particular types of polymer as bonding material has therefore to be inferred from the reduction in the strength of aggregates caused by various chemical treatments. Stefanson (1968) noted that extraction of waxes and lipids with non-polar liquids did not affect the stability of soil aggregates. Besides this the only chemical treatment so far developed which attacks semi-specifically a particular type of polymer present in soil, namely the polysaccharides, is oxidation with sodium periodate.

Mehta et al. (1960) measured the increase in the slaking of aggregates after periodate oxidation. Whereas synthetic aggregates made with added polysaccharides fell to pieces, surface aggregates from a forest soil containing some carbonate and from two rendzinas were unaffected. These authors concluded that polysaccharides were not important bonding agents. However, Greenland, Lindstrom and Quirk (1962) found that the wet strength of surface soil aggregates as measured by a Na-saturation, permeability technique could be reduced by mild oxidation with periodate. The reduction was least for permanent grassland, and there was no effect on carbonate soils.

Clapp and Emerson (1965*a*) using a qualitative Na-dispersion technique showed that periodate oxidation by itself, even under more severe conditions, was inadequate to break all the clay organic matter bonds in old grassland aggregates. This could be accomplished by an additional treatment with neutral pyrophosphate. Subsequently, quantitative dispersion of the clay present in grassland aggregates from a montmorillonitic soil was obtained by a combination of periodate oxidation and pyrophosphate (Clapp and Emerson 1965*b*), provided carbonate present was removed initially.

A quantitative Na-dispersion technique and a sequence of periodate and pyrophosphate treatments were applied by Stefanson (1971) to grassland aggregates. The combined treatment was about twice as effective as each treatment individually, but about 35 per cent of the bonds still remained after the combined treatment. This may have been due to the use of mild oxidizing conditions. The initial negative results of periodate oxidation found by Mehta et al. (1960) were therefore due to using soils containing carbonate and possibly also because one soil was under forest. Clapp and Emerson (1965*a*) found that aggregates from a forested site were almost completely resistant to the periodate, pyrophosphate treatment, whereas the corresponding grassland aggregates on the same soil type exhibited the normal complete dispersion. It seems that at least three groups of polymers are involved in linking soil particles together, one of them being periodate oxidizable materials, which are almost certainly polysaccharides and polyuronides.

Superficially there is an anomaly in suggesting that polysaccharides, which can form interlamellar complexes with montmorillonite, are restricted in soils to the external surfaces of domains. The distinction is that expanded (001) spacings are obtained by drying suspensions of the polymer and clay, whereas in soil, the polymer would have to diffuse between clay particles, not usually separated by more than a few water layers. Williams, Greenland and Quirk (1966, 1967*a*) using PVA of M.W. 10^5 (cf. dextran, M.W. 10^8)

have shown that the rate of diffusion of the polymer in a soil is very slow. They also measured the N_2 adsorption isotherms of aggregates which had absorbed PVA from solutions. The latter were interpreted as showing that due to the sorption of PVA onto the external surfaces of the domains, N_2 molecules were unable to penetrate between the individual clay particles.

BONDING OF SOIL AGGREGATES BY INORGANIC COMPOUNDS

Synthetic sesquioxides

According to Russell (1971) soils with a high content of uniformly distributed ferric iron usually have a stable structure. McIntyre (1956), for example, found a highly significant correlation between the macroporosity of surface soil aggregates and the content of clay sized Fe_2O_3 present. However, it is not easy to determine how iron actually improves structure in the field. This is partly because Al oxides often occur in conjunction with Fe oxides, but also because the techniques used to remove Fe oxides are not entirely specific. Each technique also removes some extraneous Al, including Al on exchange sites, as well as attacking to a certain extent the clay minerals present.

In synthetic systems, the precipitation of 0.5 - 1.0 per cent by weight of Al-oxide into a soil suspension, followed by drying, has been shown to reduce the swelling of the Na-form of soils (Deshpande, Greenland and Quirk 1964*a*; El Rayah and Rowell 1973). The first authors also found that Al-oxide precipitated under similar conditions, but in the absence of soil, had a positive charge of 130 m.e./100 g at pH 4 (Deshpande, Greenland and Quirk 1964*b*). The second authors, but not the first, found a similar reduction in swelling of the Na-soil after precipitating an equal quantity of Fe-oxide and drying.

Schahabi and Schwertmann (1970) synthesized Fe_2O_3 of various particle sizes and then added aliquots in suspension to a dry silty soil. After mixing, the water stability of the soil was measured, with and without an intermediate drying step. In order to increase the stability of the aggregates drying was essential and only the finest "amorphous" particles in the size range 30 - 100 μ were effective. Two per cent by weight of this material increased the percentage of water stable aggregates from 18 to 83. The smaller the oxide particles, the greater would be the positive charge per unit weight.

On the basis of these results it is suggested that provided the oxide particles carry sufficient positive charge, either Al or Fe can link the negatively charged clay particles together. Flotal (70 per cent by weight Fe alum) has been proposed by Gasperini, Alinari and Malquori (1958) as a conditioner for clay soils. Presumably the reduced dispersion and improved friability of calcareous clay soils after the addition of 5 per cent by weight Flotal (equivalent to 0.6 per cent Fe_2O_3), was due to the precipitation of finely divided Fe-oxide.

Naturally occurring oxides, carbonates and silicates

The difficulty of distinguishing the effect of Al and Fe oxides on structure has already been mentioned. In surface soil aggregates the situation is more complicated because both Al and Fe can react with organic matter present.

Deshpande, Greenland and Quirk (1968) peroxidized a number of acid soils and then extracted the soils with dithionite followed by a wash with 0.05 N HCl (McKenzie 1954). Up to 11 per cent by weight of Fe was removed in the extracts from the soils. The soils were then washed with N NaCl, the salt removed by dialysis and the soil dried. There was no increase in the swelling of cores made from the extracted soil compared with the oxidized Na-soil, suggesting that Fe was ineffective. However, these soils contained kaolin as the only clay mineral, so that little swelling of the "Na"-soil would be expected under acid conditions anyway.

Earlier, Deshpande, Greenland and Quirk (1964*a*) had carried out similar experiments on soils containing carbonate and 0.4-1.0 per cent extractable Fe. In these cases removal of Fe did cause some increase in the subsequent swelling of the Na-soil. However, extraction with 0.1N HCl was much more effective. Therefore it was suggested that Al and carbonate rather than Fe were important in bonding particles together. A complicating factor is that oxidation of a soil under alkaline conditions can lead to a large increase in positive charge when the soil is subsequently made acid. Further, this charge is also removed by acid extraction (Deshpande, Greenland and Quirk 1964*b*; Tweneboah, Greenland and Oades 1967). It probably arises from Fe and Al left after oxidation of the organic matter (fig. 2*c*). Such freshly precipitated sesquioxides could help stabilize a soil in the way already described. Thus at present the role of Fe, Al or carbonate in stabilizing surface soil aggregates has not yet been established.

Another way in which it has been suggested that clay particles are bonded together is by polymerized silica, e.g. the Keuper marl (Sherwood 1967). Indications that the friability of soil can be improved by the addition of SiO_2 have been reported by Beckwith and Reeve (1962). In the field, extreme forms of cementing by oxides, silica and carbonate do occur as hardpans. These probably develop by gradual precipitation, "engulfing" soil particles, rather than actually bonding to them (Flach et al. 1969).

SWELLING AND DISPERSION OF SOIL AGGREGATES

Dispersion

Under optimum conditions for agricultural production, surface soil aggregates should not slump to form a crust after a cycle of wetting and drying. Otherwise physical parameters such as water entry will be adversely affected. Crusting will be most severe if free water accumulates at the surface permitting the aggregates to disperse, for drying dispersed soil gives rise to greater bulk densities, and therefore soil strength, than drying flocculated soil (Sedgley 1962*a*; Coughlan, Fox and Hughes 1973*b*). Equally it is important to prevent excessive swelling of subsoil aggregates, otherwise the conducting cracks become closed off, causing waterlogging. Two main conditions under which aggregates disperse in water have been distinguished. One is the dispersion of dry aggregates, which occurs if the ESP (exchangeable sodium percentage) is sufficiently high. The other is the dispersion of wet, sheared aggregates, the minimum water content of aggregates required before this can occur being termed the water content for dispersion.

For a given soil there is a transition from the second type of dispersion to the first, as the ESP is increased. This is illustrated for subsoil aggregates from a neutral, illitic red-brown earth. The water contents for dispersion of Ca-Na and Mg-Na aggregates are shown in figure 4 plotted against the SAR (sodium adsorption ratio) of the chloride solution with which the aggregates were equilibrated initially (Bakker, Emerson and Oades 1973). SAR is plotted rather than ESP because some Na was extracted even from the Ca and Mg forms of the soil by the NH_4Cl solution used in determining ESP (Bakker, Emerson and Oades 1973). The values of the water contents for dispersion of the Ca-Na and Mg-Na soils were found to decrease rapidly with increasing SAR. The value for a given SAR when Mg was the principal cation was about half that for the Ca case. The dotted lines shown join the SAR values at the lowest content at which the soil could be uniformly remoulded, to the SAR values required for dispersion of air-dry aggregates (Bakker, Emerson and Oades 1973; Emerson and Bakker 1973). The latter corresponded to ESP values of 5 and 3 for the Ca-Na and Mg-Na aggregates respectively.

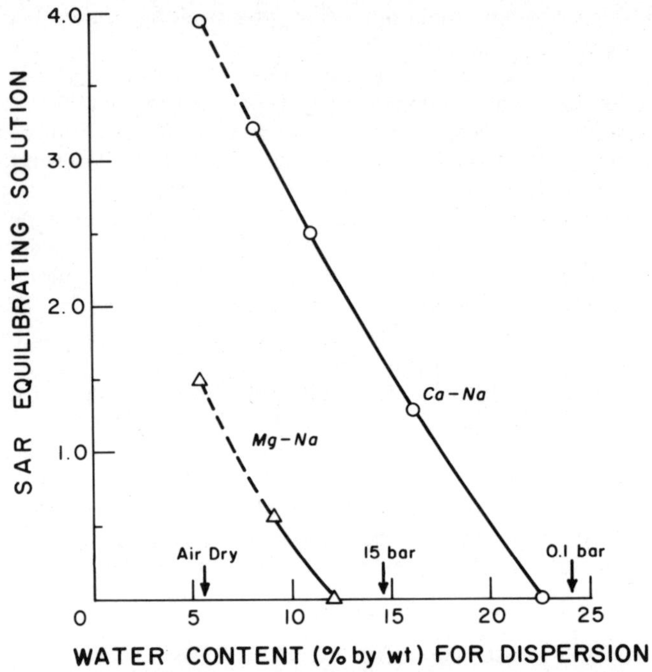

Fig. 4. The dispersion of Shepparton subsoil aggregates equilibrated with Ca-Na or Mg-Na chloride solutions of given SAR. Arrowed water contents were determined on unsheared soil.

These very low values appear to be associated with soils which contain neither free carbonate nor appreciable Al on the exchange sites (Bakker, Emerson and Oades 1973). The enhanced effect of Mg compared with Ca on the ease of dispersion of a soil appears to be a general one, e.g. Emerson (1972), but is probably of most importance in neutral soils.

As already pointed out, organic matter can hold clay particles together in unsheared aggregates against an applied osmotic stress, but will increase the repulsive forces between the particles if the organic matter bonds are sheared. Thus, the water content for dispersion of surface soils can be reduced to values appreciably below the 100 cm water content, without the necessity of the presence of exchangeable Na (Emerson and Smith 1970). For such class 3 surface soils (table 2), the shearing of the wet soil necessary for dispersion could be brought about by cultivation or stock treading under wet conditions. However, during rainfall, the mechanical action of raindrops in itself may be sufficient to cause shearing of the wetted aggregates. There is no systematic information on the lower limits of rainfall intensity and duration necessary to produce dispersion on different soil types.

Swelling

The conditions under which subsoil aggregates are exposed to free water are limited as compared with surface soils. Such conditions would occur for example if deep ripping were followed by surface ponding, or at the junction of the cultivated topsoil and undisturbed subsoil if the unsaturated conductivity of the former were greater than the saturated conductivity of the latter. Under these conditions, subsoil aggregates which disperse when immersed dry in water will also disperse in the field.

More frequently the subsoil is wet by steady infiltration. Nevertheless the suction will still be reduced to a low value, <25 cm (Marshall 1959). Aggregates which disperse in free water can swell appreciably more at low suctions compared with similar non-dispersing aggregates, i.e. with lower ESP. Some results for aggregates prepared from the same Shepparton soil as used for the previous work are shown in figure 5, where the increase in swelling at 12 cm suction was measured by the increase in water uptake (Emerson and Bakker 1973). The increase in swelling for a given ESP when Mg was the principal cation rather than Ca was most pronounced. The practical importance of such increased swelling is in the corresponding reduction in the permeability of the subsoil which would be caused.

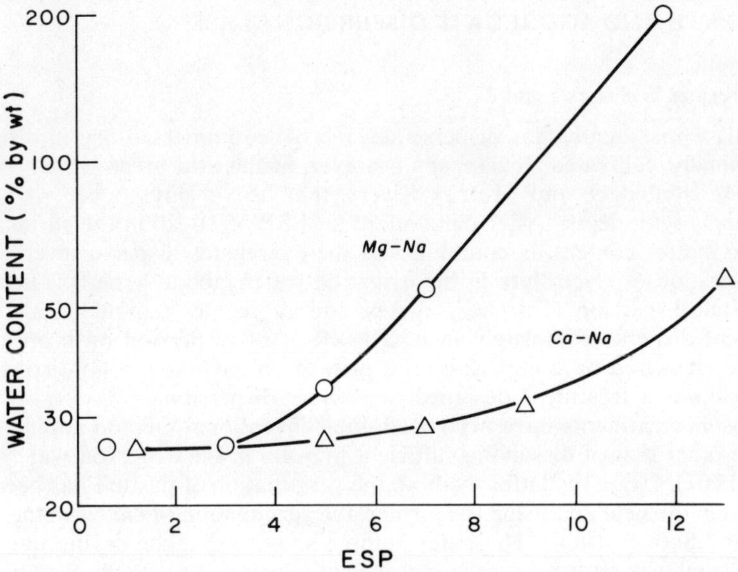

Fig. 5. The increase in water content at 12 cm suction of Shepparton subsoil aggregates with increasing ESP, the complementary ion being either Mg or Ca.

Effect of irrigation water

In the case of irrigated soils, the reaction of the aggregates to irrigation water is also important for soil physical conditions. In a study of the effect of electrolyte concentration on swelling and dispersion, Quirk and Schofield (1955) leached beds of aggregates from a surface soil with mixed $CaCl_2$-NaCl solutions of decreasing total concentration but constant SAR, so that the ESP of the aggregates was also kept approximately constant. They determined the electrolyte concentration at which a 15 per cent reduction in the permeability of the bed occurred, the "threshold concentration" and the lower concentration at which clay dispersed from the aggregates. Both concentrations were found to increase with increasing SAR, as would be expected. Rowell, Payne and Ahmad (1969) made similar measurements on a surface soil and a corresponding subsoil, and showed that the 15 per cent reduction in permeability was largely due to swelling of the aggregates.

It is not possible to obtain a universal relationship between the ESP of a soil and the minimum concentration of electrolyte needed to prevent deleterious swelling or dispersion. For a given soil, the concentration will vary with the amount of stress applied to the aggregates during the test method used, especially if the soil contains organic matter. For a given test method it will vary markedly with the soil used. McNeill and Coleman (1966), again measuring the permeability of an aggregate bed, but using as a criterion the electrolyte concentration required to cause a 25 per cent reduction in permeability rather than 15 per cent, showed that for soils in equilibrium with solutions of SAR 20 the concentration could vary from zero for an acid kaolinitic subsoil to 75 me/l for a montmorillonitic soil. Hence it seems likely that the simple dispersion and swelling tests in distilled water, outlined already for aggregates, could be usefully extended to irrigation waters. It would be appropriate for example in programming the efficient removal of exchangeable sodium from a soil by successive dilution of an initially high salt water (Reeve and Bower 1960) without the risk of causing physical damage to the soil.

CROP GROWTH AND AGGREGATE DISPERSION CLASS

Surface aggregates in classes 1 and 2

Surface soils whose aggregates disperse severely when immersed dry in distilled water are not normally cultivated. It happens however, in the arid areas which are irrigated from the Murrumbidgee and Murray Rivers, that heavy clays occur whose ESP increases sharply with depth. After cultivation, an ESP of 10-20 in the surface soil, with low organic matter content, is common and the aggregates disperse severely in water. The concentration of electrolyte in the irrigation water (about 1 me/l) is insufficient to prevent partial dispersion of dry aggregates, and aggregates remoulded at the 100 cm water content disperse completely in it. The effect of dispersion on crop growth can therefore be measured by comparing crop growth on such soils using irrigation water with and without a treatment designed to prevent dispersion.

Two gypsum treatments have been used, the conventional method adding gypsum to the soil, the other that of dissolving sufficient gypsum in the irrigation water (Davidson and Quirk 1961). Using the latter method, the germination of pasture has been shown to be improved in the field and using the former, the germination of clover in the glasshouse (Loveday and Scotter 1966). The latter authors showed that the drying out of the surface 1 cm of soil was retarded by the presence of gypsum. The mechanism by which the surface is maintained more moist is not entirely clear because of possible complications due to salt gradients (Penman 1941). Photomicrographs of dried-out surface soil show that without gypsum the soil is consolidated (Sedgley 1962*a*) and there is a complete loss of large pores to a depth of about 3 mm (Loveday and Scotter 1966). The deleterious effect of dispersion on germination is probably caused by the increased strength of this 3 mm crust, accentuated by the increased matric suction (Collis-George and Williams 1968) due to the more rapid drying out.

Aside from improved germination, the increase in the yield of crops on these soils in the field such as clover (Sedgley 1962*b*), lucerne (Loveday 1964) and cotton (Loveday et al. 1970) due to the addition of gypsum has been shown to be a result of the increase in water penetration per irrigation. Since the water content profile at the beginning of each irrigation was approximately the same for both untreated soil and the soil plus added gypsum (Sedgley 1962*b*; Loveday et al. 1970), the reduction in water entry was due to the throttling effect of the dispersed layer of surface soil. The added gypsum also caused a progressive reduction in the salt content of the subsoil (Loveday et al. 1970). This should give a long-term increase in the amount of available water with depth, but a corresponding beneficial effect on crop growth has not yet been noted.

Subsoil aggregates in classes 1 and 2

In the irrigation areas around Shepparton, in Victoria, the growth of peaches on certain red-brown earths has been unsatisfactory due to waterlogging in wet winters associated with the low permeability of the B horizon (Cockroft and Bakker 1966). Poor growth has now been found to be also strongly associated with the degree of dispersion of the B horizon aggregates in water (Bakker, Emerson and Oades 1973). It seems likely that the reduction in permeability is due to the enhanced swelling caused by an ESP in the range 5-10 together with an exchangeable Mg to exchangeable Ca ratio of 2.

Surface aggregates in class 3

The soil characteristics which are conducive to a value for the water content for dispersion being less than field capacity have already been discussed. Amongst these is a loam or silty loam texture where the clay fraction is dominantly illite. Surface soils with these textures also do not shrink appreciably on drying, so that a surface crust can be formed from the wet, dispersed soil. Such soils are common, for example, within the Great Soil Group, red brown earths. In order for surface soils of heavier texture to form similar crusts, the clay fraction has to contain proportionately more kaolinite to prevent cracking on drying and if acid, sufficient organic matter to react with Al on the exchange sites.

Experiments on the germination and growth of wheat in relation to structure have been carried out on a number of soils with surface aggregates in class 3. Some properties of four such soils, from South Australia, Victoria and Western Australia are shown in table 4. Note that the first number of the profile form indicates that the surface soils are seasonally "hard setting" (Northcote 1971). All soils disperse severely in water, after shearing wet (col. 8), despite the low ESP values. In the experiments in South Australia utilizing long-term rotation plots, the effect of structure was studied initially by comparing crop growth in wet winters and dry winters; in Victoria, structure was varied by adding gypsum and in Western Australia by adding a soil conditioner.

On a red-brown earth at Urrbrae (Litchfield 1951), the water contents at which cultivation and drilling were carried out were not measured. It was recorded that in wet winters, the soil was at or above field capacity for two-three weeks after sowing, the rainfall during this period being more than double that in dry winters. In the wet winters marked surface crusting occurred (McIntyre 1955) and germination was reduced (Millington 1959). The severity of these effects increased with the amount of cultivation during each cycle of the rotation, e.g. fallow–wheat only compared with three years pasture–wheat. Both authors found corresponding reductions in yield. On the basis of further field experiments with different levels of added N, Millington (1961) considered that the reduction in germination in wet years was due to poor soil physical conditions rather than a N deficiency. Greenland (1971) from an experiment comparing wheat growth on one of the arable rotation plots with that after freshly ploughed grassland, in which both water and N were varied, also concluded that the main cause of the reduced yield of wheat from the old arable plot was due to water loss caused by surface sealing rather than a N effect. The latter is analogous to the measured reduced water intake per irrigation previously discussed for class 1 soils.

The main physical effect of adding gypsum to a class 3 soil is to increase the value of the water content for dispersion, or as a corollary to reduce the degree of dispersion after shearing at a given water content. The latter is shown by the results in table 4 for two concentrations of $CaCl_2$, 0.3mM and 1 mM. One method of increasing the Ca in the soil solution is to add gypsum.

Table 4. Some properties of soils with surface aggregates in dispersion class 3, on which a crop response to physical conditions has been demonstrated

Soil location	Profile form*	Sample depth (cm)	pH†	ESP‡	% $<2\mu$	Water content at 100 cm suction (g/100g)	% Clay fraction dispersed from soil remoulded at 100 cm water content, immersed in		
							H_2O	0.3mM $CaCl_2$	1mM $CaCl_2$
Urrbrae, S.A.	Dr2.33	0–10	6.1	2	18	18	95	40	0
Wimmera, Vic.	Dr2.33	0–6	7.1	3	18	15	62	41	0
Goroke, Vic.	Dy3.43	0–5	5.9	3	8	9	87	48	0
Merredin, W.A.	Dr2.33	0–5	6.4	4	25	16	67	37	0
Katherine, N.T.	Gn2.14	0–5	6.8	2	34	18	65	–	0

* Northcote 1971.
† 1:5, soil water suspension.
‡ As % of CEC at pH 7.0.

Experiments on the effect of gypsum on the yield of wheat were started in Victoria in 1921 (Mullett and Scott 1923) and continued to 1933. The later results were summarized by Sims and Rooney (1965). On a red-brown earth at Salisbury in the Wimmera (Stace et al. 1968) an application of 4-5 t/ha oı gypsum before sowing wheat in a wheat–fallow–oats–fallow rotation increased the yield by 25 per cent. A residual effect was apparent after eight years, the yield of the third wheat crop being increased by 15 per cent. The untreated plots crusted badly and it was shown by hand counts that establishment was poorer; also that the gypsum-treated plots were able to withstand dry conditions longer. This is here interpreted to mean that less water was lost by run-off. At Goroke (Clarke and Blackburn 1956) on a solodized solonetz, which is a more sodium-affected Great Soil Group than red-brown earths, the addition of gypsum at the same level increased the subsequent yield of wheat by 40 per cent.

The beneficial effect of gypsum on these soils is well established. In fact it is estimated that during the three years 1963-65, cereal farmers in the Wimmera and southern mallee spread gypsum from local deposits over 45,000 ha of land. An estimate of the minimum amount of gypsum to be added to the soils in this area to maintain the concentration of Ca-ions in the soil solution above 1 mM may be made on the basis of the annual rainfall. To add this concentration to the rainfall would require about 600 kg/ha of gypsum. This is a minimum amount since time must be allowed for the gypsum present to dissolve sufficiently. However, the dressing should last for many years, as loss of salts by leaching would be small in these low rainfall areas.

At Merredin (table 4), on a solonized brown soil with kaolinite the main constituent of the clay fraction (Stace et al. 1968), the structure of the top 8 cm of soil was obviously improved by the addition of 600 kg/ha of krilium (Stoneman 1964). In the first year, a dry year, the germination of wheat was unaffected. However after adding water to ensure a crop, the yield of wheat on the treated plots was 25 per cent larger. There was a significant improvement in the germination of the subsequent crop, medic, on the treated plots.

On a quite different type of soil, a kaolinitic red earth from Katherine (table 4), severe crusting causes problems with the emergence of cotton seedlings (Arndt 1965). Previously, Arndt (1963), had shown that crusting was reduced if *subsoil* was mixed with the surface soil. The subsoil has a greater clay content, a lower organic matter content, and cannot be dispersed by remoulding wet (Emerson 1968). Similarly Greacen (1958) working on Urrbrae soil, found that deeper cultivation bringing soil of higher clay content to the surface reduced surface crusting. The water content for dispersion of the surface soil can only be increased in this way, provided the subsoil has a low ESP like the surface soil.

Another method found by Arndt (1963) of reducing crusting on this soil was to deliberately work the soil wet, so producing on drying a mulch of water-stable clods. Clods formed in this way from dispersed soil have a bulk density of 2.0 g/cm^3 instead of the value of 1.5 for normal aggregates and are resistant to slaking by water because of their very low permeability (Emerson, unpublished). This is a practical use of Vilensky's (1945) method of making water-stable aggregates, since in this case, plant roots are not required to grow through the clods.

It seems that under drying conditions the adverse effect of crusting on germination is due to the rapid increase in mechanical impedance of the crust. Under wet conditions McIntyre (1955) and Millington (1959) both ascribe reduced emergence to poor aeration caused by the pores in the crust being filled with water. This, in turn, could lead to the production of toxic levels of ethylene (Smith and Restall 1971). One additional factor under wet conditions is the visible rapid growth of algae on the surface of the soil. Bond and Harris (1964) have shown how algal strands can bond sand particles together. Whether surface impedance can be significantly increased in this way has yet to be established.

Aggregates in classes 4, 5 and 6

The water content for dispersion can be increased by the presence of finely divided carbonate in a way similar to that discussed already for applied gypsum. Soils containing class 4 aggregates can be manipulated, at least up to field capacity, without dispersion occurring when water is subsequently applied. An example of this is the Ug 5.28 soil in the irrigated area around the Namoi at Narrabri where extensive land forming is carried out for cotton growing. Such a condition would also be required for a subsoil in order to use unlined mole drainage channels, as on the Gault clay in England (Emerson 1967).

As would be anticipated, no adverse physical conditions for crop growth have been reported on class 5 (or 6) soils, since the water content for dispersion is outside the range of the normal soil water regime.

OTHER PHYSICAL FACTORS AFFECTING CROP GROWTH

Contact angle

A factor affecting water entry and movement, which is of crucial importance in the case of sands, but which had been neglected until the work of Bond (1960), is the advancing contact angle of the water on the dry soil. Apparant contact angles of over 90° have been measured by comparing the rates of capillary rise of water in dry untreated and dry ignited sand (Emerson and Bond 1963). It has been shown by Bond and Hammond (1970) how such angles can arise from the much smaller true contact angle due to the geometry of the pores (fig. 6). In the field, on deep sands, the contact angle varies

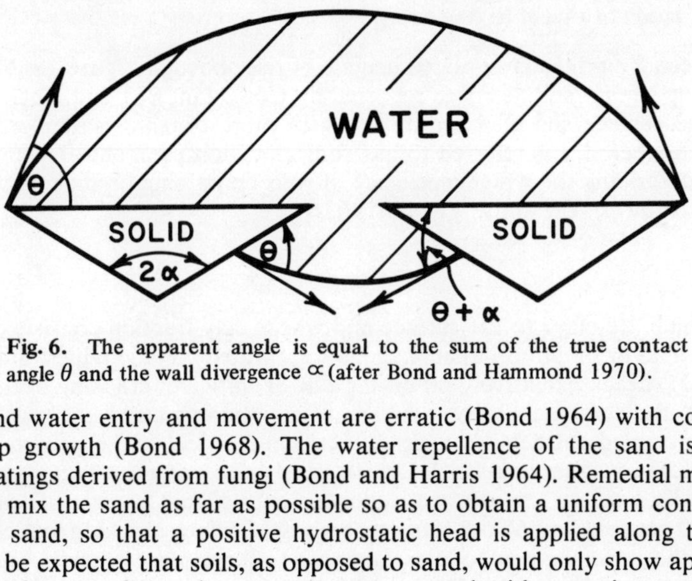

Fig. 6. The apparent angle is equal to the sum of the true contact angle θ and the wall divergence α (after Bond and Hammond 1970).

spatially and water entry and movement are erratic (Bond 1964) with correspondingly erratic crop growth (Bond 1968). The water repellence of the sand is attributed to organic coatings derived from fungi (Bond and Harris 1964). Remedial measures in the field are to mix the sand as far as possible so as to obtain a uniform contact angle and ridging the sand, so that a positive hydrostatic head is applied along the drill rows.

It would be expected that soils, as opposed to sand, would only show appreciable contact angles if most of the clay present was covered with organic matter. Quirk and Panabokke (1962*a*) found no difference in the rate of uptake of water from a source at a suction of 30 cm by dry aggregates from two plots on Urrbrae soil containing 2.5 per cent and 1.7 per cent O.M. respectively.

However, measurements have been made (Emerson, unpublished) of the rates of rise of water and alcohol, following Emerson and Bond (1963), in columns of aggregates from the surface soil of a krasnozem (0-10 cm, Beerik series, Blackburn and Leslie 1958). This soil was chosen because it contains kaolin as the only clay mineral, appreciable organic matter (8.5 per cent) and shows negligible swelling (<2 per cent by volume). Assuming that alcohol gives a zero contact angle, the apparent contact angle

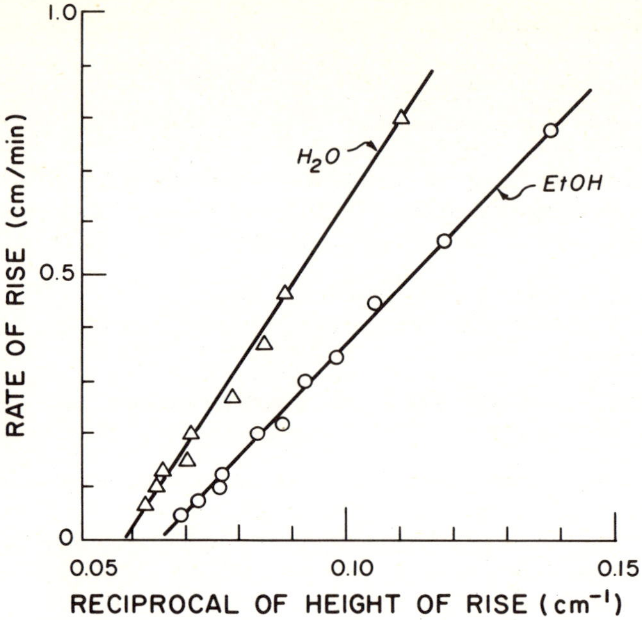

Fig. 7. The rate of rise into vertical columns of air-dry, 0.25–0.50 mm surface aggregates of a krasnozem under an applied head of 10 cm of H_2O or EtOH.

for water deduced from the extrapolated heights of rise shown in figure 7 is 68°. The difficulty of wetting similar soil has been reported by Coughlan, Fox and Hughes (1973*a*). In the field, the surface soil of a krasnozem with more organic matter present has a structure, which when dry is referred to as "snuffy" (Thompson and Beckmann 1959). Another term describing the water repellence of soils containing organic matter and appreciable iron oxide is "drummy" (Pizer 1962).

Cracking

Soil profiles which are clays to the surface and whose clay fraction consists dominantly of fine-grained illite or montmorillonite have low saturated conductivities. Such Ug soils (Northcote 1971) crack extensively on drying out. If sufficient cracking occurs before a crop wilts, then the cracks could be a means of supplying water rapidly to the plants. Fox (1964) has carried out detailed measurements in the field of the bulk density of the undisturbed 15-20 cm layer of a Ug 5.13 soil (Waco), covering the water content range 25-60 per cent by weight. The average water contents of this layer at pF values of 4.2 and 2.0 were 30 per cent and 50 per cent respectively. Surface cracking was observed on all sites when the water content of this layer was 40 per cent or less.

The relation of cracking on this Waco soil to water entry has been studied by Swartz (1966). He flooded a growing wheat crop with water when the soil had dried out sufficiently to form extensive voids, but the water content was still above wilting point, e.g. 32 per cent at 10-15 cm depth. In twenty minutes, sufficient water was absorbed to a depth of 120 cm to complete the three-dimensional swelling of the soil (Fox 1964). Swartz also showed that once three-dimensional swelling was complete, the rate of water absorption was then very slow, about 1.10^{-5} cm/sec. The former result is in marked contrast to those of Stirk (1954) who deduced from ring infiltrometer measurements on a number of Ug soils that rapid flow via cracks was not important until the pF of the soil was higher than the wilting point (>4.5). However, the soil was deliberately first cultivated to a depth of 5-8 cm and "the curves of water entry tend to reflect the effectiveness of the mulch in preventing water flow via cracks" (*sic*). Loveday (1964) in experiments with lucerne on thirty clay soils also reported that irrigation water largely entered through the cracks.

Subplasticity

From the definition of subplastic soils, the clay aggregates in these subsoils can be broken down by prolonged, vigorous working of the wet soil in the hand. It has been found that only a small percentage of the clay present is dispersed by older conventional methods for determining particle size distribution, such as high speed stirring of a soil-water suspension to which calgon and NaOH are added (Hutton 1955). To achieve full dispersion, it is necessary to apply severe mechanical stresses to the particles in suspension, such as by ultrasonics (McIntyre, private communication). The bonding mechanism is not known, although crystal growth has been suggested (Norrish 1972).

In practice in the field, these clays may be regarded as gravels with correspondingly good drainage properties (Butler 1955), enabling horticultural crops to be grown (van Dijk 1961). Subplastic clays have also been reported by engineers, e.g. Keuper marl (Dumbleton and West 1966). The aggregation can be used to advantage for foundations, but local variations in the degree of subplasticity make it difficult to extrapolate laboratory results to the field (Birch 1965).

WATER CONTROL STRUCTURES AND AGGREGATE DISPERSION CLASS

Farm dams

For a farm dam to hold water satisfactorily, it is usually necessary to reduce the hydraulic conductivity of the compacted floor to less than 1.10^{-7} cm/sec. This implies that the clay fraction of the soil has to be manipulated so that there are no continuous pores left of diameter greater than 10μ approximately. For subsoils with aggregates in dispersion classes 1 to 4, this can be accomplished by compaction at a water content near field capacity (soil groups A and B, Emerson 1960). It is not possible with subsoil aggregates in class 6 and subsoils with class 5 aggregates are borderline (soil groups C and D, Emerson 1960). In classes 6 and 5, the net attractive forces between the clay particles which cause flocculation are also sufficiently strong apparently to reverse particle displacements produced by shearing.

Another cause of excessive seepage through a dam floor is the formation of partially irreversible shrinkage cracks when a dam is emptied and the compacted soil dries out. Cracks may also develop in dam walls due to differential movement as when a dam is filled and the soil becomes saturated (Burton 1964). Such defects are most likely to occur on the first filling of dams constructed under too dry conditions for adequate compaction (Fraser 1961). However, they can also occur on dams constructed with apparently careful compaction control (Cole and Lewis 1960). Normally cracks are filled as the surrounding soil rapidly wets, and then slumps. However, if this soil disperses in the water with which a dam is filled, the soil can slough off as it slowly wets, and a piping failure results (Emerson 1959*b*; Aitchison, Ingles and Wood 1963). In practice it has been found that piping is a hazard if a layer of the soil in the dam wall consists of aggregates in class 1 (e.g. Rallings 1965).

Contour banks

Piping failures occur in contour banks constructed of class 1 aggregates in a similar way to that discussed already for dams (Emerson and Northcote 1964). With the limited size of structure, cracking is a more serious problem, since the depth of drying of clay soils under Australian conditions is about 180 cm (Aitchison and Holmes 1953). Therefore on Ug soils (class 4 aggregates), failure of contour banks is a problem (Ritchie 1964). Grass strips can be used on such soils instead of banks on slopes <5 per cent (Berndt and Swartz 1972). A mulch of suitably sized, stable material, either of surface soil or gravel, could also be used to reduce the rate of water loss from the underlying clay soil (Farrell, Greacen and Gurr 1966), and hence the incidence of cracking.

Two soil profile forms on which piping failures in dams or contour banks are common are Dy3.4 and Dr2.3 (Emerson and Northcote 1964). These are also the profiles on which tunnelling was first reported in the field by Downes (1946). It occurs on slopes where a defect, such as decomposed tree root, allows rapid entry of surface run-off into the dispersing subsoil. A similar type of defect has been shown to be responsible for sink-holes in mallee country (Stapledon 1970).

Engineering soil tests

Piping failures of the walls of farm dams are a common occurrence, whereas slip circle failures are extremely rare. Dam sites are often in gullies, with heterogeneous soil patterns and correspondingly varied ESP (e.g. Ingles 1964). Therefore some simple dispersion test is required, the cost of detailed engineering tests being prohibitive. Ritchie (1963) suggested measuring the percentage of the clay fraction released by shaking for a short time in water only, while Reilly (1970) measured the percentage of clay remaining in suspension after a given time. It is considered that the simplest test as given here for class 1 aggregates is preferable, since it obviates any mechanical action, or the necessity of taking into account the amount of salt present initially in the soil.

The shear characteristics of Na-affected soil with its in situ flocculating concentration of soluble salts differ little from those of the corresponding Ca-soil, e.g. the liquid limit of Na-montmorillonite in N NaCl is the same as that of Ca-montmorillonite in water (Warkentin and Yong 1962). Thus standard engineering tests do not detect those soils whose properties will change markedly with a change in the salt concentration of the soil solution. In general when attempting to specify stress paths for clay soils (Aitchison 1973), it may be necessary to take into account solute suction if the soil aggregates are in classes 1 or 2.

CONCLUSIONS

Soil structure in its broadest sense deals not only with particle arrangement and stability. It is also concerned with how the arrangements affect transport processes in soils and macroscopic soil properties, such as adhesion and swelling. The various interactions involved in determining particle arrangement, including organic matter, exchangeable cations, pH and clay minerals have been shown to be complex. Therefore attempts are made to develop simple tests which integrate most of these interactions for a particular transport process. For example, the dispersion tests described here in relation to the qualitative prediction of the transport of water. At the same time it is realized that the ultimate goal is the quantification of individual interactions. The recent work at Wageningen involving the calculation and measurement of the air pressures developed within aggregates during wetting is a good example.

In this review, soil structure has been mainly related to water, since in Australia water is usually the limiting factor in crop production. Australia, however, is not only the driest continent, but also one of the saltiest. The area of soils where it is recognized that the transport of water is adversely affected by the presence of exchangeable Na (in conjunction with Mg) is being continually extended, e.g. Northcote and Skene (1972), based on field observation, and more particularly using the dispersion tests outlined here. The old lower limit of ESP of 15 suggested by Richards (1954) is now agreed to be far too high.

In temperate climates, the levels of exchangeable Na are usually too low for unsheared soil to show excessive swelling or dispersion in water. The main structural damage to soils is caused by shearing the soil when it is too wet, either by cultivation or grazing animals (e.g. Boekel 1958, U.K. Min. Agric. Rep. 1970). The classical method of improving the water content range over which a soil can be safely worked is by a few years under grass (e.g. Low 1972). This has limited application in the wheat-growing areas of Australia because of the lack of persistence of the N_2 fixing legume in the grass sward, e.g. for medic, one year only. The method is also limited by economics in its application to the main problem soils in temperate climates, e.g. in United Kingdom the continuously cultivated silty soils (Russell 1971).

Deleterious crusting in the field has been shown to be associated with a low value for the water content of dispersion relative to field capacity. Further, the value of this water content for an illitic subsoil was significantly reduced by an ESP of only 2 or 3. It seems likely that there is considerable scope for reducing the dispersion of wet sheared soil in subsequent rainfall by the application of calcium (or higher valency) salts. This is not so much to displace any exchangeable sodium present as to maintain the concentration of Ca-ions in the soil solution. Calcium carbonate could be used instead of gypsum, provided time is given for the carbonate to be reprecipitated as clay sized particles so as to increase its solubility.

The aim of these additions would not be to emulate the very strong structure characteristic of old grassland, since agricultural crops such as wheat are not very sensitive to structure, but to ensure that very adverse structural conditions such as severe crusts and plough pans do not develop. It would also help to preserve the weak, ephemeral aggregates produced by the decomposing residues of a short pasture break. At the same time the supply of nutrients from the residues, on which the wheat crop depends, may be more sustained.

REFERENCES

Aitchison, G.D. (1973). 3rd Int. Conf. on Expansive Soils, Haifa, Israel, Paper No. 1.
Aitchison, G.D., and Holmes, J.W. (1953). *Aust. J. Appl. Sci.* 2, 257-66.
Aitchison, G.D., Ingles, O.G., and Wood, C.C. (1963). 4th Aust. N.Z. Conf. Soil Mech. and Found. Eng. 275-80.
Alexandrova, L.N., Yurlova, O.V., and Lobitskaya, L.V. (1964). *Trans. 8th Int. Congr. Soil Sci.* 3, 109-16.
Arndt, W. (1963). CSIRO Aust. Div. Land Res. Reg. Surv., Tech. Memo 64/3.
———. (1965). *Aust. J. Soil Res.* 3, 55-68.
Aylmore, L.A.G., and Quirk, J.P. (1960). *Nature* (*Lond.*) 183, 1752-3.
———. (1962). *Proc. Nat. Clay Conf.* 9, 104-30.
Bakker, A.C., Emerson, W.W., and Oades, J.M. (1973). *Aust. J. Soil Res.* 11, 143-50.
Beckwith, R.S., and Reeve, R. (1962). 3rd Aust. Conf. Soil Sci. Canberra, Paper No. 110.
Brendt, R.D., and Swartz, G.L. (1972). Symp. on Swelling Clay Soils, Univ. of New England. 125-38.
Birch, N. (1965). Abstr. Proc. Geol. Soc. Lond. No. 1621, 47.
Blackburn, G., and Leslie, T.I. (1958). CSIRO Aust. Soil Publ. No. 12.
Blackmore, A.V. (1973). *Aust. J. Soil Res.* 11, 75-82.
Boekel, P. (1958). Int. Symp. on Soil Structure, Ghent, 363-8.
Bolt, G.H., and Koenigs, F.F.R. (1972). Proc. Symp. Soil Conditioning, Ghent, 955-73.
Bond, R.D. (1960). CSIRO Aust. Div. Soils Rep. 10/60.
———. (1964). *Aust. J. Soil Res.* 2, 123-31.
———. (1968). *Trans. 9th Int. Congr. Soil Sci.* 1, 339-47.
Bond, R.D., and Hammond, L.C. (1970). *Soil Crop Sci. Soc. Fla. Proc.* 30, 308-15.

Bond, R.D., and Harris, J.R. (1964). *Aust. J. Soil Res.* 2, 111-22.
Brewer, R., and Blackmore, A.V. (1956). *Aust. J. Appl. Sci.* 7, 59-68.
Brewer, R., and Haldane, A.D. (1957). *Soil Sci.* **84**, 301-9.
Burton, J.R. (1964). Coll. on Failure of Small Earth Dams, CSIRO Aust. Div. Soil Mech., Paper No. 13.
Butler, B.E. (1955). *J. Aust. Inst. Agric. Sci.* 21, 239-49.
Cashen, G.H. (1959). *Faraday Soc. Trans.* 55, 477-86.
Childs, E.C. (1940). *Soil Sci.* **50**, 239-52.
Clapp, C.E., and Emerson, W.W. (1965*a*). *Soil Sci. Soc. Amer. Proc.* 29, 127-30.
———. (1965*b*). *Soil Sci. Soc. Amer. Proc.* 29, 130-4.
———. (1972). *Soil Sci.* **114**, 210-6.
Clapp, C.E., Olness, A.E., and Hoffmann, D.J. (1968). *Trans. 9th Int. Congr. Soil Sci.* 1, 627-34.
Clarke, A.R.P., and Blackburn, G. (1956). CSIRO Aust. Div. Soils Rep. 15/55.
Clarke, G.B., and Marshall, T.J. (1947). *J. Coun. Sci. Ind. Res. Aust.* **20**, 165-75.
Cockroft, B., and Bakker, A.C. (1966). *J. Aust. Inst. Agric. Sci.* 32, 292-3.
Cole, D.C.H., and Lewis, J.G. (1960). 3rd Aust. N.Z. Soil Mech. and Found. Eng. 93-98.
Collis-George, N., and Laryea, K.B. (1972). *Aust. J. Soil Res.* **10**, 15-24.
Collis-George, N., and Williams, J. (1968). *Aust. J. Soil Res.* **6**, 179-92.
Coughlan, K.J., Fox, W.E., and Hughes, J.D. (1973*a*). *Aust. J. Soil Res.* **11**, 65-73.
———. (1973*b*). *Aust. J. Soil Res.* **11**, 133-41.
Davidson, J.L., and Quirk, J.P. (1961). *Aust. J. Agric. Res.* 12, 100-10.
Deshpande, T.L., Greenland, D.J., and Quirk, J.P. (1964*a*). *Nature* (*Lond.*) 201, 107-8.
———. (1964*b*). *Trans. 8th Int. Congr. Soil Sci.* 3, 1213-25.
———. (1968). *J. Soil Sci.* 19, 108-22.
Dettmann, M.G. (1958). *J. Soil Sci.* 9, 306-15.
Dettmann, M.G., and Emerson, W.W. (1959). *J. Soil Sci.* **10**, 215-27.
Downes, R.G. (1946). *J. Coun. Sci. Ind. Res. Aust.* 19, 283-310.
Downes, R.G., and Leeper, G.W. (1939). *Proc. Roy. Soc. Vict.* 52, 1-20.
Dumbleton, M.J., and West, G. (1966). *Clay Minerals* 6, 179-93.
Edwards, A.P., and Bremner, J.M. (1967). *J. Soil Sci.* 18, 47-63.
El Rayah, H.M.E., and Rowell, D.L. (1973). *J. Soil Sci.* 24, 137-44.
El-Swaify, S.A., and Emerson, W.W. (1975). *Soil Sci. Soc. Amer. Proc.* 39, 1056-63.
Emerson, W.W. (1954). *J. Soil Sci.* 5, 235-50.
———. (1956). *J. Agric. Sci.* 47, 117-21.
———. (1959*a*). *J. Soil Sci.* **10**, 235-44.
———. (1959*b*). CSIRO Aust. Div. Soils Tech. Memo 4/59.
———. (1960). Arid Zone Tech. Conf. Warburton Vol. 1. Paper No. 38.
———. (1962). *J. Soil Sci.* 13, 40-5.
———. (1963). *J. Soil Sci.* 5, 233-50.
———. (1964). *Aust. J. Soil Res.* 2, 211-7.
———. (1967). *Aust. J. Soil Res.* 5, 47-57.
———. (1968). *Trans. 9th Int. Congr. Soil Sci.* 1, 617-26.
———. (1971). *J. Soil Sci.* 22, 50-9.
———. (1972). Symp. on Swelling Clay Soils, Univ. of New England, 39-44.
Emerson, W.W., and Bakker, A.C. (1973). *Aust. J. Soil Res.* **11**, 151-7.
Emerson, W.W., and Bond, R.D. (1963). *J. Soil Res.* 1, 9-16.
Emerson, W.W., and Grundy, G.M.P. (1954). *J. Agric. Sci.* 47, 117-21.
Emerson, W.W., and Northcote, K.H. (1964). Coll. on Failure of Small Earth Dams, CSIRO Aust. Div. Soil Mech. Paper No. 20.
Emerson, W.W., and Raupach, M. (1964). *Aust. J. Soil Res.* 2, 46-55.
Emerson, W.W., and Smith, B.H. (1970). *Nature* (*Lond.*) 228, 453-4.
Farrell, D.A., Greacen, E.L., and Gurr, C.G. (1966). *Soil Sci.* 102, 305-13.
Farrell, D.A., Greacen, E.L., and Larson, W.E. (1967). *Soil Sci. Soc. Amer. Proc.* 31, 445-50.
Flach, K.W., Nettleton, W.D., Gile, L.H., and Cady, J.G. (1969). *Soil Sci.* 107, 442-53.
Fox, W.E. (1964). *Soil Sci.* 98, 307-16.
Fraser, A.M. (1961). Inst. Engs. Aust. Brisbane Div., Tech. Paper 2, 2.
Gasperini, M., Alinari, E., and Malquori, A. (1958). *Agrochimica* 2, 54-6.
Geoghegan, M.J. (1950). *Trans. 4th Int. Congr. Soil Sci.* 1, 198-201.
Greacen, E.L. (1958). *Aust. J. Agric. Res.* 9, 129-37.
———. (1960). *Trans. 7th Int. Congr. Soil Sci.* 1, 256-64.
Greene-Kelly, R. (1964). *Clay Min. Bull.* 5, 393-400.
Greenland, D.J. (1963). *J. Colloid. Sci.* 18, 647-64.
———. (1971). *Soils Fertil.* 34, 237-51.
Greenland, D.J., Lindstrom, G.R., and Quirk, J.P. (1962). *Soil Sci. Soc. Amer. Proc.* 26, 366-71.
Greenland, D.J., and Parfitt, F.L. (1970). *Soil Sci. Soc. Amer. Proc.* 34, 862-5.
Griffith, A.A. (1924). Proc. 1st Int. Congr. Applied Maths. 55-63.

Haines, W.B. (1927). *J. Agric. Sci.* 17, 264-90.
Henin, S. (1938). "Etude physico-chimique de la stabilite structurale des terres", Paris.
Herzog, A. (1964). Coll. on Mechanisms of Soil Stabilization. CSIRO Aust. Div. Soil Mech. Paper p. 2.
Hubbell, D.S., and Chapman, J.E. (1946). *Soil Sci.* 62, 271-81.
Hutton, J.T. (1955). CSIRO Aust. Div. Soils Rep. 11/65.
Ingles, O.G. (1964). Coll. on Failure of Small Earth Dams, CSIRO Aust. Div. Soil Mech., Paper No. 37.
———. (1970). *Aust. J. Soil Res.* 8, 81-95.
Ingles, O.G., and Metcalf, J.B. (1972). "Soil Stabilization", Butterworths. London.
Jeanes, A. (1966). Encycl. Polymer Sci. and Tech. 806-24.
Litchfield, W.A. (1951). CSIRO Aust. Div. Soils, Rep. 2/51.
Loveday, J. (1964). *Aust. J. Soil Res.* 2, 96-110.
Loveday, J., and Pyle, J. (1973). CSIRO Aust. Div. Soils, Tech. Paper No. 15.
Loveday, J., Saunt, J.E., Fleming, P.M., and Muirhead, W.A. (1970). *Aust. J. Exp. Agric. Anim. Husb.* 10, 313-24.
Loveday, J., and Scotter, D.R. (1966). *Aust. J. Soil Res.* 4, 55-68.
Low, A.J. (1972). Chem. Ind. (Lond.) 373-8.
McEwen, M.B., and Pratt, M.I. (1957). *Faraday Soc. Trans.* 53, 535-41.
McIntyre, D.S. (1955). *Aust. J. Agric. Res.* 6, 797-803.
———. (1956). *J. Soil Sci.* 7, 302-6.
McKenzie, R.C. (1954). *J. Soil Sci.* 5, 167-72.
McNeill, B.L., and Coleman, N.T. (1966). *Soil Sci. Soc. Amer. Proc.* 30, 308-12.
Marshall, T.J. (1959). Commonw. Bur. Soils Tech. Commn. No. 50, CAB, Harpenden.
Marshall, T.J., and Quirk, J.P. (1950). *Aust. J. Agric. Res.* 1, 266-75.
Mehta, N.C., Streuli, H., Muller, M., and Deuel, H. (1960). *J. Sci. Food Agric.* 11. 40-7.
Middleton, H.E. (1930). U.S. Dep. Agric. Tech. Bull. No. 178.
Millington, R.J. (1959). *Aust. J. Agric. Res.* 10, 497-4.
———. (1961). *Aust. J. Agric. Res.* 12, 397-408.
Mullett, H.A., and Scott, P.R. (1923). *J. Dep. Agric. Vict.* 21, 583-97.
Norrish, K. (1972). *Proc. 1972 Int. Clay Conf. (Madrid)* 2, 3-14.
Northcote, K.H. (1971). "A Factual Key for the Recognition of Australian Soils", Rellim, Adelaide.
Northcote, K.H., and Skene, J.K.M. (1972). CSIRO Aust. Soil Publ. No. 27.
Panabokke, C.R., and Quirk, J.P. (1957). *Soil Sci.* 83, 185-95.
Penman, H.L. (1941). *J. Agric. Sci.* 31, 454-65.
Pizer, N.H. (1962). *J. Sci. Food Agric.* 13, 391-5.
Quirk, J.P. (1950). *Aust. J. Agric. Res.* 1, 276-84.
Quirk, J.P., and Panabokke, C.R. (1962*a*). *J. Soil Sci.* 13, 60-70.
———. (1962*b*). *J. Soil Sci.* 13, 71-81.
Quirk, J.P., and Schofield, R.K. (1955). *J. Soil Sci.* 6, 163-78.
Rallings, R.A. (1965). Water Res. Found. Aust. Bull. No. 10.
Reeve, R.C., and Bower, C.A. (1960). *Soil Sci.* 90, 139-44.
Reilly, L.A. (1970). Proc. Symp. Soils and Earth Structures in Arid Climates, Inst. Eng. Aust. pp. 16-9.
Richards, L.A. (Ed.) (1954). U.S. Dep. Agric. Handb. No. 60.
Ritchie, J.A. (1963). *J. Soil Conserv. Serv. N.S.W.* 19, 111-29.
———. (1964). Coll. on Failure of Small Earth Dams, CSIRO Aust. Div. Soil Mech. Paper No. 26.
Rovira, A.D., and Greacen, E.L. (1957). *Aust. J. Agric. Res.* 8, 659-73.
Rowell, D.L., Payne, D., and Ahmad, N. (1969). *J. Soil Sci.* 20, 176-88.
Ruehrwein, R.A., and Ward, D.W. (1952). *Soil Sci.* 73, 485-92.
Russell, E.W. (1971). *J. Soil Sci.* 22, 137-51.
Schahabi, S., and Schwertmann, U. (1970). *Z. Pflanzenernahrung Bodenk.* 125, 193-204.
Sedgley, R.H. (1962*a*). *Soil Sci.* 94, 357-65.
———. (1962*b*). 3rd Aust. Conf. Soil Sci. Canberra. Paper No. 55.
Sherwood, P.T. (1967). *Q. J. Eng. Geol.* 1, 47-55.
Sims, H.J., and Rooney, D.R. (1965). *J. Dep. Agric. Vict.* 63, 401-9.
Smith, K.A., and Restall, S.W.F. (1971). *J. Soil Sci.* 22, 430-43.
Stace, H.C.T., Hubble, G.D., Brewer, R., Northcote, K.H., Sleeman, J.R., Mulcahy, M.J., and Hallsworth, E.G. (1968). "A Handbook of Australian Soils" Rellim, Adelaide.
Stapledon, D.H. (1970). Proc. Symp. Soils and Earth Structures in Arid Climates, Inst. Eng. Aust. pp. 62-71.
Stefanson, R.C. (1968). *Trans. 9th Int. Congr. Soil Sc.* 2, 395-402.
———. (1971). *Aust. J. Soil Res.* 9, 33-41.
Stirk, G.B. (1954). *Aust. J. Agric. Res.* 5, 279-90.
Stoneman, T.C. (1964). *Aust. J. Exp. Agric. Anim. Husb.* 4, 363-6.
Stroosnyder, L., and Koorevaar, P. (1972). Proc. Symp. Soil Conditioning, Ghent, 1095-106.
Swartz, G.L. (1966). *Qld. J. Agric. Anim. Sci.* 23, 408-22.
Taylor, R.M., and Norrish, K. (1966). *Clay Minerals* 6, 127-42.

Thompson, C.H., and Beckmann, G.C. (1959). CSIRO Aust. Div. Soils, Soils and Land Use Series, No. 28.
Tiulin, A.T. (1928). *Perm. Agric. Exp. Sta.* 2, 77.
Tweneboah, C.K., Greenland, D.J., and Oades, J.M. (1967). *Aust. J. Soil Res.* 5, 247-62.
U.K. Min. Agric. Fish, and Food Rep. (1970). "Modern Farming and the Soil". HMSO, London.
van Dijk, D.C. (1961). CSIRO Aust. Div. Soils, Soils and Land Use Series, No. 40.
Vilensky, D.G. (1945). "Aggregation of Soil, Its Theory and Practical Application". Academy Science, Moscow.
Wada, K. (1961). *Amer. Min.* 46, 78-91.
Warkentin, B.P., and Yong, R.N. (1962). *Clays and Clay Minerals* 9, 210-8.
Williams, B.G., Greenland, D.J., and Quirk, J.P. (1966). *Aust. J. Soil Res.* 4, 131-43.
———. (1967*a*). *Aust. J. Soil Res.* 5, 77-83.
———. (1967*b*). *Aust. J. Soil Res.* 5, 85-92.

CHAPTER 6

Inorganic chemical properties

C.H. Williams and J.D. Colwell

INTRODUCTION

Australia has a wide range of soils, the result of a great diversity of parent materials and soil-forming environments. Although most of the soil groups have counterparts in other parts of the world (Prescott 1931; Hosking 1935; Stephens 1949) there are some unique features which can be attributed to unusual features of the geological history of the continent. The more important of these include limited uplift of the continent, restriction of Pleistocene glaciation to small areas of the southeastern portion of the continent and limitation of post-Tertiary volcanic activity to comparatively small areas along the east and southeast coasts (Stephens and Donald 1958). As a consequence dissection has been retarded and there has been only limited exposure of basement rock to weathering. In addition only small areas of moranic and volcanic parent materials have been available for soil formation. The overall result has been the development of many strongly leached, highly weathered soils of considerable age, together with younger soils which have been formed on erosional material which has been derived from them, such as B-horizon remnants and transported sedimentary deposits—parent materials which have already passed through earlier weathering cycles. These features have all favoured a general depletion of nutrients from the soils.

Many soils in the semi-arid regions have also been exposed to the influence of cyclic salts, resulting in large areas of solodized soils and in southern Australia calcium carbonate, much of it loessial (Crocker 1946), has also played an important part in soil development. All these factors have combined to give a high proportion of soils low in organic matter and deficient in certain nutrients, particularly phosphorus.

The semi-arid soils in Australia which have been successfully used for cropping vary widely in chemical and physical properties. Soils range from the lateritic podzolics of Western Australia, deficient in phosphorus, nitrogen and trace elements, to the highly fertile black earths of northern New South Wales and southeastern Queensland, formed on Tertiary basalt, many of which have been cropped for seventy or more years without need for any fertilizers. Important also are the brown solonized soils, the red-brown earths and the grey and brown soils of heavy texture of southern Australia which are of low to moderate fertility.

The main nutrient deficiencies are phosphorus and nitrogen, other nutrients being important only in limited areas such as the "light lands" of Western Australia where the trace elements copper, zinc and molybdenum are also deficient. Until recently little use has been made of nitrogen fertilizers, the deficiency being met initially by bare fallowing to exploit the limited soil nitrogen reserves and, later, by the use of leguminous pasture leys. Single superphosphate, containing 9-10 per cent P (sometimes with trace elements), was almost the sole fertilizer used until the late 1960s when local manufacture of high analysis fertilizers including double superphosphate and ammonium phosphates commenced.

CHEMICAL FEATURES OF SEMI-ARID SOILS

Traditionally soils are classified on the basis of their physical morphology and soil units so defined are characterized by chemical analyses. Thus when the soils of Australia were first described and classified, attempts were made to define general chemical features. Podzolics were recognized to be acidic and leached, black earths as alkaline, krasnozems to have high phosphorus sorption capacities, and so on. General chemical characterizations of particular soils in selected regions were also carried out as, for example, for the soils of the South Australian mallee (Prescott and Piper 1932), the red-brown earth soil group in South Australia (Piper 1938), and the black earth group in northern New South Wales and Queensland (Hosking 1935). Chemical characterizations of soils were also considered an essential part of soil surveys. These characterizations were, however, uncoordinated on an Australia-wide basis, and although characterizations were reasonably thorough, vast areas of soils remain for which there are no chemical data. In any case these characterizations of the past provide statistically biassed estimates of soil properties because most analyses were on soil profiles that had been carefully selected to "represent" certain types of soil. Such characterizations have become comparatively rare in recent years following a general decrease in soil survey activity and pedological studies.

A chemical characterization of semi-arid soils is currently being undertaken as part of the National Soil Fertility Project (Hallsworth 1969) and an example of the types of statistical chemical characterizations that are now possible for some regions has been given by Colwell (1970). Large numbers of soil profiles can be sampled and analyzed, using modern field and laboratory equipment, and the data then statistically processed by computer. It is thus now possible to estimate mean profile trends for soil units and to indicate profile variability within units, as illustrated for pH and $NaHCO_3$-soluble P analyses in figures 1 and 2 for four Great Soil Groups in southern New South Wales. Differences in features of profile trends between soil units can also be tested by standard analysis of variance procedures, thus defining the uniqueness of various soil features. This type of characterization has however not yet been attempted on a national scale.

The most general and ambitious chemical characterization of Australian soils so far attempted was carried out by the C.S.I.R.O. Division of Soils for the occasion of the ninth International Congress of Soil Science (Adelaide, 1968). The data from this study are reported in the *Handbook of Australian Soils* (Stace et al. 1968). The results summarized below indicate an enormous variability in most soil properties of soil groups when considered on an Australia-wide basis, and there is an obvious need for further detailed studies to enable more specific characterizations. The magnitude and variability of the analyses recorded in this book are represented diagramatically in figures 3 to 12 for the Great Soil Groups found in semi-arid regions. For these diagrams the surface analysis is presented on a horizontal scale and the overall trend in analyses in the underlying surface metre of soil is represented by the arrows ↘ ↙ ↓ indicating respectively an overall clear increase, a decrease, or no change, with depth in the surface metre of soil.

Soil pH

Soil pH is measured by a variety of equilibration procedures, the most common in Australia being for a 1:5 soil:water ratio allowing 1-2 hours for equilibrium. Other soil:water ratios are often used however, particularly a 1:2.5 ratio, a soil-water paste and also a 0.1 N $CaCl_2$ solution instead of water. In general the pH values obtained by these different procedures are highly correlated and there seems no evidence of practical advantages particularly favouring any one procedure. The pH values for narrower soil:solution ratios are usually lower, the pH for a 1:2.5 suspension being of the order

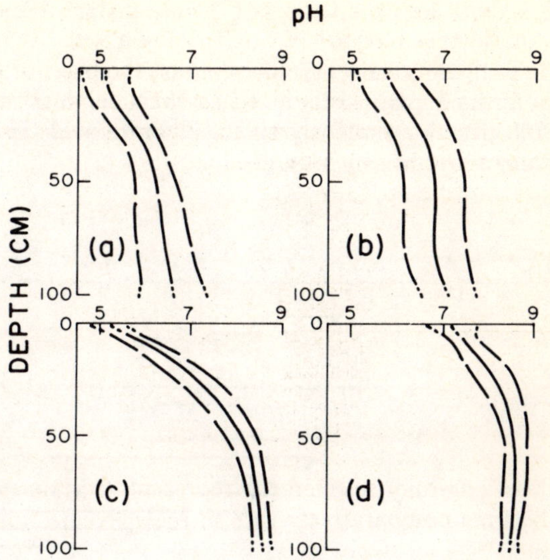

Fig. 1. Mean trend of pH in the profiles of four Great Soil Groups in southern New South Wales: (*a*) podzolic, (*b*) red earth, (*c*) red-brown earth, (*d*) grey clay, together with 5 per cent confidence intervals. Reprinted, by permission, from *Aust. J. Soil Res.* 8, 230, 231.

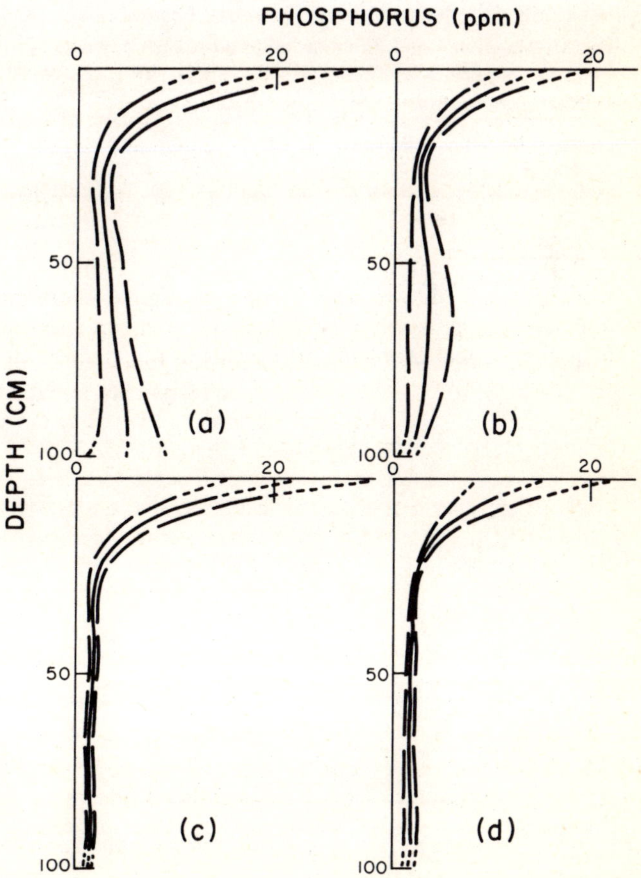

Fig. 2. Sodium bicarbonate – soluble ,phosphorus. Reprinted by permission, from *Aust. J. Soil Res.* 8, 230, 231.

0.3 units less than for a 1:5 suspension. Weak salt suspensions also give lower pH values, values for 0.1 N $CaCl_2$ being about 1.0 units less than for an aqueous suspension.

The pH values (1:5) for semi-arid soils reported by Stace et al. (1968) are represented in figure 3. pH mostly increases with depth though exceptions are not uncommon. Some soil groups are clearly typically more acid than others though values can cover wide ranges within any one group.

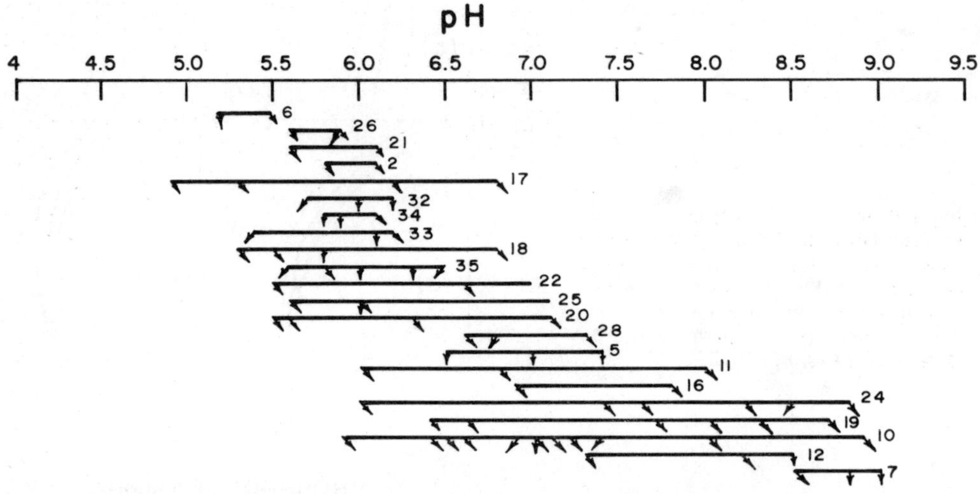

Fig. 3. Magnitude and direction of trends of pH down soil profiles of Great Soil Groups recorded by Stace et al. (1968). Numbers refer to Soil Groups: 2 = alluvial soils, 5 = siliceous sands, 6 = earthy sands, 7 = grey, brown and red calcareous soils, 10 = grey, brown and red clays, 11 = black earths, 12 = rendzinas, 16 = solonetz, 17 = solodized solonetz and solodic soils, 18 = soloths, 19 = solonized brown soils, 20 = red-brown earths, 21 = non-calcic brown soils, 22 = chocolate soils, 23 = brown earths, 24 = calcareous red earths, 25 = red earths, 26 = yellow earths, 28 = euchrozems, 32 = red podzolic soils, 33 = yellow podzolic soils, 34 = brown podzolic soils, 35 = lateritic podzolic soils.

Soluble salts

Most of the semi-arid soils contain negligible amounts of salt in their surface horizons and salt problems are mostly due to rise in salt from deeper horizons due to flooding or irrigation. Local problems also occur in the vicinity of springs or seepage points on hill-

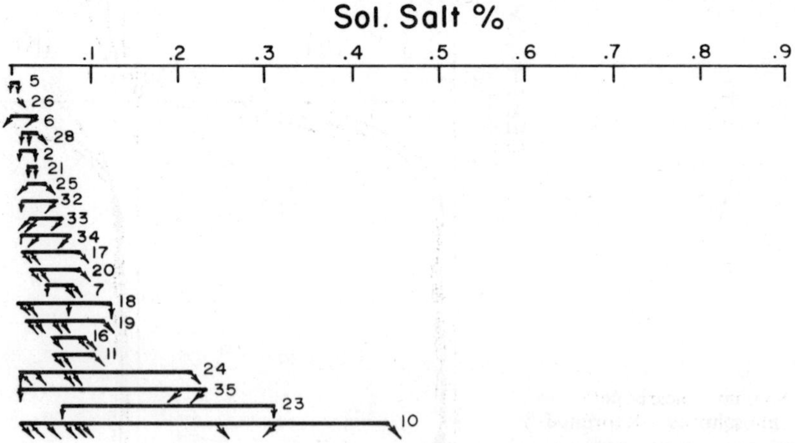

Fig. 4. Soluble salts (ppm) – as for figure 3.

sides. The highest values reported by Stace et al. (1968) (fig. 4) are for members of the calareous red earths, brown earths, lateritic podzolic and the grey, brown and red clay soil groups.

The distribution of salt in the landscape can be expected to be related to factors such as topographic situation, rainfall, drainage, accession of cyclic salt, type of soil, and this has been demonstrated in various localities (e.g. Prescott 1931; Downes 1954; Rowan and Downes 1963; Bettenay, Blackmore and Hingston 1964; Wagner 1957; Logan 1958; Coke 1958). However, for most of Australia there is insufficient data for generalizations. In a particularly comprehensive study of salt and salt effects in Australian soils, Northcote and Skene (1972) examined numerous data but this served mainly to demonstrate the general lack of information. They state in this respect that "a search of the literature for such data invariably ends in frustrations". Nevertheless, they were able to prepare a map of saline and sodic soils in Australia suggesting that most soils are dominated by chlorides (saline soils) or have been affected by salts (sodic soils). The map is based on observed association of profile morphology with saline, sodic and alkaline soil properties, but they emphasize that it should be regarded as tentative because of the general lack of chemical data.

Free calcium carbonate

Many of the soils contain free calcium carbonates in their lower horizons and some, notably the gilgaied black earths and the grey and brown clays contain calcium carbonate concentrations to the surface (fig 5).

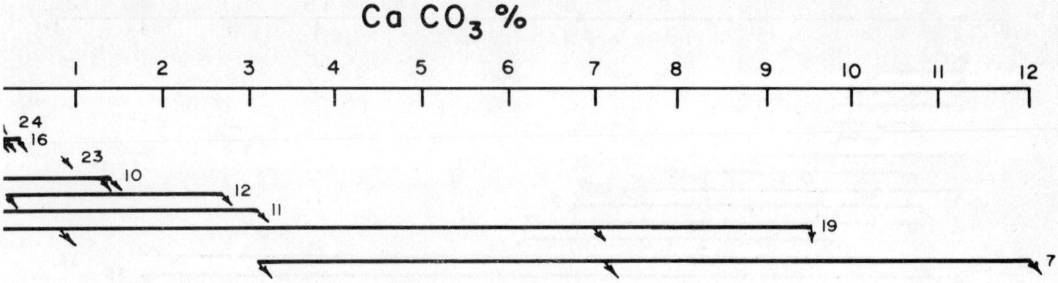

Fig. 5. Free $CaCO_3$ (per cent) – as for figure 3.

Organic matter

The organic matter content of soils is largely determined by climate and land-use history, tending to decrease with decrease in average annual rainfall, with increase in average temperature and with prolonged cropping. In some areas it is also affected by parent material and nutrient content. Most of the semi-arid soils (fig. 6) contain less than 6 per cent organic matter in the surface horizons, higher amounts generally being restricted to soils found in moister and cooler environments, and to uncultivated soils.

Exchangeable cations

The exchangeable cations (Na, K, Ca, Mg) of soils reported by Stace et al. cover wide ranges (figs. 7-10) and show few consistent differences between soils. The levels of the nutrients K and Mg are in general well above levels associated with nutrient deficiencies, especially in subsurface horizons, i.e. above about 0.2 me per cent K and 2.0 me per cent Mg (Min. Agric. Fish. Food Tech. 1967; Holford 1966; Hawkins 1967; Piper and de Vries 1964).

An unusual feature of many soils, noted by Prescott (1931), is an approximate equivalence of exchangeable magnesium and calcium, as in solonized brown soils of South Australia. Prescott attributed this feature to the accession of magnesium salts to the soil as cyclic salt.

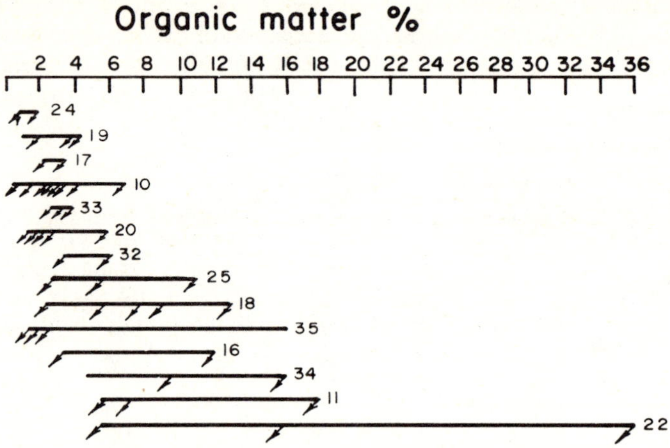

Fig. 6. Organic matter (per cent) – as for figure 3.

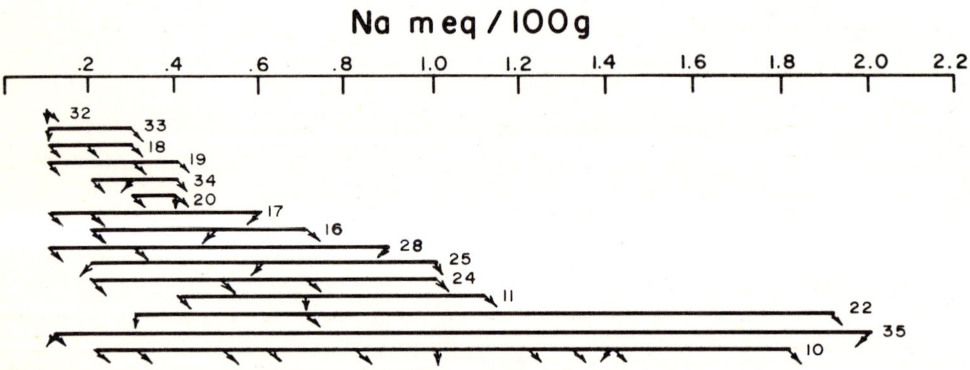

Fig. 7. Exchangeable Na (me per cent) – as for figure 3.

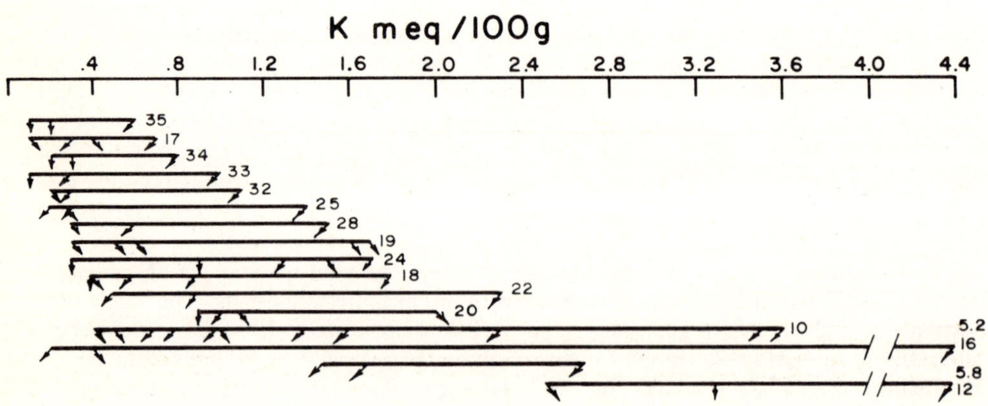

Fig. 8. Exchangeable K (me per cent) – as for figure 3.

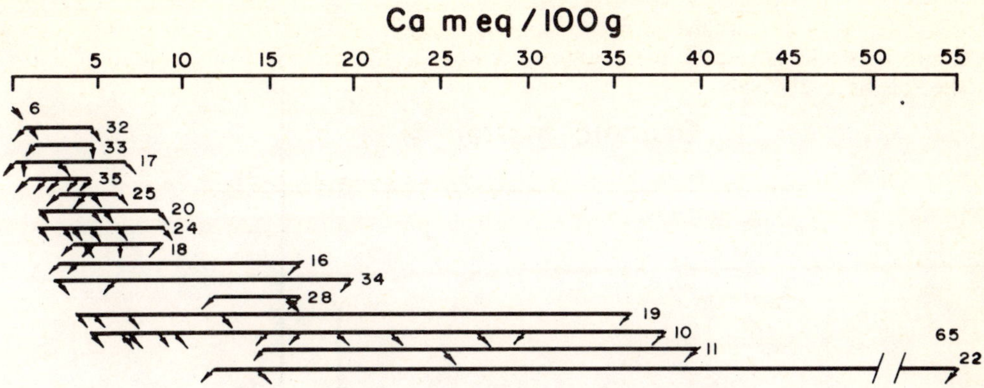

Fig. 9. Exchangeable Ca (me per cent) – as for figure 3.

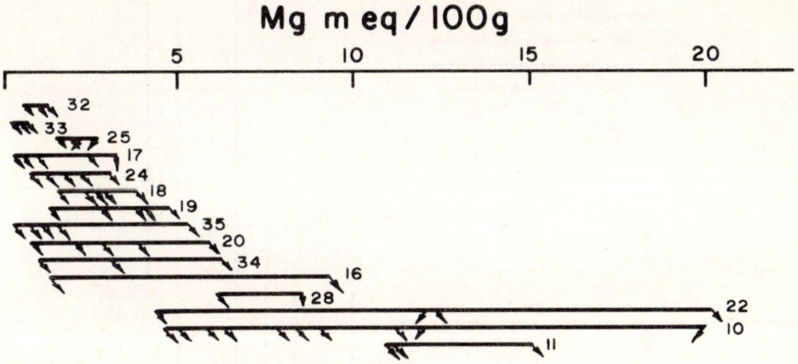

Fig. 10. Exchangeable Mg (me per cent) – as for figure 3.

Available phosphorus

The most widely used methods of analyses used to estimate phosphorus fertilizer requirements are extractions by 0.01 N H_2SO_4 (Kerr and Von Stieglitz 1938) and by 0.5 M $NaHCO_3$ (Colwell 1963). The $NaHCO_3$ analyses reported by Stace et al. are represented in figure 11 and indicate levels ranging from extreme deficiency (< 10 ppm P) to ample sufficiency (> 40 ppm P). Most soils are deficient in phosphorus as indicated by levels less than about 40 ppm P, though the interpretation of these analyses depends on other soil properties particularly pH (Colwell and Esdaile 1968).

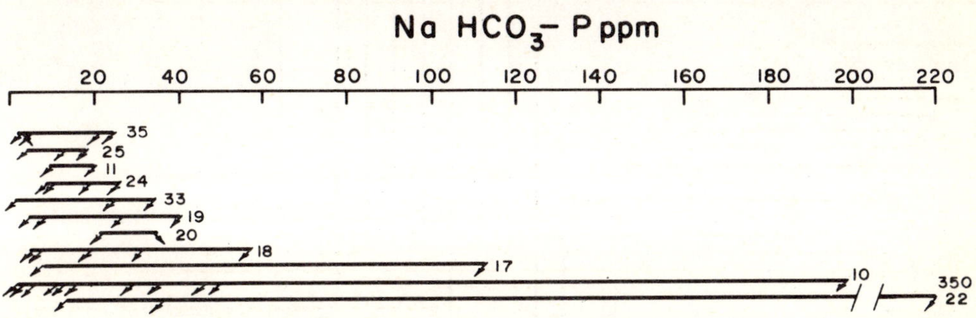

Fig. 11. Sodium bicarbonate soluble P (ppm) – as for figure 3.

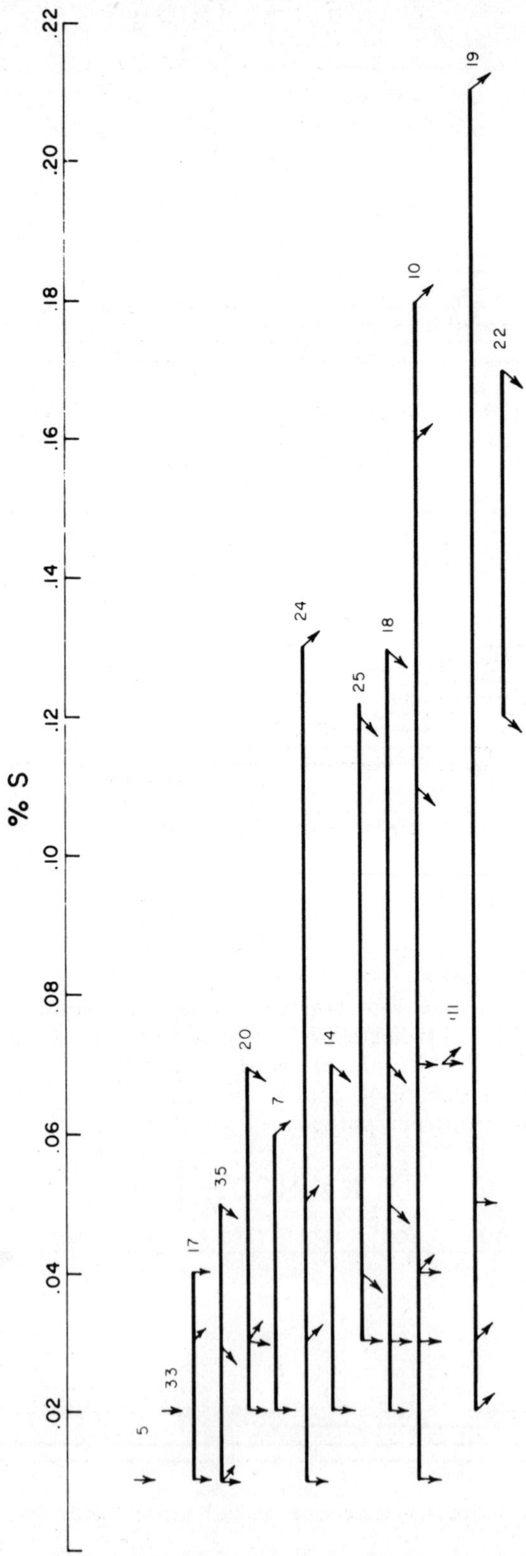

Fig. 12. Total sulphur, per cent S – as for figure 3.

Total sulphur

The amounts of combined sulphur in the surface horizons (fig. 12) vary considerably and usually reflect the organic matter status of the soil. The trends with depth vary depending on the accumulation or otherwise, of inorganic sulphate or calcium carbonate in the lower horizons.

Trace elements

Although trace element deficiencies are common in Australian soils, few analyses have been published to indicate levels of either total or available nutrient, or their distribution in the profile. Such information as is available is largely confined to particular soil types which are not necessarily representative of all soils. Table 1 summarizes data for total

Table 1. Total copper and zinc in solonetz profiles from South Australia

(Mean of six profiles. Data from Coaldrake 1951)

Horizon	Copper ppm	Zinc ppm
A	0.66	1.50
	0.60	1.37
B	1.19	13.2
	3.25	13.2

copper and zinc in six solonetz profiles of low copper and zinc status from South Australia (Coaldrake 1951). These indicate low values in the sandy surface horizons but higher concentrations, especially of zinc, in the heavy clay B-horizons. Table 2 presents

Table 2. Total cobalt, copper, molybdenum and manganese in red-brown earth profiles from South Australia

(Mean of fourteen profiles. Data from McKenzie 1957)

Horizon	Cobalt ppm	Copper ppm	Molybdenum ppm	Manganese ppm
A_1	5.7	30	3.1	500
A_2	6.7	31	3.2	440
B_1	11.0	55	4.0	470
B_2	9.0	55	3.9	480

mean values for total cobalt, copper, molybdenum and manganese in fourteen red-brown earth profiles from South Australia (McKenzie 1957). These soils are well supplied with each of these nutrients and no responses have been reported to any of them. Molybdenum content was correlated with the amount of clay in the surface soils but cobalt and copper were correlated with clay only in individual profiles. The high values for manganese are consistent with the results of Piper (1938) who reported high values (160-750 ppm) for reactive manganese (Mn extracted by hydroquinone in ammonium acetate pH 7) in these soils. Reactive manganese was highest in the surface and decreased with depth. Tiller, Honeysett and de Vries (1972) reported values for total zinc varying from 4 to 183 ppm in a wide range of Australian soils. Soils with the highest pH generally had the highest zinc contents, irrespective of soil group.

NUTRIENT LEVELS

Generally the levels of all major nutrients have been affected by the cultural and fertilizer practices imposed. Changes have been brought about by the accumulation of residual fertilizer phosphorus, the depletion of phosphorus in unfertilized soils, or as a result of changes in the amounts of soil organic matter following its mineralization during cropping or its accumulation under leys. The latter may lead directly to changes in the amounts of the organic matter components, carbon, nitrogen, sulphur and phosphorus, and to changes in the cation exchange capacity with consequent effects upon the amounts of the exchangeable cations.

Changes typical of those resulting from long-term cropping without leys in southern Australia are given in tables 3 and 4. Trends generally favouring the accumulation of nutrients in the surface soil often occur under pasture (Williams and Donald 1957). The effects of various cultural practices on such changes are discussed in chapter 13.

Factors influencing the nutrient level of soils in semi-arid areas are discussed below.

Table 3. Mean carbon, nitrogen, sulphur, and phosphorus in 55 pairs of cultivated and adjacent uncultivated soils, 0–10 cm depth, Ariah Park district of New South Wales*

				Phosphorus (ppm)		
	Carbon (%)	Nitrogen (%)	Sulphur (%)	Total	Organic	Extractable inorganic †
Uncultivated	1.64	0.094	0.0128	222	59	25
Cultivated	1.12	0.066	0.0092	223	49	32
Difference	0.52	0.028	0.0036	–1	10	–7

* Data from Williams and Lipsett 1961.

† Sum of inorganic phosphorus extracted by 2.5 per cent acetic acid – 1.0 per cent 8-hydoxyquinoline and 0.1 N sodium hydroxide (Williams 1950*a*).

Table 4. Mean exchangeable cation contents, cation exchange capacity, and pH of 55 pairs of cultivated and adjacent uncultivated soils 0–10 cm depth, Ariah Park district of New South Wales*

	Sodium	Potassium	Calcium	Magnesium	Sum	Cation exchange capacity	pH
Uncultivated (m–equiv./100 g)	0.122	1.12	6.07	2.11	9.42	11.24	6.39
Cultivated (m–equiv./100 g)	0.126	0.95	5.29	1.71	8.08	9.53	6.27
Difference (m–equiv./100g)	–0.004	0.17	0.78	0.40	1.34	1.71	0.12

* Data from Williams and Lipsett 1961.

Nitrogen

Fallowing (whereby soil is cultivated and kept bare for long periods before sowing a crop) has been widely practised throughout the semi-arid regions with obvious benefits in increased yields. The effect is mostly attributed to the conservation of water from rain received during the fallow period, to the mineralization of soil organic nitrogen to nitrate and its accumulation in surface layers of the soil, and to interaction effects of these two factors. Other factors such as the control of weeds, pests and diseases may also contribute under some circumstances but, from a nutrient point of view, mineralization of nitrogen is of major importance. The evolution of fallowing in the semi-arid areas of Australia is discussed in chapter 11.

Leaching of nitrate

Nitrate nitrogen may be leached from surface horizons into deeper subsoil horizons, although in southern Australia at least, little seems to be leached beyond the root zone and lost (Prescott and Piper 1930; Richardson and Fricke 1931; Storrier 1962, 1965*c*). In southern New South Wales, Williams and Lipsett (1961) estimated that the nitrogen removed in crops could largely account for the total loss of nitrogen from soils during a period of 50-60 years of cropping. On the "light land soils" of Western Australia however, the high intensity of the winter rainfall and the low water-holding capacity of the soils may result in considerable leaching of nitrate (Halse et al. 1969). In northern Australia intense rains in the summer-autumn may also leach nitrate to depths of 150 cm or more—well beyond the root zone of some crop species (Wetselaar and Norman 1960). Crops may sometimes recover from initial nitrogen deficiencies, induced by leaching, when roots reach the nitrogen in the lower horizons, as the plants develop (Halse et al. 1969).

In addition to possible losses of nitrogen by leaching there is evidence of gaseous loss, as N_2O and N_2, by denitrification processes. Burford and Millington (1968), Burford and Greenland (1970), McGarity (1961) and McGarity and Myers (1968) have demonstrated denitrification losses from red-brown earths and solodized solonetz soils under field conditions in South Australia, New South Wales and Queensland. Insufficient information is available as yet, however, to assess these losses accurately, although Burford and Millington's (1968) results on red-brown earths suggest only small losses as N_2O.

Yield depression by nitrogen

High levels of available nitrogen in soils following clover pastures can lead to a phenomenon termed "haying off" (Colwell 1963; Storrier 1962, 1965*a*; Dann 1969). The effect is ascribed to excessive vegetative growth in the early development stages of the crop, due to the high nitrogen level and a consequent exhaustion of available soil water before the crop has matured. Storrier (1965*b*) also considers tiller mortality due to greater competition for light a contributing factor. In some cases, however, yield depression at high soil nitrate levels may be due to inadequate molybdenum supply (Lipsett and Simpson 1971). The effect can be aggravated by the application of nitrogen fertilizers, and in some cases reduced by moderate applications of phosphorus fertilizer.

Phosphorus

Phosphorus status

Australian soils in general have low phosphorus levels, total or available, compared with soils of other countries. Wild (1958) compared total phosphorus levels, as estimated with boiling hydrochloric acid, and his results are summarized in table 5. The low level of available phosphorus is indicated by the wide need for phosphorus fertilizer throughout most agricultural areas of Australia. This generally low level may be due to great age of Australian soils compared with other countries and the consequent extreme leaching that has occurred (Wild 1958, 1961). The parent rocks of Australian soils are also often low in phosphorus content and this, too, may explain the low levels in the soils (Beadle 1962).

Australian cereal crops produced in semi-arid regions frequently have low phosphorus contents compared with crops from other countries (table 6) and this undoubtedly reflects the general low level in the soils. Only Queensland wheat (0.38 per cent P), grown mainly on black earths and their associated soils, which have been derived from phosphorus-rich basalts, has a mean phosphorus content comparable with overseas values.

Table 5. Phosphorus content of Australian soils*

Great Soil Group	Phosphorus % P
Lateritic podzolic	0.007
Groundwater podzols	0.010
Solods and solodic soils	0.010
Podzols and podzolic soils	0.015
Solonetz	0.016
Red earths and lateritic red earths	0.020
Desert loams	0.026
Grey and brown heavy soils	0.027
Solonized brown soils	0.028
Red-brown earths	0.028
Swamp soils	0.042
Black earths	0.087
Krasnozems	0.134
Average	0.020

* From Wild 1958.

Table 6. Phosphorus content of Australian wheat grain compared with other areas

Region	Number of samples	Range % P	Mean	Reference
Western Australia	36	0.18–0.37	0.25	Underwood and Snook 1935 Snook 1939
South Australia	45	0.16–0.38	0.25	Woodroffe and Williams 1953
Victoria	30	0.18–0.36	0.27	Dadswell 1935
Southern New South Wales	55	0.17–0.45	0.28	Williams and Lipsett 1961 Colwell 1963 Lipsett 1969
Queensland		0.15–0.42	0.38	Brunnich 1926
United States	310	0.15–0.54	0.40	Beeson 1941
Utah			0.36	Greaves 1940
Mean United States			0.43	Morrison 1938
United Kingdom			0.38	Wood and Woodman 1933
United Kingdom			0.41	Marshall and Halnan 1932

Some increase in crop phosphorus results from the use of fertilizer but the level is usually determined mainly by the available phosphorus status of the soil. Thus, for example, the application of 10 kg P/ha to red-brown earths (Woodroffe and Williams 1953) increased the phosphorus content of wheat grain by less than 0.004 per cent P and this is small compared with the differences resulting from differing levels of available phosphorus in these soils (fig. 13).

Residual value of phosphatic fertilizers

The amount of phosphorus removed by crops is frequently much smaller than that applied as fertilizer, so that fertilizer residues should accumulate in soils. In South Australia where phosphate fertilizer has been generally applied at rates of order 10 kg P/ha, residues have led to substantial increases in the level of available phosphorus (Williams 1950*b*; Woodroffe and Williams 1953). Similar increases are reported for Western Australia (Teakle and Cariss 1943), and Victoria (Sims 1953; Sims and Rooney 1956). In New South Wales, however, the rates of application of fertilizer have been much

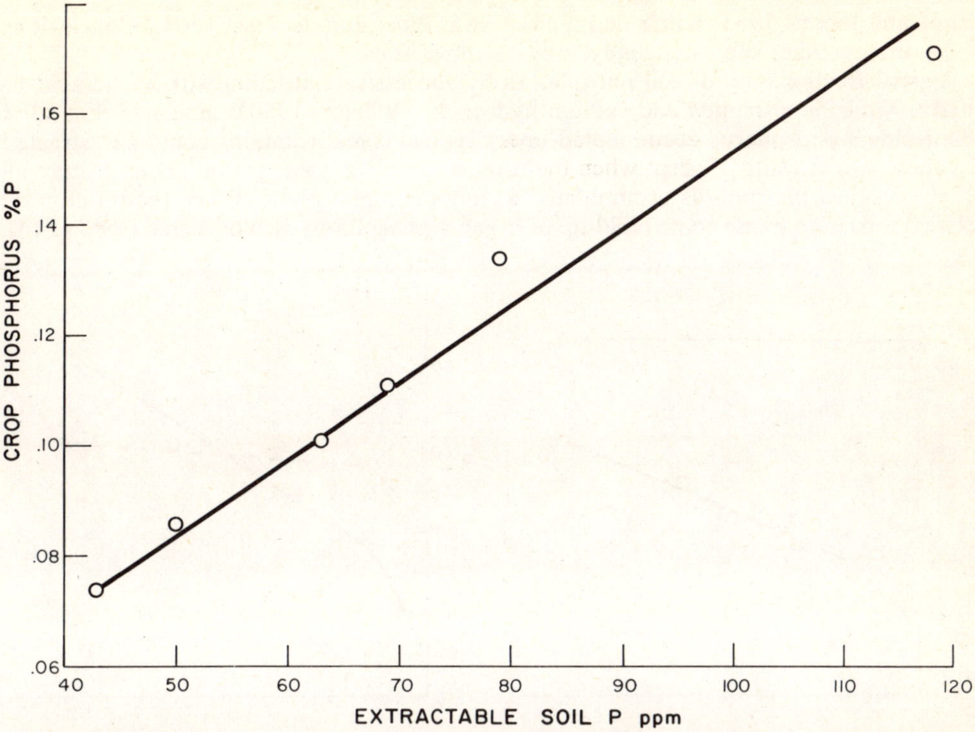

Fig. 13. Relationship betweeen extractable soil phosphorus (inorganic phosphorus extracted by acetic acid and sodium hydroxide [Williams 1950*a*]) and the phosphorus content of wheat plants grown on red-brown earths.

lower (5 kg P/ha or less) and barely sufficient to maintain the original soil phosphorus levels (Williams and Lipsett 1961).

The majority of Australian soils of the semi-arid region show no serious problems of phosphate fixation and substantial improvements in the levels of plant-available phosphorus have resulted from the accumulation of these fertilizer residues. For example, in Western Australia, soils which had received 55-70 kg P/ha as superphosphate with previous crops could be cropped for at least five years without further fertilizing (Teakle and Cariss 1943). Similarly Piper and de Vries (1964) showed in South Australia that wheat could be grown for five years without fertilizer on land that had received 260 kg P/ha during the preceding twenty-three years. However, with the higher yields now being obtained in most cereal districts, greater losses are occurring in crop produce so that fertilizer residues may be less important. A 3,000 kg/ha wheat crop, for example, will remove 7.5-12.0 kg P/ha, suggesting a need for heavier applications of fertilizer (Lipsett 1969). In contrast, on the better soils of northern New South Wales and Queensland, where crops have been grown for many years without fertilizer, appreciable depletion of available phosphorus has occurred and responses to fertilizer phosphorus are now being obtained on previously unfertilized soils (Colwell and Esdaile 1968). The level of available phosphorus in some soils is still sufficient however for many more years of exploitation.

Phosphorus fractions in soils

Detailed chemical studies on soil phosphorus have been largely confined to the red-brown earths and brown solonized soils of southern Australia (Williams 1950*b*; Williams and Lipsett 1961; Kurtz and Quirk 1965; Piper and de Vries 1964). Conclusions from these studies may not apply fully to other soils.

Partial fractionation of soil phosphorus by successive extraction with acetic acid-8-hydroxyquinoline solution and sodium hydroxide (Williams 1950*b*) indicated that all of the residual phosphorus accumulated under typical cereal rotations could be extracted by these two reagents. Except when the rotation involved some period under pasture all of the residual phosphorus accumulated as inorganic phosphate. When the rotation included a pasture phase some build-up of organic phosphorus also occurred (see fig. 14).

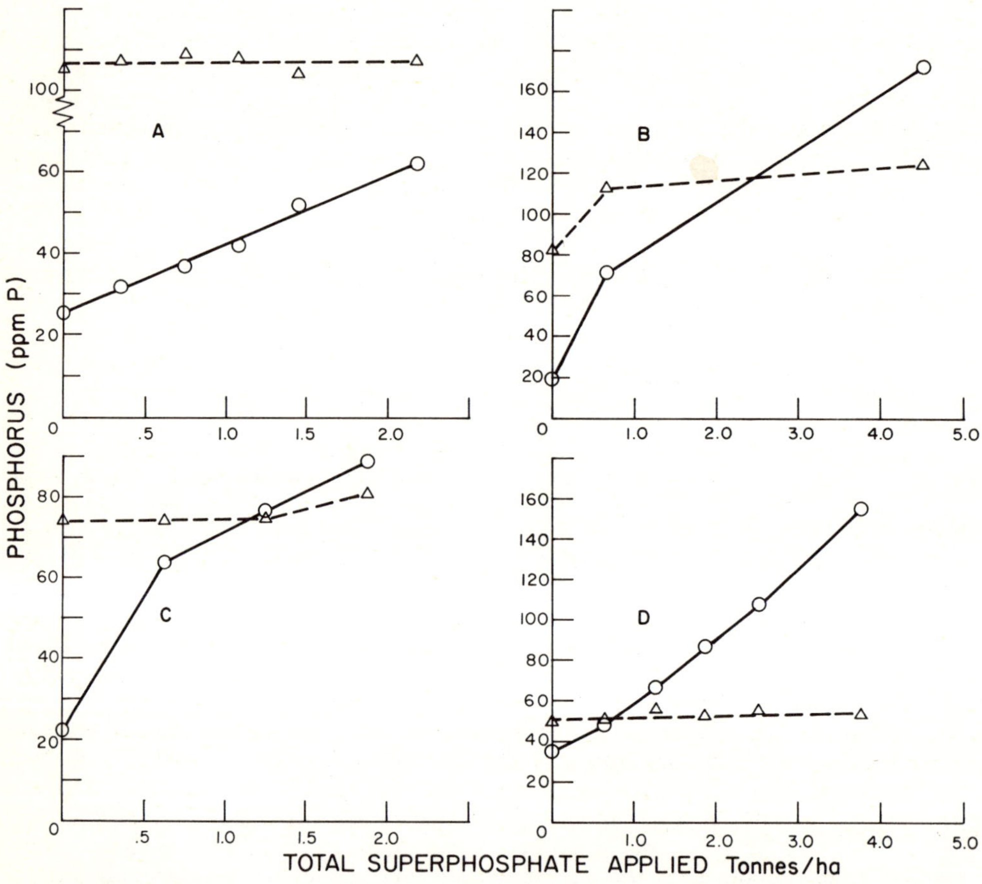

Fig. 14. Phosphorus, ppm P, extracted by acetic acid-sodium hydroxide (Williams 1950*a*) from soils with the same cropping histories for different rates of superphosphate application.

A. Adelaide	Soil pH 6.5	Fallow, wheat.	
B. Adelaide	Soil pH 6.5	Continuous pasture.	O — O Inorganic P
C. Booborowie	Soil pH 6.5	Fallow, wheat, pasture.	Δ — Δ Organic P
D. Appila	Soil pH 8.8	Fallow, wheat.	

These two extractants are considered to extract mainly calcium, iron and aluminium phosphates (Williams 1950*a*) so that the results for these two groups of soils suggest that the residual phosphate is accumulated largely in these forms. Fractionation studies using the procedure of Chang and Jackson (1957) on red-brown earths in South Australia (Kurtz and Quirk 1965) showed that all of the residual phosphate had accumulated in the fractions extracted by ammonium fluoride, sodium hydroxide and sulphuric acid, also suggesting that it is retained largely as calcium, iron and aluminium phosphates. Kurtz and Quirk (1965) results indicate that similar mechanisms could also be involved in the retention of phosphate by lateritic podzolic soils.

Extraction of soils with acetic acid-8-hydroxyquinoline and sodium hydroxide (Williams 1950*b*) indicated a relationship between soil pH and the relative amounts of residual phosphate extracted by the two reagents. At high pH relatively more of the residual phosphate was extracted by the acetic acid reagent, suggesting, as would be expected, that in the more alkaline soils a greater proportion of the phosphate was retained as calcium phosphate.

Organic phosphorus generally accounts for 20-40 per cent of the total phosphorus in the red-brown earth and brown solonized soils of South Australia (Williams 1950*b*) but pot culture experiments have indicated that its natural rate of mineralization is extremely slow and that very little is likely to be released for plant uptake during crop growth (Williams 1950*c*). Under the conditions of the pot culture experiment wheat plants derived at least 95 per cent of their phosphorus from inorganic sources. In red-brown earths and associated soils of the southern New South Wales wheat belt, organic phosphorus accounted for an average of approximately 25 per cent of the total soil phosphorus (Williams and Lipsett 1961). Comparison of cropped with contiguous virgin soils indicated that after cropping for 50-60 years an average of 17 per cent of the total organic phosphorus (equivalent to about 14 kg P/ha) had been mineralized. In the same soils approximately 30 per cent of the organic carbon, nitrogen and sulphur was mineralized during this period. This also indicates that the contribution of the organic phosphorus to the overall available phosphorus supply is generally small.

Leaching of phosphorus

Leaching of fertilizer phosphorus from the surface horizons of some light soils under pasture has been reported in southern Australia (Hingston 1959; Russell 1960; Loneragan and Paton 1960; Ozanne, Kirton and Shaw 1961). In Western Australia Ozanne, Kirton and Shaw (1961) have reported losses from the surface 10 cm of pasture soils ranging from 17 to 80 per cent of the phosphate from a single application of superphosphate under rainfall of 550 mm per annum. It seems likely that similar losses could occur in this type of soil under cereal crops. However, the extent of such losses and their importance under cereal cultivation have not been assessed. It seems possible however that the recovery of leached phosphate by cereal crops could be considerably greater than is the case with shallow-rooted pasture species.

Yield depression by phosphorus

Excessive levels of soil phosphorus may depress yields, in a similar fashion to excess soil nitrogen, as indicated by yield depressions that may result from the application of superphosphate (Colwell 1963). These depressions occur in years in which rainfall is low for the later stages of crop development, and may be termed "haying off" as for nitrogen yield depression. Such effects from phosphorus are, however, less frequent than for nitrogen.

Phosphorus toxicity, causing a grey to white necrosis of the tips of wheat and oat crops, has been reported for light-textured soils in Western Australia. Rossiter (1955) reported these symptoms with a glasshouse study and Loneragan, Carroll and Snowball (1966) and Bhatti and Loneragan (1970 *a,b*) reported phosphorus injury to wheat and oat seedlings on such soils under field conditions. Significant losses in cereal yields due to phosphorus toxicity, under field conditions, have however not been demonstrated and it seems doubtful that phosphorus toxicity will occur under normal field conditions even on the poorest of sand soils (Rossiter, private communication).

Sulphur

Although sulphur deficiency is widespread in Australian soils, and there have been many reports of its occurrence, particularly in pastures, there have been no important reports of responses by cereals. Sims (1959), using rock phosphate as a sulphur-free source of phosphorus, obtained no response to sulphur by wheat at several localities in Victoria; Rudd (1970), using sodium pyrophosphate, obtained no response at four localities in South Australia, and Lipsett and Williams (1971), using triple superphosphate, did not obtain responses at any of fourteen sites in southern New South Wales. Holford (1971) refers to similar observations on wheat crops in the northern New South Wales wheat belt.

Any likelihood of sulphur deficiency in the cereal-growing soils of southern Australia has almost certainly been prevented by the widespread application at seeding of single superphosphate containing over 11 per cent sulphur. In most cases this superphosphate would provide more than adequate sulphur to meet the crop requirement and balance removals in crop produce (Williams and Lipsett 1961).

The most important sources of plant-available soil sulphur are soluble sulphate, adsorbed sulphate and organic sulphur. The latter however, like nitrogen, must be mineralized to inorganic sulphate before it can be utilized by plants. The N:S ratio in soil organic matter is usually about 8:1, whereas the critical ratio required for the healthy growth of cereal plants is about 17:1. It could thus be expected that, provided nitrogen and sulphur are mineralized in proportions similar to those occurring in the soil organic matter, sulphur is unlikely to limit the growth of crops unless nitrogen fertilizer is applied. Lipsett and Williams (1971) have shown that in wheat soils in southern New South Wales mineralization of soil organic matter does indeed provide adequate sulphur for crops grown without nitrogen fertilizer. In addition appreciable amounts of readily available sulphate sulphur are present in the subsoil of many soils and these may frequently be utilized by deep-rooting cereal species.

Use of nitrogen fertilizer may increase the crop requirement for sulphur by increasing yields and it is significant that in most cases where crop responses to sulphur have been reported overseas these have been obtained in the presence of nitrogenous fertilizer. In Australia, Crofts and Blair (1966) have reported responses to sulphur by nitrogen-fertilized oats grown on a basaltic soil on the central tablelands of New South Wales. In many of the cereal-growing soils of southern Australia however, subsoil sulphate would still provide adequate sulphur even for crops fertilized with nitrogen.

The use of sulphur-containing fertilizer may still be desirable on clover leys in some cases however, because soil organic matter is accumulated, rather than mineralized, during the ley and because shallower root penetration by the pasture species may prevent full use of subsoil sulphate. Table 7 (Williams and Lipsett, unpublished) presents data showing significant responses by pasture leys to application of sulphur fertilizer in the presence of adequate phosphate at three of five sites in southern New South Wales.

Table 7. Response of subterranean clover-grass leys in southern New South Wales to sulphur (15 kg sulphate sulphur applied per ha)

Locality	Pasture yield kg/ha		Significance
	No sulphur	Sulphur applied	
A Cootamundra	2,160	3,850	*
B Ariah Park	4,120	3,980	n.s.
C Ariah Park	5,560	6,620	*
D Boorowa	3,210	4,150	*
F Harden	7,810	8,100	n.s.

* Response significant at 5 per cent level.
n.s. Response not significant.

Calcareous soils may often contain appreciable quantities of sulphur as a sulphate "impurity" in the calcium carbonate (Williams and Steinbergs 1964). This however is quite unavailable to plants and contributes little, if anything, to the available sulphur supply to the crop.

Potassium, calcium and magnesium

In general Australian cereal soils are well supplied with available potassium, calcium and magnesium and there have been no reports of crop responses to these nutrients. However, potassium responses by the shallow-rooted clover leys have been reported on some of the "light soils" in the cereal-growing districts of Western Australia (Carroll 1961). Calcium as a component of superphosphate has been applied to most cultivated soils in amounts considerably in excess of crop needs but it is unlikely that this has influenced crop yields. Application of gypsum to heavy textured soils in the Wimmera district of Victoria has resulted in substantial increases in wheat yields and lasting beneficial residual effects (Sims and Rooney 1965). These increases result however from improved physical conditions, prevention of crusting and improved water entry into the soil and not from nutrient effects.

Trace elements

Most of the information concerning the occurrence or cause of trace element deficiencies has been derived from agronomic studies, involving mainly field and pot culture experiments, or from plant analysis. Much of this information relates to soils outside the semi-arid region. However several broad generalizations on factors likely to have contributed to the occurrence of these deficiencies are possible.

Firstly the intense weathering and leaching to which many soils, or their parent materials, have been subjected has probably resulted in very low nutrient contents (near-absolute deficiency). Limited numbers of analyses indicate the likelihood of this in the case of copper and zinc (Baldwin and Crocker 1941; Coaldrake 1951) but its general importance has still to be confirmed by further analytical data.

Secondly, under conditions of intense leaching and weathering some nutrient elements have been immobilized into chemical forms of very low plant-availability. Oertel and Prescott (1944), for example, showed that the concentration of molybdenum in the ironstone gravels of Australian soils was considerably higher than that in the associated soil suggesting that molybdenum had concentrated with iron during weathering. The molybdenum in the gravel is now totally unavailable to plants. Immobilization of molybdenum in this way has undoubtedly contributed to the poor molybdenum status of the lateritic podzolic soils. Zinc also appeared to have been concentrated in the gravel to some degree but copper and manganese had not (Oertel and Prescott 1944). Taylor, McKenzie and Norrish (1964) and Taylor and McKenzie (1966) have also indicated the immobilization of cobalt and, to a smaller extent, zinc and molybdenum in the minerals birnesite and lithiophorite contained in manganiferous nodules and deposits in a wide range of Australian soils. The association of zinc with these minerals, however, may be largely due to adsorption rather than to occlusion (Tiller, Honeysett and de Vries 1972).

Thirdly, the high proportion of calcium carbonate in many soils, including aeolian calcareous sands, the brown solonized soils and black earths has limited the availability of several trace elements, especially copper, zinc and manganese. In acid soils adsorption of molybdate by iron oxides is also probably an important factor contributing to the occurrence of molybdenum deficiency in many soils since liming frequently releases available molybdenum and may often eliminate the deficiency.

Wells (1959) in a study of the possible influence of soil-forming processes and parent materials on the occurrence of nutrient deficiencies in pasture legumes in South

Australia concluded that on soils in which podzolizing processes had been dominant manganese, copper and zinc deficiencies were likely whereas where laterization or solonization had prevailed manganese deficiency was less likely and molybdenum deficiency more likely. On calcareous soils responses to copper, zinc and manganese were possible but molybdenum deficiency was unlikely.

In the semi-arid regions responses by cereal crops to copper, zinc, manganese and molybdenum have been recorded on many soils, especially in South Australia and Western Australia (tables 8-10). In all cases the soils are also deficient in phosphorus so that their development for crop production has involved the application of phosphatic fertilizer. It has thus proved relatively simple to correct trace element deficiencies by their application as additives in superphosphate.

Table 8. Response of cereal crops to application of copper and zinc

Soil designation	pH	State	Crop	Yield kg/ha				Reference
				Nil	+Cu	+Zn	+Cu, Zn	
Calcareous sand	7.8	S.A.	Wheat	320	1,760			Riceman and Donald 1938
			Oats	34	205			
			Barley	890	3,920			
			Rye	1,230	1,170			
Yellow podzolic	5.5	W.A.	Oats	0	3,070			Teakle, Thomas and Turton 1941
Lateritic podzolic	5.5	W.A.	Wheat	100	630			Teakle, Turton and Throssell 1940
Podzolic sand	7.0	S.A.	Wheat	240	700			King 1972
Solodized solonetz	8.4			1,050	1,380			
Brown sandy loam	8.3			1,040	1,140			
Calcareous sand	8.7			700	1,010			
Lateritic solod	6.4			15	1,900			
Granitic gneiss solod	5.4			1,810	3,810			
Alluvial solod	5.8			2,260	2,620			
Grey clay	8.2	Vic.	Wheat	1,960		2,590		Millikan 1938
Calcareous loam	8.1	S.A.	Wheat	1,260		1,400		King 1972
Solonetz	6.5	S.A.	Oats	1,620		2,150		Riceman 1945
Black earth	8.2	Qld.	Maize	1,970		4,430		Hibberd 1970
Calcareous sand	>8	W.A.	Wheat	820	820	780	1,010	Teakle 1942
Lateritic yellow earth	5.5	W.A.	Oats	400		1,420	1,520	Dunne, Smith and Cariss 1949
Lateritic podzolic	6.0			750		1,340	1,440	
Lateritic podzolic	6.0			710		1,100	1,290	
Calcareous sand	8.5	W.A.	Wheat	1,160	1,390	1,390	2,090	Dunne and Throssell 1948

The most important areas of trace element deficient soils are the "light lands" (generally acid sandy soils) of the Western Australian cereal-growing districts, extensive areas of which have been shown to be deficient in copper and zinc (Teakle 1942; Toms 1958; Gartrell 1969*a*), while more recently, many have been shown to be also deficient in molybdenum (Gartrell 1966, 1969*b*).

In many cases two or more element deficiencies occur on the same soil (Teakle 1942; Dunne and Throssell 1948; Riceman 1945; Higgs and Burton 1955; Stephens and Donald 1958; Gartrell 1969*a*), and on some of these soils different crop species have responded differently to the correction of the deficiencies. Wheat, for example, generally responds more to copper and oats more to zinc (Toms 1958). On many of the copper and zinc deficient "light" soils of Western Australia application of zinc in the absence of copper has depressed yields especially of wheat (Toms 1958). Similar depressions in

Table 9. Response of cereal crops to application of molybdenum and nitrogen

Soil group	pH	State	Crop	Yield kg/ha				Reference
				Nil	+Mo	+N	+N,Mo	
Lateritic yellow earth	5.7	W.A.	Wheat	1,210	1,200	1,260	1,510	Gartrell 1966
Lateritic yellow earth	5.7		Wheat	630	770	800	1,030	
Lateritic yellow earth	5.7		Oats	680	770	550	800	
Yellow earth	6.0		Wheat	480	650	450	730	
Brown podzolic	5.3	N.S.W.	Wheat	1,910	1,910	1,600	1,850	Lipsett and Simpson 1971
Brown podzolic	5.4	N.S.W.	Wheat	1,630	2,420	1,260	1,990	Lipsett and Simpson 1973

Table 10. Response of cereal crops to application of manganese

Soil group	pH	State	Crop	Yield kg/ha		Reference
				No Mn	+Mn	
Rendzina	7.6	S.A.	Oats	0	2,960	Samuel and Piper 1928
Calcareous lateritic	6.5	W.A.	Wheat	910	1,460	Wild 1934
Grey calcareous	>8	S.A.	Barley	590	1,080	Scott 1932
				570	1,650	
Calcareous sand	8.7	S.A.	Wheat	910	1,460	King 1972

yield due to the application of zinc in the absence of copper have been observed in South Australia by King (1972).

Substantial responses to the application of zinc have been reported in flax and linseed crops on many soils (Adam and Piper 1944; Millikan 1946; Cass Smith and Harvey 1948; Duncan 1967). These crops are more susceptible to zinc deficiency than cereals and the deficiency has frequently been reported on soils on which cereals show no deficiency. The deficiency in flax occurs most frequently on black alkaline soils of heavy texture (black earths) but it may also occur on lighter textured soils including some red-brown earths and brown solonized soils, and in Western Australia it occurs on acid sandy soils (lateritic podzols).

In many cases the responses to manganese, copper and zinc reflect low availability associated with high pH and high calcium carbonate but in the acid sandy soils the deficiencies undoubtedly reflect acute deficiency, the effects of intense leaching and low nutrient contents of the parent materials.

On the acid soils a single application of copper, zinc or manganese is usually sufficient to overcome the deficiency, at least for many years, and care must be exercised to avoid excess application, but on alkaline and calcareous soils immobilization of copper, zinc or manganese commonly occurs so that further applications are necessary to maintain an adequate level of available nutrient (Higgs and Burton 1955; Toms 1958; Fitzpatrick 1962; Sims and Rooney 1956). On grey calcareous soils of Yorke Peninsula, South Australia, for example, Higgs and Burton (1955) found that manganese applied to barley crops had little residual value for subsequent crops (table 11) and considered the application of manganese necessary for every crop for several years at least. On similar soils Reuter, Heard and Alston (1973) found that manganese added as manganous sulphate was rapidly immobilized and that a combination of soil and foliar application was necessary for best results.

In most cases molybdenum deficiency appears to be largely due to unavailability of the soil molybdenum as a result of low pH and the presence of substantial quantities of free sesquioxides rather than to an absolute deficiency of molybdenum since liming

Table 11. Response of barley crops grown on grey calcareous soil, pH 8.6, to the application of manganese
(Data from Higgs and Burton 1955)

Treatment	Yield kg/ha	
	1951	1953
No Mn	580	410
Mn applied with each crop	1,410	930
Mn applied to first crop only	1,450	440

releases sufficient molybdenum to meet plant requirements on the deficient soils of the Western Australian region (Gartrell, private communication).

In cereals molybdenum deficiency is often inversely related to nitrogen status (Freney and Lipsett 1965), responses to molybdenum being more pronounced under conditions of high nitrate than when the nitrate supply is low. Indeed response to molybdenum is often only obtained in the presence of nitrogenous fertilizers and in some cases application of fertilizer nitrogen may depress yield in the absence of adequate molybdenum (Lipsett and Simpson 1971, table 9). Although molybdenum deficiency is wide-spread in Australian soils and its importance for the growth of leguminous pasture has long been known (Anderson 1942) it is only comparatively recently that its importance to cereal crops has been recognized (Gartrell 1966; Lipsett and Simpson 1971). It seems likely that earlier recognition of the deficiency may have been inhibited by the low nitrogen status prevailing in most soils and it is of interest to note that responses by wheat reported by Lipsett and Simpson (1971) were obtained on soils which had previously carried subterranean clover showing no signs of molybdenum deficiency.

REFERENCES

Adam, D.B., and Piper, C.S. (1944). *J. Dep. Agric. S. Aust.* **47**, 422-6.
Anderson, A.J. (1942). *J. Aust. Inst. Agric. Sci.* **8**, 73-5.
Baldwin, J.G. and Crocker, R.L. (1941). *Trans. Roy. Soc. S. Aust.* 65, 263-75.
Beadle, N.C.W. (1962). *Aust. J. Agric. Res.* 13, 434-42.
Beeson, K.C. (1941). U.S. Dep. Agric. Misc. Publ. No. 369.
Bettenay, E., Blackmore, A.V., and Hingston, F.J. (1964). *Aust. J. Soil Res.* 2, 187-210.
Bhatti, A.S. and Loneragan, J.F. (1970*a*). *Agron. J.* 62, 288-90.
———. (1970*b*). *Aust. J. Agric. Res.* 21, 881-92.
Brunnich, J.C. (1926). *J. Dep. Agric. Qld.* 26, 180-21.
Burford, J.R., and Greenland, D.J. (1970). *Proc. 11th Int. Grassl. Congr.* 458-61.
Burford, J.R., and Millington, R.J. (1968). *Trans. 9th Int. Congr. Soil Sci.* 2, 505-11.
Carroll, M.D. (1961). *J. Dep. Agric. West. Aust.* 2, 457-8.
Cass Smith, W.P., and Harvey, H.L. (1948). *J. Dep. Agric. West. Aust.* 25, 136-42.
Chang, S.C., and Jackson, M.L. (1957). *Soil Sci.* **84**, 133-44.
Coaldrake, J.E. (1951). Coun. Sci. Ind. Res. Aust. Bull. No. 266.
Coke, F. (1958). Catchment salting in Victoria, Soil Conserv. Auth. Vict., Melbourne.
Colwell, J.D. (1963). *Aust. J. Exp. Agric. Anim. Husb.* 3, 51-61.
———. (1970). *Aust. J. Soil Res.* **8**, 221-38.
Colwell, J.D., and Esdaile, R.J. (1968). *Aust. J. Agric. Res.* **6**, 105-20.
Crocker, R.L. (1946). Coun. Sci. Ind. Res. Aust. Bull. No. 193.
Crofts, F.C., and Blair, G.J. (1966). *J. Aust. Inst. Agric. Sci.* 32, 142.
Dadswell, I.W. (1935). *Aust. J. Exp. Biol. Med. Sci.* 13, 33-42.
Dann, P.R. (1969). *Aust. J. Exp. Agric. Anim. Husb.* **9**, 626-9.
Downes, R.G. (1954). *Aust. J. Agric. Res.* 5, 448-64.
Duncan, O.W. (1967). *Qld. J. Agric. Anim. Sci.* 24, 301-7.
Dunne, T.C., Smith, H.G., and Cariss, H.G. (1949). *J. Dep. Agric. West. Aust.* **26**, 75-82.
Dunne, T.C., and Throssell, G.L. (1948). *J. Dep. Agric. West. Aust.* 25, 43-6.
Fitzpatrick, E.N. (1962). *J. Dep. Agric. West. Aust.* 3, 323-27.
Freney, J.R., and Lipsett, J. (1965). *Nature (Lond.)* **205**, 616-7.
Gartrell, J.W. (1966). *Nature (Lond.)* **209**, 1050.

———. (1969*a*). West. Aust. Dep. Agric. Tech. Bull. No. 3.
———. (1969*b*). *Aust. J. Exp. Agric. Anim. Husb.* 9, 432-6.
Greaves, R.E. (1940). Utah Agric. Exp. Stn. Circ. No. 113.
Hallsworth, E.G. (1969). *J. Aust. Inst. Agric. Sci.* 35, 78-89.
Halse, N.J., Greenwood, E.A.N., Lapins, P., and Boundy, C.A.P. (1969). *Aust. J. Agric. Res.* 20, 987-98.
Hawkins, C.A. (1967). *Agric. Gaz. N.S.W.* 78, 167-70.
Hibberd, D.E. (1970). *Qld. J. Agric. Anim. Sci.* 27, 89-94.
Higgs, E.D., and Burton, M.V. (1955). *J. Dep. Agric. S. Aust.* 59, 114-21.
Hingston, F.J. (1959). *J. Aust. Inst. Agric. Sci.* 25, 209-13.
Holford, I.C.R. (1966). *Aust. J. Exp. Agric. Anim. Husb.* 6, 409-17.
———. (1971). *Aust. J. Agric. Res.* 22, 879-84.
Hosking, J.S. (1935). *Trans. Roy. Soc. S. Aust.* 59, 168-200.
Kerr, H.W., and Von Stieglitz, C.R. (1938). Qld. Bur. Sugar Exp. Stn. Tech. Comm. No. 9.
King, P.M. (1972). Aust. Cereal Agron. Res. Newsletter No. 7, 28-30.
Kurtz, L.T., and Quirk, J.P. (1965). *Aust. J. Agric. Res.* 16, 403-12.
Lipsett, J. (1969). *J. Aust. Inst. Agric. Sci.* 35, 260-2.
Lipsett, J., and Simpson, J.R. (1971). *J. Aust. Inst. Agric. Sci.* 37, 348-51.
———. (1973). *Aust. J. Exp. Agric. Anim. Husb.* 39, 563-6.
Lipsett, J., and Williams, C.H. (1971). *Aust. J. Exp. Agric. Anim. Husb.* 11, 59-63.
Logan, J.N. (1958). *J. Soil Conserv. Serv. N.S.W.* 14, 229-42.
Loneragan, J.F., Carroll, M.D., and Snowball, K. (1966). *J. Aust. Inst. Agric. Sci.* 32, 221-3.
Loneragan, J.F., and Paton, D.F. (1960). *Aust. J. Agric. Res* 11, 524-9.
McGarity, J.W. (1961). *Plant Soil.* 14, 1-21.
McGarity, J.W., and Myers, R.J.K. (1968). *Soil Sci. Soc. Amer. Proc.* 32, 812-7.
McKenzie, R.M. (1957). *Aust. J. Agric. Res.* 8, 246-52.
Marshall, F.H.A., and Halnan, E.T. (1932). "Physiology of Farm Animals". Cambridge Univ. Press, Cambridge.
Millikan, C.R. (1938). *J. Dep. Agric. Vict.* 36, 409-16.
———. (1946). *J. Dep. Agric. Vict.* 44, 69-74.
Min. Agric. Fish. Food Tech. (London). (1967). Bull No. 19.
Morrison, F.D. (1938). "Feeds and Feeding", 20th Edn. Morrison Publishing Co., New York.
Northcote, K.H., and Skene, J.K.M. (1972). CSIRO Aust. Soil Publ. No. 27.
Oertel, A.C., and Prescott, J.A. (1944). *Trans. Roy. Soc. S. Aust.* 68, 173-6.
Ozanne, P.G., Kirton, D.J., and Shaw, T.C. (1961). *Aust. J. Agric. Res.* 12, 409-23.
Piper, C.S. (1938). *Trans. Roy. Soc. S. Aust.* 62, 53-100.
Piper, C.S., and de Vries, M.P.C. (1964). *Aust. J. Agric. Res.* 15, 234-72.
Prescott, J.A. (1931). Coun. Sci. Ind. Res. Aust. Bull. No. 52.
Prescott, J.A., and Piper, G.R. (1930). *J. Agric. Sci.* 20, 517-31.
———. (1932). *Trans. Roy. Soc. S. Aust.* 56, 118-47.
Reuter, D.J., Heard, T.G., and Alston, A. (1973) *Aust. J. Exp. Agric. Anim. Husb.* 13, 434-9.
Riceman, D.S. (1945). *J. Coun. Sci. Ind. Res. Aust.* 18, 336-48.
Riceman, D.S., and Donald, C.M. (1938). Coun. Sci. Ind. Res. Aust. Pamph. No. 78.
Richardson, A.E.V., and Fricke, E.F. (1931). *J. Dep. Agric. S. Aust.* 35, 57-86.
Rossiter, R.C. (1955). *Aust. J. Agric. Res.* 6, 1-14.
Rowan, J.M. (1971). Salting on dry land farms in north-western Victoria (Soil Conserv. Auth. Vict.) Publ. No. T.C. 7.
Rowan, J.M., and Downes, R.G. (1963). A study of the land in north-western Victoria (Soil Conserv. Auth. Vict.), Publ. No. T.C. 2.
Rudd, C.L. (1970). Dep. Agric. S. Aust. Exp. Rec. No. 5, 28-31.
Russell, J.S. (1960). *Aust. J. Agric. Res.* 11, 902-26.
Samuel, G., and Piper, C.S. (1928). *J. Dep. Agric. S. Aust.* 31, 696-705, 789-99.
Scott, R.C. (1932). *J. Dep. Agric. S. Aust.* 35, 771-80.
Sims, H.J. (1953). *J. Aust. Inst. Agric. Sci.* 19, 89-98.
———. (1959). *J. Dep. Agric. Vict.* 57, 759-65.
Sims, H.J. and Rooney, D.R. (1956). *J. Dep. Agric. Vict.* 54, 499-507.
———. (1965). *J. Dep. Agric. Vict.* 63, 401-9.
Snook, L.C. (1939). *J. Dep. Agric. West. Aust.* 16, 283-7.
Stace, H.C.T., Hubble, G.D., Brewer, R., Northcote, K.H., Sleeman, J.R., Mulcahy, M.J., and Hallsworth, E.G. (1968). "A Handbook of Australian Soils" Rellim, Adelaide.
Stephens, C.G. (1949). *J. Soil Sci.* 1, 123-49.
Stephens, C.G., and Donald, C.M. (1958). *Adv. Agron.* 10, 167-256.
Storrier, R.R. (1962). *Aust. J. Exp. Agric. Anim. Husb.* 2, 185-92.
———. (1965*a*). *Aust. J. Exp. Agric. Anim. Husb.* 5, 310-6.
———. (1965*b*). *Aust. J. Exp. Agric. Anim. Husb.* 5, 317-22.

———. (1965*c*). *Aust. J. Exp. Agric. Anim. Husb.* 5, 323-8.
Taylor, R.M., and McKenzie, R.M. (1966). *Aust. J. Soil. Res.* **4**, 29-40.
Taylor, R.M., McKenzie, R.M., and Norrish, K. (1964). *Aust. J. Soil. Res.* 2, 235-48.
Teakle, L.J.H. (1942). *J. Dep. Agric. West. Aust.* **19**, 242-53.
Teakle, L.J.H., and Cariss, H.G. (1943). *J. Dep. Agric. West. Aust.* **20**, 1-28.
Teakle, L.J.H., Thomas, I., and Turton, A.G. (1941). *J. Dep. Agric. West. Aust.* **18**, 70-86.
Teakle, L.J.H., Turton, A.G., and Throssell, G.L. (1940). *J. Dep. Agric. West. Aust.* 17, 161-73.
Tiller, K.J., Honeysett, J.L., and de Vries, M.P.C. (1972). *Aust. J. Soil Res.* **10**, 165-82.
Toms, J. (1958). *J. Dep. Agric. West. Aust.* 7, 197-203.
Underwood, E.J., and Snook, L.C. (1935). *J. Dep. Agric. West. Aust.* 12, 326-35.
Wagner, R. (1957). *J. Soil Conserv. Serv. N.S.W.* 13, 33-40.
Wells, C.B. (1959). *J. Aust. Inst. Agric. Sci.* 25, 273-81.
Wetselaar, R., and Norman, M.J.T. (1960). *Aust. J. Agric. Res.* 11, 693-704.
Wild, A. (1958). *Aust. J. Agric. Res.* **9**, 193-204.
———. (1961). *Aust. J. Agric. Res.* 12, 286-99.
Wild, A.S. (1934). *J. Dep. Agric. West. Aust.* 11, 223-5.
Williams, C.H. (1950*a*). *J. Agric. Sci.* **40**, 233-42.
———. (1950*b*). *J. Agric. Sci.* **40**, 243-56.
———. (1950*c*). *J. Agric. Sci.* **40**, 257-62.
Williams, C.H., and Donald, C.M. (1957). *Aust. J. Agric. Res.* **8**, 179-89.
Williams, C.H., and Lipsett, J. (1961). *Aust. J. Agric. Res.* 12, 612-29.
Williams, C.H., and Steinbergs, A. (1964). *Plant Soil.* 21, 50-62.
Wood, T.B., and Woodman, H.E. (1933). Min. Agric. (London) Bull. No. 48, 8th Edn.
Woodroffe, K., and Williams, C.H. (1953). *Aust. J. Agric. Res.* **4**, 127-50.

CHAPTER 7

Biochemical properties: carbon and nitrogen metabolism

J.M. Oades and J.N. Ladd

INTRODUCTION

An understanding of the nature, distribution and dynamics of carbon and nitrogen compounds in soils is particularly relevant in semi-arid regions of Australia where both organic matter levels and fertility tend to be low. In cereal-growing areas of southern Australia, legumes are grown to maintain or improve the nitrogen status of soils. At the same time carbon contents are increased with beneficial effects on soil structure.

Factors influencing the nitrogen content of soils have been discussed by Stevenson (1965). Jenny (1930, 1941, 1961) has indicated that climate and vegetation mainly determine soil carbon and nitrogen levels. The concentrations of organic carbon and nitrogen tend to be related and decrease with depth, most of the organic-N occurring within 60 cm of the surface. C:N ratios of surface soils typical of those in semi-arid regions of Australia range from 11 to 15.

In this chapter we deal only briefly with studies on the chemical properties of organic carbon and nitrogen compounds in soils or soil fractions since these have been the subject of several more comprehensive reviews (Dubach and Mehta 1963; Bremner 1965; Felbeck 1965; Flaig 1966; Bremner 1967, 1968; Paul 1970; McLaren and Peterson 1967; McLaren and Skujins 1971; Schnitzer and Khan 1972). Moreover, in general terms the chemistry of organic materials appears to be similar in soils from different regions of the world and data currently available give little indication that the organic materials in Australian soils possess unique characteristics. We have attempted to assess studies, based largely on the use of ^{14}C and ^{15}N, which have aimed to follow the flux of soil carbon and nitrogen through organic matter fractions in order to define their relative contributions to carbon dioxide release and to the mineral nitrogen pool. Limitations of the chemical fractionation approach to studies of the dynamics of carbon and nitrogen in soils are discussed and our views on fractionation schemes of more biological significance are presented.

NATURE OF ORGANIC CARBON AND NITROGEN COMPOUNDS IN SOILS

The nature of soil humus has exercised the minds of chemists for almost two hundred years and the chemical approach to studies of organic materials in soils has remained dominant, although recognition of the major role of microorganisms in the transformation of soil organic compounds is evident in the shift towards more biological and biochemical studies, especially over the last decade.

The chemical approach may be divided according to two major procedural differences, viz. (1) the approach of the purist chemist, initiated by Schreiner and Shorey (1908), in which individual compounds, belonging to many distinct organic chemical groups, were identified and measured, and (2) the classical approach, based on alkaline extraction of organic materials from soils. This procedure was initiated in 1786 by Achard and developed by Sprengel, Berzelius, Mulder and subsequently Oden, to give a

Fig. 1. The classical fractionation scheme for organic materials in soils

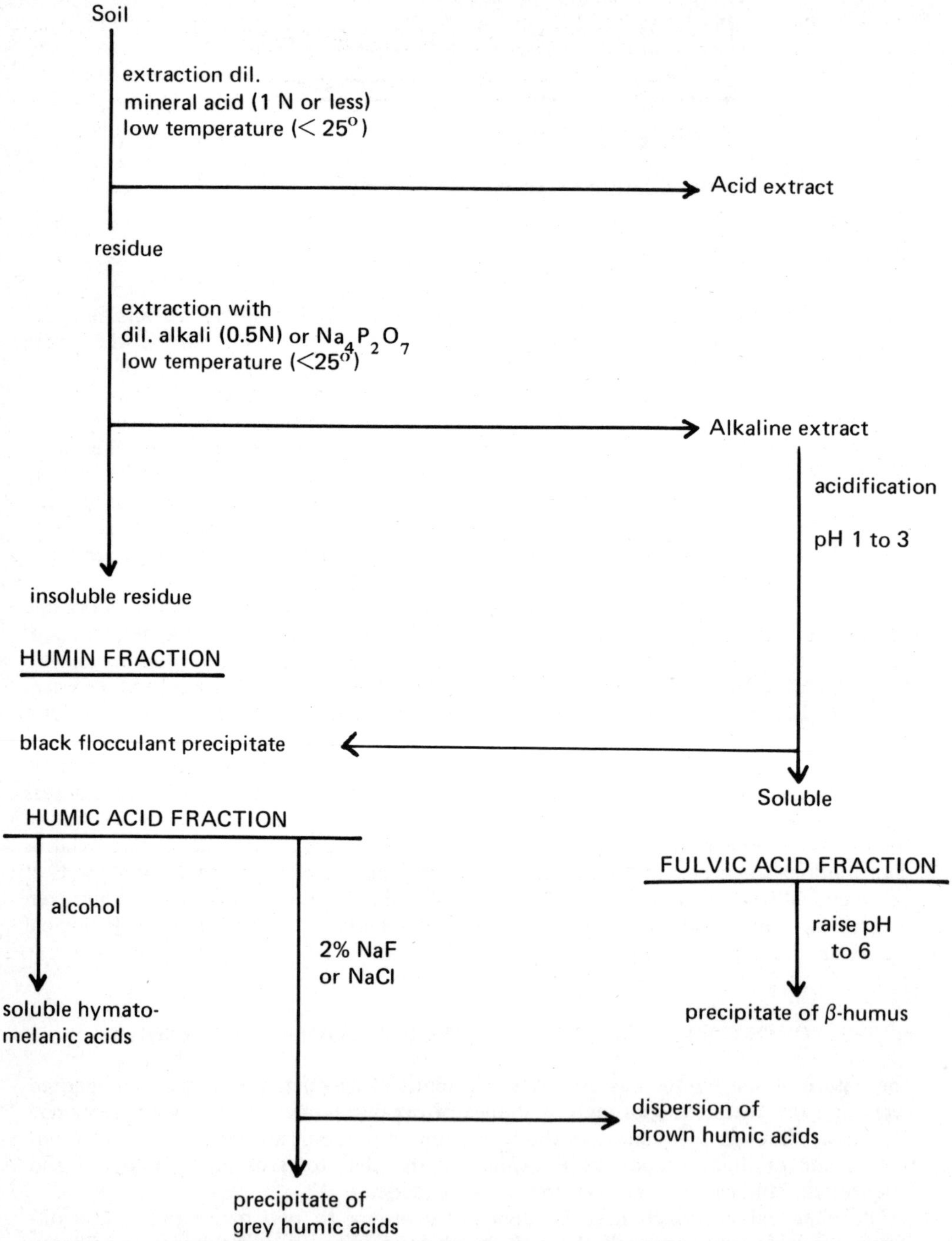

fractionation scheme similar to that shown in figure 1 (Russell 1961; Kononova 1966).
Refinements were later applied by German workers, e.g. Springer, who classified materials into humic and non-humic materials and divided humic acids into grey and

brown fractions based on their susceptibility to flocculation by electrolytes (Dubach and Mehta 1963). Pretreatment of soils by mineral acids and the use of sodium pryophosphate in place of or with sodium hydroxide are more recent modifications now used extensively.

Many of the humic materials of soils cannot be described in precise chemical terms, either because existing techniques are inadequate for the separation and characterization of the pure compounds or perhaps because no two humic molecules are exactly alike. Most current chemical studies apply some form of the classical approach followed by as much precise organic chemistry as possible depending upon the aims of the investigator.

Carbon compounds

The defined chemical "non-humic" substances can be grouped into carbohydrates, lipids, organic acids, polyphenols, nitrogenous components, etc. Each group of compounds can, to some extent, be considered and studied in isolation, and requires its own specialized methodology.

(*a*) Carbohydrates usually represent from about 10 to 20 per cent of the C in soil. They serve as a major substrate for heterotrophic soil organisms and are involved in the stabilization of soil structure. Several reviews, Gupta (1967), Swincer, Oades and Greenland (1969), Finch, Hayes and Stacey (1971), and Greenland and Oades (1975), indicate the recent interest in soil carbohydrates and exemplify investigation of a group of well defined compounds, i.e. the sugars as they occur in soil. Figure 2 shows the aldose composition of 12 Australian soils and also the distribution of the sugars in fractions of different densities (Oades 1972).

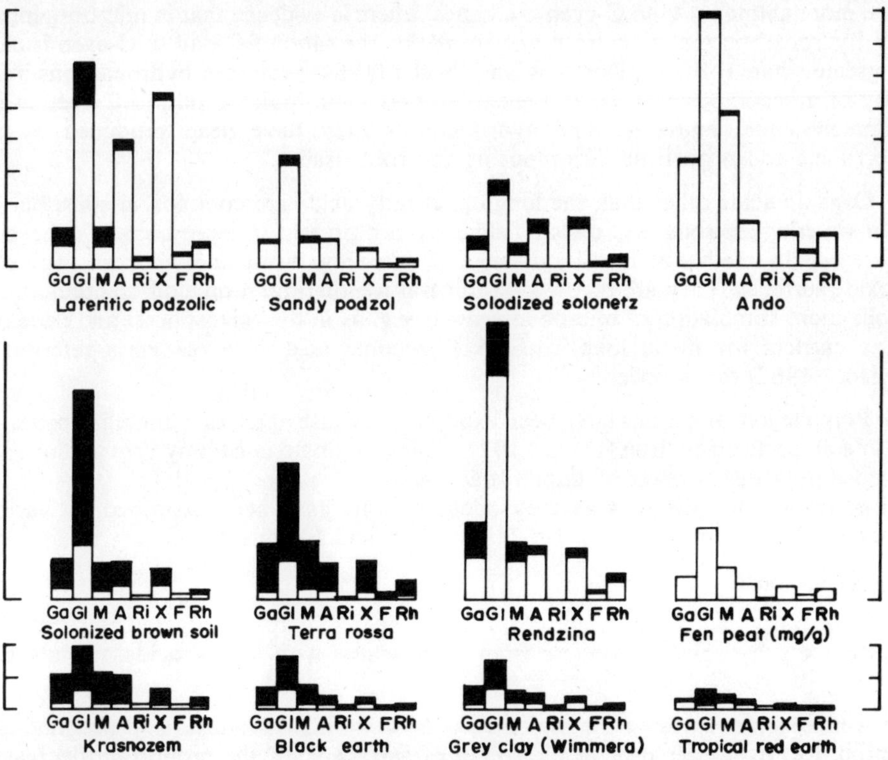

Fig. 2. The distribution of monosaccharides between light (white) and heavy (black) fractions of soils. Each unit on the vertical scale represents 50 mg sugar/100 g soil. Ga, galactose; Gl, glucose; M, mannose; A, arabinose; Ri, ribose; X, xylose; F, fucose; and Rh, rhamnose.

Glucose is the dominant aldose and differences in the relative amounts of the aldoses are associated with glucose, arabinose and xylose, which are the major sugars in plant materials. Galacturonic and glucuronic acids, galactosamine and glucosamine are always associated with the aldoses along with minor amounts of deoxy and O-methyl sugars.

In aerobic soils most of the sugars are present in complex, heterogeneous macromolecules which are polydisperse (M.W. range from <1,000 to >100,000). Some are easily extracted using N HCl; others are strongly associated with inorganic colloids.

(*b*) Lipids occur in all soils but because there is a wide variety of chemicals included in this group, there is no single method for determination of lipid content. In most mineral soils, lipids probably account for about 5 per cent of the organic-C. This percentage increases with increasing soil acidity to about 15 per cent in podzols and even higher values have been reported for acid peats (Stevenson 1966; Morrison 1969).

The term lipids is generally considered to cover waxes, fatty acids, hydrocarbons, and other minor components. The refined chromatographic procedures currently used in lipid chemistry have been applied to soil lipids, e.g. Butler, Downing and Swaby (1964); Himes and Bloomfield (1967); Morrison and Bick (1967). Waxes so far isolated from soils are generally similar to those of plants and animals and consist mainly of straight chain compounds with a predominance of acids and alcohols with an even number of carbon atoms (C-even) in the range C_{12} to C_{30}.

Lipid extracts contain a range of acidic materials including long chain fatty acids. Again C-even acids are dominant with chain lengths in the C_{20} to C_{30} region. Hydrocarbons on the other hand show the reverse situation as also exhibited by plant materials. n-Alkanes with an odd number of carbon atoms (C-odd) in the range C_{25} to C_{35} are more abundant than C-even n-alkanes. There is evidence that in microorganisms and in hydrocarbon materials from ancient rocks, the ratio of C-odd to C-even is close to or greater than 1. Butler, Downing and Swaby (1964) examined hydrocarbons in the vicinity of a decomposing root of a eucalyptus tree and found C-odd to C-even in approximately equal amounts. The hydrocarbons may have been produced by the heterotrophic soil organisms decomposing the root tissues.

(*c*) Organic acids other than the long chain fatty acids are common in soils, but except in waterlogged soils, e.g. paddy fields, are not present in large quantities because they are usually readily utilized by a range of microorganisms and will react with clay and oxide surfaces. They are involved in mineral weathering, profile differentiation, e.g. podzolization, stimulation of microbial growth, e.g. as in the rhizosphere, and probably also as carriers for metal ions, e.g. 2-ketogluconic acid. The reader is referred to Stevenson (1967) for a review.

(*d*) Polyphenolic materials have been examined because again they mobilize metal cations in soil, particularly iron (Davies 1971). They originate in canopy drip and are also released during degradation of humic materials.

Other groups of chemicals as they occur in soils have been described by various authors (McLaren and Peterson 1967; McLaren and Skujins 1971).

Nitrogen compounds

The nature of organic-N compounds in soils has been mainly investigated by identifying and estimating hydrolytic products from soils heated with strong acids, usually 6 N HCl.

(*a*) Amino acids represent the major class of identifiable products and account, as a minimum, for 20-40 per cent of the soil-N of surface soils, the proportion decreasing with depth in the profile. Most of the amino acids are typical of those derived from proteins and are considered to be released mainly from the cleavage of peptide bonds but the occurrence of "non-protein" amino acids has been established. Free amino acids occur in soils in only trace amounts. Qualitatively, the amino acid compositions of different soil hydrolysates are similar (table 1) although differences in their relative concentrations, especially those of the "non-protein" amino acids, have been recorded.

Table 1. Distribution of amino acid-N in soil hydrolysates

Amino acid	N in amino acid as % of total amino acid-N	Amino acid	N in amino acid as % of total amino acid-N
Basic	19.1	Neutral	66.4
Lysine	10.1	Alanine	11.0
Arginine	3.6	Glycine	10.9
Histidine	2.0	Serine	8.2
Ornithine	3.4	Threonine	7.3
		Valine	6.4
Acidic	14.5	Leucine	5.0
Glutamic acid	6.3	Isoleucine	3.4
Aspartic acid	8.2	Proline	4.7
		Hydroxyproline	1.2
		Tyrosine	1.2
		Phenylalanine	1.8

SOURCE. Adapted from Bremner 1965.

(*b*) Amino sugars, glucosamine and galactosamine are present in soils as high molecular weight compounds. They account for 5-10 per cent of the total-N of surface soils, the proportion increasing with depth in the profile to a maximum in the B horizon.

(*c*) Purine and pyrimidine bases, derived probably from nucleic acids, account for less than 1 per cent of soil organic-N. Few investigations have aimed specifically at estimating the proportion of soil-N in the form of nucleic acids and their hydrolysis products, and the methods available are less satisfactory than those employed for amino acids and hexosamines. Hydrolysis with hot HCl (as commonly employed in soil hydrolysis) leads to extensive deamination of some of the bases (Vischer and Chargaff 1948; Bendich 1955). Under milder acidic conditions often employed for hydrolysis of nucleic acids, adsorption of the purine bases to the soil residue occurs, leading to a considerable underestimation of these bases when hydrolysates alone are analyzed (Parsons and Ladd, unpublished data).

Alkaline extracts of soil remove both added ribonucleic acid and deoxyribonucleic acid, the former being hydrolysed to nucleotides subsequently determined in the fulvic acid fraction (Adams, Bartholomew and Clark 1954), whereas DNA is alkali-stable and appears in the humic acid fraction on subsequent acidification (Anderson 1958, 1961, 1967). An examination of soil fulvic acids for nucleotides and of acid-hydrolysed humic acids for DNA bases tended to confirm that nucleic acid-N accounted for only small percentages of soil total-N, although the amounts exceeded those calculated (Anderson 1958) to be present in an assumed microbial biomass.

(*d*) Approximately 25 per cent of the organic-N of most surface soils is not solubilized after hydrolysis with 6 N HCl and, together with almost 30 per cent of the organic-N of the hydrolysates, remains unidentified. The total extent to which amino acid-N, amino sugar-N and nucleic acid-N may be underestimated because of either degradation, artefact formation or adsorption to insoluble soil residues during hydrolysis, and the extent to which soil-N may be protected from acid hydrolysis by soil minerals have not been accurately assessed. In addition, there has been much conjecture on the occurrence in soils of complex, organic, nitrogenous derivatives, hypothesized to be formed either by the reaction of phenolic compounds (lignins, tannins, quinones) with ammonia, amino acids, peptides or proteins or by the reaction of carbohydrates with amino compounds. The feasibility of the various hypotheses has been discussed by Bremner (1965, 1967). Direct, substantive evidence that these nitrogenous organic complexes contribute significantly to the total-N of the soil is lacking.

Humic acid, fulvic acid and humin fractions

Since about 1960 much progress has been made in the chemical characterization of two of the three most important fractions widely recognized and used around the world, viz. the humic acid fraction, the fulvic acid fraction and the humin fraction.

Humic acid fraction

(*a*) Molecular weight and shape. Humic acids are generally recognized as dark brown to black polydisperse, heterogeneous condensates with molecular weights ranging from about 5,000 to more than 100,000 (Kononova 1966; Lindquist 1967; Butler and Ladd 1969; Ladd 1969; Swift,Thornton and Posner 1970; Swift and Posner 1971). Electron microscopy, ultracentrifugation and viscosity measurements suggest that the particles are roughly spherical (40 to 80 Å radius) although other shapes have been proposed (Flaig and Beutelspacher 1968; van Dijk 1971). More recent suggestions are that the humic acid polymer is linear (Flaig 1969) and in fact may be a random coil (Cameron et al. 1972). Most of this work involves humic acids purified by reprecipitation until only a minimum "ash" is present. The effect of multivalent cations on the configuration and on the disposition of humic acids in soils is unknown.

(*b*) Functional groups. Humic acids possess a variety of functional groups with a range of pK values. Cation-exchange capacities, measured at pH 7, are of the order of 300 me/100 g, i.e. about one charge site per 70 $Å^2$, which is similar to the charge density on inorganic colloids. The acidity is largely due to carboxyl and phenolic groups although other groups also exist, e.g. alcoholic hydroxyl, carbonyl, quinone carbonyl (Theng and Posner 1967; Stevenson and Butler 1969; Felbeck 1971; Schnitzer 1971; Schnitzer and Khan 1972). Most of the functional groups are associated with the periphery of an open or "spongy" molecule and probably are largely associated with aliphatic materials.

(*c*) Elemental analysis. Elemental analyses for humic acids usually fall in the range C 50-60 per cent, O 30-40 per cent, H 3-5 per cent and N 2-5 per cent. Some of the C, O and H is involved in the functional groups, but most is present in the more resistant and older (Campbell et al. 1967) core or chain of humic acids. Reviews on the nature of the many and various phenolic, aromatic and polycyclic aromatic compounds which have been isolated from humic acids have been written by Flaig (1966), Felbeck (1971), Haworth (1971) and Schnitzer and Khan (1972). Many of these compounds occur in lignin of plant materials but others are of microbial origin (Hurst and Burges 1967; Piper 1972). The polycyclic aromatics may be the source of the high free-radical contents of humic acids (Steelink and Tollin 1967).

(*d*) Spectroscopic characteristics. These have been reviewed by Schnitzer (1971). Generally humic acids show featureless absorption spectra in the visible region, rising to a maximum in the UV region due to the presence of single and polycyclic aromatics, e.g. polynuclear quinones. High extinction values or absorbance in the 400-465 nm region are supposed to reflect a high degree of aromatic condensation and vice versa, and the ratio of absorbances at 465 nm and at 665 nm (E_4:E_6 ratio) has been used extensively to characterize humic materials, particularly in Soviet literature. As the C:H ratio increases and condensation increases, the E_4:E_6 ratio should increase (Kononova 1966). Thus, values obtained when dealing with well-humified materials such as from a chernozem should be higher than those from podzolic soils. Butler and Ladd (1969) obtained higher values for such a ratio in the lower molecular weight fractions, which was substantiated by Swift, Thornton and Posner (1970). Thus, E_4:E_6 ratios cannot be used for comparisons of humic materials from different soils unless a particular molecular weight fraction is studied.

Duchaufour and Jacquin (1964) and Flaig (1967) have shown that grey humic acids which are precipitated by 2 per cent sodium halide contain more C, are more aromatic and show higher absorbance than brown humic acids which are more aliphatic and better dispersed. Anderson (1972) presented evidence that the extinction values given by humic materials at 280 nm are a good index of aromaticity. There was a high correlation between the absorbance at this wavelength and the C:H ratio, the acid hydrolysability, the free radical content and other chemical properties.

Japanese workers classify humic acids into types A, B, P and R_p based on spectral characteristics (Kumada et al. 1967). A-type humic acids absorb strongly at 600 nm, have high C contents and high C:H and C:N ratios, and are supposedly strongly humified. Aromaticity decreases $A > B \simeq P > R_p$ (Kumada and Matsui 1970). P-type humic acids are common in cool climates and absorption at 615, 570 and 450 nm is due to a green fungal metabolite (Kumada and Hurst 1967; McGrath 1967; Sato and Kumada 1967; Lowe and Tsang 1970).

(*e*) Nitrogenous components. Bremmer (1965, 1967), in summarizing the literature, states that amino acid-N and amino sugar-N accounted for 30-50 per cent and 3-10 per cent respectively of the nitrogen of most humic acids. Small amounts of purines and pyrimidines, probably derived from DNA, are also present. Humic acids extracted with alkali (0.5 N NaOH) contain more N and higher percentages of acid-hydrolysable-N and amino acid-N than do humic acids extracted from the same soils with neutral sodium pyrophosphate solutions. Nevertheless, amino acid compositions of humic acids obtained with either extractant from a variety of soils are similar, and the small quantitative variations which do occur are associated mainly with "non-protein" amino acids.

Studies (Ladd and Butler 1966; Piper and Posner 1968; Butler and Ladd 1969; Swift and Posner 1972*a*) of humic acids extracted from soils of southern Australia and fractionated according to molecular size have shown that the N content and the proportion of N accounted for as amino acid-N decrease with the molecular weight of the humic compounds present and may be due to hydrolysis of peptide-bound components of decreasing chain length.

The general assumption that acid hydrolysis of humic acids releases amino acids by cleavage of peptide bonds has been supported by studies (Scharpenseel and Krausse 1962; Ladd and Brisbane 1967) demonstrating that proteolytic enzymes (papain, pronase) liberate up to 39 per cent of the acid-hydrolysable amino acid-N as free amino acids. Further, Simonart, Batistic and Mayaudon (1967), commencing with an alkali-extracted humic acid, have isolated by chromatography in a phenol solvent a humoprotein fraction containing 11 per cent N, which accounted for 15 per cent of the N of the original source material. A second fractionation procedure, based upon phenol extraction of humic material, dialysis, paper chromatography in a phenolic solvent and acetone precipitation, yielded a proteinaceous fraction containing 14.8 per cent N.

Although the percentage of humic acid-N extracted as protein-N was low, the purification procedure is the most successful of any applied and indicates that the extracted protein was originally linked by hydrogen bonding within the humic acid complex.

More recently, Biederbeck and Paul (1973) showed that the fraction of a soil humic acid solubilized by phenol was predominantly proteinaceous, whereas the phenol-insoluble fraction was largely aromatic by nature. They concluded that at least one-third of the humic acid-N was accounted for as amino-rich compounds of molecular weight greater than 50,000 which were combined reversibly by hydrogen bonding to predominantly aromatic, humic constituents.

Peptide or protein components of humic acids not separated by phenol treatment may be bonded covalently, in which case the failure to obtain large yields of purified material by any method is explicable. Covalent bonding of α- and ϵ-amino groups of amino acids, peptides and proteins to aromatic structures has been hypothesized to occur during the formation of humic acids by the oxidative polymerization of quinones (see reviews by Bremner 1965, 1967; Flaig 1963; Kononova 1966). N-phenyl amino acids are relatively resistant to hot acid hydrolysis (Haider, Frederick and Flaig 1965; Ladd and Butler 1966) but are readily cleaved with amino acid release when conditions favour their oxidation to the quinone-imine intermediates. Such a mechanism has been proposed by Piper and Posner (1972) to explain the observation that alkaline hydrolysis of humic acids releases amino acids that are resistant to acid hydrolysis.

Fulvic acid fraction

The name fulvic acid was originally used for the pale yellow-orange-brown compounds found in the acidic supernatant liquid after precipitation of the humic acid fraction from an alkaline soil extract (derivation Latin *fulvus* = orange).

In terms of total-C, this fraction is less significant than the humic acid fraction in neutral to alkaline soils, but is dominant in many acidic soils and is the main component of the humic B horizons of podzols. The fraction contains a very wide range of different chemicals including low molecular weight acidic substances produced by microorganisms during anaerobic incubation and organic acids, as reviewed by Stevenson (1967). Many plant materials also appear in this fraction, e.g. the polyphenolics, as reviewed by Dragunov and Murzakov (1970) and Davies (1971).

The fulvic acids are mainly low molecular weight (<10,000) materials with less C and more O than humic acids (Kononova 1966; Butler and Ladd 1971; Shurygina et al. 1971; Stevenson and Goh 1971). They possess a higher content of functional groups and are involved in complexing and mobilizing metals in acidic soils (Schnitzer 1968). Fulvic acids from the Bh horizon of a Canadian podzol have been examined in detail by Schnitzer and coworkers. The fraction examined was free of N and carbohydrates and was of low molecular weight. Elemental and functional group analysis gave the following formula: $C_{21}H_{12}(COOH)_6(OH)_5(CO)_2$. The fraction reacted with various metals. Subsequently the material was further fractionated and shown to be a number of compounds of different molecular weights.

A proposal for the structure of fulvic acid and descriptions of chemical methods employed in the identification of fragments obtained from fulvic acid preparations by both degradative and non-degradative techniques is described by Schnitzer and Khan (1972).

A major component of the fulvic fraction is the polysaccharide fraction which represents by far the greater portion of the carbohydrates extracted by NaOH from soils (Swincer, Oades and Greenland 1968*a*, 1968*b*, 1969). Quantitative differences exist in the content and extractability of the carbohydrates in different soils but qualitatively the monosaccharide composition of different fulvic acids is similar. Glucose represents about one-third of the sugars in fulvic fractions from a range of soils, and in addition to galactose, mannose, arabinose and xylose significant amounts of rhamnose, fucose and ribose are also present. In two podzolic soils mannose was the most abundant sugar, probably reflecting a dominantly fungal population. The polysaccharides in the fulvic fraction are heterogeneous and polydisperse and are always associated with uronic acids, amino sugars and amino acids (Oades 1972).

Amino acid-N and amino sugar-N generally account for 20-30 per cent and 8-10 per cent respectively of fulvic acid-N. The amino acids are typical of those also present in humic acids (Bremner 1965, 1967). Khan and Sowden (1972) found a similar distribution of amino acids in fulvic acid hydrolysates from a solonetz, solod and chernozem but the contents of basic amino acids were less than those of hydrolysates of the corresponding humic acid. Amino sugars were present in the three fulvic acid hydrolysates, the glucosamine:galactosamine ratios being 1:1 in all preparations. Swincer, Oades and Greenland (1968*a*) found that a partly purified fulvic acid from a red-brown earth contained 2.1 per cent N, of which amino acid-N accounted for about 45 per cent. The amino acid distribution was similar to that reported by Khan and Sowden (1972). Upon further fractionation of the fulvic acid, Oades and Swincer (1968) demonstrated that amino acids were mainly concentrated in the low molecular weight fraction ($< 4{,}000$), whereas amino sugars were in relatively constant proportions in all fractions, including that of molecular weight $>100{,}000$. Indications were obtained that fulvic acids of soils sampled in the winter contained a higher proportion of amino acids than those of samples taken in other seasons, which may reflect differences in the extent of the metabolism or ease of extractability of microbially derived proteins according to field conditions.

Humin fraction

The humin fraction often comprises a major portion of the soil organic materials, particularly in soils of arid regions. Because it is intractable, less information is available than for the other fractions. Limited data (Arshad and Lowe 1966; Kononova 1966) obtained after destruction of the silicate minerals using HF/HCl mixtures suggest that the humin fraction is more strongly humified and possibly of higher molecular weight than the extractable humic fraction. Such molecules are supposedly firmly linked to the inorganic colloids.

Although some of the compounds of the humic fraction may be similar in nature to humic acids, it is becoming increasingly clear that plant components make up a significant proportion of humin carbon. The more efficiently plant fragments can be removed from a soil by flotation on a heavy liquid prior to chemical extraction, the smaller is the humin fraction. Addition of ^{14}C-labelled plant components to soil shows conclusively that plant material is largely insoluble in NaOH.

^{14}C dating of organic-matter fractions has shown that humin fractions are generally young fractions (Campbell et al. 1967; Scharpenseel, Ronzani and Pietig 1968), and Costin and Polach (1969) showed that this is in part due to alkali-insoluble plant residues. Perraud, Nguyen and Jacquin (1971) and Oades (1972) have shown that plant fragments encrusted with inorganic materials are present in the humin fraction. Other components known to occur in the humin fraction are charcoal and fungal hyphae (Shields 1972; Carter and Oades, unpublished). All three sources of organic material have wide C:N ratios and account for the wide C:N ratios of most humin fractions.

The proportion and nature of the nitrogenous components of the humin fraction will be dependent partly on the efficiency with which plant fragments are removed prior to extraction of the soil and partly on the efficiency of the extractant to dissolve particulate, cellular compounds and those bound to clay and organic surfaces. Addition of ^{15}N-labelled microbial cells to soils has demonstrated that cellular-N becomes part of the humin-N fraction, at a time when little decomposition of the added material has occurred (Chu and Knowles 1966; Knowles and Chu 1969).

McGill (1971) has examined two soils, a fine sandy loam and a clay loam, from semi-arid areas of southern Saskatchewan. Their distribution of nitrogen, both unlabelled and labelled with $^{15}NH_4{}^{15}NO_3$-N, previously immobilized in the field at least two years before sampling, was determined. The proportion of soil-N accounted for in the humin fraction was 33 per cent and 40 per cent for the sandy and clay soils respectively when extracted successively with NaOH, then $Na_2S_2O_4$ solutions. However humin-N represented 55 per cent and 63 per cent of the soil-N respectively when an NaOH-$Na_4P_2O_7$ solution was used as extractant. The humin fraction of each soil contained similar proportions of unlabelled and labelled nitrogen after extraction with the NaOH-$Na_4P_2O_7$ solution, but extractions involving the use of sodium dithionite solutions preferentially removed labelled material. Humin-N was hydrolysed to a greater extent than humic acid-N and amino acid-N accounted for 27-39 per cent of the total humin-N.

DISTRIBUTION OF ORGANIC CARBON AND NITROGEN COMPOUNDS IN SOILS

Macroorganic matter

The most obvious organic matter in soil is in the form of large plant fragments. Such material has been termed macroorganic matter and, although there is no agreement on a precise definition, Barley (1955) and Garwood, Clement and Williams (1972) have suggested that 0.25 mm should represent the lower size limit.

Macroorganic matter is plant material in the early stages of decomposition. Total phytomass production varies from 0.3 to 50kg/m^2, from the extremes of tropical desert to humid tropical forest. Total phytomass for cereal crops varies from 0.1 to 1 kg/m^2/annum and for leys 1.2 kg/m^2/annum (Bazilevich and Rodin 1968; Pershina and Avdeyeva 1968; Garwood, Clement and Williams 1972). From 20 to 80 per cent of the phytomass is in root systems, depending on the plant community, and for agricultural crops about two-thirds of the plant materials consist of root residues. It is known that one-fourth to one-half of the root residues decompose annually, even with perennials (Jenkinson 1964; Dahlmann 1968; Ignatenko, Krillova and Ponyatovskaya 1968). Because of the rapid decomposition of plant components once they die, the composition of the macroorganic matter will be dependent on the season and much of the material present at any particular time will represent relatively resistant plant components. The C:N ratios of macroorganic matter are usually >20:1 and the percentage nitrogen contents have been correlated with mineralizable-N, cereal yield and nitrogen uptake by the crop (Heard 1965).

Fractions of different particle size

In the last decade increasing interest has been shown in separating soils into various size fractions by physical means only, and determining the nature of organic materials in the fractions. Edwards and Bremner (1964, 1967) showed that ultrasonic energy can be used to disperse soils in water, a treatment considered unlikely to cause major alterations to the organo-mineral components. Soviet workers have used trituration of soil pastes to achieve similar results. Results show that organic materials are not always predominantly associated with the clay fraction, but in some soils are concentrated in particles 1 to 20 μm in diameter, and occasionally in larger particles. The distribution of C and N within the size fractions depends on the soil type but from data now available it is possible to make certain generalizations.

In chernozemic and closely related soils, e.g. chestnut soils or other calcareous clay soils, the fine silt to coarse clay fractions contain the highest contents of organic materials (Grati, Sinkevich and Klesch 1965; Pokotilo 1967; Trofimenko and Kizyakov 1967; Kyuma, Hussain and Kawaguchi 1969; McKeague 1971; Watson and Parsons 1974; Ladd and Parsons, unpublished). This trend is obvious even within the <2 μm fraction (Arshad and Lowe 1966; Dudas and Pawluk 1969/70). Watson and Parsons (1974) found most of the soil-N to be in silt-sized particles although, due to the quantitative dominance of the <1 μm fraction in such soils, the fine clay is generally associated with more organic matter than the coarser fractions. In other soils, particularly those which have undergone some leaching, the <2 μm or finer clay fractions are richest in C and N although significant amounts of organic materials are present in coarser particles up to about 20 μm. Above this size the organic matter content falls rapidly (Alexsandrova, Yurlova and Lobitskaya 1964; Lichmanova 1962; Kuzmin 1969). In particles >20 μm, plant fragments have often been recognized and occasionally separated using density fractionations (Kyuma, Hussain and Kawaguchi 1969; McKeague 1971).

In other soils there is a steady increase in the content of organic matter as the particle size decreases, e.g. Ando soils (Kyuma, Hussain and Kawaguchi 1969; Satoh and Yamane 1972*a*) and a red-brown earth (Turchenek and Oades 1974).

Some of the 1-20 μm particles rich in organic matter have been shown to be aggregates in which clay particles are cemented together by organic materials, the process probably involving clay-metal cation-organic material bonding which is destroyed to some extent by the use of dilute mineral acid or sodium pyrophosphate. Aggregates 1-5 μm have been shown in some cases to be enriched in humic acids which could be removed by pyrophosphate, but not by sodium hydroxide (Kyuma, Hussain and Kawaguchi 1969) or an acid pretreatment, before sodium hydroxide extraction (Trofimenko and Kizyakov, 1967). Evidence that such aggregates may be old and cause stabilization of the organic materials present in them was obtained by Campbell et al. (1967), who showed that Ca, Fe and Al humates were the oldest fractions based on ^{14}C-dating techniques. The position of organic materials within and around microaggregates is important in the prevention of dispersion of clay and possibly the slaking of aggregates by affecting the entry of water into pores (see chapter 5).

With some exceptions, the extraction of organic materials from particle-size fractions is more efficient the smaller the particles, i.e. the amount of humin decreases with particle size. This is correlated with the ratio of C in the humic acid fraction to C in the fulvic acid fraction. In most soils studied this ratio decreases with particle size and fulvic acids tend to be associated with the finer clay fractions. Such trends suggest that the larger humic materials are associated with and influenced by plant components in the coarser particles and that the soluble fulvic materials are the result of oxidative degradation of the larger humic components through microbiological processes. This type of process was suggested by Schnitzer and Hoffman (1965) and Swift, Thornton and Posner (1970), the latter showing the similarity of spectral characteristics of high molecular weight humic acids and those of lignin.

The use of E_4:E_6 ratios and Δ log k values on extracts from the different particle sizes has failed to produce data from which generalizations can be drawn. In some cases the results suggested an increase in condensation of the humic materials as the particle size decreased, and in other cases the reverse or no change. In view of the work of Butler and Ladd (1969) and Swift, Thornton and Posner (1970), interpretation of spectrophotometric work in terms of condensation should be viewed with caution. Condensation has also been confused with "humification".

The N content of the organic materials increases with decrease in particle size and to a greater extent than for C, such that with the odd exception (e.g. Domke 1969) the C:N ratio decreases with particle size. McKeague (1971) showed that in a range of soils the C:N ratio changed from 26:1 to as low as 8:1 as the particle size changed from >50 μm to <2 μm. Chichester (1969) obtained similar results, demonstrating a C:N ratio of 20:1 for particles ranging from 53 to 2,000 μm and 10.7:1 for particles <0.25 μm diameter. Satoh and Yamane (1972*a*) showed a change from 27:1 to 13:1 for the C:N ratio as particle sizes from 53-105 μm down to <0.1 μm were examined. Swift and Posner (1972*a*, 1972*b*) obtained C:N ratios from 30:1 to 13:1 as the particle size investigated changed from 1-2 mm to <0.2 μm. The exceptions which show a silt-size fraction with the lowest C:N ratio may be explained by aggregation.

Hydrolysis of the particle-size fractions with 6N HCl released a greater proportion of the N bound to clay than that bound to larger particles (Kyuma, Hussain and Kawaguchi 1969). Generally, however, the percentages of hydrolysate-N accounted for as ammonium-N, hexosamine-N and amino acid-N showed no consistent trend with the size of the particles hydrolysed.

McGill (1971) determined the distribution of N in two chernozemic soils from the semi-arid regions of southern Saskatchewan. The fractionation procedure involved preliminary extractions of the soil with a 0.1M NaOH-0.1M $Na_4P_2O_7$ solution, followed by ultrasonic treatment of suspensions of the residue in distilled water and differential centrifugation. Table 2 (adapted from McGill 1971) shows that 28-34 per cent of the

Table 2. Distribution of nitrogen in soil extracts and in sedimentation fractions of different particle sizes

Fraction	Percentage of soil-N in fraction		Percentage of fraction-N hydrolysed by 6N HCl		Percentage of hydrolysate-N accounted for as amino acid-N	
	Soil A*	Soil B*	Soil A	Soil B	Soil A	Soil B
Extract						
Humic acid	16.9	23.7	67	59	45	42
Fulvic acid	15.4	17.9			22	21
Sediments						
<0.04 μm	34.2	28.3	92	88	46	42
0.04–0.2 μm	4.9	4.4	100	96	34	26
0.2–2 μm	12.4	10.1	84	82	42	36
>2 μm	11.1	12.1	88	88	42	35

* Soil A, Sutherland clay loam. Soil B, Bradwell fine sandy loam.

soil-N was present in a fraction containing material not sedimented by centrifugation at 12,000*g* for 1 hour and calculated to be of a particle size <0.04 μm. Further centrifugation of this fraction at 40,000*g* for 1 hour sedimented material containing 71 per cent of the fraction-N from the Sutherland soil and 34 per cent of the fraction-N from the Bradwell soil. The ash contents of the fraction, < 0.04 μm, approximate 50-60 per cent and the organic-N was considered intimately associated with fine mineral colloids, probably sesquioxides (Anderson 1972). Acid hydrolysis released a greater proportion of the N associated with particulate fractions than that present in the humic acid extracts. However, there were no obvious trends in the percentages of total-N and amino acid-N in hydrolysates of the fractions in relation to their particle size.

Fractions of different density

Densimetric fractionations have been limited due to the problem of dispersing soils in suitable liquids, which are either a concentrated electrolyte solution (e.g. Thoulet's solution) or an organic liquid (e.g. bromoform) or are extremely poisonous and expensive (e.g. thallous formate). The method has potential for separation of both plant and microbial tissues from heavier inorganic components. The crudest fractionation should allow separation of "free" organic materials from the clay-organic complex and both of these from "free" inorganic materials. Most of the earlier work was Soviet (Lein 1940; Khan 1959) or French (Hénin and Turc 1949, 1950; Monnier, Turc and Jeanson-Luusinang 1962; Monnier 1965; Duthien and Chretien 1966; Lefebvre-Drouet and Meriaux 1966). The method was further developed by Greenland and Ford (1964) and Ford, Greenland and Oades (1969) and has been applied to particle-size separates including clay fractions by Satoh and Yamane (1972*b*) and Turchenek and Oades (1974).

The "light fractions" usually represent a small portion of the soil by weight and yield up to a tenfold concentration of organic materials compared with the whole soil. However, certain soils contain significant quantities of poorly crystalline inorganic colloids which have a low density, e.g. Ando soils (Oades 1972; Satoh and Yamane 1972*b*), so that the use of liquids of density 2.0 cannot be successfully applied to all soils for concentration and separation of free organic materials.

The C content is greatest in the lightest fractions, whether the technique is applied to the whole soil or the clay fraction. The light fraction (S.G. < 2.0) is intermediate between plant materials and humic materials with respect to C:N ratios, the amount of humic material present, the negative charge due to functional groups and its carbohydrate composition (Ford 1968; Oades 1972). The light fractions also contain microbial components (Oades 1974). Fungal hyphae contribute very significantly to this fraction (Carter and Oades, unpublished) and Ascomycetes have been grown from such fractions, indicating that fungal spores are present in the light fraction (Dr. J.H. Warcup, personal communication).

The amount and proportion of the soil-N accounted for in the light fraction depend upon soil type and agronomic history (Greenland and Ford 1964). Densimetric fractionation of fourteen soils by the technique, modified to include a surfactant, demonstrated that the percentage of soil-N in the light fraction ranged from 4 per cent to 79 per cent. A detailed comparison of the nitrogenous components of the light fraction and of a humic acid preparation from Urrbrae loam under permanent pasture showed that 88 per cent and 75 per cent of the total-N of each respective fraction was solubilized by hydrolysis with 6N HCl. Of the hydrolysate-N, 95 per cent of that from the light fraction was accounted for as amino acid-N (70.6 per cent), hexosamine-N (4.4 per cent) and amide + NH_4^+ - N (20.3 per cent), whereas only 60.6 per cent of the humic acid hydrolysate-N could be accounted for by these components. Some differences were noted in the proportions of individual amino acids in the two hydrolysates but overall similar amounts of neutral (55-60 per cent), basic (25 per cent) and acidic (15-20 per cent) amino acids were present. The gross composition of the light fraction approximated that of plant material more closely than that of soil humic extracts (Ford and Greenland 1968).

MINERALIZATION OF CARBON AND NITROGEN COMPOUNDS IN SOILS

Turnover of soil organic matter

The amounts of organic materials added to soils annually represent about 5 per cent of the organic matter content of the soils. The annual gains of C and N in soils close to equilibrium are offset by mineralization of some C and N of both the added and native organic materials. Various equations have been used to describe this system in general terms (Jenny 1941; Woodruff 1949; Hénin, Monnier and Turc 1959; Bartholomew and Kirkham 1960; Russell 1964), and the reader is referred to Clarke and Russell (chapter 13, this volume) for more detailed discussions of the applicability of the different mathematical models to soil organic matter turnover.

Rates of change of organic matter or nitrogen in soil have been shown to approximate the simple first-order rate equation.

$$\frac{dX}{dt} = A - rX$$

where X = the organic matter or nitrogen content of a given mass of soil at time t
A = the amount of organic matter or nitrogen added to the given mass of soil in unit time
r = the fraction of X decomposed per unit time.

Jenkinson (1966*a*) has discussed the limitations of this model, pointing out that its validity rested upon the assumptions that the rate of addition of organic matter remained constant and that the decomposition rate was independent of the amount of material added and of the chemical and physical nature of the organic material present. Values for the constants A and r have been estimated by Ford (1968) for some Australian soils. Greenland (1971) has used the above model to describe the changes in N content with time of Urrbrae loam under a fallow-wheat rotation, together with the probable changes occurring when the plots were put down to continuous pasture and a rotation of four years of pasture then two years of wheat.

Stability of soil organic matter

The biological stability of soil organic-C and -N cannot be attributed entirely to the nature of their chemical structures since much of the soil organic-C and about 40 per cent of the soil organic-N can be accounted for as compounds derivable from substances readily attacked by microorganisms. Rather, their stability is considered due to the existence of or development of conditions within the soil microenvironment which limit the opportunities for reaction between enzyme and substrate.

Evidence relating to theories that soil organic-N is stabilized by adsorption to clay minerals or reaction with resistant organic constituents (lignin, polyphenolic heterocondensates, tannins) has been reviewed by Estermann and McLaren (1959), Bremner (1965), Greenland (1965), Bremner (1967, 1968), Mayaudon (1968) and Swaby (1968). More recent studies supporting the role of clay minerals in stabilizing soil organic matter include those of Colom and Wolcott (1967), Sϕrensen (1969, 1972*b*), Craswell, Saffigna and Waring (1970) and Craswell and Waring (1972*a*, 1972*b*). Generally the bound substrates are regarded as being of high molecular weight (usually proteins) or as becoming part of a high molecular weight complex. Thus, their stability would depend upon the extent to which their reaction with inorganic or organic soil constituents rendered them inaccessible to the appropriate extracellular enzymes.

Presumably some material may be so bound that, although it becomes insoluble, it remains partly vulnerable to enzyme attack provided that free enzyme is present in the immediate vicinity. Such a situation may occur during the lysis of cells or due to *de novo* enzyme synthesis by microorganisms in the presence of an available energy source, e.g. added plant material. The enzymes themselves, however, would be subject to similar reactions which bind and insolubilize protein substrates. Hence any attack on bound, but otherwise accessible, substrates would seem likely to be temporary, and would decline rapidly after exhaustion of the energy source.

The activities of the bound enzymes may be greatly influenced by several factors, including the number and type of bonds involved in the formation of the enzyme complexes and their proximity to the active sites of the enzymes. By analogy with the results obtained with synthetic organic polymer-enzyme complexes (reviewed by Goldman, Goldstein and Katchalski 1971), formation of clay-enzyme or organo-enzyme complexes in soils may decrease their specific activities, especially toward substrates of high molecular weight. Those adsorbed enzymes which do retain some capacity to hydrolyse high molecular weight substrates are themselves presumably potentially vulnerable to attack by free proteolytic enzymes, formed when conditions favour enzyme synthesis. By contrast, adsorbed enzymes, which as a result of complex formation retain their activities towards low molecular substrates only, may be expected to be relatively stable since they are inaccessible to attack by free proteolytic enzymes.

Thus in soil, whilst a portion (probably small) of the adsorbed extracellular enzymes may be potentially active against high molecular weight substrates and a portion of the adsorbed nitrogenous substrates may be potentially accessible to biological attack, these potentials would not be realized if both enzyme and substrate remain physically separated. Unless conditions change to displace either enzyme or substrate from the complexes, their participation in nitrogen transformation may be confined to reactions with newly synthesized materials following availability of an energy source. In this context, Broadbent and Norman (1946) and Broadbent (1947) showed that decomposition of soil organic matter and mineralization of soil-N are apparently stimulated by addition of plant material. This and subsequent literature describing the priming action of added plant material or C sources on the decomposition of native soil organic matter has been reviewed by Jenkinson (1966*b*), who also pointed out that the magnitude of the calculated priming action on soil-C oxidation may in several cases be subject to error from a number of sources. Jansson (1958, 1971) too has cautioned that the apparent increased rates of soil-N mineralization following decomposition of labelled plant material may be misinterpreted as a priming effect, whereas an alternative explanation may lie in a partial equilibration of labelled and unlabelled N during mineralization-immobilization reactions. The extent to which stability of soil organic carbon and nitrogen is due either to inaccessibility of substrates as a result of complex formation or adsorption reactions or to inadequate concentrations of the appropriate free extracellular enzymes has yet to be determined. Part of the soil-C and -N resides in cellular material and although microbial biomass has been calculated to account for only a few per cent of the total soil organic matter (Russell 1961; Jenkinson 1966*c*; Clark and Paul 1970), accurate estimates are difficult to obtain. The relative stabilities of the C and N components of microbial cells, living or dead, active or senescent, have not been determined. Microbial cells when added to soils are rapidly decomposed (Chu and Knowles 1966; Mayland and McIntosh 1966; Knowles and Chu 1969) but cell walls, especially those of melanic fungi, are more resistant than cytoplasmic constituents (Bloomfield and Alexander 1967; Wagner 1968; Hurst and Wagner 1969).

Decomposition of organic carbon compounds in soil

Extent and rates of decomposition

Studies with ^{14}C-labelled substrates have shown that most organic compounds are decomposed to some extent when added to soils. The more readily available the ^{14}C substrate the more completely is the ^{14}C mineralized, but even with a simple natural substrate such as glucose some of the ^{14}C remains as an integral part of the soil organic matter.

The amount of C from simple organic compounds which is immobilized during the incubation of ^{14}C-labelled materials in soils is variable. For acetate, 20 and 11 per cent was immobilized after 10 weeks' incubation (McGill 1971; Sørensen and Paul 1971; Shields 1972). More results are available for glucose, where the extremes are 45 and 11 per cent. The times of incubation have varied considerably but it is clear that the proportion of ^{14}C immobilized from ^{14}C-glucose differs from one soil to another and from one incubation to another. There are results which suggest that a high clay content favours immobilization (Jansson 1960; Shields 1972), but other factors are also involved. Oxygen supply during the early stages of the incubation could be a dominant factor and possibly the available surface area in the soil could be more important than clay content. C:N ratios and pH are other soil factors involved.

There are differences in the speed of mineralization of added ^{14}C. An initial lag, as occurs after addition of labelled acetate, probably indicates an adjustment of the microbial enzyme complex to a compound not normally found in soil, or in other cases to a change of pH.

Detailed examination during the early stages of incubation, including measurement of the release of both $^{14}CO_2$ and native $^{12}CO_2$ from the soil, have shown that a stimulation of mineralization of native carbon, i.e. a positive priming effect, occurred after 1 to 2 days depending on how quickly the mineralization commenced. This was followed by a second priming effect 2 or 3 days later (Keefer and Mortensen 1963; Shields 1972; Ladd and Paul 1973). It was postulated that the second priming effect may be due to the synthesis of a different suite of enzymes, possibly produced by a secondary microbial population.

The rates of decomposition of added ^{14}C-labelled compounds can be recognized to consist of two and sometimes three discrete phases (Shields 1972; Ladd and Paul 1973). In the first phase the efficiency of conversion of ^{14}C to microbial components was 60-65 per cent, in accord with Payne (1970) for culture systems. Ladd and Paul (1973) found that the rate of disappearance of glucose during the active oxidation phase was calculated to be equivalent to a half-life of 0.22 days. ^{14}C-labelled microbial metabolites formed from glucose were oxidized less readily, exhibiting (from semi-log plots) half-lives of approximately 4 days during the 1.5-20-day incubation period, and of 190 days upon entering a more stabilized phase (after 30 days), when 33 per cent of the glucose-derived carbon still remained in the soil. Little information is available on the turnover of individual organic materials formed during the early stages of incubation although it is well established that the ^{14}C is rapidly redistributed in amino acids, sugars and other organic compounds.

Distribution of ^{14}C in chemical fractions following decomposition of ^{14}C-glucose in soils

During the incubation of ^{14}C-labelled glucose in soils, labelled products are introduced into all fractions so far obtained by chemical techniques. Simonart and Mayaudon (1958), Wagner (1968), Persson (1968), Igel (1969) and Oades (1974) have shown that the distribution of ^{14}C in the classical fractions suggests that they are all significantly involved in receiving microbial tissue and metabolites. Simonart and Mayaudon (1958) also showed that ^{14}C was lost from all these fractions with time, indicating a much more rapid turnover of the C of the new microbial materials than that of the native C. However, differences in stabilities of the new microbial materials in the various classical fractions could not be detected. Thus, a secondary microbial product is not "humified" because it occurs in the humic acid or humin fraction.

Persson (1968) investigated this aspect in some detail and used a relatively mild acid treatment (0.5N H_2SO_4) as the first extractant, followed by hot 0.5N NaOH, which left an insoluble residue or humin fraction. ^{14}C in these three fractions was determined at various times during the incubation. Although the amounts of total-C extracted by acid (10 per cent) and alkali (75 per cent) did not change in either a neutral or an acidic soil during the incubation, the distribution of ^{14}C was different in the two soils and changed with time. The acid-soluble ^{14}C decreased in 28 days to less than half the original values. The NaOH-soluble ^{14}C gradually increased with time. ^{14}C in the residue changed little with time in the neutral soil but increased to almost three times the original value in the acidic soil, which was thought to be due to fungi. Further fractionation of the NaOH extracts and a separate pyrophosphate extraction did not yield any fraction in which ^{14}C was concentrated. Flux of ^{14}C was greatest through the acid extract.

A proximate analysis scheme was also tested and involved the following successive treatments: Soxhlet extraction with ethanol/benzene; reflux with 0.55N H_2SO_4; reflux with 3.5N H_2SO_4; extraction with NaOH. ^{14}C reached a maximum in the ethanol/benzene extract after 3 to 4 days, then decreased slowly, falling to low levels at 28 days. Two-thirds of the ^{14}C in the soil was extracted by refluxing in 0.55N H_2SO_4 but one-third of the soil-C was also extracted so that the fraction contained considerable quantities of native carbon. The ^{14}C in this fraction and in that dissolved using 3.5N H_2SO_4 increased to a maximum and then declined. ^{14}C in the acid-resistant fraction slowly increased with time and remained constant and was present either in fulvic acid or hymatomelanic acid fractions extracted by NaOH. Very little label was found in humic materials or the final insoluble residue. Persson showed that the stronger the acid hydrolysis before NaOH extraction the less contaminated with fresh secondary material were the residue and humic acid fraction. Successively increasing the concentration of acid to a final 6N H_2SO_4 hydrolysis yielded subsequent humic and residue fractions containing least ^{14}C. Jenkinson (1965) and Oades (1974) have also shown that dilute mineral acid extracted materials with high specific activities.

Chemical fractionations of ^{14}C-labelled organisms before and after addition to soil

The addition of ^{14}C-glucose to soil is a means of labelling the biomass and recent metabolites. The addition of labelled microbes directly to soil yields information of the fate of individual organisms and different components, e.g. cell walls of a particular organism. The data of Simonart and Mayaudon (1963) make it abundantly clear that the classical chemical scheme fractionates microbial tissue and that the presence of soil alters this fractionation.

alcohol	Soluble in 80%	Fulvic fraction	Humic fraction	Humin fraction
Aspergillus	48.7	9.8	6.8	34.7
Aspergillus in soil	33.4	8.5	13.5	44.6
Azotobacter	64.5	35.5		0
Azotobacter in soil	21.5	60.7		7.6

There was a marked shift towards insoluble fractions when the fractionation was done after the labelled organisms were added to soil. During subsequent incubation, ^{14}C was lost from all fractions but particularly the soluble and fulvic fractions, and the humin fraction eventually contained a vastly increased proportion of the label.

There is convincing evidence that fungal material is less soluble than bacterial components and is concentrated in humin fractions. Wagner (1968) quoted a fractionation of *Aspergillus* which showed 54 per cent in the humin fraction. Zeller et al. (1966) showed that after 30 days' incubation of labelled mycelium from *Aspergillus* more than 60 per cent of the label resided in the humin fraction. Carter and Oades (unpublished) grew fungi for 10 days on agar containing ^{14}C-labelled glucose and allowed the hyphae to grow through a bed of sterilized soil aggregates placed immediately above but not in contact with the agar. The aggregates plus labelled hyphae were fractionated as shown in the table.

Fraction	*Rhizoctonia solani*	*Stachybotrys atra*	*Ceratobasidium*
Light fraction	46.7	25.0	20.9
Acid extract	12.4	13.5	32.7
Fulvic fraction	3.7	9.6	7.8
Humic fraction	8.6	14.4	8.8
Humin fraction	28.5	37.5	29.7

Most of the hyphal material either floated off as recognizable hyphal fragments at density 2.0 or was insoluble. Highest specific activities were found in the light fraction and acid extracts.

Wagner (1968) and Hurst and Wagner (1969) showed that both cell wall and cytoplasm of hyaline fungi were readily utilized when incubated in soil but cell wall components of a melanic fungus showed considerable resistance to decomposition. A number of soils, including acidic soils and some soils of arid regions, have high humin components due in part to resistant, insoluble fungal materials. It has been known for 30 years that fungi produce such materials (Pinck and Allison 1944).

Incubation of ^{14}C-labelled compounds and well defined fractions in soils

Cellulose (Mayaudon and Simonart 1959*b*; Igel 1969), hemicelluloses (Mayaudon and Simonart 1959*a*; Sϕrensen 1963, 1966), starch (Cheshire, Mundie and Shepherd 1969) and dextran (Oades and Wagner 1971) all decompose in a similar manner to glucose. Cellulose, which is insoluble in NaOH, may appear in the humin fraction before incubation. Decomposition of the carbohydrates leads to a distribution of ^{14}C in all the classical fractions but only low amounts in the humin fraction. Amino acids released from humic acids by acid hydrolysis have been shown to be labelled.

When labelled lignin was added to soil, it became part of the humic acid fraction and to a lesser extent the humin fraction (Mayaudon and Simonart 1959*b*; Sϕrensen 1963; Simonart 1964). The decomposition of lignin was slow although mineralization reached 30 per cent in 60 days. Both Sϕrensen (1963) and Jenkinson (1971) point out that the extraction procedures used to obtain the lignin possibly render the material less accessible to organisms than might be the case in the natural state. Mayaudon and Simonart (1959*b*) showed that vanillin, syringaldehyde and *p*-hydroxybenzaldehyde, when added to soil, were distributed in a manner similar to lignin. They point out that lignin and aromatics become part of the acid-insoluble humic fraction whereas most of the nitrogen-containing materials become part of the acid-hydrolyzable humic fraction, although some carbon from phenolics, including lignin, may also be incorporated in hydrolyzable components, presumably protein and carbohydrate. Plant pigments such as chlorophyll b and γ-carotene behaved like the aromatics and became acid-insoluble humic and humin components, only slowly utilized by microorganisms.

It is now clear from the work of Simonart and Mayaudon (1961), Mayaudon and Simonart (1963), Mayaudon (1968), Oberlander and Roth (1968) and Sauerbeck and Fuhr (1968) that protein material of plant or microbial origin becomes part of the humic acid fraction due to association of the protein with humic components.

Incorporation of ^{14}C into identifiable compounds during decomposition of labelled substrates in soils

When microorganisms utilize ^{14}C substrates in soil, some of the ^{14}C is rapidly incorporated into amino acids, sugars and phenolics. ^{14}C has been found in a wide range of amino acids with alanine, glycine, glutamic acid and lysine, i.e. those formed readily in the citric acid cycle, possessing the highest specific activities (Wagner 1968; Sϕrensen 1969). Similar results were obtained when labelled glycine or acetate was used as substrate (Ivarson and Stevenson 1964; Stevenson and Ivarson 1964; Mayaudon and Simonart 1965; Sϕrensen and Paul 1971).

Keefer and Mortensen (1963) showed rapid labelling of sugars during incubation of labelled alfalfa in a soil. Decomposition of ^{14}C-glucose and other carbohydrates leads to rapid labelling of glucose, galactose and mannose, which account for 80 per cent or more of the ^{14}C in aldoses. Xylose and arabinose become labelled lightly, if at all (Cheshire, Mundie and Shepherd 1969, 1971; Oades and Wagner 1970, 1971; Oades 1974). A large proportion of the pentoses in several soils has been shown to be in plant materials and the available evidence indicates that the plant components are the major source of pentoses in soils. During incubation with a ^{14}C substrate, rhamnose and fucose have sometimes become labelled and sometimes have not incorporated ^{14}C.

The labelling of polysaccharides in soils (Chahal 1968) indicated that differences between those of fungal and microbial origin can be detected. Studies of ^{14}C in aldoses in a series of extracts showed that a microbial glucan can be isolated in a light fraction ($<$ 2.0 S.G.) and that the most highly labelled sugars are extracted by cold N HCl and are markedly concentrated in low molecular weight fractions according to gel chromatography on Sephadex G-10 (Oades 1974).

Incorporation of ^{14}C into phenolics has been demonstrated by Simonart and

Mayaudon (1966) and Sinha (1972*a*, 1972*b*). No doubt it could be found in many other simple compounds which form part of larger molecules in microbial tissue and metabolic products.

Decomposition of labelled plant materials added to soils

Plant materials contain proteins, carbohydrates and lignin, and when subjected to the classical fractionation procedures for soil organic matter yield a large insoluble "humin fraction" but also significant soluble and fulvic fractions (Oberlander and Roth 1968; Sauerbeck and Fuhr 1968). Mixing the plant materials with soil resulted in sorption of some components, causing increases in both humic and humin fractions (cf. microbial tissue p. 142). Neither plant nor microbial tissues are "humified" just because they appear in humic and humin fractions. ^{14}C was lost from all fractions during incubation of labelled plant components in soil. The percentage of ^{14}C in insoluble materials fell from 74 to 19 per cent during incubation and, in spite of absolute losses of ^{14}C from the fulvic and humic fractions, these latter fractions were enriched in labelled carbon. Evidence was obtained for interaction of plant components with soil during the NaOH extraction. Persson (1968) showed that concentrated acid treatments can render certain easily extractable components insoluble, i.e. a series of acid extractions of increasing concentration extracted more ^{14}C than a single treatment of the most concentrated acid used.

Bearing in mind this possibility of artefact formation during both acid and alkaline extraction and considering the complexity of plant material, it is not surprising that even with the use of ^{14}C the decomposition of plant material in soil can only be interpreted in a general way. Microbial degradation of plant tissue leads to synthesis of microbial tissue and metabolic products, which in turn become substrates for further successive microbial activity until the materials remaining are oxidized relatively slowly.

Studies on the decomposition of plant-C in soil are limited (Fuhr and Sauerbeck 1966; Oberlander and Roth 1968; Jenkinson 1971; Shields 1972; Shields and Paul 1973). This work has shown that, despite differences in climate and in the type of plant material and soil, about one-third of the plant material was retained in the soil after 1 year and about one-fifth after 4 or 5 years. The decomposition of the added material was initially retarded in soils with low pH or soils growing plants but both these effects were transient. Growing plants could lead to drier conditions, which definitely retard decomposition of plant materials on the soil surface (Shields 1972; Shields and Paul 1973). Decomposition of roots may initially be slow compared with above-ground phytomass, but the overall rate of decomposition was similar for both roots and tops (Dahlman and Kucera 1965; Dahlman 1968; Jenkinson 1971; Ignatenko, Krillova and Ponyatovskaya 1968).

Laboratory experiments using ^{14}C-labelled plant materials have given similar results (Chekalov and Illuvieva 1962; Sφrensen, 1963, 1966; Chahal, Mortensen and Himes 1966; Sauerbeck 1968). Approximately one-half of the ^{14}C in plant material remained in the soil after 100 days.

The immobilized ^{14}C was rapidly distributed between the classical fractions, but it is not clear whether the ^{14}C was in the original plant material or new microbial products (Sφrensen 1963, 1966; Jenkinson 1965; Neunylov and Khavkina 1968). As the incubation proceeded, ^{14}C was lost from the ethanol/water and fulvic acid fractions and the ratio of ^{14}C in humic acid to fulvic acid increased. ^{14}C in the insoluble residue also decreased as shown by the field data.

Persson (1968) showed that during the incubation of ^{14}C-labelled straw, extraction by reflux in 0.55N H_2SO_4 removed ^{14}C-labelled materials which were not primary straw materials but rather microbial materials derived from the straw. The microorganisms utilized cellulose or straw tissue soluble in ethanol/benzene, to produce materials dissolved by the reflux in acid. Acid-resistant materials were not affected during 2 years' incubation. Cold 0.5N H_2SO_4 extracted significant proportions of straw-C initially, but after 6 months the ^{14}C in this fraction decreased and then reappeared, presumably as microbial tissue (cf. the fate of the hemicellulose fraction, Russell 1961). From both work on incubation of ^{14}C-glucose and ^{14}C-plant material, Persson concluded that most microbial material in soil was extracted by a reflux in 0.55N H_2SO_4. Plant components resistant to 6N HCl treatments were concentrated in humic and humin fractions. By pre-extracting the ^{14}C-labelled straw using various acid treatments, Persson showed that incubation of the extracted straw quickly led to incorporation of ^{14}C into secondary products subsequently dissolved by the various acid treatments.

Decomposition of organic nitrogen compounds in soil

Availability of organic-N

Bartholomew (1965) has reviewed the evidence indicating that in soils, microorganisms have a dominant role in transforming the nitrogenous organic constituents of plant residues. The participation of plant enzymes, e.g. hydrolytic enzymes, in the overall decomposition process may assume some importance especially in the early stages of breakdown of incorporated fresh residues (Chibnall 1939). However it would seem that a high proportion, if not all, of the plant organic-N is rapidly transformed in soils and that the changes are mainly affected by microbial action. Hence the more immediate products of the decomposition process are mainly organic-N compounds of microbial origin, together with excess mineral-N, if any, not required for microbial syntheses. Subsequent mineralization of that portion of the plant-N immobilized either within microbial cells or as extracellular microbial metabolites is a relatively slow process. Factors which influence both the rates and extent of immobilization of plant residue-N and also the availability of the N of organic matter, including immobilized-N, have been reviewed by Bartholomew (1965) and Scarsbrook (1965).

Although immobilized nitrogenous material is considerably less susceptible to further decomposition than that of the original plant and animal residues, nevertheless it has been shown that recently immobilized-N is, for a considerable time, more available than the remainder of the soil organic-N.

Using ^{15}N-labelled immobilized material, Jansson (1958) found that the proportion of label in the ammonium-N formed during the remineralization phase considerably exceeded that of labelled organic-N originally present. The preferential availability of the labelled organic-N continued for some months. Jansson concluded that mineral-N was derived in soils from a small active fraction, accounting for up to 10-15 per cent of the total soil-N and which he identified with the heterotrophic microbial biomass. Later, Jansson (1963, 1968) showed in long-term pot experiments that the availability to oat crops of ^{15}N-labelled plant residues and of microbial tissue declined each year but continued to be greater than that of the remaining soil-N for several years. Thereafter the availability of labelled and unlabelled sources was similar. Their biological half-life after several years of equilibration was about 40 years.

Experiments, particularly by Broadbent (Broadbent and Nakashima 1965, 1967; Broadbent 1968*a*. 1968*b*) and Legg (Legg and Allison 1967; Legg et al. 1971), have clearly demonstrated the progressive decline in the availability of immobilized or plant-residue-N as a function of time. Broadbent and Nakashima (1967) calculated availability ratios, defined as the percentage of labelled-N in a crop divided by the percentage of labelled-N in the soil at the onset of crop growth. Availability ratios determined for successive cuttings of sudan grass grown on loam soils declined progressively, ranging from 9.2 to 11.9 for the second cutting to 1.5 to 1.9 for the eighth (Broadbent and Nakashima 1967). Legg et al. (1971) reported that availability ratios determined for successive oat crops approached a value of approximately 2. Thus, under their conditions, where all of the plant material was returned to the soil after each cropping, the labelled-N (from immobilized and fresh plant residue sources) remained consistently twice as available as the indigenous soil-N.

Changes in the nature of organic nitrogen components of soils during nitrogen mineralization

Attempts have been made to define the relative importance of various organic nitrogenous components as sources of inorganic-N and to determine the extent to which their net mineralization rates and percentage decomposition are affected by their distribution and state of occurrence in soils. Comparisons have been made of the distribution and availability of ^{15}N-labelled and unlabelled nitrogenous compounds of different soil fractions throughout periods of net mineralization. Experiments may be of short duration (1-3 months), when the greatest relative changes in availability ratios are observed, or may be continued for several years.

(*a*) Availability of organic-N of chemically defined compounds. Of the various organic-N compounds formed in soil during immobilization of mineral-N, those comprising the alkali non-distillable, acid-hydrolysable fraction appear quantitatively to be the most important (Cheng and Kurtz 1963; Stewart, Porter and Johnson 1963*a*; Chu and Knowles 1966; Simpson and Freney 1967; Broadbent 1968*a*, 1968*b*; Danneberg, Haunold and Kaindl 1968; Miki and Mori 1968; Knowles and Chu 1969; McGill 1971). Amino acid-N accounts for much of the N of this fraction.

Subsequent mineralization leads to decreased concentrations of all forms of soil organic-N, both of those constituents formed by recent immobilization and those less available components previously present. In each case the acid-hydrolysable amino acid-N fraction is the major contributor to the inorganic-N pool (Stewart, Porter and Johnson 1963; Keeney and Bremner 1964, 1966; Mayland and McIntosh 1966; Broadbent 1968*a*, 1968*b*; Freney and Simpson 1969; Harada and Hayashi 1968; Miki and Mori 1968; Moore and Russell 1970; Isirimah and Keeney 1973) but in moderate to long term experiments at least, there is no clear indication that this fraction exhibits consistently a greater biological availability than other chemical components (Keeney and Bremner 1964, 1966; Broadbent 1968*a*, 1968*b*). However, a preferential availability of the non-distillable, acid-hydrolysable-N (estimated as acid-hydrolysable amino acid-N in some cases) has been demonstrated after shorter mineralization periods (Stewart, Johnson and Porter 1963; Chu and Knowles 1966; Miki and Mori 1968; Freney and Simpson 1969) or especially when the soils were subjected to alternate wetting and drying treatments to promote mineralization (Harada and Hayashi 1968; Moore and Russell 1970).

McGill (1971) demonstrated that recently immobilized, amino sugar-N was more available subsequently than amino acid-N, the initial decline being associated with the disappearance of a primary fungal population formed during the metabolism of the supplied acetate energy source. Nevertheless the greater susceptibility of the amino sugar-N fraction continued to be evident throughout the whole incubation period (268 days). Also Isirimah and Keeney (1973) reported that during a period (6 months) of mineralization of some organic soils, the greatest percentage decline in various organic-N fractions was exhibited by the amino sugars and by unidentifiable N components of acid hydrolysates.

The nitrogenous fraction, resistant to acid hydrolysis, generally exhibits an availability comparable with that of other fractions, although some exceptions have been reported, e.g. Freney and Simpson (1969) and Isirimah and Keeney (1973). The nitrogenous components of the acid-insoluble fraction are unknown but may in part consist of compounds derived from perhaps cellular sources which become adsorbed during acid hydrolysis, for example nucleic acid bases and their decomposition products. The extent to which the apparent availability of the N of the acid-resistant fraction is affected by such artefacts is unknown. However, it is of interest to contrast the situation generally observed with soil incubation experiments with that of Danneberg, Haunold and

Kaindl (1968) and Danneberg (1971*a*, 1971*b*), who followed changes in the labelled organic-N levels during periods of net immobilization and remineralization accompanying the aerobic decomposition of plant material in the absence of soil. During the first 10 days' incubation, a small amount of the labelled-N is incorporated into the non-hydrolysable fraction, from which it is concluded that humified nitrogenous materials are formed relatively rapidly. During the period of subsequent remineralization (up to 180 days), the level of organic-N decreases, due mainly to a decline in the acid-hydrolysable "protein" component, but the level of non-hydrolysable organic-N remains constant. Although comparisons may well be misleading, the results support the possibility that in mineral soils the availability of the organic-N of the fraction not hydrolysable by hot 6N HCl may be overestimated as a result of adsorption processes taking place during hydrolysis. It is of interest in this context that the results of Isirimah and Keeney (1973), demonstrating the comparative unavailability of the nitrogen of the acid-resistant fraction, were obtained with highly organic soils, where the adsorption of hydrolysis products may be anticipated to be less than that obtained with mineral soils.

(*b*) Availability of organic nitrogen of the classical humic fractions. The classical procedure of fractionating soils into extractable humic and fulvic acids and non-extractable humin components has been applied in several studies of the changes in the distribution of immobilized-N and N of incorporated plant residues. It is recognized that the results are difficult to interpret since the fractions will contain not only humified nitrogenous compounds, i.e. compounds bonded as an integral part of the humic molecule, but also materials co-extracted from cells and cellular debris. Most of the investigations have involved frequent sampling of soils during relatively short incubation periods (up to 14 days from the commencement of immobilization of labelled inorganic-N) when the greatest relative changes in the levels of labelled organic-N take place. Freytag and Igel (1970) showed that, during immobilization of labelled NH_4^+-N in soil, label was rapidly (within 24 hours) incorporated into all fractions, including the acid-resistant components of extracted humic acids.

McGill (1971) showed that during net mineralization of ^{15}N-labelled immobilized-N in a heavy clay chernozemic soil, marked changes occurred in the concentrations of labelled organic-N in the fulvic and humic acid fractions obtained by previously extracting with 0.1M neutral sodium pyrophosphate. By using a reagent milder than alkali, the labelled compounds in the extracts were thought to consist of microbial extracellular metabolites and lytic products and not to be derived from intact cells. However, the preliminary drying of the soil samples at 60°C may have led to subsequent extraction of some intracellular material. McGill found that the labelled organic-N of the fulvic acid fraction rapidly declined in amount and contributed significantly to the mineral-N pool. Less labelled-N was present in the humic acid fraction but it appeared to be no less available than that in the fulvic fraction. Nitrogenous material further extracted from the humic acid with phenol by a similar technique to that of Simonart, Batistic and Mayaudon (1967) was more readily mineralized than the residue, presumably held in the humic acid complex by other than hydrogen bonding.

Wojcik-Wojtkowiak (1972), studying the decomposition of ^{15}N-labelled straw in soil, found that up to 67 per cent of the straw-N was present in the alkali-insoluble humin fraction within 14 days of incorporation of the plant material, and that this fraction showed the greatest decline in organic-N during N mineralization. Labelled-N was present in both humic and fulvic acids, whether extracted with alkali or sodium pyrophosphate solutions. The proportions of ^{15}N-label in the preparations varied according to the extractant but, for a given extractant, generally showed only small changes during the 14-112 day incubation period.

In conclusion, chemical fractionation procedures, involving measurement either of the nitrogenous components of the classical fractions (fig. 1) or of identifiable organic compounds, give no consistent indication of the nature of the more readily available organic-

N pool in soils.

An alternative approach to the use of classical extractants has been to use milder reagents which selectively remove nitrogenous compounds differing in their nature or in the manner in which they are bound within the soil. Extracts whose N contents are more highly correlated with the capacity of the unextracted soil to mineralize N might be anticipated to more closely reflect changes in the size of the readily decomposable N pool. The N availability of soils and soil fractions of different particle size was found to be highly correlated both with the percentage N extractable by autoclaving with 0.01M $CaCl_2$ and with the percentage N in the NaOH-distillable portion of the extracts (Chichester 1969; Stanford 1968, 1969; Stanford and DeMar 1969). However, such extracts were found to remove similar proportions of both labelled and unlabelled nitrogenous compounds from soil throughout a period in which eight successive oat crops were grown, the total plant material being returned to the soil after each crop (Legg et al. 1971). Yet the labelled nitrogenous sources in the soil were clearly more available to plants as determined by availability ratios (Broadbent and Nakashima 1967) for each cropping period, and to microorganisms as determined by the relative percentages of labelled and unlabelled N mineralization during aerobic incubation tests. The extractant therefore was unselective from the standpoint of biologically available N.

Change in the distribution of organic nitrogen components of soils during nitrogen mineralization

(*a*) Availability of organic nitrogen of fractions of different particle size. Clay soils from virgin and cultivated sites in the Darling Downs, a major cereal cropping area of Queensland, were fractionated by Craswell, Saffigna and Waring (1970) into aggregates ranging in diameter from > 5 mm to < 0.1 mm. Aggregates of size 0.5-1 mm diameter were of highest N content, which decreased progressively in aggregates of greater and smaller size. Mineralization rates, determined by net ammonium release during incubation under waterlogged conditions, demonstrated that the average availability of the organic-N of the aggregates increased with decreasing aggregate size. The trend, observed in all soils, was inexplicable in terms of differential aeration effects within the aggregates, C:N ratios, or amounts of free organic matter of high C:N ratio present in the aggregates. Rather, the size of the aggregates is considered to affect directly the proportion of the organic-N inaccessible to microbial action. The same trend was observed by Kereszteny, Szabo and Teuk (1963) and Seifert (1964) using aerobic nitrification rates as a measure of mineralizability.

The results are consistent with earlier studies by Bremner and colleagues (Waring and Bremner 1964; Edwards and Bremner 1967), who showed the N mineralization rates were increased by grinding soils to finer particle size. Reaggregation of finely ground ($<$ 300 mesh) particles by a moistening and drying treatment depressed mineralization rates. A similar treatment applied to larger ($<$ 10 mesh) particles did not affect their subsequent rates of mineralization.

Craswell and Waring (1972*a*) showed that grinding caused the greatest increases in N mineralization rates with soils of high clay content, especially those with montmorillonite clays. Increased rates of nitrification and oxygen uptake in virgin and cultivated cracking clay soils occurred soon after commencing incubation of the ground soils (Craswell and Waring 1972*b*). The results support the contention that the organic matter is readily metabolizable when conditions render it accessible to the soil microflora.

Chichester (1969) fractionated soil suspensions by sonic vibration to obtain components ranging in particle size from 2,000 to < 0.25 μm. Again, the percentage of N in the fractions mineralized during 6 weeks of anaerobic incubation increased with decreases in particle size. Comparisons were later made of the distribution and availability of labelled- and total-N of fractions from the same soil into which ^{15}N had been previously incorporated by successive cycles of cropping and plant-residue return (Chichester 1970). Labelled material was present in all particle-size fractions and chemical fractions prepared therefrom, but the proportion of label decreased in those materials requiring greater periods of ultrasonic vibration for dispersion. Labelled material was mineralized more readily.

(*b*) Availability of organic-N of fractions of different density. Greenland and Ford (1964) have shown that the release of nitrate-N during 14 days' aerobic incubation of Urrbrae loam was augmented by the inclusion of light fraction material, consisting of partially humified plant fragments separated from soils by densimetric techniques. Net nitrate release from incubation mixtures containing added light fraction material of relatively high C:N ratio (23.1:1) separated from continuous wheat plots was less than from mixtures containing material of lower C:N ratio (11.5:1) separated from the same soil under continuous pasture. Assuming that the additional nitrate-N was formed entirely from the nitrogenous sources of the added light fraction, the releases expressed as percentages of the input nitrogen are 1.5 and 2.2 respectively. Thus, the availability of the light fraction nitrogen did not differ appreciably from that of the total-N of the unamended soil, 2 per cent of which is mineralized during the incubation period (Ford 1968). An amendment of a solution of an alkali-extracted humic acid (C:N ratio, approximately 12:1) depressed nitrate release from the native soil organic-N, possibly due to inhibition of enzymes involved in their degradation (Ladd and Butler 1968*a*, 1968*b*).

In a later study (Ford and Greenland 1968), unamended soils were aerobically incubated for 4 weeks and the changes in the total mineral-N and light fraction-N determined. The net loss of light fraction-N accounted for 25-60 per cent of the net gain of mineral-N, although the amounts of light fraction-N originally present accounted for only 7-23 per cent of the total soil-N. Thus, in these experiments a clear indication is given that the light fraction-N is more available than the remainder of the soil organic-N. However, the possibility cannot be excluded that both physical and chemical changes in the light fraction occurred during incubation, leading not only to mineralization of some of the light fraction-N but also to transfer to the denser "humified" component on subsequent fractionation. The use of ^{15}N-labelled plant components would be a distinct advantage in resolving this doubt. Field experiments showed that seasonal changes in light fraction-N were relatively small and not clearly related to changes in the capacity of the soils to release mineral-N in aerobic incubation experiments.

FRACTIONATION OF SOILS INTO BIOLOGICALLY SIGNIFICANT COMPONENTS

The biomass and other recognizable fractions

In attempting to define those organic fractions which contribute most of the available C and N pools in soils, it is desirable to use a fractionation procedure which allows direct analysis of recognizable biological entities (Jansson and Persson 1968). Pertinent sources of mineralizable material would include intact plant, animal and microbial cells, particulate cell debris, and extracellular compounds, either free in the soil solution, loosely (ionically) bonded to surfaces of the soil matrix, or bonded more strongly and enmeshed more deeply in complexes with soil organic and inorganic colloids.

The activities of plant enzymes may affect directly the distribution of plant C and N within these pools before the soil heterotrophic microflora have an opportunity to cause further changes (Chibnall 1939; Swaby and Ladd 1962, 1966; Gallus and Jennings, 1971). Nevertheless, microorganisms play a most important role in the processes of transformation of C and N compounds in soils (Bartholomew 1965) and mineralization of both C and N depends on the size of the active microbial population in addition to the chemical nature and accessibility of the substrates being utilized. However, in only a few instances have attempts been made to fractionate soils into even the crudest biological entities, e.g. cellular and extracellular components.

Little is known of the effects of soil properties on either (*a*) the proportions of C and N which become immobilized in intact microbial cells and extracellular materials (bound or free) or (*b*) the relative rates of mineralization from each of these pools.

Jenkinson (1966*c*) provided indirect evidence that the intact microbial biomass constitutes an important component of the labile C and N pools in soils. Irradiation or fumigation of soils, followed by inoculation and incubation, caused the release of CO_2 and NH_4^+ considered to have originated in the tissues of killed microorganisms.

Based on the size of the flush of decomposition after exposure to $CHCl_3$ vapour, Jenkinson calculated that 2.3-3.5 per cent of the soil organic-C was accounted for in the biomass. Nevertheless biomass-C accounted for 31-39 per cent of the residual C of plant material incorporated 1 year previously and 19 per cent of that of plant material after 4 years of decomposition.

Shields (1972) and Shields et al. (1973) found that following incubation of ^{14}C-labelled glucose in the field, about 60 per cent of the labelled-C was present in the soil as microbial cells plus metabolites after 7 days' incubation and about 40 per cent after 87 days. From measurements of total bacterial numbers and fungal hyphae lengths, microbial biomass was calculated to account for one-third of the labelled-C after 7 days. However, based on the procedure involving pretreatment of the soil with chloroform vapour, about three-quarters of the labelled-C present after 87 days was calculated to be in the biomass. The residual material was less susceptible to priming than native organic matter. Shields considered that in his system the use of chloroform had led to an overestimate of the biomass. McGill (1971) used the same soil in a laboratory investigation of changes in microbial biomass following addition of acetate and ammonium sulphate. Computer simulation studies indicated that C and N present as microbial biomass accounted for only about 10 per cent of that present as extracellular metabolites.

The results of Jenkinson (1966*c*) and Shields (1972) indicate that a substantial portion of recently immobilized C in field soils becomes part of the microbial biomass. Unlike Shields, Jenkinson (personal communication) found that the magnitude of the biomass as determined by pretreatment of soils with chloroform was similar to that obtained by methods involving direct counts of microbial cells of different sizes. Estimates of biomass based on the amounts of adenosine triphosphate extractable from soils (Doxtader 1969; Ausmus 1973; Holm-Hansen 1973*a*, 1973*b*) may be useful in indirectly corroborating measurements obtained by other methods.

Biological significance of chemical extracts

It is clear that the classical fractionation scheme takes no account of microbial tissue or recent metabolites and, as Jenkinson (1971) has pointed out, the distribution of ^{14}C in fulvic, humic and humin fractions is relatively uniform and gives no indication that the availability of compounds within these fractions may differ up to twentyfold. A further disadvantage in using extractants such as alkali is that the extracted material would be derived in part from several sources, e.g. part decomposed plant fragments, the microbial biomass and clay organic complexes. Thus there is no way of deciding whether, for example, the hydrogen-bonded protein isolated from a humic acid fraction (Simonart, Batistic and Mayaudon 1967) may not have been derived from material which was not humified in the soil.

Summarizing the work of Persson (1968), Jansson and Persson (1968) described a procedure which had a chemical basis but which was intended to be biologically justified. They suggested that the basis of such a scheme should be treatments involving gradually increasing acid concentrations up to a final concentration of 6N prior to any alkaline extraction. Reflux with 0.55N H_2SO_4 was considered to dissolve most of the biomass and recent metabolites. Cold alkaline extraction should be used to avoid extraction of lignin, which could then be extracted by a hot alkaline treatment, leaving only highly resistant materials.

There is convincing evidence to show that a gentle acid treatment (up to 1M at 0-20°) is a most useful pretreatment to any chemical extraction scheme. Apart from the fact that it increases the subsequent extraction of a range of soil organic chemicals (humic materials, Choudhri and Stevenson 1957; polysaccharides, Swincer, Oades and Greenland 1968*a*; organic phosphorus, Steward and Oades 1972), it also extracts materials of relatively high specific activity from soils recently incubated with a ^{14}C-labelled substrate (Persson 1968; Chekalov and Illuvieva 1962; Chahal, Mortensen and Himes 1966; Jenkinson 1968*a*; Oades 1974).

The use of more concentrated acid to remove fresh secondary products from humic and humin fractions has been demonstrated (Mayaudon and Simonart 1958*b*; Simonart and Mayaudon 1961; Søφrensen 1966; Wagner 1968) and was used by Haworth (1971) as a pretreatment to obtain the so-called "resistant core" of humic materials. The treatment will remove protein and polysaccharides from humic and humin fractions. Thus, mineral acids appear to have a role as both mild and vigorous hydrolytic agents in differentiating between "new" and "old" organic fractions in soils.

The use of milder, more specific extractants may also have some advantages. Aqueous buffers have been used to extract enzymes from soils (Bartha and Bordeleau 1969; Ladd 1971; Burns, El-Sayed and McLaren 1972) so presumably some non-enzymic proteins may also be extracted under these conditions. Ladd (1971) showed that an extracted, peptide-hydrolysing enzyme may be separated to a great extent from co-extracted humic compounds by precipitating the latter with Ca^{++}. McGill (1971) and Shields (1972) found that the organic-N and C of fulvic acids, extracted from soils with neutral sodium pyrophosphate solution, were comparatively available and important contributors to the mineral pool. However, even with relatively mild reagents, intracellular constituents may also be extracted if the soil sample is previously dried or subjected to preliminary dispersion by sonic vibration.

Jenkinson (1968*a*, 1968*b*) has discussed the merits of various extractants in tests for potentially available N in soils and the conditions under which the tests may best apply. Uptake of N by rye grass in pots was best correlated with barium hydroxide-extractable carbohydrate and with sodium bicarbonate-extractable nitrogen. During a period of immobilization and remineralization of nitrate-^{15}N in a sandy loam, Ladd and Paul (1973) recorded changes in the distribution of acid-hydrolysable amino acid-^{15}N in fractions of different particle size and in both high and low molecular weight fractions obtained by extracting with 0.1M sodium bicarbonate. During rapid microbial growth, nitrate-^{15}N was immobilized and the contents of acid-hydrolysable amino acid-^{15}N rose markedly. However, upon entering the phase of decline of viable bacterial numbers, during which microbial metabolites were the main energy sources, there was initially a rapid decline in the amounts of extractable protein. A secondary transient increase in the concentration of bicarbonate-extractable protein-^{15}N after 4-5 days' incubation coincided approximately with the rise and fall of net proteolytic activity of the soil. Since appreciable amounts of proteolytic activity are also extractable, it seemed possible that extracellular proteolytic enzymes might account for much of the extractable protein, newly synthesized or released during this secondary phase. If so, the results support the suggestion of Søφrensen (1969, 1972*a*) that extracellular enzyme-N may contribute measurably to the total-N immobilized as proteins. The concentrations of ^{15}N-labelled, extractable amino acids and peptides remained relatively constant during this early period but then decreased rapidly as the rates of oxidation of microbial metabolites decreased. The major changes in the levels of bicarbonate-extractable, acid-hydrolysable amino acid-

^{15}N took place before substantial net mineralization of labelled compounds occurred. Ladd and Paul (1973) took the view that the mineralization by microorganisms of soil organic matter required that it be converted to soluble, low molecular weight compounds, which thus constitute a small but important organic pool in the overall process. The extraction conditions were considered sufficiently mild to remove loosely adsorbed or free compounds from the soil with minimal damage to intact microbial cells.

Specific chemical fractions

Of the various chemical fractions examined, acid-hydrolysable amino acid-N is quantitatively the major contributor to the mineral-N pool, although there is no consistent trend to indicate a greater availability of this fraction compared with that of others. When a particular chemical fraction has been shown to be more readily decomposed, there is generally no supporting evidence to explain the observation in terms of changes in some recognizable biological entity. As an exception, McGill (1971) found that amino sugar-N, recently formed by immobilization of inorganic-N, was subsequently mineralized more readily than other fractions. The relatively rapid initial decrease in amino sugar-N was accompanied by the disappearance of a primary fungal population formed during the initial immobilization period.

Changes in the concentration of nucleic acid-N during net mineralization in soils have not been reported even though it is known that during rapid microbial growth, nucleic acid-N may account for up to 30 per cent of the cell nitrogen. The proportion decreases to approximately 10 per cent as the growth rate declines, due mainly to decreases in the concentration of ribose nucleic acid (Herbert, Phipps and Strange 1972; Ribbons 1972). It would be of considerable interest to examine the fate of the RNA-N and the extent to which the decline in concentration coincides with the onset of the net mineralization phase. Further, nucleic acid-N is found in very low amounts in soils. This may be due partly to limitations in the procedures used for its detection or partly to its greater availability than that of say protein-N when released from cellular material. Loss of nucleic acid-N from the biomass may thus be disproportionately greater than that of other constituents.

Physically defined fractions

Fractionation of soil into biologically significant entities requires that the application of chemical techniques be preceded by some form of physical separation. Plant remnants contribute significantly to the total C and N of soils and may be removed by sieving or flotation techniques (Barley 1955; Ford, Greenland and Oades 1969; Roulet et al. 1963; Swincer, Oades and Greenland 1968*a*; Garwood, Clement and Williams 1972).

Further separation of plant fragments, intact microorganisms including fungal hyphae, and cell wall fragments from each other and from sorbed extracellular material is highly desirable but difficult to achieve. Their separation requires an input of energy and a major problem is to determine how vigorous the physical treatment shall be. If whole cells are to be preserved the treatments must be gentle but separation of the components will be incomplete. More complete dispersion can be obtained using ultrasonic energy, which will disrupt an indeterminate number of cells giving rise to artefacts in the form of soluble and sorbed cytoplasmic components and cell wall fragments.

Studies by Chichester (1970), McGill (1971) and Ladd and Paul (1973) have shown that the relative concentrations of soil N and immobilized ^{15}N-labelled N differ in fractions of different particle size. Although a general labelling of all forms of soil-N occurs during immobilization, the proportion of label is less in those materials requiring more extensive ultrasonic treatment for dispersion (Chichester 1970). This observation, together with the fact that N mineralization rates are inversely related to particle or aggregate size, indicate the desirability, before dispersion, of separating material occurring on the surfaces of microaggregates from that present within the particles. Hattori (1969) has partially separated microflora on the surfaces and within wider pores of aggregates from those within narrower pores. The former populations exhibit greater responses to the addition of a soluble nitrogenous substrate, glycine, and are more sensitive to drying subsequently. Hattori's procedure may be applicable in some soils as a preliminary for separating a more readily available nitrogenous fraction, i.e. the fraction containing N recently immobilized in a microbial biomass plus metabolic products, from older, less available sources.

Bae, Cota-Robles and Casida (1972) and Bae and Casida (1973) described an exhaustive centrifugal washing of dispersed soils and fractionation in a density gradient. A fraction was obtained in which microbial cells were concentrated in numbers equivalent to those of the viable cells as determined by plating techniques. Lammers (1971) has used similar techniques to obtain microorganisms from natural waters.

Paul and his colleagues, especially McGill and Shields, have examined in detail the transformations of ^{14}C- and ^{15}N-labelled materials during their decomposition in soils in both field and laboratory studies. Soils were fractionated by a combined chemical and physical technique based on that described by Anderson (1972). Following extraction of soil samples with sodium pyrophosphate solutions, the sodium-saturated soils were dispersed in water by ultrasonics and particulate fractions of different sizes were obtained by centrifugation.

During the metabolism of added ^{14}C-acetate and ^{15}N-ammonium sulphate, immobilized labelled C and N became distributed amongst pyrophosphate-extracted humic and fulvic acids and in fractions of particle size ranging from $> 0.2\ \mu m$ to $< 0.04\ \mu m$. McGill (1971) found that during the early stages of net remineralization of N, the greatest decline in organic-^{15}N occurred in the extracts (especially the fulvic acid and the phenol-soluble portion of the humic acid), whereas the organic-^{15}N content in particles $> 0.2\ \mu m$ continued to increase to a maximum of about 45 per cent of the total. Later, after 12-91 days' incubation, the organic-^{15}N concentration of the $> 0.2\ \mu m$ fraction decreased sharply apparently leading to continually increasing levels of inorganic-^{15}N. Approximately 20-25 per cent of the immobilized-^{15}N was present in material of particle size $< 0.04\ \mu m$. The N of this fraction showed only minor fluctuations throughout the net mineralization phase, although early in this period a small decline in concentrations was observed, which contrasted with the increased ^{15}N concentration in the $> 0.2\ \mu m$ fraction obtained at the same time.

The Canadian workers also demonstrated a close relationship between the turnover of C and N in the various organic fractions. Not only did the evolution of carbon dioxide closely parallel N immobilization and mineralization, but also, throughout the incubation, the concentrations of ^{14}C-carbon and ^{15}N-nitrogen were highly correlated within the fractions, especially within the humic acid and fractions $< 0.04\ \mu m$ and $> 0.2\ \mu m$ in diameter. Changes in the immobilized ^{14}C and ^{15}N in the soil fractions were also related to changes in the microbial populations in the soil.

McGill and Shields postulated that, after fractionation, cell walls and other particulate debris sedimented in the fraction of particle size $> 0.2\ \mu m$, which represented a relatively resistant pool during the early stages of mineralization but which later became a major source for the steady release of carbon and nitrogen. A significant portion of the immobilized C and N also resided in the fraction of particle size $< 0.04\ \mu m$, and was considered to represent cytoplasmic material adsorbed to fine, inorganic colloids following cell lysis.

The main difficulty in interpretation lies in deciding the extent to which the observed distribution is due to natural processes of cell growth and lysis during the incubating period and that due to cell disruption from the drying and sonication methods employed for soil dispersion. Thus artefact formation may affect conclusions regarding the relative availability of carbon and nitrogen in the above pools.

In any case, immobilized C and N in the fraction of particle size > 0.2 μm will not be accounted for exclusively by cell wall debris. Possibly the observed mineralization from this pool may be due to the preferential decomposition of other microbial components following the death of the intact cells also present, and thus this particle size fraction may represent a pool where relatively resistant cell wall components eventually accumulate. This question may be resolved only after more intensive fractionation, possibly by density gradient centrifugation, to separate intact cells, particular cell wall material and compounds bound in clay-organic complexes.

McGill and Shields also examined in detail the decomposition of ^{14}C- and ^{15}N-labelled plant material over two or four growing seasons at several field sites. Much of the labelled-N initially mineralized was derived in all soils from the relatively rapid degradation of the fulvic acid fractions. In all soils also, incorporation of ^{15}N-label into the fraction of particle size > 0.20 μm reached a maximum after the first season and net decomposition of this fraction was evident only after depletion of more readily available nitrogenous sources in the fulvic acid fraction. Most of the labelled organic-N was accounted for in particles > 0.2 μm and < 0.04 μm in diameter, the latter also slowly releasing its labelled-N during the later stages of decomposition.

The results of the Canadian workers have been of particular interest not only because their investigations have included field experiments in a semi-arid environment but also because of their intensive application of a fractionation technique such that changes in recently immobilized C and N in soil organic fractions may be interpreted as changes in recognizable biological entities.

REFERENCES

Adams, A.P., Bartholomew, W.V., and Clark, F.E. (1954). *Soil Sci. Soc. Amer. Proc.* 18, 40-6.

Alexandrova, I.V. (1968). Pochvovedenie No. 8, 71-8.

Alexandrova, L.N., Yurlova, O.V., and Lobitskaya, L.V. (1964). *Trans. 8th Int. Congr. Soil Sci.* 3, 109-16.

Anderson, D.W. (1972). Ph.D. Thesis, Univ. of Saskatchewan, Saskatoon.

Anderson, G. (1958). *Soil Sci.* 86, 169-74.

———. (1961). *Soil Sci.* 91, 156-61.

———. (1967). In "Soil Biochemistry". vol. 1 (Eds. A.D. McLaren and G.H. Peterson). pp. 67-90. Marcel Dekker, New York.

Arshad, M.A., and Lowe, L.E. (1966) *Soil Sci. Soc. Amer. Proc.* 30, 731-5.

Ausmus, B.S. (1973). *Bull. Ecol. Res. Comm. (Stockholm)* 17, 223-34.

Bae, H.C. and Casida, Jr., L.E. (1973). *J. Bacteriol.* 113, 1462-73.

Bae, H.C., Cota-Robles, E.H., and Casida, Jr., L.E. (1972). *Appl. Microbiol.* 23, 637-48.

Barley, K.P. (1955). *Agron. J.* 47, 145-6.

Bartha, R., and Bordeleau, L. (1969). *Soil Biol. Biochem.* 1, 139-43.

Bartholomew, W.V. (1965). In "Soil Nitrogen". (Eds. W.V. Bartholomew and F.E. Clark). pp. 285-306. Amer. Soc. Agron., Madison.

Bartholomew, W.V., and Kirkham, D. (1960). *Trans. 7th Int. Congr. Soil Sci.* 3, 471-7.

Bazilevich, N.I., and Rodin, L.E. (1968). In "Methods of Productivity Studies in Root Systems and Rhizosphere Organisms". pp. 4-7. U.S.S.R. Acad. Sci. Nauka.

Bendich, A. (1955). In "The Nucleic Acids". vol. 1. (Eds. E. Chargaff and J.N. Davidson). pp. 81-136. Academic Press, New York.

Biederbeck, V.O., and Paul, E.A. (1973). *Soil Sci.* 115, 357-66.

Bloomfield, B.J., and Alexander, M. (1967). *J. Bacteriol.* 93, 1276-80.

Bremner, J.M. (1965). In "Soil Nitrogen" (Eds. W.V. Bartholomew and F.E. Clark). pp. 93-149. Amer. Soc. Agron., Madison.

———. (1967). In "Soil Biochemistry". vol. 1 (Eds. A.D. McLaren and G.H. Peterson) pp. 19-66. Marcel Dekker, New York.

———. (1968). In "Organic Matter and Soil Fertility". *Pontif. Acad. Sci. Scr. Varia.* 32, 143-93.

Broadbent, F.E. (1947). *Soil Sci. Soc. Amer. Proc.* 12, 246-9.

———. (1968*a*). In "Organic Matter and Soil Fertility". *Pontif. Acad. Sci. Scr. Varia.* 32, 61-82.

———. (1968*b*). In "Isotopes and Radiation in Soil Organic Matter Studies". pp. 131-42. IAEA, Vienna.
Broadbent, F.E., and Nakashima, T. (1965). *Soil Sci. Soc. Amer. Proc.* 29, 55-60.
———. (1967). *Soil Sci. Soc. Amer. Proc.* 31, 648-52.
Broadbent, F.E., and Norman, A.G. (1946). *Soil Sci. Soc. Amer. Proc.* 11, 264-7.
Brydon, J.E., and Sowden, F.J. (1959). *Can. J. Soil Sci.* 39, 136-43.
Burns, R.G., El-Sayed, M.H., and McLaren, A.D. (1972). *Soil Biol. Biochem.* 4, 107-8.
Butler, J.H.A., Downing, D.T., and Swaby, R.J. (1964). *Aust. J. Chem.* 17, 817-19.
Butler, J.H.A., and Ladd, J.N. (1969). *Aust. J. Soil Res.* 7, 229-39.
———. (1971). *Soil Biol. Biochem.* 3, 249-57.
Cameron, R.S., Thornton, B.K., Swift, R.S., and Posner, A.M. (1972). *J. Soil Sci.* 23, 394-408.
Campbell, C.A., Paul, E.A., Rennie, D.A., and McCallum, K.J. (1967). *Soil Sci.* **104**, 217-24.
Chahal, K.S. (1968). In "Isotopes and Radiation is Soil Organic Matter Studies". pp. 207-217. IAEA, Vienna.
Chahal, K.S., Mortensen, J.L., and Himes, F.L. (1966). *Soil. Sci. Soc. Amer. Proc.* **30**, 217-20.
Chahal, K.S., and Wagner, G.H. (1965). *Soil Sci* **100**, 96-103.
Chekalov, K.I., and Illuvieva, V.P. (1962). Pochvovedenie No. 5, 40-50.
Cheng, H.H., and Kurtz, L.T. (1963). *Soil Sci. Soc. Amer. Proc.* 27, 312-6.
Cheshire, M.V., Mundie, C.M., and Shepherd, H. (1969). *Soil Biol. Biochem.* 1, 117-30.
———. (1971). *J. Soil Sci.* 22, 222-36.
Chibnall, A.C. (1939). "Protein Metabolism in the Plant". Yale Univ. Press, New Haven.
Chichester, F.W. (1969). *Soil Sci.* **107**, 356-63.
———. (1970). *Plant Soil.* 33, 437-56.
Choudhri, M.B., and Stevenson, F.J. (1957). *Soil Sci. Soc. Amer. Proc.* 21, 508-13.
Chu, J.P.H., and Knowles, R. (1966). *Soil Sci. Soc. Amer. Proc.* **30**. 210-3.
Clark, F.E., and Paul, E.A. (1970). *Adv. Agron.* 22, 375-435.
Clement, C.R., and Williams, T.E. (1964). *J. Agric. Sci.* **63**, 377-83.
———. (1967). *J. Agric. Sci.* **69**, 133-8.
Colom, J., and Wolcott, A.R. (1967). *Plant Soil.* **26**, 261-8.
Costin, A.B., and Polach, H.A. (1969). Atom. Energy, Sydney 12, No. 4, 13-17.
Craswell, E.T., Saffigna, P.G., and Waring, S.A. (1970) *Plant Soil.* 33, 383-92.
Craswell, E.T., and Waring, S.A. (1972*a*). *Soil Biol. Biochem.* **4**, 427-33.
———. (1972*b*). *Soil Biol. Biochem.* **4**, 435-42.
Dahlman, R.C. (1968). In "Methods of Productivity Studies in Root Systems and Rhizosphere Organisms". pp. 11-21. U.S.S.R. Acad. Sci. Nauka.
Dahlman, R.C., and Kucera, C.L. (1965). *Ecology* **46**, 84-9.
Danneberg, O.H. (1971*a*). *Bodenkultur.* 22, 264-78.
———. (1971*b*). Österreichische Studiengesellschaft für Atomenergie, Publication LA-8/1971.
Danneberg, O.H., Haunold, E., and Kaindl, K. (1968). In "Isotope Studies on the Nitrogen Chain". pp. 89-97. IAEA, Vienna.
Davies, R.I. (1971). *Soil Sci.* **111**, 80-5.
van Dijk, H. (1971). In "Soil Biochemistry". vol. 2 (Eds. A.D. McLaren and J.J. Skujins) pp. 16-35. Marcel Dekker, New York.
Domke, K. (1969). *Albrecht-Thaer-Arch.* **13**, 241-6.
Doxtader, K.G. (1969). *Bacteriol Proc.* **69**, 14.
Dragunov, S.S., and Murzakov, B.G. (1970). *Sov. Soil Sci.* 3, 220-5.
Dubach, P., and Metha, N.C. (1963). *Soils Fertil.* 26, 293-300.
Duchaufour, P., and Jacquin, F. (1964). *C.R. Hebd. Séances Acad. Agric. Fr.* **50**, 376-87.
Dudas, M.J., and Pawluk, S. (1969/70). *Geoderma.* 3, 5-17.
Duthien, C., and Chrêtien, J. (1966). Sci. Sol., No. 2, 15-28.
Edwards, A.P., and Bremner, J.M. (1964). *Can. J. Soil Sci.* **44**, 366.
———. (1967). *J. Soil Sci.* **18**, 64-73.
Estermann, E.F., and McLaren, A.D. (1959). *J. Soil Sci.* **10**, 64-78.
Felbeck, G.T., Jr. (1965). *Adv. Agron.* **17**, 327-68.
———. (1971). In "Soil Biochemistry". vol. 2 (Eds. A.D. McLaren and J.J. Skujins). pp. 36-59. Marcel Dekker, New York.
Finch, P., Hayes, M.H.B., and Stacey, M. (1971). In "Soil Biochemistry". vol. 2 (Eds. A.D. McLaren and J.J. Skujins). pp. 257-319. Marcel Dekker, New York.
Flaig, W. (1966). In "The Use of Isotopes in Soil Organic Matter Studies". FAO/IAEA Symposium, Brunswick-Volkenrode. pp. 103-27. Pergamon Press, Oxford.
———. (1967). In "Studies about Humus" (Eds. B. Novak and V. Rypacek). Trans. Int. Symp. "Humus et Planta IV". pp. 81-112.
———. (1969). *Landbauforsch Volkenrode.* **19**, 53-66.
Flaig, W., and Beutelspacher, H. (1968). In "Isotopes and Radiation in Soil Organic Matter Studies". pp. 23-30. IAEA, Vienna.
Ford, G.W. (1968). Ph.D. Thesis, Univ. of Adelaide, Adelaide.

Ford, G.W., and Greenland, D.J. (1968). *Trans. 9th Int. Congr. Soil Sci.* 2, 403-10.
Ford, G.W., Greenland, D.J., and Oades, J.M. (1969). *J. Soil Sci.* 20, 291-96.
Freney, J.R., and Simpson, J.R. (1969). *Soil Biol. Biochem.* 1, 241-51.
Freytag, H.E., and Igel, H. (1970). *Abrecht-Thaer-Arch.* **14**, 495-507.
Fuhr, F., and Sauerbeck,D. (1968). In "Isotopes and Radiation in Soil Organic Matter Studies". pp. 241-50. IAEA, Vienna.
Gallus, H.P.C., and Jennings, A.C. (1971). *Aust. J. Biol. Sci.* 24, 747-53.
Garwood, E.A., Clement, C.R., and Williams, T.E. (1972). *J. Agric. Sci.* **78**, 333-41.
Goldman, R., Goldstein, L., and Katchalski, E. (1971). In "Biochemical Aspects of Reactions on Solid Supports". (Ed. G.R. Stark). Academic Press, New York.
Gorbunov, N.I., Erokhina, G.A., and Shchurina, G.N. (1971). Pochvovedenie No. 7, 117-28.
Grati, V.P., Sinkevich, Z.A., and Klesch, F.I. (1965). Pochvovedenie No. **10**, 72-81.
Greenland, D.J. (1965). *Soils Fertil.* 28, 521-32.
———. (1971). *Soils Fertil.* **34**, 237-51.
Greenland, D.J., and Ford, G.W. (1964). *Trans. 8th Int. Congr. Soil Sci.*, 3, 137-48.
Greenland, D.J., and Oades, J.M. (1975). In "Soil Components". Vol. 1 Organic Components (Ed. J.E. Gieseking). Springer-Verlag, Berlin.
Gupta, U.C. (1967). In "Soil Biochemistry". vol. 1 (Eds. A.D. McLaren and G.H. Peterson). pp. 91– 118. Marcel Dekker, New York.
Haider, K., Frederick, L.R., and Flaig, W. (1965). *Plant Soil.* 22, 49-64.
Harada, T., and Hayashi, R. (1968). *Soil Sci. Plant Nutr.* **14**, 13-9.
Hattori, T. (1969). Soil Biol. No. 11, 30-1.
Haworth, R.D. (1971). *Soil Sci.* **111**, 71-9.
Heard, A.J. (1965). *J. Agric. Sci.* **64**, 329-34.
Hénin, S., Monnier, G., and Turc, L. (1959). *C.R. Acad. Sci. (Paris)* **248**, 138-41.
Hénin, S., and Turc, L. (1949). *C.R. Hebd. Séances Acad. Agric. Fr.* 35, 41-4.
———. (1950). *Trans. 4th Int. Congr. Soil Sci.* **1**, 152-4.
Herbert, D., Phipps, P.J., and Strange, R.E. (1972). In "Methods in Micro-biology". vol. 5B (Eds. J.R. Norris and D.W. Ribbons). pp. 209-344. Academic Press, London.
Himes, F.L., and Bloomfield, C. (1967). *Plant Soil.* 26, 383-4.
Holm-Hansen, O. (1973*a*). In "Estuarine Microbial Ecology" (Ed. L.H. Svetenson). Univ. of South Carolina Press, Columbia.
———. (1973*b*). *Bull. Ecol. Ecol. Res. Comm. (Stockholm)* 17, 215-22.
Hurst, H.M., and Burges, N.A. (1967). In "Soil Biochemistry". vol. 1 (Eds. A.D. McLaren and G.H. Peterson). pp. 260-86. Marcel Dekker, New York.
Hurst, H.M., and Wagner, G.H. (1969). *Soil Sci. Soc. Amer. Proc.* 33, 707-11.
Igel, H. (1969). *Albrecht-Thaer-Arch.* 13, 267-82.
Ignatenko, I.V., Krillova, V.P., and Ponyatovskaya, V.M. (1968). In "Methods of Productivity Studies in Root Systems and Rhizosphere Organisms". pp. 66-71. U.S.S.R. Acad. Sci. Nauka.
Isirimah, N.O., and Keeney, D.R. (1973). *Soil Sci.* **115**, 123-9.
Ivarson, K.C., and Stevenson, I.L. (1964). *Can. J. Microbiol.* **10**, 677.
Jansson, S.L. (1958). *Lantbrukshogsk. Ann.* 24, 101-361.
———. (1960). *Trans. 7th Int. Congr. Soil Sci.* 2, 635-42.
———. (1963). *Soil Sci.* 95, 31-7.
———. (1966). In "The Use of Isotopes in Soil Organic Matter Studies". FAO/IAEA Symposium, Brunswick-Volkenrode. pp. 283-96. Pergamon Press, Oxford.
———. (1968). In "Festskrift til H.L. Jensen". pp. 49-60. Gadgaard Nielson, Lemvig.
———. (1971). In "Soil Biochemistry". vol. 2 (Eds. A.D. McLaren and J.J. Skujins). pp. 129-66. Marcel Dekker, New York.
Jansson, S.L., and Persson, J. (1968). In "Isotopes and Radiation in Soil Organic Matter Studies". pp. 111-24. IAEA, Vienna.
Jenkinson, D.S. (1964). In "Experimental Pedology". (Eds. E.G. Hallsworth and D.V. Crawford). pp. 199-207. Butterworths, London.
———. (1965). *J. Soil Sci.* **16**, 104-15.
———. (1966*a*). In "The Use of Isotopes in Soil Organic Matter Studies". FAO/IAEA Symposium, Brunswick-Volkenrode. pp. 187-97. Pergamon Press, Oxford.
———. (1966*b*). In "The Use of Isotopes in Soil Organic Matter Studies". FAO/IAEA Symposium, Brunswick-Volkenrode. pp. 199-208. Pergamon Press, Oxford.
———. (1966*c*). *J. Soil Sci.* **17**. 280-302.
———. (1968*a*). *J. Soil Sci.* **19**, 25-39.
———. (1968*b*). *J. Sci. Food. Agric.* **19**, 160-8.
———. (1971). *Soil Sci.* **111**, 64-70.
Jenny, H. (1930). Mo. Agric. Exp. Stn. Res. Bull. No. 152.
———. (1941). "Factors of Soil Formation". McGraw-Hill, New York.
———. (1961). Mo. Agric. Exp. Stn. Res. Bull. No. 765.

Keefer, R.F., and Mortensen, J.L. (1963). *Soil Sci. Soc. Amer. Proc.* 27, 156-60.
Keeney, D.R., and Bremner, J.M. (1964). *Soil Sci. Soc. Amer. Proc.* 28, 653-6.
———. (1966). *Soil Sci. Soc. Amer. Proc.* 30, 714-9.
Kereszteny, B., Szabo, I., and Tenk, A. (1963). *Agrartud. Kozl.* 6, 19-24.
Khan, D.V. (1959). *Sov. Soil Sci.* No. 1, 7-14.
Khan S.V., and Sowden, F.J. (1972). *Can. J. Soil Sci.* 52, 116-8.
Knowles, R., and Chu, J.P.H. (1969). *Can. J. Microbiol.* 15, 223-8.
Kononova, M.M. (1966). "Soil Organic Matter". 2nd Edn. Pergamon Press, Oxford.
Kumada, K., and Hurst, H.M. (1967). *Nature* (*Lond*). 214, 631-3.
Kumada, K., and Matsui, Y. (1970). *Soil Sci. Plant Nutr.* 16, 250-5.
Kumada, K., Sato, O., Ohsumi, Y., and Ohta, S. (1967). *Soil Sci. Plant Nutr.* 13, 151-8.
Kuzmin, V.A. (1969). Pochvovedenie, No. 6, 3-11.
Kyuma, K., Hussain, A., and Kawaguchi, K. (1969). *Soil Sci. Plant Nutr.* 15, 149-55.
Ladd, J.N. (1969). *Soil Sci.* 107, 303-6.
———. (1971). *Soil Biol. Biochem.* 4, 227-37.
Ladd, J.N., and Brisbane, P.G. (1967). *Aust. J. Soil Res.* 5, 161-71.
Ladd, J.N., and Butler, J.H.A. (1966). *Aust. J. Soil Res.* 4, 41-54.
———. (1968*a*). *Aust. J. Soil Res.* 7, 241-51.
———. (1968*b*). *Aust. J. Soil Res.* 7, 253-61.
Ladd, J.N., and Paul, E.A. (1973). *Soil Biol. Biochem.* 5, 825-40.
Lammers, W.T. (1971). In "Water and Water Pollution Handbook". vol. 2 (Ed. L. Ciaccio). pp. 593-638. Marcel Dekker, New York.
Lefebvre-Drouet, E., and Meriaux, S. (1966). Sci. Sol. No. 1, 53-63.
Legg, J.O., and Allison, F.E. (1967). *Soil Sci. Soc. Amer. Proc.* 31, 403-6.
Legg, J.O., Chichester, F.W., Stanford, G. and Demar, W.H. (1971). *Soil Sci. Soc. Amer. Proc.* 35, 273-6.
Lein, Z.Y. (1940). Pochvovedenie No. 10, 41-57.
Lichmanova, A.I. (1962). Pochvovedenie No. 6, 58-69.
Lindquist, I. (1967). *Acta Chem. Scand.* 21, 2564-6.
Lowe, L.E., and Parasher, C.D. (1971). *Can. J. Soil Sci.* 51, 136-7.
Lowe, L.E., and Tsang, W.C. (1970). *Can. J. Soil Sci.* 50, 456-7.
McGill, W.B. (1971). Ph.D. Thesis, Univ. of Saskatchewan, Saskatoon.
McGrath, D. (1967). *Nature* (*Lond*). 215, 414.
McKeague, J.A. (1971). *Can. J. Soil Sci.* 51, 499-505.
McLaren, A.D., and Peterson, G.H. (1967). "Soil Biochemistry". vol. 1. Marcel Dekker, New York.
McLaren, A.D., and Skujins, J.J. (1971). "Soil Biochemistry", vol. 2. Marcel Dekker, New York.
Mayaudon, J. (1966). In "The Use of Isotopes in Soil Organic Matter Studies". FAO/IAEA Symposium, Brunswick-Volkenrode. pp. 259-69. Pergamon Press, Oxford.
———. (1968). In "Isotopes and Radiation in Soil Organic Matter Studies". pp. 177-88. IAEA, Vienna.
Mayaudon, J., and Simonart, P. (1958*a*). *Plant Soil.* 9, 376-80.
———. (1958*b*). *Plant Soil.* 9, 381-4.
———. (1959*a*). *Plant Soil.* 11, 170-5.
———. (1959*b*). *Plant Soil.* 11, 181-92.
———. (1963). *Ann. Inst. Pasteur* (*Paris*). 105, 257-66.
———. (1965). *Suppl. Ann. Inst. Pasteur* (*Paris*) 109, No. 3, 224-34.
Mayland, H.F., and McIntosh, T.H. (1966). *Soil Sci. Soc. Amer. Proc.* 30, 606-9.
Miki, K., and Mori, T. (1968). *Tokai-Kinki Natl. Agric. Exp. Stn. Bull.* 17, 67-73.
Monnier, G. (1965). *Ann. Agron.* (*Paris*) 16, 327-400 and 471-534.
Monnier, G., Turc, L., and Jeanson-Luusinang, C. (1962). *Ann. Agron.* (*Paris*) 13, 55-63.
Moore, A.W., and Russell, J.S. (1970). *Aust. J. Soil Res.* 8, 21-30.
Morrison, R.I. (1969). In "Organic Geochemistry". (Eds. G. Eglinton and M.T.J. Murphy). pp. 558-75. Longman, Springer-Verlag, Berlin.
Morrison, R.I., and Bick, W. (1967). *J. Sci. Food Agric.* 18, 351-5.
Mundra, M.C., Bhandari, G.S., and Srivastava, O.P. (1973). *Geoderma.* 9, 27-33.
Mutatkar, V.K., and Wagner, G.H. (1967). *Soil Sci. Soc. Amer. Proc.* 31, 66-70.
Neunylov, B.A., and Khavkina, N.V. (1968). Pochvovedenie No. 2, 103-8.
Oades, J.M. (1972). *Aust. J. Soil Res.* 10, 113-26.
———. (1974). *Trans. 10th Int. Congr. Soil Sci.* 3, 93-100.
Oades, J.M., and Swincer, G.D. (1968). *Trans. 9th Int. Congr. Soil Sci.* 3, 183-92.
Oades, J.M., and Wagner, G.H. (1970). *Geoderma.* 4, 417-23.
———. (1971). *Soil Sci. Soc. Amer. Proc.* 35, 914-7.
Oberlander, H.E., and Roth, K. (1968). In "Isotopes and Radiation in Soil Organic Matter Studies". pp. 251-61. IAEA, Vienna.
Paul, E.A. (1970). *Recent Adv. Phytochem.* 3, 59-104.
Payne, W.J. (1970). *Annu. Rev. Microbiol.* 324, 17-52.
Perraud, A., Nguyen Kha, and Jacquin, F. (1971). *C.R. Hebd. Seances Acad. Paris* 272, 1594-7.

Pershina, M.N., and Avdeyeva, L.A. (1968). In "Methods of Productivity Studies in Root Systems and Rhizosphere Organisms". pp. 154-7. U.S.S.R. Acad. Sci. Nauka.
Persson, J. (1968). *Lantbrukshogsk. Ann.* **34**, 81-2117.
Pinck, L.A., and Allison, F.E. (1944). *Soil Sci.* **57**, 155-61.
Piper, T.J. (1972). Ph.D. Thesis, Univ. of Western Australia, Perth.
Piper, T.J., and Posner, A.M. (1968). *Soil Sci.* **106**, 188-92.
———. (1972). *Plant Soil.* **36**, 595-8.
Pokotilo, A.S. (1967). *Sov. Soil Sci.* 1854-62.
Ribbons, D.W. (1972). In "Methods in Microbiology". vol. 3A (Eds. J.R. Norris and D.W. Ribbons). pp. 297-304. Academic Press, London.
Roulet, N., Dubach, P., Muller-Vonmoos, M., and Deuel, H. (1963). *Z. Pflanzernähr. Düng. Bodenkd.* **101**, 210-4.
Russell, E.J. (1961). "Soil Conditions and Plant Growth". 9th Edn. Longmans Green, London.
Russell, J.S. (1964). *Nature (Lond).* **204**, 161-2.
Sato, O., and Kumada, K. (1967). *Soil Sci. Plant Nutr.* **13**, 121-2.
Satoh, T., and Yamane, I. (1972*a*). *J. Sci. Soil Manure, Japan.* **43**, 41-5.
———. (1972*b*). *J. Sci. Soil Manure, Japan.* **43**, 61-5.
Sauerbeck, D. (1968). In "Isotopes and Radiation in Soil Organic Matter Studies". pp. 57-66. IAEA, Vienna.
Sauerbeck, D., and Fuhr, F. (1968). In "Isotopes and Radiation in Soil Organic Matter Studies". pp. 3-11. IAEA, Vienna.
Scarsbrook, C.E. (1965). In "Soil Nitrogen". (Eds. W.V. Bartholomew and F.E. Clark). pp. 481-502. Amer. Soc. Agron., Madison.
Scharpenseel, C., Ronzani, C., and Pietig, F. (1968). In "Isotopes and Radiation in Soil Organic Matter Studies". pp. 67-72. IAEA, Vienna.
Scharpenseel, H.W., and Krausse, R. (1962). *Z Pflanzernähr. Düng. Bodenkd.* **96**, 11-34.
Schnitzer, M. (1968). *Trans. 9th Int. Congr. Soil Sci.* **1**, 635-44.
———. (1971). In "Soil Biochemistry", vol. 2 (Eds. A.D. McLaren and J.J. Skujins). pp. 60-95. Marcel Dekker, New York.
Schnitzer, M., and Hoffman, I. (1965). *Geochim. Cosmochim. Acta.* **29**, 859-70.
Schnitzer, M., and Khan, S.V. (1972). "Humic Substances in the Environment". Marcel Dekker, New York.
Schreiner, O., and Shorey, E.C. (1908). *J. Amer. Chem. Soc.* **30**, 1295-307.
Seifert, J. (1964). *Folia Microbiol.* **9**, 115-20.
Shields, J.A. (1972). Ph.D. Thesis, Univ. of Saskatchewan, Saskatoon.
Shields, J.A., and Paul, E.A. (1973). *Can. J. Soil Sci.* **53**, 297-306.
Shields, J.A., Paul, E.A., Lowe, W.E., and Parkinson, D (1973). *Soil Biol. Biochem.* **5**, 753-64.
Shurygina, E.A., Larina, N.K., Chubarova, M.A., and Kononova, M.M. (1971). Pochvovedenie No. 6, 35-44.
Simonart, P. (1964). *Ann. Inst. Pasteur (Paris)* **107**, 7-20.
Simonart, P., Batistic, L., and Mayaudon, J. (1967). *Plant Soil.* **27**, 153-61.
Simonart, P., and Mayaudon, J. (1958). *Plant Soil.* **9**, 367-75.
———. (1961). Pedologie Symp. Intern. 2 Appl. Sci. mid Ped. 91-103.
———. (1963). *Ann. Inst. Pasteur (Paris)* **105**, 257-66.
———. (1966). In "The Use of Isotopes in Soil Organic Matter Studies". FAO/IAEA Symposium, Brunswick-Volkenrode. pp. 245-58. Pergamon Press, Oxford.
Simonart, P., Mayaudon, J., and Batistic, L. (1959). *Plant Soil.* **11**, 176-80.
Simpson, J.R., and Freney, J.R. (1967). *Aust. J. Agric. Res.* **18**, 613-23.
Sinha, M.K. (1972*a*). *Plant Soil.* **36**, 283-93.
———. (1972*b*). *Plant Soil.* **36**, 295-99.
Sϕrensen, H. (1963). *Soil Sci.* **95**, 45-51.
———. (1965). *Nature (Lond.)* **208**, 97-8.
———. (1966). In "The Use of Isotopes in Soil Organic Matter Studies". FAO/IAEA Symposium, Brunswick-Volkenrode. pp. 271-4. Pergamon Press, Oxford.
———. (1967). *Soil Sci.* **104**, 234-41.
Sϕrensen, L.H. (1969). *Experientia*, **25**, 40-1.
———. (1972*a*). *Soil Biol. Biochem.* **4**, 245-55.
———. (1972*b*). *Soil Sci.* **114**, 5-11.
Sϕrensen, L.H., and Paul, E.A. (1971). *Soil Biol Biochem.* **3**, 173-80.
Stanford, G. (1968). *Soil Sci.* **106**, 345-51.
———. (1969). *Soil Sci.* **107**, 323-8.
Stanford, G., and DeMar, W.H. (1969). *Soil Sci.* **107**, 203-5.
Steelink, C., and Tollin, G. (1967). In "Soil Biochemistry" vol. 1 (Eds. A.D. McLaren and G.H. Peterson). pp. 147-72. Marcel Dekker, New York.
Stevenson, F.J. (1965). In "Soil Nitrogen". (Eds. W.V. Bartholomew and F.E. Clark). pp. 1-42. Amer. Soc.

Agron., Madison.
———. (1966). *Amer. Oil Chem. Soc.* 43, 203-10.
———. (1967). In "Soil Biochemistry" vol. 1 (Eds. A.D. McLaren and G.H. Peterson). pp. 119-46. Marcel Dekker, New York.
Stevenson, F.J., and Butler, J.H.A. (1969). In "Organic Geochemistry". (Eds. G. Eglinton and M.T.J. Murphy). pp. 534-57. Longman, Springer-Verlag, Berlin.
Stevenson, F.J., and Goh, K.M. (1971). *Geochim. Cosmochim. Acta.* 35, 471-83.
Stevenson, I.L. and Ivarson, K.C. (1964). *Can. J. Microbiol.* **10**, 139-42.
Steward, J.H., and Oades, J.M. (1972). *J. Soil Sci.* 23, 38-49.
Stewart, B.A., Johnson, D.D., and Porter, L.K. (1963). *Soil Sci. Soc. Amer. Proc.* 27, 656-9.
Stewart B.A., Porter, L.K., and Johnson, D.D. (1963). *Soil Sci. Soc. Amer. Proc.* 27, 302-4.
Swaby, R.J. (1968). In "Organic Matter and Soil Fertility". *Pontif. Acad. Sci. Scr. Varia.* 32, 583-613.
Swaby, R.J., and Ladd, J.N. (1962). Trans. Int. Soc. Soil Sci. Soil Conference, New Zealand, Commissions IV and V, pp. 3-8.
———. (1966). In "The Use of Isotopes in Soil Organic Matter Studies". FAO/IAEA Symposium, Brunswick-Volkenrode. pp. 153-9. Pergamon Press, Oxford.
Swift, R.S., and Posner, A.M. (1971). *J. Soil Sci.* 22, 237-49.
———. (1972*a*). *J. Soil Sci.* 23, 50-7.
———. (1972*b*). *Soil Biol. Biochem.* 4, 181-6.
Swift, R.S., Thornton, B.K., and Posner, A.M. (1970). *Soil Sci.* **110**, 93-9.
Swincer, G.D., Oades, J.M., and Greenland, D.J. (1968*a*). *Aust. J. Soil Res.* 6, 211-24.
———. (1968*b*). *Aust. J. Soil Res.* 6, 225-35.
———. (1969). *Adv. Agron.* **21**, 195-235.
Theng, B.K.G., and Posner, A.M. (1967). *Soil. Sci.* **104**, 191-201.
Trofimenko, K.I., and Kizyakov, Yu.E. (1967). Pochvovedenie No. 2, 82-90.
Turchenek, L.W., and Oades, J.M. (1974). *Trans. 10th Int. Congr. Soil Sci.* 2, 65-72.
Vischer, E., and Chargaff, E. (1948). *J. Biol. Chem.* **176**, 715-34.
Wagner, G.H. (1968). In "Isotopes and Radiation in Soil Organic Matter Studies". pp. 197-205. IAEA, Vienna.
Waring, S.A., and Bremner, J.M. (1964). *Nature* (*Lond.*) 202, 1141-2.
Watson, J.R., and Parsons, J.W. (1974). *J. Soil Sci.* 25, 1-8.
Wojcik-Wojtkowiak, D. (1972). *Plant Soil.* 36, 261-70.
Woodruff, C.M. (1949). *Soil Sci. Soc. Amer. Proc.* **14**, 208-12.
Zeller, A., Oberlander, H.E., Roth, K., and Stadler, I. (1966). In "The Use of Isotopes in Soil Organic Matter Studies". FAO/IAEA Symposium, Brunswick-Volkenrode. pp. 275-9. Pergamon Press, Oxford.

Part C

Soil-plant mechanisms and models

CHAPTER 8

Mechanisms and models of water transfer

E.L. Greacen

INTRODUCTION

In the first section of this book, which deals with the properties of semi-arid soils, a review is given of the level of our understanding of soil water relations. The movement and retention of water in non-swelling soils has been described using the physics and mathematics of moisture potential, hydraulic conductivity and diffusivity theory. The development of these ideas is given as they are presently being extended, again by soil physicists in this country, to cover the case of swelling soils. This sets out the fine detail, as it were, of our knowledge of the behaviour of water in soil; it forms the basis of our understanding of the processes of water supply to the plant.

Crop yield depends largely on a continuous supply of water to the photosynthesizing leaves. Plants are able to withstand long periods of moderate water stress but the processes determining yield slow down; at moderate stress (—4 bar) plant turgor is decreased and vegetative growth stops; at higher stress (—9 to —12 bar) assimilation for grain-filling is restricted by stomatal closure and a general decline in the rate of biochemical reactions (Boyer 1973). In a semi-arid environment the supply of water by rain is not continuous, and successful cropping must rely on the performance of the soil as a catching and storing device for this rain.

That water is an important factor in crop production scarcely needs documenting, but it is rewarding to determine even from broad general data the extent to which yield is determined by water supply. Using a simple statistical model Cornish (1950) has done this for yield of wheat in South Australia. He described the water supply factor by three variates: rainfall for the three periods April through May, June through August, and September through November, for the years 1896-1944, and he related these to crude wheat yields for the whole of South Australia. His conclusions are worthy of quotation: "In 267 'hundreds',* out of a total of 296, 50 per cent or more of the observed variation in yield is expressible in terms of the rainfall variates and time, a fact sufficient to excite remark when it is recalled that all other causes of variation without exception are included in the remaining 50 per cent or less."

This result is even more exciting when we consider that any effect of the soil itself in modifying the water factor through storage or availability is not included in this value. These effects can be considerable. Nix and Fitzpatrick (1969) developed a water balance model which included the soil in the water system but merely as a simple fixed reservoir. On a range of sites throughout the Queensland wheat belt they were able to ascribe 66 per cent of yield variability to a water stress index; for individual varieties but in different crops the stress index accounted for 80 per cent of the variability. With an improved model Greacen and Hignett (1976) for well-documented South Australian wheat sites increased this to 86 per cent.

Considering the gross simplification of the water supply system in the above cases relative to the understanding we have of the various processes on the microscale level,

*A hundred is the terminology used in South Australia for a subdivision of a county, roughly 259 km^2.

described in chapter 4, one can be hopeful of obtaining solutions to many of the problems in soil-plant-water relations. On the other hand the soil water regime is determined by many complex factors and their interactions: infiltration with changing soil surface structure, re-distribution of infiltered water with the serious complication of hysteresis. These are difficult problems in one dimension even for the computer. Add to these water removal by the growing plant with a second or third space dimension, and it becomes obvious that considerable simplifications and limitations to scale and extent are required for a useful description of field situations.

This is largely the subject matter of this section of the book. We propose to outline the present state of our understanding of mechanisms of the particular processes, infiltration, drainage and evaporation involving water and crop roots in soil and our competency in describing these and their interactions in numerical models. Since we are concerned here with describing the soil water regime as a factor in crop production, the initial emphasis is on computational models; later analytical models will be discussed as they are used to investigate particular mechanisms in detail. Consideration will be given to the possibilities of refining and simplifying these in the further development of predictive simulation models.

WATER BALANCE MODELS

Early computational models of the soil water regime, of which that of Fitzpatrick and Nix referred to above was a development, were based largely on the water balance concept. The trend has been to test and incorporate new ideas and so develop successive water balance models. In the earliest form as developed by Penman (1949) the soil was considered as a fixed frictionless reservoir; water storage, expressed as surface inches of water, was determined by the water-holding capacity of the root zone, with an extra allowance of one inch of water for upward movement from the soil below. Upper and lower limits of the water-holding capacity were the field capacity and the wilting point water contents respectively. The qualification "frictionless" was an acceptance of Veihmeyer's concept (Marshall 1959) that water was freely available to the plant over the full range of available water.

Fleming (1964) demonstrates the operations involved in a simple water balance model. The estimated evaporation for the chosen time interval is subtracted from the rainfall for that interval and this amount is then added to the soil water to give the soil water storage for the next interval. In Penman's model the input requirements for the two main variables were the monthly losses from the soil reservoir by evaporation and the monthly gains from rainfall. Loss by evaporation from grass, E_g, was shown to vary seasonally between 0.6 and 0.8 E_{tank}, the loss from a free water surface. Penman (1948) had presented a method for calculating E_{tank} from meteorological data, this calculated value being referred to as the potential evaporation, E_o.

Penman's water balance model is limited, not unexpectedly, in several respects. While use of the month as the time interval had advantages in smoothing short-term deviations in E_g relative to E_{tank} and E_o, there can be a serious loss in discrimination when short-term processes are important, e.g. as water supply at a critical stage of growth. Also, it is now thought that evapotranspiration decreases significantly even at moderate plant water stress. In irrigated areas or under dry land conditions, where local meteorological data are available, water balance models are ideally run on a daily basis (Fleming 1964; Baier and Robertson 1966). However where meteorological data are sparse, a reasonable simulation of the soil water regime even in a semi-arid environment can be made with a time interval of one week (Slatyer 1960*b*; Fitzpatrick and Nix 1969).

A weekly interval is usually satisfactory for gradual processes, e.g. the development of different stages of plant growth, but the response of the model is severely dampened to disturbances where the relaxation time is of the order of a day or less. These limitations become evident when the order of performing the operations is reversed. Rose et al. (1972) found that incrementing water storage with weekly rainfall before taking off the loss by evaporation underestimated cumulative evaporation and yield. At the other extreme, processes of infiltration and water movement between soil layers can require time intervals of less than one hour and in this case variable time steps must be used for economic computing.

Potential evaporation

The concept of potential evaporation used in Penman's model has generally been retained (Cowan 1965). By combining a simple energy balance equation, relating evaporation and the available radiant energy, with an expression giving evaporation as a function of the vapour pressure gradient, the Dalton equation, Penman was able to estimate evaporation from a free water surface. The required meteorological data are mean air temperature, mean vapour pressure, mean wind speed and mean duration of sunshine. Although obtained for sunken pan data in England Penman's equation has been used successfully for several widely spaced stations in Australia, Katherine and Yass (Slatyer 1960*a*, 1960*b*), Griffith (Fleming 1964), and Aspendale (McIlroy and Angus 1964).

The analysis has been further developed to cover evaporation from different surfaces (Penman 1952). Several important parameters, hidden for simplicity in constants in Penman's original equation, depend on the properties of the evaporating surface. The most obvious of these are the reflection coefficient or albedo (taken by Penman as 0.05 for water) in the energy balance term, and the transfer coefficient, h, in the vapour pressure term, which depends in part on the roughness of the surface; h is usually determined empirically, although an analytical expression has been given for it by Tanner and Fuchs (1968).

In an attempt to include the effect of varying surface properties and conditions some of these parameters have been retained in more recent but similar forms of the combination formula. That of McIlroy (Slatyer and McIlroy 1961), which is finding more frequent use (McIlroy and Angus 1964; Holmes and Watson 1967; Dunin and Costin 1970), is written

$$LE_o = \frac{S}{S+\gamma}(R_n - G) + h(D_z - D_o), \qquad (1)$$

where E_o is the potential evaporation and L is the latent heat of vaporization. On the right-hand side the first term gives the radiant energy available for evaporation, modified slightly by temperature through $S/S+\gamma$, R_n is the net radiation and G is the soil heat flux. The second term covers surface geometry effects on wind through the transfer coefficient h and the effects of the humidity of the environment through D_z; $D_z - D_o$ is the difference in wet bulb depression between screen height and the evaporating surface. McIlroy and Angus (1964) claim that any large-scale oasis effect is absorbed in the second term, provided that on-site data and a matching h is used for the crop and the locality.

Because of differences in albedo between vegetation and a water surface (0.25 for crops versus 0.05 for water) net radiation can be 25 per cent higher for water than for grass. Under conditions of low advection, European environments, E_g as noted earlier lies about 0.8 E_{tank}. Generally though, under Australian conditions, while Penman's equation describes tank evaporation surprisingly well, E_g is often about 5 per cent higher than E_o. This is generally explained on the failure of the second term to account fully for advection and surface roughness. As h is empirically determined the problem of fulfilling the requirement of non-limiting water supply and the overall complexity of vegetation evaporation makes it suspect.

This complexity of the vegetation surface and the overall empiricism of some of the terms has inclined people towards Penman's original method of using evaporimeters, which have a well defined surface, as a reference. But this practice too has disadvantages. In evaporation from water the G term becomes relatively significant compared to the crop and although monthly or even weekly totals are correlated, there is little short-term (hourly or daily) correlation between E_g and E_{tank}. Conversion factors between types of vegetation and evaporimeters and even between types of evaporimeters can be strongly location- and season-specific and any significant water balance project requires laborious and well-controlled calibration to determine them. For crop and pasture evaporation, conversion factors are best measured by means of some form of lysimeter (Rose et al. 1972) but here again strict control of geometry is important; also, replication is likely to be essential as cropped areas are plagued by hot spots (Holmes and Watson 1967) and thermal funnels (Byrne and Rose 1972).

While the above methods estimate potential evaporation with reasonable confidence, when one is dealing with a recently developed region the meterological records are often so limited that recourse has to be made to simple expressions for evaporation based on the saturation deficit (Prescott 1934). This approach works because there is a high correlation between saturation deficit and net radiation (Fleming 1972), the two terms in equation (1). Fitzpatrick (1963) has shown that E_{tank} is related to vapour pressure and a modified temperature value over much of the continent of Australia. In figure 1 from data by Hignett (1973), the relationship between Class A pan evaporation and this modified saturation deficit is given for a restricted region of Australia, the South Australian wheatlands.

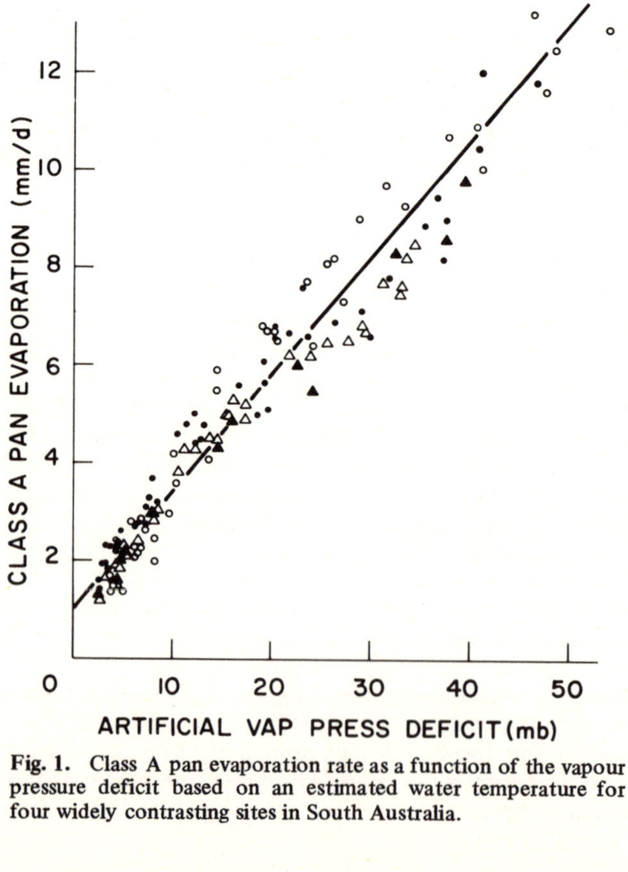

Fig. 1. Class A pan evaporation rate as a function of the vapour pressure deficit based on an estimated water temperature for four widely contrasting sites in South Australia.

Potential evapotranspiration with crop development

The original idea of potential evaporation of a pasture, which applied to a uniform and well-covered turf, has been extended to the idea of potential evapotranspiration for the crop at different stages of development. Where Penman allowed for a winter and summer season for pasture, Slatyer considered four stages of crop development (for sorghum in the Northern Territory of Australia). These stages were emergence, establishment, period of maximum growth, and finally maturation with a crop factor, f, of 0.4, 0.6, 0.8 and 0.2 respectively, based on Penman's coefficient for turf and his estimates of bare soil evaporation. For water supply not limiting this gives the potential evapotranspiration E_t of the crop as $E_t = fE_o$. This coefficient, in one value, takes account of the leaf area of the crop, evaporation from the soil, and changes in structure and morphology of the crop.

The stage of development of the crop can sometimes be predicted from various meteorological parameters, e.g. Fitzpatrick and Nix (1969) for day-neutral varieties of wheat found that accumulated potential evaporation, which correlates closely with total radiation and temperature over the winter growing season, predicted anthesis better than day-degrees summation or photothermal units. However French (personal communication) finds that this approach does not work for the early growth stages of wheat in South Australia.

Actual evaporation

While the practice of first estimating potential crop evaporation has been retained, in more recent water balance models there have been significant departures from the original Penman model in estimating actual evaporation loss. Where Penman originally assumed that full potential evaporation was maintained until the water supply in the root zone was exhausted, i.e. water content less than wilting point, present models modulate evaporation loss according to the soil water content. Monteith (1965) accounted for the reduction in evaporation below potential by incorporating a stomatal water pathway into the combination formula. In a similar approach McIlroy(1966) first calculates potential evaporation E_t, using equation (1) and then subsequently modifies it according to soil water content as

$$E_a = E_t/[1 + \frac{\gamma}{S+\gamma}(\frac{h}{h_i})] \qquad (2)$$

The internal transfer coefficient, h_i, of the soil-root system depends on soil water content as $h_i = C(\theta - \theta_c)$, where C and θ_c are constants for the particular system and θ_c is the critical water content beyond which there is no significant uptake. These parameters can be determined experimentally with a particular soil-crop system. Dunin and Costin (1970) give an example of their evaluation in a water use study on pastures.

The potential transmission concept based on the conductance of the soil-root system, initially demonstrated by Denmead and Shaw (1962) (fig. 2), has been fully developed and amplified by Fleming (1964, 1972) as a method for modulating potential evaporation in relation to soil water content and the potential evaporation rate. The potential transmission rate, E_m (Fleming 1964), is the maximum rate at which the soil conductivity, the root concentration, and the plant properties allow water to be supplied to the plant evaporating surface. Hydraulic conductivity decreases rapidly with decreasing soil water content and E_m decreases accordingly. Cowan (1965) has given a detailed analysis of the process.

Thus there is an upper limit of crop evaporation, E_m, the maximum transmission rate; while E_t, the potential evaporation for the particular growth stage, is less than E_m, actual evaporation E_a will equal E_t. As the soil store is depleted, E_m falls, or, if E_t becomes greater than E_m due to a weather change, then E_a will fall below E_t and become equal to E_m. This is referred to as the falling rate phase. Fleming (1964) has summarized the

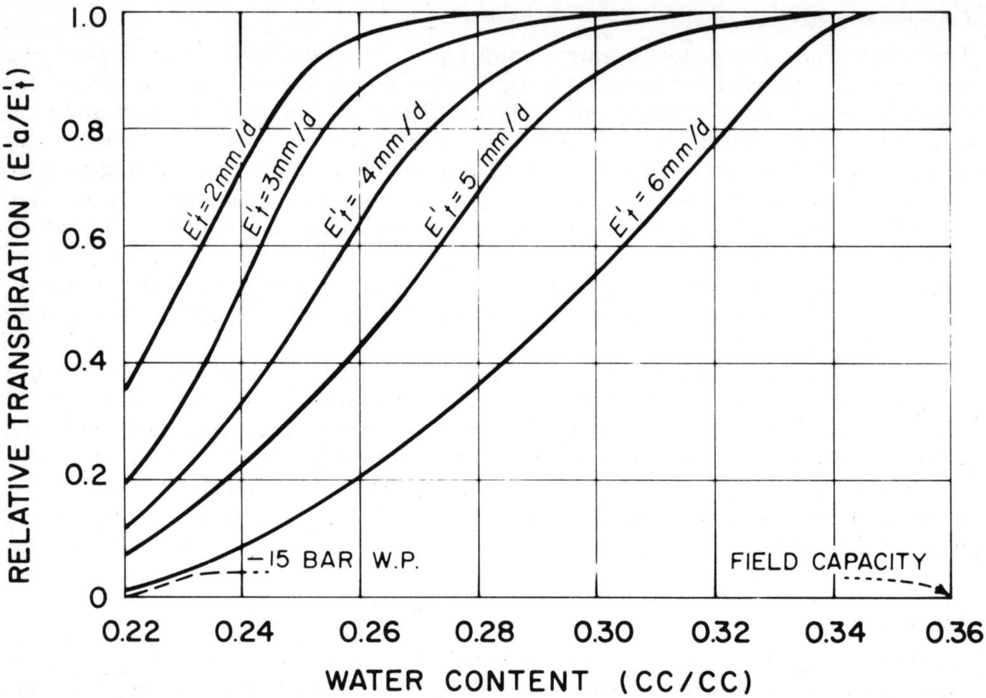

Fig. 2. Relative transpiration rate for corn as a function of the volumetric water content, θ, and the potential transpiration rate. E' refers to transpiration loss only, as the soil surface was covered (after Denmead and Shaw 1962).

Table 1. Table of soil moisture factor P

Soil moisture regime		Range of free water evaporation E_o in mm/day			
	% Range	0–3.0	3.1–4.0	4.1–6.0	6.1 and above
A	100–75	1.0	1.0	1.0	1.0
B	74–50	1.0	1.0	0.8	0.6
C	49–25	1.0	0.7	0.5	0.35
D	24–0	0.5	0.3	0.25	0.15

process as $E_a = PE_t$, where P is the soil water factor with a maximum value of 1 (table 1). P is seen to depend on E_o and the available soil water content.

Table 1 is based on the assumption that E_m at field capacity is always greater than E_o, and there is considerable evidence that this assumption is reasonable for many crops. But maximum rates of uptake for wheat roots have been measured at ~0.002 cm^3/cm length of root per day (Greacen and Hignett 1976; Walter and Barley 1974) so that with limited roots lengths (< 200 cm/cm^2) E_m can restrict E_a even at high soil water potentials.

MOVEMENT OF WATER IN SOIL

The hydrologic parameters of soils

Solutions to the continuity equation (equation [6] in chapter 4) form the basis for realistic modelling of the movement of water in soil. The required soil parameters are the water characteristic (the suction-water content curve) and the conductivity-water content function. As early as 1911 Green and Ampt stressed the importance of characterizing soils and modelling water movement by means of potential and conductivity parameters, and this remains true today. Effective description of soil-water processes depends on a sufficient description of these parameters. Methods of measurement have been discussed fully in Klute (1972) and will not be dealt with here, except to refer the reader to a field method for conductivity based on water balance study described by Rose, Stern and Drummond (1965), a comparison of field and laboratory determinations by Greacen, Ponsana and Forrest (1974), and core sampling techniques described by McIntyre (1974).

For analytical laboratory models measurements are usually made over the complete range of water content. The techniques are lengthy and demand expertise. For field soils variability is high and less precise methods usually suffice: for a red-brown earth subsoil the coefficient of variation for values of water content at a given suction is 10 per cent (Greacen, Ponsana and Forrest 1974). Nielsen, Biggar and Corey (1972) report a variation of 20 per cent for a supposedly "uniform" alluvial soil, giving an uncertainty of field water content θ at a particular potential of 0.05 g/cc. Rogowski (1972) suggests that values of the coefficient of variation of 15 per cent at -15 bar and 10 per cent at the wet end (air-entry) are to be expected for uniform soils. These criteria set 10 per cent as a minimum value of error for the estimate of water storage capacity from a reasonable sampling effort.

Variability in measured values of conductivity is higher than that of water content. The uncertainty in water content reported by Nielsen, Biggar and Corey above, 0.05 g/cc, sets an uncertainty of an order of magnitude on conductivity in the value of conductivity itself; this stresses the need for caution when inferring conductivity from measured values of soil water suction. Referring to variability in hydraulic conductivity, K, Rogowski suggests an antilog standard deviation of two (2), i.e. 68 per cent of the measured values of K to lie between 2K and K/2, as a criterion for uniformity. Thus, depending on the magnitude of K, this criterion sets the lower attainable limit of error in the estimation of loss by drainage from the water store.

Various models have been proposed for calculating conductivity from the soil water characteristic. In view of field variability these give a sufficient description of the conductivity function when they are matched at some measured value, the measured saturated conductivity being the point commonly used for matching. For example the model of Marshall (1959) modified by Green and Corey (1971) gives K as a function of θ from the $\psi(\theta)$ curve as

$$K(\theta)_i = K_o \frac{\sum_{j=i}^{m} [(2j + 1 - 2i)\psi_j^{-2}]}{\sum_{1}^{m} [2j + 1 - 2i)\psi_j^{-2}]} \tag{3}$$

where m is the selected number of water content classes, K_o is the saturated conductivity, and ψ is the matric potential of the soil water.

Rogowski (1971) goes one step further and models the water characteristic to be used in Marshall's model. Using two measured values of water content, one, θ_{15} at -15 bar and the other θ_e at air entry suction, ψ_e, he gives

$$\psi = \psi_e - 1 + \exp[(\theta - \theta_e)/\alpha] \tag{4}$$

where for $\theta < \theta_e$

$$\alpha = (\theta_{15} - \theta_e)/\ln(\psi_{15} - \psi_e + 1) \tag{5}$$

Compared to other models this gives a more realistic representation of the pore sizes between field capacity and saturation which dominate the drainage term of the water balance. Its performance in a red-brown earth subsoil is demonstrated in figure 3 based on data by Greacen, Ponsana and Forrest (1974).

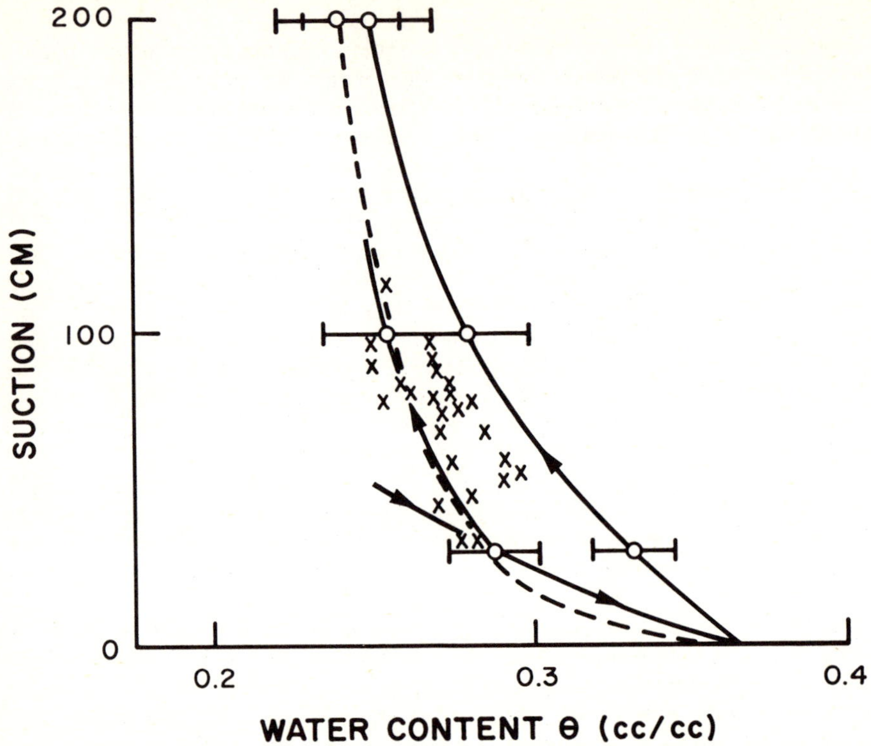

Fig. 3. Water characteristic of a red-brown earth subsoil determined in the laboratory on natural cores (-o-) and from field tensiometry (x). Curve (- - -) was calculated by Rogowski (1971) model using two points, $\Psi = -200$ and $\Psi e = 0$.

Other more empirical models of the water characteristic have been developed for analytical or computational convenience, the most common of which is the simple linear log-log relationship for the K(θ) and the $\psi(\theta)$ functions as $\psi = a\theta^{-m}$, or alternatively an exponential relation for the conductivity function as $K = ae^{b\theta}$.

Infiltration

Infiltration as a process modifying rainfall input has been largely ignored in water balance work concerned.with crop production. Recent models simulating the soil water regime have used the "instantaneous overflow" system: infiltration is considered to be non-limiting; rainfall for the time interval, day or week, is added directly to soil storage until the maximum water content is reached, after which any excess is considered as "overflow" to the next layer or discarded as drainage when the storage is filled. Slatyer (1960*a*) justifies this approach in the Southern Tableland region of New South Wales, where in summer, when soil water is low, even heavy falls of rain produce only one-quarter as much run-off as do rains in winter and spring when the soil store is full. In this region rainfall is distributed uniformly throughout the year. In a summer-rainfall sub-tropical region of central Queensland, Fitzpatrick and Nix (1969) found that the same "overflow" model predicted soil water stored at the end of the fallow period.

In the above regions run-off may be small but, while the overflow system continues to be widely used, in general there is a need for a more definitive treatment of infiltration. Slatyer (1960*a*) stresses the possible importance of limiting infiltration giving rise to localized concentration of water, particularly on sites of uneven slope; Rose et al. (1972)

entertain doubts regarding non-limiting infiltration on sites in tropical Australia; Arndt (personal communication) has developed a method for measuring rainfall acceptance on Australian wheat-land soils. These workers stress the need for a more critical assessment of the importance of infiltration in agronomic hydrology.

In chapter 4 of this review a solution to the continuity equation is given for the vertical infiltration process (Philip 1957*b*) as

$$i = St^{1/2} + At \qquad (6)$$

and differentiating,

$$I = \tfrac{1}{2}St^{-1/2} + A \qquad (7)$$

where I is the instantaneous infiltration rate, i is cumulative water infiltrated depending on time t, for a condition of full supply. S and A are soil parameters, the sorptivity S being a function of the initial soil water content.

A model simulating run-off (Chapman 1968) uses this equation to obtain the mean infiltration rate for the selected time interval; the values of S and A are reset after any interval when the supply is not sufficient to meet the infiltration rate. He suggested that values of S and A be obtained from the soil water characteristic and the conductivity function determined in the laboratory on samples of the surface soil. Talsma (1969) has recently described a more direct method for measuring these parameters. To measure sorptivity a standard 30 cm diameter infiltrometer ring is used in the field to obtain the initial rapid intake of water when the process is almost completely sorptivity dominated. Conductivity is measured on the core obtained by breaking the ring out of the ground.

This particular application of Philip's equation has several objections (Fleming 1973). As the time interval is reduced the rate becomes infinitely high and the infiltration behaviour depends on the time interval chosen. Furthermore the strong dependence of sorptivity on water content accents the problem of assigning a value to the depth of the surface soil, since current models assume a uniform instantaneous distribution of infiltered water throughout the surface layer.

In contrast to the time-dependent theoretical models for infiltration, models have been developed along the lines of the empirical Horton (1933) model in which a constant infiltration capacity is reached, but with the difference that the rate is expressed as a function of the amount of water infiltrated; Holtan (1961) uses the expression

$$F = a(W_m - W)^{1.4} + F_c \qquad (8)$$

where F is the infiltration rate, a is an index of surface-connected porosity, $(W_m - W)$ is the available water storage remaining in the upper soil zone and F_c is the hydraulic conductivity. Boughton (1966) has used a similar water content dependent approach in a model designed to make use of daily rainfall data. Rainfall acceptance is limited to a potential daily infiltration rate based on the expression

$$F = F_c + (F_o - F_c)e^{-bW} \qquad (9)$$

where F_c is the daily infiltration rate at saturation, F_o is the rate at zero soil water storage W, and b is a constant.

The water content dependent approach has considerable advantage in computation and in handling variable rainfall intensity input, and both models described above have been used with some success. Nevertheless there is considerable difficulty in evaluating the different parameters, in particular the depth and thus the water content of the upper soil zone wetted by the rainfall input.

Because of the empirical nature of the above models, the inclination is to use Philip's equation or to develop new models based on the differential equation for water movement in soils. Solutions of the finite differential form of equation (6) of chapter 4 have been developed for computers (Richtmyer 1957; Molz and Remson 1970) and used successfully for a wide range of boundary conditions: limited supply for infiltration (Rubin and Steinhardt 1963), surface crusting effects (Hillel and Gardner 1970), layered profiles (Hanks and Bowers 1962). The complexity of these models precludes them from use in agronomic computational models, but attempts are being made to incorporate their physical reality into simpler layered models, and to use the more sophisticated model for checking the validity of the simplification.

Following this approach the number of soil layers in water balance models has been increased. In place of the two-layer models (Chapman 1968; Greacen and Hignett 1976) multi-layered or segmented models have been developed in which the soil water characteristic and the conductivity function for each layer controls the rate of movement of water in it. Infiltration is modelled by assuming a saturated water content (Dutt, Shaffer and Moore 1972) or a zero water potential (Warrick, Nielsen and Biggar 1971) at the soil surface. Problems arise in the choice of the time interval (Hanks and Bowers 1962) but these and other problems inherent in segmented models will be discussed later.

The simple overflow approach to infiltration and indeed the classical infiltration process assumed in multi-layered models are both based on water movement in an idealized uniform porous medium; their application to field situations cannot be accepted without question. Current work using applied (Blake, Schlichting and Zimmermann 1973) or natural (Allison and Hughes 1974) tritium as a tracer shows that wetting of the profile during infiltration need not occur progressively from the surface layer downwards, but can occur simultaneously throughout the profile via the macrostructure of the soil. This concept of the infiltration process is contained in the surface-connected porosity term, a, in Holtan's equation above. While it has the same form as Philip's analytical expression it is a complete contradiction in concept. Even in fine-structured soils, water repellence (Bond 1968) or the presence of surface crusts or deeper restrictions to flow reduce the chances of idealized soil wetting (Raats 1973).

Redistribution

Movement of water within the profile following gain by precipitation and loss by drainage and evaporation is important agronomically in that it determines the availability of water and the movement and availability of certain plant nutrients. Development of multi-layered models has been directed mainly towards simulating this process of redistribution. Hysteresis in the soil water characteristic presents considerable difficulty even in uniform soil (see chapter 4) but hysteresis is usually ignored in computational models on the grounds that field variability and crop variability are usually such that error introduced by neglecting hysteresis is relatively unimportant. This simplification should be treated with caution however in analytical models.

Multi-layered models simulating transfer processes in soil have been developed by Dutt, Shaffer and Moore (1972) and Molz and Remson (1970). The basic technique has been discussed at length by de Wit and van Keulen (1972), but it will be useful to describe them briefly here and to consider what use is being made of them in current water balance models.

The soil profile is divided into a number of layers of thickness χ, each considered as homogeneous with respect to hydrologic properties. Neglecting gravity for reasons of simplicity, the flow of water between two adjacent layers is calculated from the difference in water content, $\Delta\theta$, and the mean diffusivity $\bar{D}$ as

$$d\theta_n/dt = \bar{D}(\theta_n - \theta_{n+1})/(\tfrac{1}{2}\chi_n + \tfrac{1}{2}\chi_{n+1}).$$

These flows are applied over the time interval to give the water contents for the next interval. $\bar{D}$ is the effective or average diffusivity between the layers. There are a number of ways in which $\bar{D}$ can be obtained and the problem is to decide which to use. The arithmetic mean can be taken as $[D(\theta_n) + D(\theta_{n+1})]/2$, or $\bar{D}$ can be taken as the diffusivity at the average water content $(\theta_n + \theta_{n+1})/2$; Passioura and Cowan (1968) use the diffusivity at the geometric mean water content $\theta_n{}^{1/2}\theta_{n+1}{}^{1/2}$. Another alternative is to treat the diffusivities of the two layers in the same way as electrical conductivities as $1/\bar{D} = (1/D_n) + (1/D_{n+1})$, i.e. a harmonic mean. Philip has used a weighted mean diffusivity giving higher weighting to the wetter zone where flow is greatest,

$$\bar{D} = (D_n\theta_n + D_{n+1}\theta_{n+1})/(\theta_n + \theta_{n+1}).$$

Using a diffusivity-water content function from Hanks and Bowers (1962) de Wit and van Keulen (1972) show that these different methods of averaging can give forty fold differences in the computed flow rates, ranging from the lowest for the harmonic mean to the highest at the weighted mean value. To resolve this problem they simulated the infiltration process for this soil using a more precise model based on a solution of the difference form of the basic equation for water movement and compared the result with that obtained by the multi-layered model using the different methods of averaging. They show for the particular soil that the arithmetic average closely followed by the weighted average gave the closest approximation. It is likely however that there can be no one best method of averaging applicable to all soils since this will be determined by the diffusivity-water content relationship. This is an obvious area requiring further research effort in modelling transfer processes.

In recently developed layered models redistribution has been treated either by the simple overflow process or is controlled as above in some way by the diffusivity and conductivity functions. Chapman (1968) moves water between the two layers at the rate

$$Q = a_o(\theta_n - a_1\theta_{n+1} - a_2)\exp(a_3\theta_{n+1}) \qquad (10)$$

where the term in the brackets represents the difference in potential between the two layers in terms of θ, and the exponential term the conductivity of the lower soil layer. This layer is used quite arbitrarily as being the more stable. Paltridge (1970) on the other hand chooses the diffusivity of the upper one of any two layers as an approximate control of redistribution. Ross (1971) treats the two layers as conductances in series and obtains the overall conductance g_{n+1} as $1/g_{n+1} = (\chi_n/2K_n) + (\chi_{n+1}/2K_{n+1})$, i.e. the harmonic mean referred to above.

Thus the flow from layer n to $n+1$ is

$$Q_n = a\,g_{n+1}(\psi_n - \psi_{n+1} + \tfrac{1}{2}(\chi_n + \chi_{n+1}))] \qquad (11)$$

where ψ is the matric potential of the soil water;
the time derivative of water content then is

$$d\theta_n/dt = (Q_n - Q_{n+1} - U_n)/\chi_n \qquad (12)$$

where U_n is the root extraction term for the layer n.

It is obvious if the time step, Δt, is too large, the change in θ_n given by $(d\theta_n/dt)\Delta t$ could increase the value above the maximum allowable "field capacity" value and would oscillate. This is overcome by an "overflow" approach which passes any excess input through to the next layer. Stability in this particular model is also improved by the use of the harmonic mean conductivity which tends to underestimate conductivity. The problem of time interval and stability has been considered at some length by de Wit and van Keulen.

No one of these models has been tested exhaustively but it is obvious that there is a continued stress on simplification in the face of field error, cutting corners wherever possible, but striving to retain the basic physical principles.

Drainage

In simple water balance models the assumption that infiltration is non-limiting and all rainfall is accepted by the surface layer prohibits in effect useful simulation of the drainage process. "Overflow" models assign a maximum water-holding capacity to the profile or soil layer and rainfall input in excess of this is considered as loss by drainage. In real life since sorptivity decreases markedly with increasing water content, much of this drainage loss is in fact surface drainage or run-off. The maximum water-holding capacity approach, an extension of the field capacity concept, is essentially a simplification of the drainage process: for values of θ greater than field capacity, $\theta_{f.c.}$, K approaches infinity; for $\theta \leqslant \theta_{f.c.}$, K is assumed to be equal to zero. This situation is closely approached in coarse-textured soils where underdrainage becomes negligible shortly after infiltration ceases; but in the case of finer textured soils significant drainage can continue for prolonged periods.

It is also worth pointing out that with a continuous steady but restricted infiltration, e.g. through a surface crust, θ will increase throughout the profile until the increased conductivity is able to cope with the input; in this case the effective water-holding capacity is increased. The same mechanism of adjusting "field capacity" would operate if the infiltration input were intermittent either with or without loss by evaporation, an equilibrium water content profile being set up between the infiltration and drainage rates.

To overcome the limitations of these storage models, particularly where a simulation of drainage is required, e.g. in plant nutrient or salt-leaching situations, various drainage models have been developed on the basis of Darcy's equation applied to the whole profile as a unit,

$$v = -K(d\psi/dz + 1) \qquad (13)$$

where v is a soil water flux (cm/day) and K is the mean hydraulic conductivity (cm/day). In a uniform soil during drainage $d\psi/dz$ is close to zero and the column drains uniformly under unit hydraulic gradient. Black, Gardner and Thurtell (1969) arrive at the solution that the drainage flux at the lower boundary is a function of the content of stored water, which Wilcox (1960) and Ogata and Richards (1957) had found experimentally. For a covered lysimeter containing Plainfield sand they found that

$$v = 0{\cdot}35e^{0{\cdot}7(W-15{\cdot}0)} \qquad (14)$$

where W is the stored water in cm depth of water.

This simplified treatment of drainage was developed further by Davidson et al. (1969). Using an exponential $K(\theta)$ function, $K = K_o e^{a(\theta-\theta_o)}$ they arrive at an expression for drainage at depth $x = L$ as

$$v_L = K_o/(1 + aK_o t/L) \qquad (15)$$

This model predicted drainage where conditions of uniformity were met in Yolo clay loam, and reasonably well in a highly variable silty clay profile where the required condition of equal drainage from all points in the profile was met and K_o was taken as a mean value for the whole profile. In a loamy sand profile where the condition of equal drainage did not hold the performance was poor. Both of these models were developed for covered soil but were found to apply where moderate evaporation was allowed. They would not be expected to apply under high evaporation from a fully transpiring crop.

Drainage has been modelled successfully using numerical solutions to the flow equation (Wang and Lakshminarayana 1968) and consequently there is a move to develop models based on the conductivity and potential function in place of the empirical field capacity concept. Chapman (1968) using a two-layer model defines a field capacity value and water in excess of this in the lower layer is counted as direct drainage, i.e. a simple overflow system; below field capacity he assumes that drainage occurs at a rate proportional to the unsaturated conductivity. Greacen and Hignett (1976) again in a two-layer model use an exponential $K(\theta)$ function based on Rogowski's (1971) model above as an option to the field capacity model. In the more sophisticated multi-layer models single-valued functions of D and K with θ are employed and the lower boundary condition is set at a constant high water content simulating conditions above a static water table. This allows movement of water both upward and down through the lower boundary (Dutt, Shaffer and Moore 1972).

Evaporation from bare soil

Water is lost from the soil by direct evaporation from the surface both during the fallow period and during the final stages of cultivation preparatory to seeding. In semi-arid regions sound cultivation practice is based on methods aimed at minimizing this loss. French (1963) has found that in South Australia from 2 to 20 per cent of the rainfall over a 10-month period may be conserved under fallow.

Recent attempts to model this component of the water balance, for example on a weekly time interval budget, set criteria which determine the ratio of actual evaporation from the soil, E_s, to potential evaporation. The criteria are based on the amount of rainfall received during the interval and are evaluated by trial and error to fit measured values of soil water content (Fitzpatrick and Nix 1969).

The physics of evaporation from a bare soil surface is reasonably well established. From solutions of the flow equation (Philip 1957*a*; Gardner 1959) three stages can be distinguished: (i) the wet or constant rate phase, where E_s is the same as that for a saturated soil surface, (ii) at intermediate water contents, the falling rate phase, where E_s is almost independent of E_o and depends on the physical properties of the soil and the soil water distribution and (iii) for dry soil, where thermal effects are dominant but in terms of magnitude are often negligible.

Using a simple solution of the flow equation, neglecting the effect of gravity and assuming constant diffusivity, Black, Gardner and Thurtell (1969) arrive at an expression for E_s as

$$E_s = (\theta_i - \theta_o)(\bar{D}/\pi t)^{1/2} \tag{16}$$

where θ_i and θ_o refer to the initial and surface soil water contents. Since $\bar{D}$ depends on θ, a weighted mean diffusivity for desorption is used, as

$$\bar{D} = \int_{\theta_o}^{\theta_i} \omega D(\theta) d\theta \Big/ \int_{\theta_o}^{\theta_i} \omega\, d\theta, \tag{17}$$

where ω is the weighting function. For desorption Crank (1956) gives $\omega = (\theta_i - \theta_o)^{0 \cdot 85}$.

Cumulative evaporation from Plainfield sand, obtained by integrating equation (17) and using laboratory values for $\bar{D}$ and θ_i, gave a value of 7·5 cm compared to a lysimeter-measured value of 7·9 cm. The data do not fit the linear time$^{1/2}$ relationship exactly but the prediction is surprisingly accurate. Hingston et al. (personal communication) obtained an equally good performance with the same model on several Western Australian sands with the still further approximation of using the Plainfield sand parameters for their soils. It appears that cumulative evaporative loss from coarse-grained sand is strongly conservative.

In contrast to sandy soils where the initial wet phase is brief enough to be ignored and evaporation control is within the soil proper, for medium-textured soils with relatively high water transmission rates there is considerable wet-phase evaporation where the evaporation potential of the environment is rate controlling. For these soils Gardner and Gardner (1969) found that a solution of the flow equation with an exponential diffusivity function effectively described both the prolonged initial constant rate phase and the falling rate phase of evaporation. They simulated mulching effects by controlling the evaporation rate E and concluded that mulching must reduce E below potential evaporation by 50 per cent or more to be effective. Gardner and Hillel (1962) give an expression for predicting the end of the constant rate phase based on the total water content of the profile. When W of a soil evaporating at E reaches a value equal to the expression below, the constant rate phase is complete:

$$W = \frac{L}{a} \ln(1 + EaL/2D_o) \tag{18}$$

L is the depth of the soil column and a is a constant from the chosen $D(\theta)$ function for the soil.

The analysis of evaporation from bare soil described above applies to a uniform profile. The problem is made more difficult in agricultural soils by the presence of the surface cultivated layer. In multi-layered models this layer may be treated as a distinct surface layer with peculiar conductivity and water characteristics.

But the basic hydrologic properties of cultivated soils are not readily established and are frequently transient in nature. Furthermore, in coarse-tilled soil, pore spaces can be large enough to allow significant transfer of water vapour by air turbulence within the soil (Farrell, Greacen and Gurr 1966). These problems are often avoided empirically in water balance models for particular localities by measuring the evaporation from soil surfaces under different management practices. In figure 4 evaporation is given for a

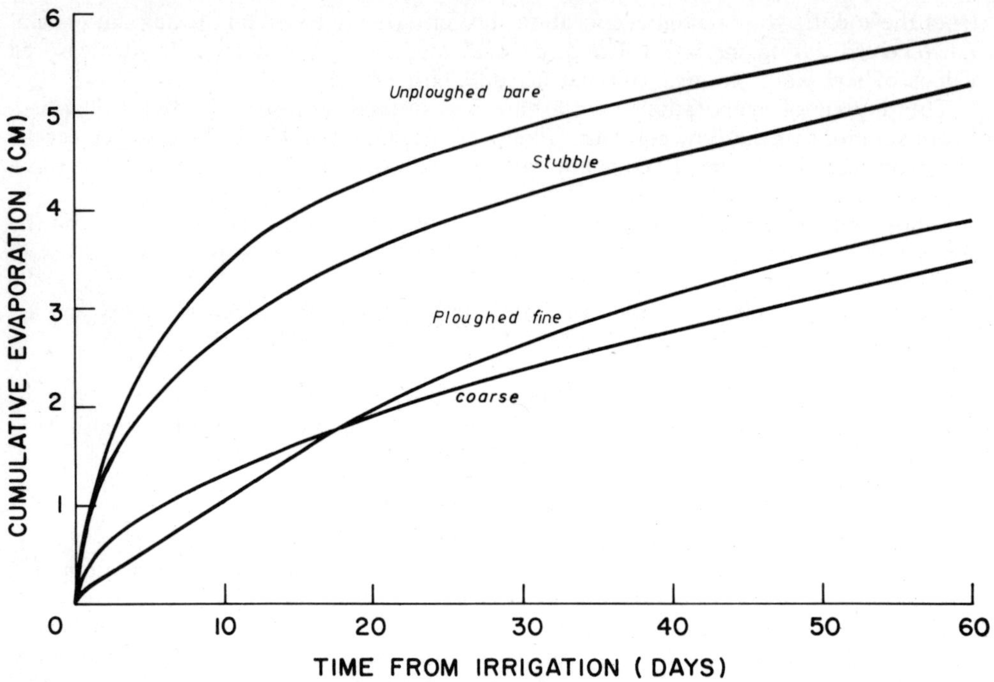

Fig. 4. Cumulative evaporation versus time from a red-brown earth under different surface conditions.

red-brown earth loam, under coarse and fine tilth, and bare soil undisturbed (Greacen and Hignett 1976). Different drying-rate phases are apparent and can be specified at particular levels of accumulated evaporation.

THE UPTAKE OF WATER BY PLANTS

The processes of water movement and distribution through the soil profile have been considered in response to changes in water potential resulting from infiltration and drainage events. These have been treated as problems in one-dimensional flow with the soil properties uniform in the horizontal plane and gradients occurring only in the vertical direction. Extraction of water from the soil by crop plants, when not widely spaced, can also be treated effectively on the macro scale as a one-dimensional vertical flow process but considerations of defining control mechanisms require analysis at a lower level if effective models of the transpiration process are to be developed.

Transpiration involves the flow of water through the soil to the root surface, through the plant to the site of evaporation in the leaves and then as vapour to the outside air. The rate of flow can be expressed in terms of the water potential and the hydraulic resistance of the plant and the soil as

$$E = \frac{\Psi_r - \Psi_l}{R_p} = \frac{\psi_s - \psi_r}{R_s} \tag{19}$$

where E is the transpiration rate, Ψ and ψ are the total water potential and the matric potential, i.e. total water potential minus the osmotic component; the subscripts s, r and l refer to the soil, the root-soil interface and the sites of evaporation within the leaves. R_s and R_p are soil and plant resistances respectively. Equation (19) implies that E is determined by the soil and plant properties, but in a fully turgid plant transpiration is more often determined by the environment as expressed by the potential transpiration rate, E_t;

under these conditions potential gradients throughout the system, and hence operative values of the hydraulic parameters of the soil and plant, are determined by E_t.

As conditions become less than ideal either due to soil drying or to an increased demand in E_t, transpiration is regulated by the stomates in response to the decreasing leaf water potential. Under these conditions the properties of the soil-plant system determine E through their effect on Ψ_l.

Control mechanisms in transpiration

The relationship between stomatal closure and leaf water potential is not fully understood. It is commonly assumed that there is a narrow critical range of Ψ_l over which the stomates close (Lawlor 1972; Begg et al. 1964; Slatyer 1967); results of Gardner and Ehlig (1963) show a pronounced drop in transpiration at Ψ_l near −4 bar for pepper and −12 bar for cotton; sunflower and trefoil are intermediate. Kanemasu and Tanner (1969) present data showing an almost on-off effect of stomatal resistance in beans at Ψ_l from −11 to −12 bar. On the above basis a water use model developed by Cowan (1965) postulates a critical leaf water potential Ψ_w, set generally at −15 bar, above which the stomates remain fully open; Ψ_w is later re-defined by Slatyer (1967) as that potential at which stomatal closure reduces transpiration to a degree which prevents further decline in Ψ_l. The critical leaf water potential is used to define the supply function E_w of the soil-plant system as

$$E_w = (\Psi_s - \Psi_w)/(R_s + R_p) \tag{20}$$

E_w expresses the ability of the soil-plant system to supply water in response to full demand, Ψ_w. It depends on the soil water potential, Ψ_s, and the resistance to flow in the soil and plant pathways R_s and R_p. Plant resistance R_p is taken as constant and is evaluated using equation (20) for the conditions that R_s is zero, i.e. the soil is at field capacity, and that Ψ_l is equal to Ψ_w, the plant is at the point of wilting.

The supply function provides the conceptual basis widely used in the modelling of transpiration; E_w corresponds to Fleming's maximum transmission rate, E_m, described earlier. For an assumed steady state condition of transpiration, a condition which occurs rarely if ever, and considering mean daily rates, these models give rise to characteristic transpiration rate curves with an initial constant rate phase followed by a falling-off phase after Ψ_w is reached. In the more realistic situation when diurnal fluctuations are taken into account models based on this on-off stomatal control at Ψ_w = −15 bar

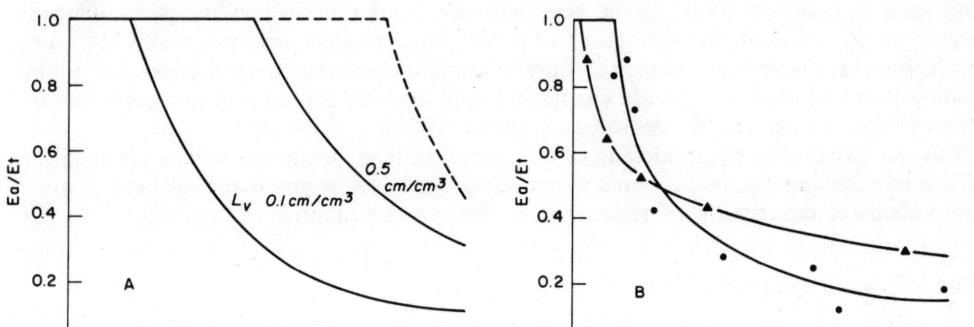

Fig. 5a. Relative evaporation versus soil water potential for (i) an assumed steady state condition (- - -) and (ii) for diurnally fluctuating E_t with root concentrations of L_v equal to 0.1 and 0.5 cm/cm^3 (Cowan 1965).

Fig. 5b. Relative evaporation for pot-grown perennial rye (▲) (Lawlor 1972) and sugar beet (●) (Biscoe 1972) as a function of soil water potential.

predict restricted transpiration rates at soil water potentials as high as −2 bar or more, but only under conditions of extremely low root density (see figure 5*a*, based on Cowan 1965).

The response of stomates to leaf water potential is however more complex than this simple one-step Ψ_w model. Lawlor (1972) has shown that a change in Ψ_l at quite high potentials, −2 and −4 bar, reduces stomatal aperture from 2·8 to 1·0 μm in perennial rye grass with little further change as Ψ_l approaches −16 bar. This suggests effects of soil water potential on transpiration beginning at −1 bar. Biscoe (1972) found that in sugar beet stomatal resistance was sensitive to changes in Ψ_l, and hence Ψ_s, over the full range of Ψ_l from −4 to −14 bar ($\Psi_s \simeq -1$ to −10 bar). He showed that stomatal response to Ψ_l depended on the rate at which moisture stress was applied, and suggested that results which support the critical leaf water potential concept might be so explained. Stomatal resistance depends not only on the turgor pressure of the guard cells but also on that of the surrounding epidermal cells; the leaf is able to osmo-regulate, although only slowly, to maintain turgor pressure against reduced leaf water potential. Biscoe suggests that instances where stomates fail to regulate transpiration until a relatively low value of Ψ_l is reached are associated with a change in Ψ_l too rapid to allow osmoregulation.

In figure 5*b* relative evaporation is given as a function of soil water potential for perennial rye grass (from Lawlor 1972) and for sugar beet (from Biscoe 1972). E is seen to fall sharply with the first decrease in Ψ_s and is close to its minimum value at Ψ_s near −5 bar. The curves indicate a continuous control of evaporation by the stomates regulated by leaf water potential.

While stomatal transpiration appears to be regulated at high water potentials, regulation is not completely effective, and there may even be considerable transpirational loss following stomatal closure; the data of Gardner and Ehlig (1963), referred to above, show a continued transpiration rate as high as 20 per cent E_t at leaf water potentials approaching −20 bar.

Visser (1966) has developed a model which includes stomatal regulation of transpiration using an assumed exponential relation between stomatal resistance r_s and leaf water potential of the form $1/r_s = a - b|\Psi|^l$ where a, b and l are constants describing stomatal action. This represents an advance, but, in common with the steady state models, plant resistance between the root surface and the surfaces of the stomatal cavities is set constant and equal to its minimum measurable value, and the critical resistance is assumed to be in the soil pathway. Under these conditions Ψ_l depends largely on Ψ_r and both the stomatal and steady state models yield essentially the same result. But there is strong evidence to show that plant resistance to liquid flow, i.e. not including stomatal resistance, is not constant, but in osmotic solutions at low potentials increases exponentially as Ψ_l decreases (Lawlor 1973).

Further advance in the modelling of this aspect of plant water use awaits clarification of the relationship between stomatal resistance and leaf water potential, and a more comprehensive description of resistance to flow in the plant.

Soil resistance in uptake of water by plants

The form of the flow equation which applies to the flow of water through the soil to a cylindrical root is given as

$$\partial\theta/\partial t = \frac{1}{r}\frac{\partial}{\partial r}\left(rD\frac{\partial\theta}{\partial r}\right) \tag{21}$$

where r is the radial distance from the central axis of the root. Details of approximate solutions of this equation for different problems may be obtained from Gardner (1960) and Cowan (1965); simplifications involve the assumption of a constant average diffusivity, D, and the approximation of a continually decreasing water content by a series of steady state situations. Solutions for the steady rate case and steady state case have been checked by Passioura and Cowan (1968) against a more exact numerical solution and the agreement is reasonable, with the steady rate solution being slightly superior.

The constant rate of change of soil water content with time Q($cm^3 cm^{-3} s^{-1}$) due to root extraction using the steady state solution is given as

$$Q = d\theta/dt = (\theta_s - \theta_1)\bar{D}/\alpha \tag{22}$$

where θ_s is the water content of the bulk of the soil, θ_1 is the water content at r_1, the root radius, and $\bar{D}$ is the average diffusivity taken as

$$D(\theta_s^{1/2}\theta_1^{1/2}); \quad \alpha = \frac{r_2^4}{2(r_2^2 - r_1^2)} \ln(r_2/r_1) - \frac{3r_2^2 - r_1^2}{8};$$

r_2 is the average distance between roots and is equal to $(1/\pi L_v)^{1/2}$; L_v is the rooting density in cm^{-2}.

Equation (22) can be expressed in terms of soil water potential and hydraulic conductivity as $Q = (\psi_s - \psi_1)\bar{K}/\alpha$, in which form it gives results generally similar to those given by Gardner's model:

$$q = (\psi_s - \psi_1)2\pi\bar{K}/\ln(r_2/r_1) \tag{23}$$

In this case the flow rate q is the flow rate in $cm^3 s^{-1}$ per cm length of root and equals $\pi r_2^2 Q$.

These models have been used directly (Paltridge 1970) or in a simplified form (Ross 1971) as root extraction terms. Gardner (1964) assumes that resistance to flow in the plant pathway is negligible and from equation (23) obtained for salt-free soil

$$Q = d\theta/dt = (\Psi_l - \Psi_s)/I_s \tag{24}$$

where I_s is the impedance to water movement in the soil and is equal to $1/B\bar{K}L_v$; B includes all the terms in equation (23) not individually mentioned. For a layer of thickness χ and depth Z below the surface this becomes

$$Q = d\theta/dt = B\bar{K}L_v(\Psi_l - \Psi_s - Z)\chi \tag{25}$$

In these models Ψ_l is taken as Ψ_w, and is usually set at −15 bar. In effect this gives the supply function E_{w_i}, the maximum rate of uptake for the ith layer of the model. Actual uptake is taken as the maximum rate $\sum_{i=1}^{n} E_{w_i}$ if $E_a \leqslant E_t$ or as a fraction $E_{w_i}/\sum_{i=1}^{n} E_{w_i}$ of E_t if $\sum_{i=1}^{n} E_{w_i} > E_t$.

It will be appreciated that these models are based implicitly on the critical leaf water potential concept and allow no regulation of E_a by the water potential of the system for $\Psi_l > \Psi_w$. The assumption is made that extraction rates at well below maximum rates are directly proportional to maximum rates.

This assumption is not unreasonable in drying soil. Soil conductivity decreases rapidly with decreasing potential, so that decreases in ψ_r with resultant increase in hydraulic gradient cause only diminishing increase in flux. Lang and Gardner (1970) show by experiment, as originally predicted by Whisler, Klute and Millington (1970), that maximum uptake rate by a root is given by

$$q_{max} = \frac{2\pi a}{\ln(r_2/r_1)(n-1)|\psi_s|^{n-1}} \tag{26}$$

and that this is approached rapidly with decreasing Ψ_l. They showed for several soils that E_a for cotton was not increased by decreasing Ψ_l.

Failure of E_a to increase with increase in E_t in the above experiments does not provide conclusive proof however that q_{max} as determined by soil resistance was approached. Indeed using known values of total root length the authors showed that the measured flux per cm length of root was 200-fold less than the calculated flux. This result is not uncommon; in wheat, measured maximum rates of q averaged over the whole profile or over a single layer are 100-fold less than that calculated using the above model (Walter and Barley 1974; Greacen and Hignett 1976) and in perennial rye Lawlor (1972) measured maximum q values of 0·006 to 0·007 cm^2/day compared to 0·1 cm^2/day commonly assumed in water uptake models (Gardner 1960; Cowan 1965).

Consideration of similar results has thrown doubts on the relative importance of R_s, the resistance to flow in the soil surrounding individual roots, compared to resistance in the plant pathway, R_p (Newman 1969). From table 2 based on equation (23) for a soil with a conductivity function of the form $K = a|\psi|^{-n}$, where $n > 1$, R_s becomes significant at several bars below zero for q values ~0·1 cm² day⁻¹. Newman (1969) argues however that, because of the great total length of roots under most field crops, values of q must be several orders of magnitude below this, and that in general R_s will not be significant or will remain below R_p until near or beyond the wilting point. Although he cautions that these conclusions are based on several questionable assumptions, he claims in a later paper to show that many results purporting to demonstrate appreciable soil resistance in the rhizosphere have been incorrectly interpreted, and in some instances even show that R_s is negligible.

Table 2. Values of hydraulic potential at the root surface to provide the transpiration rate, e_t, for different values of total root length L_A(cm/cm²) and soil water potential ψ_s

($K = 187|\psi_s|^{-2.07}$)

Transpiration rate E_t = 0.50 cm/day

L_A	E_t/L_A	Soil water potential (−bar)							
(cm−1)	(cm² day⁻¹)	0.1	0.5	1.0	2.0	4.0	8.0	12.0	16.0
		Potential at root surface (−bar)							
10	0.050	0.10	0.53	1.13	2.58	7.19	29.35	52.96	103.04
20	0.025	0.10	0.52	1.06	2.26	5.16	14.39	26.48	45.30
40	0.013	0.10	0.51	1.03	2.12	4.53	10.33	17.74	27.10
80	0.006	0.10	0.50	1.01	2.06	4.25	9.05	14.50	20.66
160	0.003	0.10	0.50	1.01	2.03	4.12	8.50	13.15	18.11

Transpiration rate E_t = 1.00 cm/day

L_A	E_t/L_A	Soil water potential (−bar)							
(cm−1)	(cm² day⁻¹)	0.1	0.5	1.0	2.0	4.0	8.0	12.0	16.0
		Potential at root surface (−bar)							
10	0.100	0.10	0.57	1.29	3.60	26.26	51.52	141.91	303.82
20	0.050	0.10	0.53	1.13	2.58	7.19	29.35	52.96	103.04
40	0.025	0.10	0.52	1.06	2.26	5.16	14.39	26.48	45.30
80	0.013	0.10	0.51	1.03	2.12	4.53	10.33	17.74	27.10
160	0.006	0.10	0.50	1.01	2.06	4.25	9.05	14.50	20.66

The questionable assumption in Newman's argument is that the total root length is equally absorbing, whereas the evidence available indicates considerable variation along the root and with the conditions of uptake (Brouwer 1954; Tanton and Crowdy 1972). The author is not aware of any significant work where direct measurement of root resistance has been made under unsaturated conditions of moderate to low matrix potential. Root resistance under these conditions is critical to the above argument and to the question of water availability in semi-arid soils.

Root resistance in uptake of water by plants

Flow of water in a single root is conveniently considered as two component flows, a radial flow across the root into the xylem and thence an axial upwards flow in the root xylem to the above-ground parts of the plant. Evidence by Tanton and Crowdy (1972) using a lead chelate tracer supports the view that in the cortex the radial pathway is via the free space; i.e. along the cell walls of the root hairs and then across the cortical tissue proper in the walls until flow is blocked by the suberized casparian band on the radial and horizontal walls of the endodermis. The pathway is then through the symplasm of the endodermis and finally into the xylem.

Along the length of the root the radial pathway becomes increasingly restricted as suberization of the endodermis progresses with age. Tanton and Crowdy suggest that the main zone of absorption extends from 4 to 50 mm behind the apex in the case of the bean and to 100 mm in the case of cereals. Clarkson, Robards and Jackson (1972) relate calcium uptake to the per cent of unsuberized cells in barley (fig. 6). This parallels Tanton and Crowdy's results on water uptake in cereals and suggests that the main zone of uptake in the seminal axis is the apical 10 cm and in laterals the apical 4 cm.

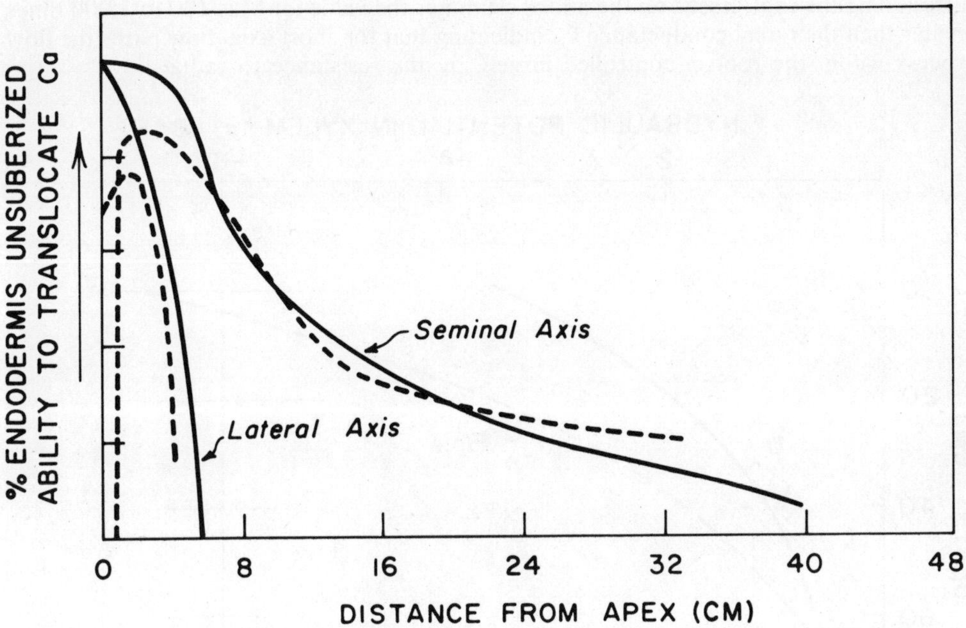

Fig. 6. Distribution of the relative ability of the root of barley to translocate calcium (- - - -) and of the percentage of unsuberized endodermal cells (——) (Clarkson, Robards and Jackson 1972).

Flow of water in the above two pathways may be expressed for an individual root member as $f = k_x d\psi_x/dz$ and $df/dz = -k_r(\Psi_s - \Psi_x)$, where f is the axial flow (cm^3s^{-1}), z is the distance along the root, ψ_x is the hydraulic potential in the xylem, Ψ_s and Ψ_x are the water potentials surrounding the root and in the xylem and k_x and k_r are the conductances for axial and radial flow (Cowan and Milthorpe 1968).

Early measurements of the flow resistance of xylem elements were made by Dixon (1914) and Emerson (1954). Using a syphon technique for the dead root axes of perennial rye, Emerson found that the flow rate tended to be less than that predicted from the measured diameter of the xylem using the Poiseuille equation for flow in a tube.

More recently Ponsana (personal communication) using 2-3 cm segments of the seminal and axial roots of wheat in a potometer showed that measured flow could vary from 0.5 to 2 times Poiseuille flow. He found that the diameter of the metaxytem varied from 50 μm near the surface to 70 μm at 50 cm depth both for field and water cultured roots. Cox (1966) applied the above flow-path equations to whole root systems (4-5 seminal axes) of young wheat plants. He obtained surprisingly low values of k_x for individual axes, 10-40 cm^3 day^{-1} bar^{-1} cm, corresponding to metaxylem elements of only 30 μm diameter.

Direct measurement of radial flow is not easy. For bean roots Brouwer (1954), using a potometer, obtained values of k_r of 0·05 cm^3 day^{-1} cm^{-1} bar^{-1}. The same order of magnitude was obtained also by direct measurement for wheat roots by Ponsana (personal communication) and by the indirect method for the whole root system by Cox (1966).

These are acceptable estimates of radial root conductance for the saturated condition, but there is considerable evidence that this property varies both with water potential and with position along the root (Brouwer 1965); furthermore there is no evidence that conductance for the saturated root is applicable to roots in drying soil. Nevertheless it is interesting to compare resistance to flow in the axial and radial pathways and in the soil surrounding the root. Based on the above estimates the value for k_x is 100 to 1,000 times greater than the radial conductance k_r, indicating that for short axial flow paths the flow process within the root is controlled largely by the resistance to radial flow.

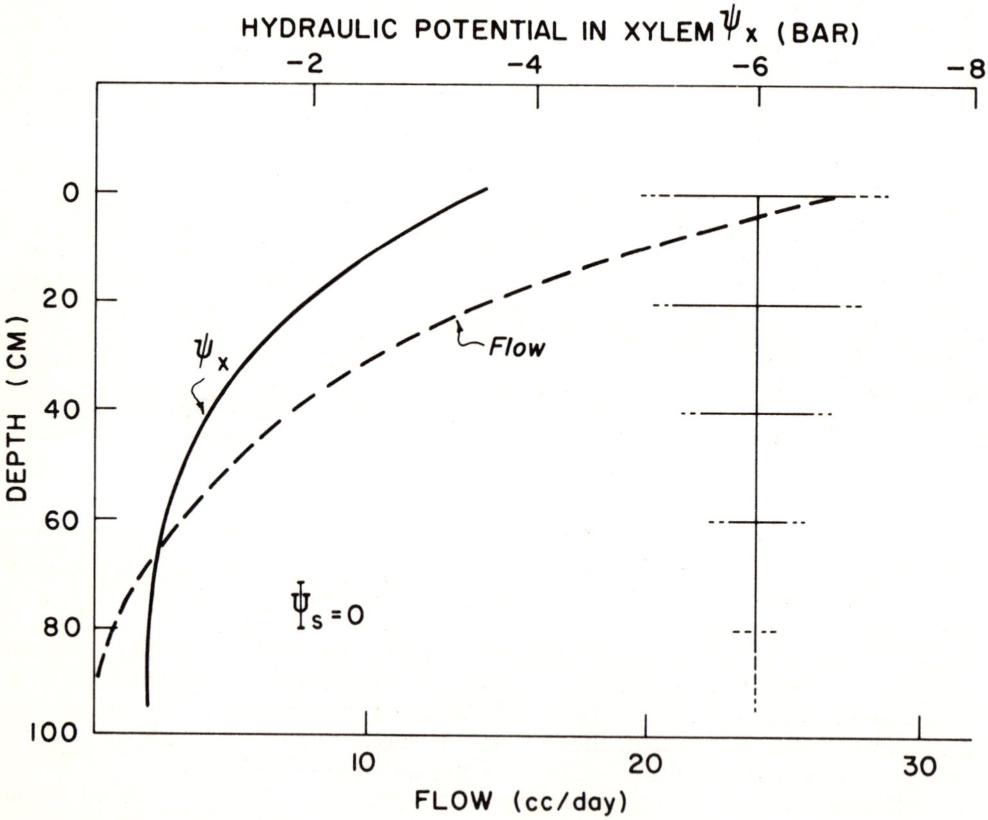

Fig. 7. Estimates of flow and hydraulic potential in a seminal axis of wheat with soil water potential equal to zero. Geometry of the root based on growth attributes given in table 1 by Lungley (1973); only 1 in 20 laterals shown.

Cox (1966) gives estimates of conductance for the soil surrounding the root. Using Gardner's expression for cylindrical flow (equation 23) and assuming a value of $r_2/r_1 = 10$ he gets $df/dx = (2\pi K/\ln 10)\Delta\psi_s$, or $k_{rs} = 2{\cdot}7$ K, where k_{rs} is the radial conductance per cm length of root for the soil surrounding the root. Typical values of k_{rs} for loam soil at matrix potentials −1, −2, −4, −8 and −16 bar are 6, 4, 0·5, 0·2 and

0·005 cm³ day⁻¹ cm⁻¹ bar⁻¹. The assumption of constant conductivity in Gardner's model is crude, but for the present purpose it errs in the right direction by underestimating soil conductance. The values of k_{rs} are to be compared directly with the value of 0·05 for measured saturated root conductance given above. Errors of estimate could be an order of magnitude, but the comparison supports the contention that soil rhizosphere resistance becomes critical only as wilting point is approached.

Notwithstanding the apparent dominance of radial resistance to flow in the plant pathway, resistance to flow in the root xylem can play a significant role in water use by plants. In figure 7 estimates are given of flow rate and hydraulic potential in the metaxylem of a seminal axis in wheat based on the above considerations. k_x for the main axis is taken as constant and equal to 250 cm^3 day^{-1} cm bar^{-1} (diameter 60 μm), although recent work in this laboratory has shown that for young roots this can be reduced by as much as 30 per cent at moderate potentials (—4 bar) presumably by mechanical compression of the xylem under the suction load; k_r has been assumed equal to 0.05 for the main axis and 0.03 cm^3 day^{-1} cm^{-1} bar^{-1} for the laterals; only the apical 10 cm in the main axis, and 4 cm in the case of the laterals, is considered to be absorbing. Considering that a decrease in Ψ_l from -1 to -4 bar can reduce transpiration twofold by stomatal control (fig. 5*b*), it is obvious that water potentials arising in the xylem due to flow can be significant. The effect is even greater than this suggests. Since we are concerned with net vertical flow through the soil, axial conductance is modified to allow for root tortuosity as t^2k_x, where t is the cosine of the angle made between the root and the vertical. Again, particular situations of root geometry and soil water can also operate to make axial conductance critical. This occurs for example when the plant has to depend on a supply of subsoil water through sparse axial roots. Passioura (1972) has demonstrated under glasshouse conditions that it is possible to manipulate this situation of axial resistance to advantage in wheat to conserve water for use during the more critical post-anthesis stage (fig. 8). But classical concepts of this aspect of the plant water

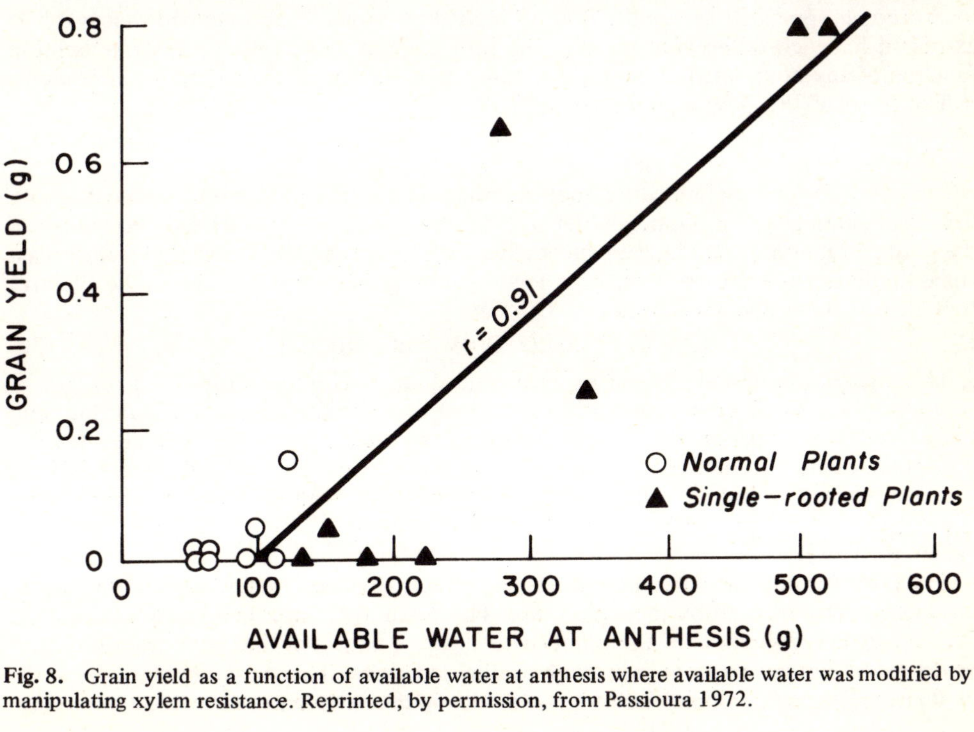

Fig. 8. Grain yield as a function of available water at anthesis where available water was modified by manipulating xylem resistance. Reprinted, by permission, from Passioura 1972.

supply, namely the xylem pathway, are currently under scrutiny. The efficiency of a xylem network at carrying water under high stress must be inversely proportional to the number of breakdowns and leaks in the system, and these are determined to some extent by soil movement during cracking and the predations of rhizophilic soil-living organisms. Recent work at this laboratory suggests that soil sterilization increases the efficiency of the wheat root system and water uptake somewhat more than it increases the extent of the root system.

Macroscopic models of water uptake

The cylindrical flow model described earlier has been used to develop water uptake terms for individual layers in multi-layered models. Gardner (1960) uses equation (6) to obtain an expression for transpiration rate for the whole soil profile consisting of *n* equal layers of thickness χ as

$$E = \sum_{i=1}^{n} \frac{d\theta_i}{dt} = B\chi \sum_{i=1}^{n} (\Psi_l - \Psi_s - Z_i)K_i L_{v_i} \tag{27}$$

Although equation (27) is based on microscopic processes occurring around individual roots it describes the extraction rate of the root system as a macroscopic process. The extraction rate $d\theta_i/dt$ is a function of the depth in the profile, the water content, root and soil parameters, and time. It can be used to obtain macroscopic or integrated parameters for the root system as a whole. For example, the integrated soil matric potential can be obtained as

$$\bar{\psi}_s = [\sum K_i L_{v_i}(\psi_{s_i} + Z_i)]/\sum K_i L_{v_i} \tag{28}$$

and the integrated conductance for the profile defined by $\bar{K} = \sum K_i L_{v_i}$. These expressions have been used effectively by Lawlor (1972) as indices of integrated water potential and conductivity in relation to water uptake and plant growth.

Where Gardner uses a water uptake term, based on the cylindrical flow model, in an attempt to incorporate microscopic processes, Molz and Remson (1970) in a strictly macroscopic approach ignore the flow to individual roots. They assumed that water is extracted for each differential layer of the root zone at some rate U, this rate being a function of any nominated variables, e.g. time, depth in the profile or soil water content.

The form of the flow equation then is

$$\frac{\partial \theta}{\partial t} = \frac{\partial}{\partial z}\left(D\frac{\partial \theta}{\partial z}\right) - \frac{\partial K}{\partial z} - U, \tag{29}$$

which is solved for θ by finite difference methods. The extraction term U can involve any degree of simplicity, e.g. Dutt, Shaffer and Moore (1972) assume a fixed percentage of the total extraction occurs in each successive layer. Alternatively, Molz and Remson assume that water is extracted from any depth z in the profile in proportion to the effective root density L(z) and the diffusivity $D(\theta)$ as

$$U(z, \theta) = E[L(z)D(\theta)/\int_0^z L(z)D(\theta)dz] \tag{30}$$

This model was tested against experimental data of Gardner (1964) for pot-grown sorghum using an effective root distribution term L(z) obtained from data (Gardner and Ehlig 1962) for birdsfoot trefoil grown in similar soil and pots. A steady state situation for uptake existed in the trefoil experiment and this allowed an easy assessment of L(z). Water content distribution computed from equation (29) with the sink term from equation (30) based on this artificial effective root distribution is compared with measured values in figure 9.

The L(z) term fits the experimental results perfectly for the first few days of the experiment but is seriously off by the eighth day. The deviation is ascribed to a change in the effective root distribution of the plant, as it is to be expected that root resistance and hence water uptake patterns will vary as the soil dries. The same conclusion can be made from the performance of Gardner's model on the same data: the extraction pattern

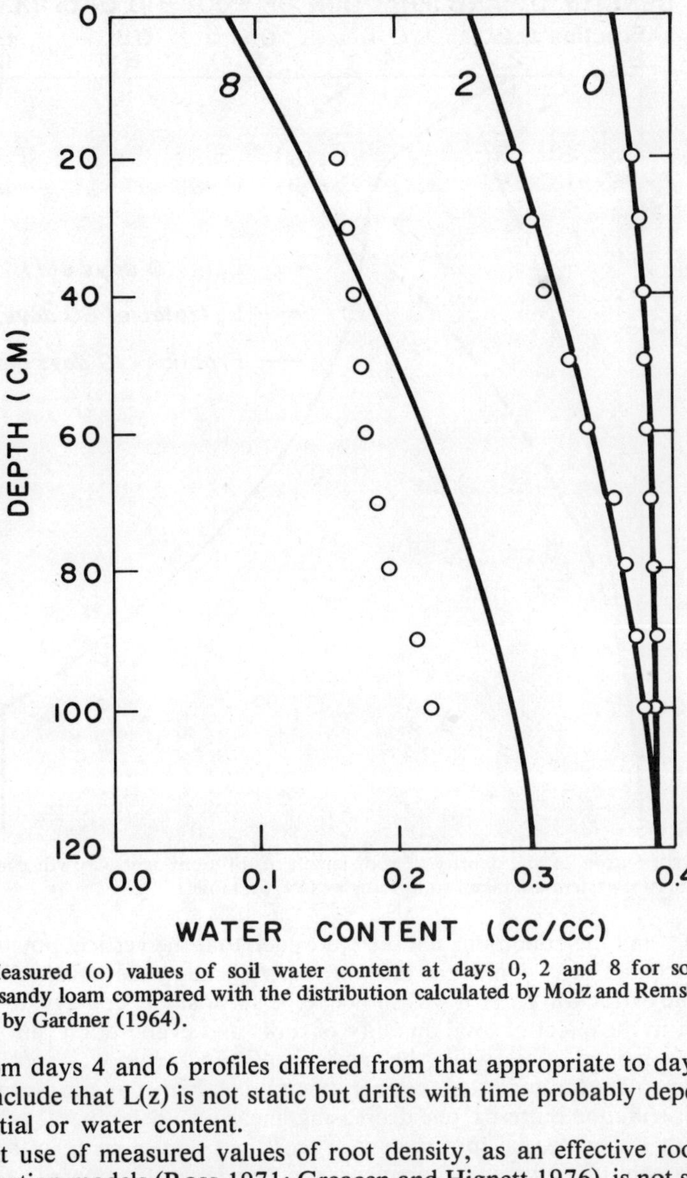

Fig. 9. Measured (o) values of soil water content at days 0, 2 and 8 for sorghum in Pachappa sandy loam compared with the distribution calculated by Molz and Remson (1970) from data by Gardner (1964).

obtained from days 4 and 6 profiles differed from that appropriate to day 8. Molz and Remson conclude that L(z) is not static but drifts with time probably dependent on soil water potential or water content.

The direct use of measured values of root density, as an effective root distribution term in extraction models (Ross 1971; Greacen and Hignett 1976), is not satisfactory. It is generally accepted that uptake is most active in young root material but the length of young roots is not directly related to total root length. This is demonstrated clearly in figure 10, where the fraction of root length composed of zones less than 10 days old is plotted as a function of depth for a simulated ideal wheat root system (Lungley 1973).

THE ROOT SYSTEM

Geometry of the root system

The root extraction model, equation (27) above, has been used by Gardner (1964) to examine the effect of root distribution and other factors on the pattern of water uptake from the soil profile. The model is limited since resistance to flow in the roots is assumed

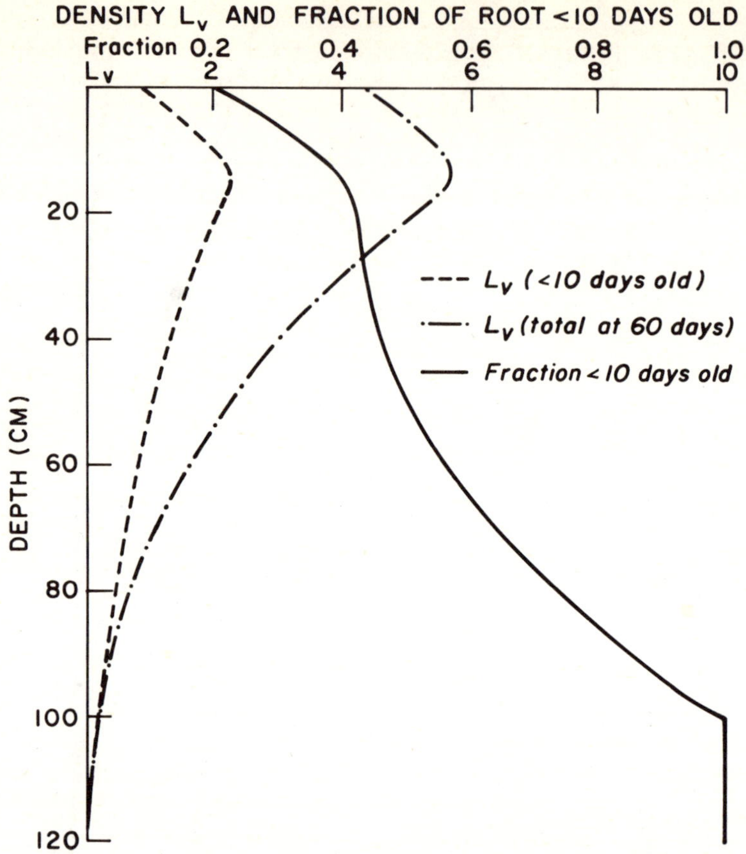

Fig. 10. Distribution of the density of root length of different ages. Growth characteristics of the root system are taken from Lungley (1973), table 1.

to be negligible, and the conditions chosen were such that the vertical potential gradient set up did not cause any significant flow of water through the soil itself.

Generally the extraction pattern was shown to be surprisingly conservative particularly with respect to the effect of total quantity of roots and evaporation rate. Soil properties had only a limited effect while the most significant factor was root distribution. Water content profiles for a constant evaporation rate of 2 cm/day are given in figure 11 for two root distribution patterns, one decreasing linearly with depth, the other exponentially. Initially, the exponential root system, with 70 per cent of roots in the top 30 cm, extracts 50 per cent of its water requirement from this layer. Compared to this the surface 30 cm in the case of the linear system has 50 per cent of the roots and supplies 40 per cent of the required water. This difference however decreases by day 6, suggesting only a moderate effect of root geometry on water uptake.

Further evidence for the conservative nature of the water extraction pattern can be inferred from other work. Danielson (1967) gives as a rule of thumb for irrigationists that 40, 30, 20 and 10 per cent of water requirements are supplied by successively deeper quarters of the root zone. Molz and Remson (1970) have applied this simple empirical rule as an extraction term in equation (29), to the data for trefoil from Gardner and Ehlig (1962) and predicted water movement as accurately as could be measured experimentally.

These results suggest that root distribution might not have a significant effect on water availability or on the extraction pattern. But they apply only to "normal" or ideal root geometry, and under field conditions in a semi-arid environment root geometry is in general not ideal. In the case of cereal crops the fraction of roots in the surface layer can be much greater than that described by the exponential relationship of figure 11; where the crop depends on current rainfall the role of this layer, compared to that of the sparsely rooted subsoil, becomes critical in the supply of water and nutrients to the crop. In figure 11*b* a root distribution pattern for wheat is given for a medium-textured soil in South Australia. The high concentration of roots in the surface layer probably reflects the depth of cultivation with the associated lower mechanical resistance to root penetration and the depth of phosphate application.

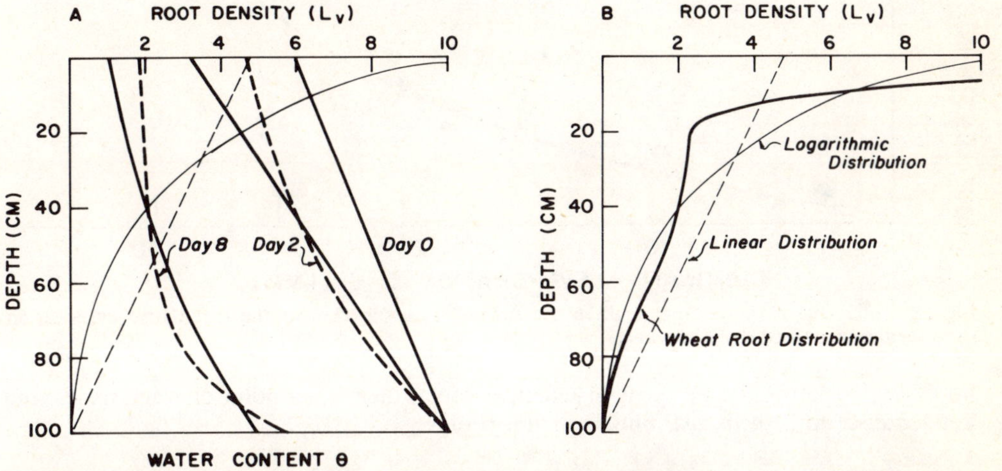

Fig. 11*a*. Water content profiles (heavy lines) for a constant evaporation rate of 2cm/day for days 0, 2 and 8 for root systems decreasing linearly (- - -) and logarithmically (——) with depth, after Gardner (1964).

Fig. 11*b*. Measured root distribution for wheat in a red-brown earth compared to two hypothetical patterns from figure 11*a*.

In a water balance model for wheat in South Australia this strongly developed root pattern has been treated effectively as a two-layer system (Greacen and Hignett 1976). The surface A layer was chosen as 0-30 cm to accommodate strong changes in texture commonly found at this depth. Even in poor seasons root length in this layer (L_A = 150 cm/cm^2) is sufficient to support relatively high evaporation rates (3 mm/day) and the soil is always dried to wilting point by the time of crop ripening. In the subsoil B layer (30-130 cm) the total length of root L_A varies with the season, increasing exponentially with the cumulative water used, $\sum E_a$ (fig. 12). For a total water use of 20 cm, values of L_A in the B layer at anthesis can be as low 50 cm/cm^2. Since the measured maximum rate of water uptake q_{max} is approximately 0·002 cm^3/cm length of root/day (Walter and Barley 1974; Greacen and Hignett 1976) the rate of water uptake from below the A layer is restricted to less than 1 mm/day. Under these conditions significant amounts of water remain untapped in the subsoil.

The explanation of the low q_{max} values is not known. Walter and Barley (1974) suggest that resistance to Poiseuille flow in the xylem tubes of the axial roots restricted water uptake when root concentration at depth was limited by planting density. Lang and Gardner (1970), in explaining similar results with cotton, do not consider xylem resistance, but suggest that only part of the root is absorbing and that resistance to radial flow in the cortex could be appreciable. Work by Boyer (1971) supports this conclusion.

Whatever the reason for the failure of crop roots to deplete subsoil reserves of water, total root length L_A is not a sufficient description of the effective absorption system. It is

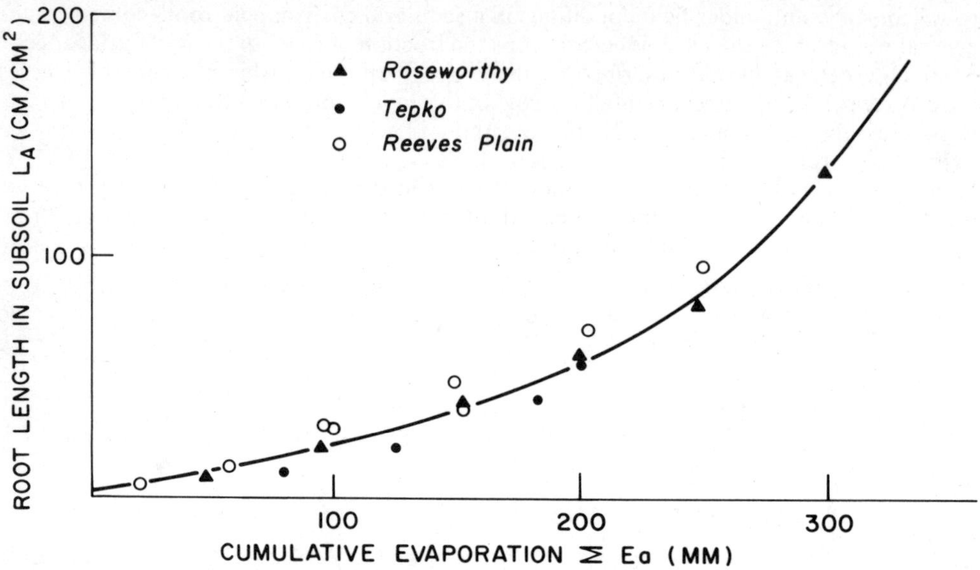

Fig. 12. Total length of root per cm^2 in the subsoil, as a function of the cumulative evaporation since seeding, for three crops of wheat.

necessary to know the per cent of effective roots, their absorption characteristics and conductance and their distribution in the profile.

The effect of the soil environment on root growth and distribution

The root systems of plants differ according to their genetic constitution but the extent to which these differences are expressed is modified by the supply of materials, nutrients, water and oxygen, and also by the temperature and mechanical properties of the soil. Even in a semi-arid environment temporary poor aeration can follow unusually heavy rainfall events and, depending on the stage of crop development, this can lead to the death of crop roots (Erickson and van Doren 1960); soil parameters and critical conditions for aeration have been dealt with by Greenwood (1968) and Eavis and Payne (1968) and these can be incorporated into layered models.

In this section we will be concerned with the influence of temperature and the mechanical properties of soil. Numerical models have been developed to simulate the morphological development of root systems (Lungley 1973; Hackett and Rose 1972); inherited characteristics, such as number of initial roots, and the orientation of members, control the type of root system in the model, while the rate of elongation and branching, modified in individual layers by the physical environment of the soil, determine its extent and distribution with time.

Root growth against mechanical resistance

In sands and finely structured soils the resistance to penetration of the soil by a metal penetrometer can be calculated from the mechanical properties of the soil and the shape of the tip of the probe (Farrell and Greacen 1966). Greacen, Barley and Farrell (1968) provide evidence that the pressure applied by the root is approximately the radial pressure, σ_r, required to expand the growth cavity cylindrically, and show that this can be estimated from the soil resistance measured with a metal penetrometer. For a finely tapered probe

$$\sigma_r \approx \sigma_n = p/(1 + \tan\phi \cot\alpha) \quad (31)$$

where α is the included semi-angle of the conical tip of the probe (= 5 degrees), ϕ is the angle of soil-probe friction, p is the pressure on the cross-sectional area of the probe, and σ_n is the force normal to the surface of the probe tip.

Resistance to penetration is determined in a complex way by the strength and compressibility properties of the soil (Farrell and Greacen 1966) and these in turn depend strongly on bulk density and water content (Barley and Greacen 1967). Mechanical resistance is far more easily obtained from penetrometer measurements than calculated from measured soil properties. In some soils p increases linearly with suction over the normal part of the range, and strength-water content relations can be described empirically by a linear log relationship (fig. 13).

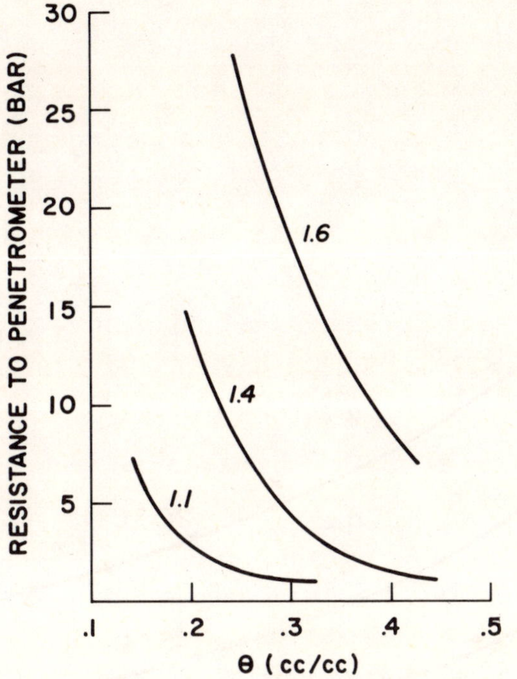

Fig. 13. Resistance to a penetrometer as a function of water content in a sandy loam soil at three bulk densities 1.1, 1.4 and 1.6 g/cm^3 (based on data by Eavis and Payne 1968).

In heavily compacted soil the lack of oxygen may limit the rate of growth (Gill and Miller 1956), and in aggregated crumb structures the air-filled porosity-aeration relationship is not unique (Greenwood 1968); but, where aeration or other factors are not limiting and soil temperature is constant, the rate of growth is inversely proportional to log p. In figure 14 the rate of growth of seminal roots of wheat in cores of sandy loam remolded to different strengths is given at 10, 15 and 20°C (Greacen and Oh, unpublished).

Effect of mechanical resistance in macroscopically structured soil

In contrast to sands and fine structured soils, i.e. where the characteristic dimension of the structure is less than that of the root, in coarse-structured soils the mechanical resistance of the structural units or peds tends to restrict root growth to regions of lower resistance rather than reduce the rate of root elongation. The restriction of roots to vertical channels and cracks in some soils is an extreme example of this. In other soils the influence of orientation of faces of higher strength on root development may be less obvious and impregnation techniques and serial thin sections of the soil may be needed to show the influence of structural features on root distribution (Melhuish and Lang 1968; Willoughby 1967). While the rate of growth within a structural unit or ped may be described by figure 14, the probability of the root tips entering the ped is determined by several factors, e.g. the mechanical state of the soil surrounding the layer and also the angle of approach to the surface of resistant peds or layers (Greacen, Barley and Farrell 1968).

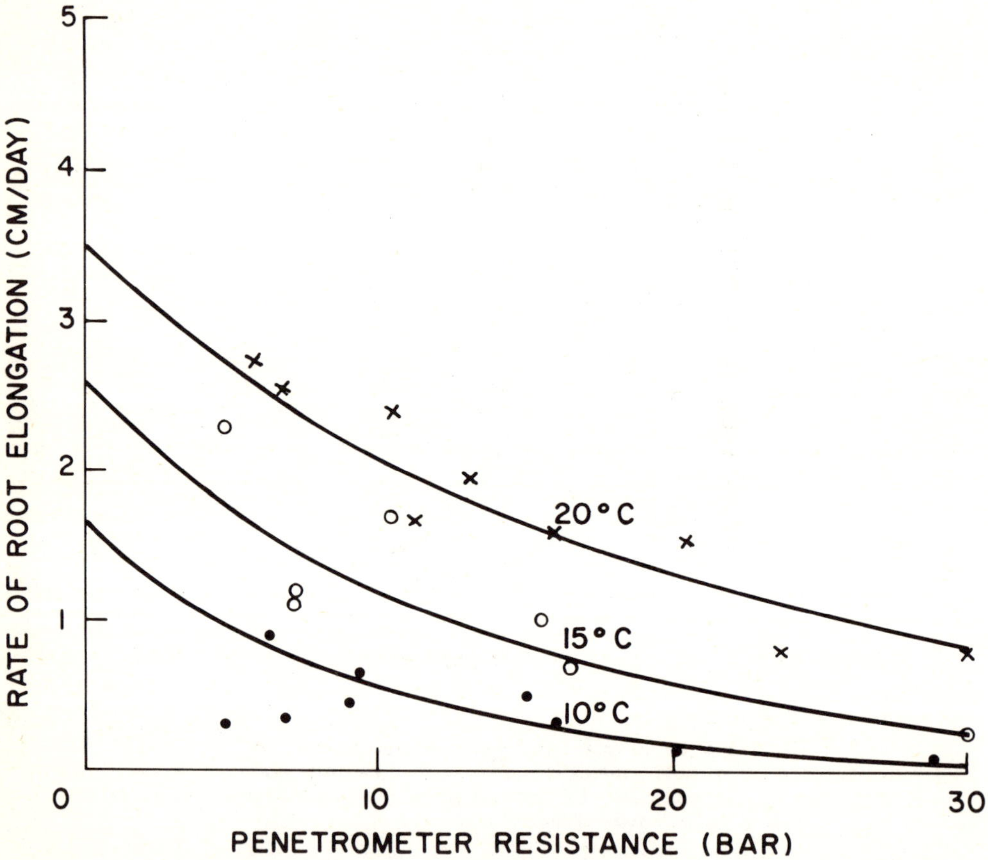

Fig. 14. The rate of elongation of the seminal roots of wheat in cores of sandy loam as a function of soil strength (penetrometer resistance) at 10°, 15° and 20°C.

The problem of quantifying the effect of soil structure on root distribution is not easy. Root distribution can be examined at the macroscopic level by excavation and pin-board techniques (Ponsana 1975), and at the microscopic level by impregnation techniques (Barley and Sedgley 1961), but to the author's knowledge, no method is available for describing soil structure quantitatively at the microscopic level and predicting the volume of soil that is available to plant roots. This situation still further complicates the problem of water availability. In an earlier section, the question was raised as to what part of the root is effective in absorbing water. Besides this must be answered the question, what part of the soil volume does the physical environment of the soil allow the root to exploit.

The physics of root growth

Useful measurements of the rate of root elongation must be obtained under conditions that are controlled for temperature and to some extent for soil strength, evaporative demand, and supply of assimilate. For convenience the radicles of germinating seedlings have been used (Eavis and Payne 1968; Barley, Farrell and Greacen 1965). This method has the advantage that the supply of assimilate is not limiting and the evaporative demand is zero. Under normal conditions, however, the plant is growing under an evaporative demand and the assimilate available for growth has to be partitioned between the plant parts, "tops and roots". These factors are likely to modify the rate of root elongation but little information is available on the extent of this modification.

The problem of simulating this aspect of plant growth has been attempted in an empirical way by Rose at al. (1972) for Townsville stylo in the north of Australia: total growth is directly related to cumulative evaporation modified by soil nutrient status; a value for the rate of elongation of the tap root is assumed from experience and partitioning of growth is made between roots and tops and between roots in the different soil layers again on the basis of earlier measurements. A predictive model based on these concepts over-estimated yield at times, but generally the agreement was promising.

In an attempt to reduce the degree of empiricism in the modelling of root growth Greacen and Oh (1972) sought an understanding of the bio-physical processes involved. They applied the concept of cell elongation, as an extension or stretching process of the cell wall under the action of turgor pressure, to the zone of elongation of the root as a whole; this is written as

$$R_g = m_r(w - w_c) \qquad (32)$$

where R_g is the rate of elongation of the root, m_r is the extensibility of the cell wall material, w is the tensile stress in the cell wall and w_c the threshold critical stress for growth. w increases with increase in hydraulic or turgor pressure P inside the cell, but it is reduced by mechanical reaction or confining pressure, σ_n, of the soil on the root as $w = P - \sigma_n$. P can be equated with the hydrostatic pressure potential, ψ_p, within the cell vacuole and together with the osmotic potential, ψ_o, makes up the total water potential of the cell, as $\Psi_c = \psi_p + \psi_o$.

To evaluate Ψ_c, and hence w, Greacen and Oh assume for simplicity that Ψ_c is equal to and in equilibrium with the water potential Ψ_s of the soil surrounding the zone of elongation, i.e. that the potential gradient required to supply water for growth is negligible. This assumption is not critical and under the conditions of their experiments, i.e. high soil water potential and zero evaporative flux, it is reasonable. Although the radial flux required to supply water to the elongating cells is not insignificant (0·05 cm^3/cm length of elongation zone per day for the pea radicle) the potential drop between the bulk of the soil and the root surface is likely to be <0·1 bar; in the root tissue itself, considering the geometry of the flux pattern and using published values of radial root resistance (Brouwer 1954; Dainty 1969), the potential drop can be shown to be of the same magnitude. Under these conditions it was possible to demonstrate that, in the short term, root elongation followed the wall pressure model.

At low soil water potentials the resistance to flow of water in the soil surrounding the root tip and in the root tissue is likely to limit the flow of water and control the rate of elongation, e.g. at −7 bar in peanuts and −12 bar in cotton (Taylor and Ratliff 1969) the growth rate of roots appears to be limited by Ψ_s. This situation of limiting flux has been dealt with analytically for individual cells (Lockhart 1965) but its application awaits more confident estimates of the permeability of root tissue.

The above model describes the effect of soil water potential and mechanical resistance of the soil on the rate of root elongation. For growth to proceed the water potential within the cells Ψ_c must be lowered enough to produce the required flux of water from the soil along the potential gradient $\Psi_c - \Psi_s$ and to increase ψ_p sufficiently to overcome soil resistance and to stretch the cell walls. The plant is able to do this by lowering its osmotic potential, within limits, with almost 100 per cent efficiency against decreasing Ψ_s, but only less efficiently against increasing σ_n; the lowest limit of this osmo-regulation in the case of pea radicles appears to be about −15 bar. Besides the recognized effect of temperature on root permeability (Brouwer 1965), both temperature and the supply of assimilate could influence root growth indirectly through their role in the biochemistry of osmo-regulation and by changing the rate of cell division.

A model for root and top growth in plants

In the previous section the rate of growth of an individual root tip has been described as it is determined by the physical environment of the soil, under conditions of no evaporative demand and with full supply of growth substances. It represents the potential growth rate of the root for that particular soil strength and water potential; when summed over the root system it describes the sink strength of the root for growth assimilates. On this basis the model suggests mechanisms through which distribution of assimilated growth substances and evaporative demand might operate to determine the growth and morphology of the plant.

In the representation of the growing root tip, given in figure 15, the principal region of water uptake in the bean root extends from just above the zone of elongation, about 4 mm from the apex. With the setting up of an evaporative flux, a gradient is established between the water potential in the xylem, Ψ_x, along the flux pathway in the cell walls, and the water potential in the bulk of the soil, Ψ_s. The extent to which the decrease in Ψ_x will affect Ψ_c in the growing tip, and eventually growth rate, as minimum ψ_o is approached, will depend on the conductivity of the root tissue between the end of the xylem and the zone of elongation, and the conductivity of the surrounding soil (see also Lawlor (1973)). Under conditions of moderate evaporative demand potential gradients across the root can be taken as less than 1 bar and the available evidence suggests that the zone of elongation does not contribute significantly to the evaporative flux (Tanton and Crowdy 1972). On the basis of this argument the direct effect of Ψ_x on the water potential in the most active growth zone would be small. Within the soil the decrease in Ψ_x will cause the water potential at the soil root interface to decrease, but again this perturbation will be confined to the region of high absorption and again its affect on growth rate should be secondary. On the other hand Ψ_s will decrease generally with water loss throughout the bulk of the soil and consequently around the growing tip. Ψ_c will respond to this decrease in Ψ_s both directly and because of the effect of Ψ_s on soil strength. This however is an effect of soil environment on root growth as treated in the previous section; the effect of evaporative demand on root growth and on the root sink strength for assimilate is secondary.

A similar demand for water and assimilate occurs in the apices of the above-ground parts of the plant. The same growth mechanism embodying cell wall extensibility can be postulated, in this case with ψ_{ol}, m_l and w_{cl} being the osmotic potential, the extensibility and critical wall stress characteristics of the plant tops. Ψ_{cl} is taken as Ψ_l the leaf water potential measured with a psychrometer. Likewise for mechanical impedance Williams (1975) has demonstrated the importance of confining mechanical stresses in the growth of leaf buds in growing shoots.

In direct contrast to the situation in root growth the water potential in the growing shoot Ψ_{cl} must be determined largely by the water potential in the leaf xylem, Ψ_{xl}, and fluctuates directly with evaporative demand. Ψ_l falls below Ψ_s to give the evaporative

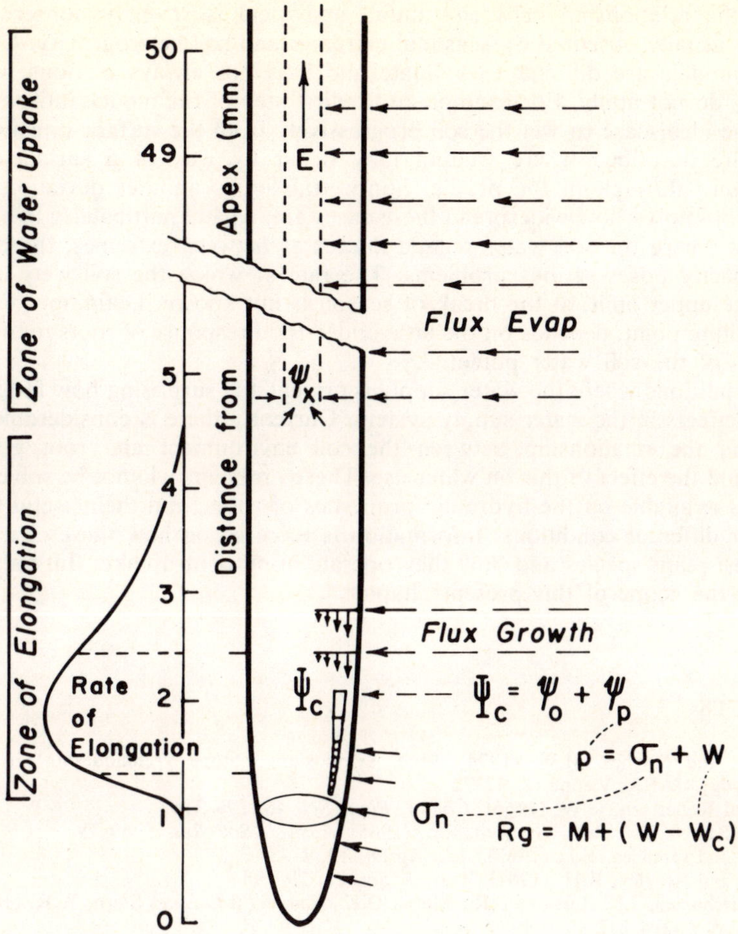

Fig. 15. Diagram of the growing root showing principal zones of absorption and elongation.

flux against the combined resistance of soil and plant. As long as the decreasing Ψ_l does not cause ψ_{ol} to approach its lower limit apical growth will continue and assimilate will be distributed according to the sink strength. As the plant comes under water stress, minimum ψ_o can be expected to obtain earlier in the shoots than in the roots and root growth will be enhanced at the expense of shoot growth. The model is consistent with the general observation that the shoot-root ratio decreases with increasing drought conditions.

OUTLOOK

The water supply factor in crop production has been discussed on the basis of the simple water balance model. Water from rainfall goes into storage from where it is taken up against varying resistance through the root system in response to the radiation load of the environment. The different mechanisms and processes involved have been described by soil physics theory for ideal media at the microscale level, and the resulting concepts have been incorporated into models of water supply in the soil-crop macrosystem.

Yield prediction is often used as a test of model performance, but in a semi-arid environment the relationship between rainfall and yield is strongly conservative and furthermore usually obscured by seasonal extremes and soil heterogeneity. Under such conditions models are difficult to validate and it is not always obvious when ideal mechanisms do not apply. For example in the first step of the model, infiltration is assumed in the ideal case to wet the soil progressively from the surface downwards; but situations are described where wetting may occur by way of a surface-connected macrostructure throughout the profile; non-wettability is another deviation from the ideal. It is not known how widespread these cases are. Again, particularly in a semi-arid environment where the soil water regime fluctuates between extremes, the concept of storage capacity poses serious problems. The rate at which the soil wets up to field capacity, the upper limit, at the break of season is not known. Definition of the lower limit, the wilting point, depends on the little understood response of roots to the different components of the soil water potential.

Under conditions where the water supply is critical it is surprising how little is known about root effects in the water supply system. Currently there is considerable research emphasis on the relationship between the soil environment and root growth and geometry, and the effect of this on water use. These problems will not be solved until information is available on the hydraulic properties of roots, and their useful life expectancy under different conditions. Information is required on how these characteristics vary between plant species and how they operate in nutrient uptake. But these matters are beyond the scope of this present chapter.

REFERENCES

Allison, G.B., and Hughes, M.W. (1974). Proc. Symposium Isotope Techniques in Ground Water Hydrology. IAEA, Vienna. 1, 57-72.

Baier, W., and Robertson, G.W. (1966). *Can. J. Plant Sci.* **46**, 299-316.

Barley, K.P., Farrell, D.A., and Greacen, E.L. (1965). *Aust. J. Soil Res.* 3, 69-79.

Barley, K.P., and Greacen, E.L. (1967). *Adv. Agron.* 19, 1-43.

Barley, K.P., and Sedgley, R.H. (1961). *Aust. J. Sci.* 23, 297-98.

Begg, J.E., Bierhuizen, J.F., Lemon, E.R., Misra, O.K., Slatyer, R.O., and Stern, W.R. (1964). *Agric. Meteorol.* 1, 294-312.

Biscoe, P.V. (1972). *J. Exp. Bot.* 23, 930-40.

Black, T.A., Gardner, W.R., and Thurtell, G.W. (1969). *Soil Sci. Soc. Amer. Proc.* 33, 655-60.

Blake, G., Schlichting, E., and Zimmermann, U. (1973). *Soil Sci. Soc. Amer. Proc.* 37, 669-72.

Bond, R.D. (1968). *Trans. 9th Int. Congr. Soil Sci.* 1, 339-47.

Boughton, W.C. (1966). Inst. Engs. Aust. Civ. Engng. Trans. CE 3, 83.

Boyer, J.S. (1969). *Science* 163, 1219-20.

———. (1971). *Crop Sci.* 11, 403-7.

———. (1973). *Phytopathology* 63, 466-72.

Brooks, R.H., and Corey, A.T. (1964). Hydrology Papers, Colorado State Univ. 3, 1-27.

Brouwer, R. (1954). Proc. K. Ned. Akad. Wet. Ser. C Biol. Med. Sci. 57, 68-80.

———. (1965). *Symp. Soc. Exp. Biol.* 19, 131-49.

Byrne, G.F., and Rose, C.W. (1972). *Agric Meteorol.* 10, 13-9.

Byrne, G.F., and Tognetti, K. (1969). *Agric. Meteorol.* 6, 151-63.

Chapman, T.G. (1968). In "Land Evaluation". (Ed. G.A. Stewart). pp. 312-23. Macmillan, Melbourne.

Clarkson, D.T., Robards, A.W., and Jackson, S.M. (1972). Agric. Res. Coun. Letcombe Laboratory Annu. Rept. 1972. 5-7.

Cornish, E.A. (1950). *Aust. J. Sci. Res. Ser. B. Biol. Sci.* 3, 178-218.

Cowan, I.R. (1965). *J. Appl. Ecol.* 2, 221-239.

———. (1968). In "Land Evaluation". (Ed. G.A. Stewart). pp. 299-311. Macmillan, Melbourne.

Cowan, I.R., and Milthorpe, F.L. (1968). In "Water Deficits and Plant Growth". Vol. 1 (Ed. T.T. Kozlowski). pp. 131-93. Academic Press, New York.

Cox, E.F. (1966). Ph.D. Thesis, Univ. of Nottingham, Nottingham.

Crank, J. (1956). "The Mathematics of Diffusion". Clarendon Press, Oxford.

Dainty, J. (1969). In "Physiology of Plant Growth and Development". (Ed. M.B. Wilkins). pp. 421-51. McGraw-Hill, London.

Danielson, R.E. (1967). In "Irrigation of Agricultural Lands" (Eds. R.M. Hogan, H.R. Haise and T.W. Edminster). pp. 390-424. Amer. Soc. Agron., Madison.
Davidson, J.M., Stone, L.R., Nielsen, D.R., and Larue, M.E. (1969). *Water Resour. Res.* 5, 1312-21.
Denmead, O.T. (1972). Possible roles of vegetation and soil in controlling evapotranspiration. ANZAAS 44th Congr. Sydney. Sect. 21: 5pp.
Denmead, O.T., and Shaw, R.H. (1962). *Agron. J.* 45, 385-90.
de Wit, C.T., and van Keulen, H. (1972). "Simulation of Transport Processes in Soils". Pudoc, Wageningen.
Dixon, H.H. (1914). "Transpiration and the Ascent of Sap in Plants". Macmillan, London.
Dunin, F.X., and Costin, A.B. (1970). I.A.S.H. UNESCO Symposium on representative and experimental basins. Wellington, N.Z. **1**, 39-54.
Dutt, G.R., Shaffer, M.J., and Moore, W.J. (1972). Ariz. Agric. Exp. Stn. Tech. Bull. No. 196.
Eavis, B.W., and Payne, D. (1968). In "Root Growth". (Ed. W.J. Whittington). pp. 315-38. Butterworths, London.
Emerson, W.W. (1954). *J. Agric. Sci.* 45, 241-5.
Erickson, A.E., and van Doren, D.M. (1960). *Trans. 7th Int. Congr. Soil Sci.* 4, 428-34.
Farrell, D.A., and Greacen, E.L. (1966). *Aust. J. Soil Res.* 4, 1-17.
Farrell, D.A., Greacen, E.L., and Gurr, C.G. (1966). *Soil Sci.* **102**, 305-13.
Fitzpatrick, E.A. (1963). *Appl. Meteorol.* 2, 780-92.
Fitzpatrick, E.A., and Nix, H.A. (1969). *Agric. Meteorol.* 6, 303-19.
Fleming, P.M. (1964). *Trans. 6th Int. Congr. Agric. Engng.* 2, 66-77.
———. (1972). The aerial energy balance in water resource modelling. ANZAAS 44th Congr. Sydney, Section 21: 21 pp.
———. (1973). Modelling of total hydrologic systems. Aust. Water Resourc. Council. Groundwater seminar; Manly Vale: 12 pp.
French, R.J. (1963). *J. Dep. Agric. S. Aust.* 67, 43-8.
Gardner, H.R., and Gardner, W.R. (1969). *Soil Sci. Soc. Amer. Proc.* 33, 192-6.
Gardner, W.R. (1959). *Soil Sci. Soc. Amer. Proc.* 23, 183-7.
———. (1960). *Soil Sci.* **89**, 63-73.
———. (1964). *Agron. J.* 56, 41-5.
———. (1968). In "Water Deficits and Plant Growth". Vol. 1 (Ed. T.T. Kozlowski). pp. 107-35. Academic Press, New York.
Gardner, W.R., and Ehlig, C.F. (1962). *Agron. J.* 54, 453-6.
———. (1963). *J. Geophys. Res.* **68**, 5719-24.
Gardner, W.R., and Hillel, D.I. (1962). *J. Geophys. Res.* **67**, 4319-25.
Gill, W.R., and Miller, R.D. (1956). *Soil Sci. Soc. Amer. Proc.* **20**, 154-7.
Greacen, E.L., Barley, K.P., and Farrell, D.A. (1968). In "Root Growth". (Ed. W.J. Whittington). pp. 256-69. Butterworths, London.
Greacen, E.L., and Hignett, C.T. (1976). CSIRO Aust. Div. Soils Tech. Paper No. 27.
Greacen, E.L., and Oh, J.S. (1972). *Nature (Lond.)* 235, 24-5.
Greacen, E.L., Ponsana, P., and Forrest, J.A. (1974). *Trans. 10th Int. Congr. Soil Sci.* **1**, 319-26.
Green, R.E., and Corey, J.C. (1971). *Soil Sci. Soc. Amer. Proc.* 35, 3-8.
Green, W.H., and Ampt, G.A. (1911). *J. Agric. Sci.* **4**, 1-24.
Greenwood, D.J. (1968). In "Root Growth". (Ed. W.J. Whittington). pp. 202-23. Butterworths, London.
Hackett, C., and Rose, D.A. (1972). *Aust. J. Biol. Sci.* 25, 669-81.
Hanks, R.J., and Bowers, S.A. (1962). *Soil Sci. Soc. Amer. Proc.* 26, 530-4.
Hignett, C.T. (1973). CSIRO Aust. Div. Soils Tech. Mem. 42/1973.
Hillel, D., and Gardner, W.R. (1970). *Soil Sci.* **109**, 69-77.
Holmes, J.W., and Watson, C.L. (1967). *Agric. Meteorol.* **4**, 177-88.
Holtan, H.N. (1961). U.S. Dep. Agric. ARS 41-51.
Horton, R.E. (1933). *Trans. Amer. Geophys. Union.* **14**, 446-60.
Kanemasu, E.T., and Tanner, C.B. (1969). *Plant Physiol.* **44**, 1547-52.
Klute, A. (1972). *Soil Sci.* **113**, 264-76.
Lang, A.R.G., and Gardner, W.R. (1970). *Soil Sci. Soc. Amer. Proc.* 34, 693-5.
Lawlor, D.W. (1972). *J. Appl. Ecol.* 9, 79-98.
———. (1973). *New Phytol.* 72, 297-305.
Lockhart, J.A. (1965). *J. Theoret. Biol.* **8**, 264-75.
Lungley, D.R. (1973). *Plant Soil.* **38**, 145-59.
McIlroy, I.C. (1966). In "Agricultural Meteorology", Vol. 2. World Meteorol. Org. and Commonw. Bur. Meteorol., Melbourne. pp. 409-31.
McIlroy, I.C., and Angus, D.E. (1964). *Agric. Meteorol.* **1**, 201-24.
McIntyre, D.A. (1974). In "Methods for Analysis of Irrigated Soils". (Ed. J. Loveday). pp. 154-66. Commonw. Bur. Soils. Tech. Commun. No. 54, CAB, Harpenden.
Marshall, T.J. (1959). Commonw. Bur. Soils Tech. Commun. No. 50, CAB, Harpenden.
Melhuish, F.M., and Lang, A.R.G. (1968). *Soil Sci.* **106**, 16-22.

Molz, R.J., and Remson, I. (1970). *Water Resour. Res.* 6, 1346-56.
Monteith, J.L. (1965). *Symp. Soc. Exp. Biol.* 19, 205-34.
Newman, E.I. (1969). *J. Appl. Ecol.* 6, 1-12.
Nielsen, D.R., Biggar, J.W., and Corey, J.C. (1972). *Soil Sci.* **113**, 254-63.
Nix, H.A., and Fitzpatrick, E.A. (1969). *Agric. Meteorol.* 6, 321-37.
Ogata, G., and Richards, L.A. (1957). *Soil Sci. Soc. Amer. Proc.* 21, 355-9.
Paltridge, G.W. (1970). *Agric. Meteorol.* 7, 93-130.
Passioura, J.B. (1972). *Aust. J. Agric. Res.* 23, 745-53.
Passioura, J.B., and Cowan, I.R. (1968). *Agric. Meteorol.* 5, 129-34.
Penman, H.L. (1948). *Proc. Roy. Soc. A* **193**, 120-45.
———. (1949). *J. Soil Sci.* **1**, 74-89.
———. (1952). The physical basis of irrigation control. 13th Int. Hort. Congr. London, 1-12.
Philip, J.R. (1957*a*). *Proc. 3rd Int. Congr. Irrig. Drain.* **8**, 125-54.
———. (1957*b*). *Soil Sci.* **84**, 257-64.
———. (1969). *Adv. Hydroscience.* 5, 215-96.
Ponsana, P. (1975). Ph.D. Thesis, Univ. of Adelaide, Adelaide.
Prescott, J.A. (1934). *Trans. Roy. Soc. S. Aust.* **58**, 48-61.
Raats, P.A.C. (1973). *Soil Sci. Soc. Amer. Proc.* **37**, 681-5.
Richtmyer, R.D. (1957). "Difference Methods for Initial Value Problems". Interscience, New York.
Rogowski, A.S. (1971). *Water Resour. Res.* 7, 1575-82.
———. (1972). *Water Resour. Res.* **8**, 1015-23.
Rose, C.S., Stern, W.R., and Drummond, J.E. (1965). *Aust. J. Soil Res.* 3, 1-9.
Rose, C.W., Begg, J.E., Byrne, G.F., Torssell, B.W.R., and Gonz, J.H. (1972). *Agric. Meteorol.* **10**, 161-85.
Ross, P.J. (1971). CSIRO Aust. Div. Soils, Notes on Soil Techniques, 25-32.
Rubin, J., and Steinhardt, R. (1963). *Soil Sci. Soc. Amer. Proc.* **27**, 246-51.
Slatyer, R.O. (1960*a*). CSIRO Aust. Div. Land Res. Reg. Surv. Tech. Paper No. 6.
———. (1960*b*). CSIRO Aust. Div. Land Res. Reg. Surv. Tech. Paper No. 13.
———. (1967). "Plant-Water Relationships". Academic Press, London.
Slatyer, R.O., and McIlroy, I.C. (1961). "Practical Microclimatology". CSIRO, Melbourne.
Talsma, T. (1969). *Aust. J. Soil Res.* 7, 269-76.
Tanner, C.B., and Fuchs, M. (1968). *J. Geophys. Res.* 73, 1299-304.
Tanton, T.W., and Crowdy, S.H. (1972). *J. Exp. Bot.* 23, 600-18.
Taylor, H.M., and Ratliff, L.F. (1969). *Soil Sci.* **108**, 113-9.
Visser, W.C. (1966). Inst. Land Water Manage. Res. Tech. Bull. No. 45, 18-44.
Walter, C.J., and Barley, K.P. (1974). *Trans. 10th Int. Congr. Soil Sci.* **1**, 150-8.
Wang, F.C., and Lakshminarayana, V. (1968). *Soil Sci. Soc. Amer. Proc.* 32, 329-34.
Warrick, A.W., Nielsen, D.R., and Biggar, J.W. (1971). *Water Resourc. Res.* 7, 1216-25.
Whisler, F.D., Klute, A., and Millington, R.J. (1970). *Soil Sci. Soc. Amer. Proc.* **34**, 382-7.
Wilcox, J.C. (1960). *Can. J. Soil Sci.* **40**, 15-27.
Williams, R.F. (1975). "The Shoot Apex and Leaf Growth". Cambridge Univ. Press, Cambridge.
Willoughby, D.R. (1967). *Aust. J. Soil Res.* 5, 21-36.

CHAPTER 9

Mechanisms and models of plant nutrition

G.D. Bowen and B. Cartwright

INTRODUCTION

Many soil, plant and climatic factors interact in crop production, and it is important to recognize this interaction when assessing possible strategies for improving the efficiency of use of soil and plant resources. Intuitive thinking alone is inadequate to forecast the consequences of many processes interacting in time and space, so that some formalization of concepts is necessary. It is improbable that any rationale could ever fully account for all the factors which affect crop growth, but the method of systems analysis does seem to us to provide a tool with which to test the wider implications of research data and hypotheses concerning the relative importance of particular soil or plant properties to crop growth.

Examples of the cases for and against modelling have been presented by Harré (1970) and Passioura (1973), respectively. Further debate is superfluous here, except to note that the degree of critical appraisal which must always be applied to the predictions of models should also be given to the generality of conclusions drawn from other investigations. These include field trials, where many parameters are neither controlled nor monitored and where weather (and soil conditions) are virtually unique to a given season, as well as experiments conducted in the artificial conditions of pots and solution-culture in glass-houses and controlled environment cabinets.

Loomis, Williams and Hall (1971) have reviewed factors affecting crop productivity and the application of modelling techniques. Here we discuss more specifically some of the factors affecting crop nutrition and examine how far current understanding is capable of supporting simulation analysis. We have concentrated mainly on approaches which involve cause and effect hypotheses rather than regression-correlation type models. A number of excellent discussions are already available on the techniques of simulation and their application to agricultural and ecological problems (Jones 1970; Jeffers 1971; Patten 1971, 1972). A series of monographs on computer simulation of processes important in agricultural and related systems is being published by the Centre for Agricultural Publishing and Documentation, Wageningen. The discussions are developed in terms of the IBM Continuous System Modelling Program (CSMP/360), but this should create only minor difficulties for users of other simulation languages. Among the discussions so far published are "Simulation of transport processes in soils" (de Wit and van Keulen 1972) and "Simulation of nitrogen behaviour in soils" (Beek and Frissel 1973).

The simulation of a complex system firstly requires a decision about the overall level of detail needed for a particular purpose. Simple models can be developed quickly to give useful approximations and to identify the most important areas of investigation where more detailed sub-models and data are required. Simplicity, however, often reduces flexibility, so that the behaviour of the model may be narrowly predetermined by the in-built relationships. Sophisticated models incorporating feedback loops and combinations of deterministic and stochastic functions, etc. may be used to analyze more complex representations with possibly greater appeal to realism.

Attempts to formally represent the factors affecting crop growth and nutrition inevitably lead to a requirement for specific data. At the present time, much of the most detailed data available has been obtained using isolated plants grown in controlled environments. As indicated above, considerable caution is required in interpreting the quantitative (and sometimes qualitative) relevance of this information to the field situation where competition between individuals for light, nutrients, water, etc. occurs, and where canopy micro-climate and many soil factors may differ from that experienced by the isolated plant.

CROP GROWTH

General growth functions

The representation of growth as a time-dependent function may be achieved by several curve-fitting methods. Williams (1964) has discussed several examples, including those of the logistic type,

$$W = \frac{K}{1-e^{k_1-k_2t}}$$

and polynomial type, e.g.

$$\log_{10} W = k_3 + k_4t + k_5t^2 + k_6t^3 \quad ....$$

where W is plant dry weight, t is time, K is the upper asymptote and the subscripted k values are constants. It is possible to obtain very good fit to experimental data with these types of formulae, but their use is mainly confined to the description of results rather than explanation, since no biological meaning can be attached to the coefficients.

Mitscherlich-type models (Russell 1972) have been used to analyze the responses of crops to nitrogen and phosphorus under a range of conditions in semi-arid environments. These models are typically of the form

$$\log (Y_{max} - Y) = \log Y_{max} - c(b+x)$$

where Y is yield, Y_{max} is the limiting maximum yield, x is the level of applied nutrient, b is the soil nutrient level at the beginning of the season, and c is a constant.

Russell (1972) discussed the application of such models for predictive purposes and their limitations, particularly under conditions of widely varying Y_{max}, for example where variation in seasonal conditions may considerably affect yield without apparently affecting fertilizer requirement. He noted that a more accurate analysis of the relationship between crop yield at the end of the season and the level of nutrient applied at the beginning of the season is likely to require a more detailed representation of the processes affecting a greater number of soil and crop parameters throughout the growth period than are usually considered in such simple models.

The growth of undefoliated *Stylosanthes humilis* pasture has been predicted very simply (Rose et al. 1972) as a function of accumulated evapotranspiration (ΣET), where ΣET was used as a integrator of the environmental factors affecting above-ground dry matter production. It was claimed that once a correlation between ΣET and dry matter production had been established for the species in question, the relationship was sufficiently robust to be applicable in seasons of quite different weather pattern. The correlation was not thought to be greatly affected by water stress, but strongly sensitive to phosphate nutrition, at least in the case of *Stylosanthes*. Because of the high frequency of phosphate deficiency in Australian soils, such relationships need to be more widely tested before the general applicability of the method can be established.

A more commonly used method for predicting plant growth has been by means of Relative Growth Rates (see models by Scaife and Smith 1973 and Brouwer and De Wit 1969). One difficulty in this approach with nutritional studies is that the relation between percentage of potential growth attained and nutrient status is affected greatly by the age of the plant or tissue. Thus Ozanne, Keay and Biddiscombe (1969) showed that at day 29 the near-maximum yield of ryegrass was obtained at 0.96 per cent phosphorus in the tops, but by day 92 maximum yield corresponded to about 0.1 per cent. Such findings arise in part from the commonly accepted practice of averaging the nutrient concentration for the whole of the tops rather than only the actively growing parts of the plant. The data of Ozanne, Keay and Biddiscombe show a similar shape of response curve for four species of clover, two grasses and two other species for relative yield (per cent of maximum) plotted against per cent phosphorus in tops for each of two times, but show great differences between times and between groups (e.g. clovers v. grasses) in actual values. These differences arise as much from variations in the distribution of leaf and stem, etc. between groups and times as from any inherent differences in physiological efficiency.

Phasic development

The time-course of phasic development (floral initiation, stem extension, anthesis, grain ripening) is affected by environmental factors including nutrition. A useful device for predicting phenological development is that of heat-sum or degree-day accumulation (Nuttonson 1955; Wang 1963). Correlations between levels of applied P and K and the number of degree-days between emergence and silking or maturity, which were also correlated with yield, in maize were observed by Peaslee, Ragland and Duncan (1971). In species which showed strong photo- or thermal control of flowering, nitrogen had little effect on phasic development, but in less sensitive plants nitrogen can modify the initiation and abundance of flowering (Heslop-Harrison 1969).

Assimilation

Dry-matter accumulation by the crop occurs mainly via the process of carbon assimilation. The rate of assimilation affects and, in turn, is affected by other components of the soil-crop system. For example, assimilation leads to an increase in total leaf area of the crop. As a result, gradients of light and carbon dioxide become established which modify the profile of further photosynthetic activity in the canopy. Redistribution of nutrients from leaves which become shaded are among other consequences which may follow from such changes. The rates at which such changes occur and may be optimized in different environments are of interest to systems analysts.

Schemes to represent the interrelationships of some of the components of a soil-crop model are shown in figures 1 and 7. The real world is more subtle than can adequately be shown in this manner. Nevertheless, such schemes do provide a minimal conceptual base on which hypotheses for the different subsystems can be developed. In addition they draw attention to the range of system components and controls which may be eliminated when an attempt is made to isolate one component for detailed observation under laboratory conditions.

Several models are available which are capable of usefully simulating leaf and crop photosynthesis for most canopies under near-optimal conditions, e.g. those of Idso and De Wit (1970), Duncan et al. (1967), Monteith (1965) (see also *Prediction and Measurement of Photosynthetic Activity*, published by the Centre for Agricultural Publishing and Documentation, Wageningen, 1970). The more comprehensive versions take account of the penetration and scattering of direct radiation and diffuse skylight through the canopy, and of the factors affecting leaf temperature and transpiration (Lommen et al. 1971).

The relationship of productivity to net assimilation rate (E) and leaf area index (LAI) has been the subject of considerable investigation, debate and simulation analysis. Two general types of situation have been documented, one in which productivity increases as

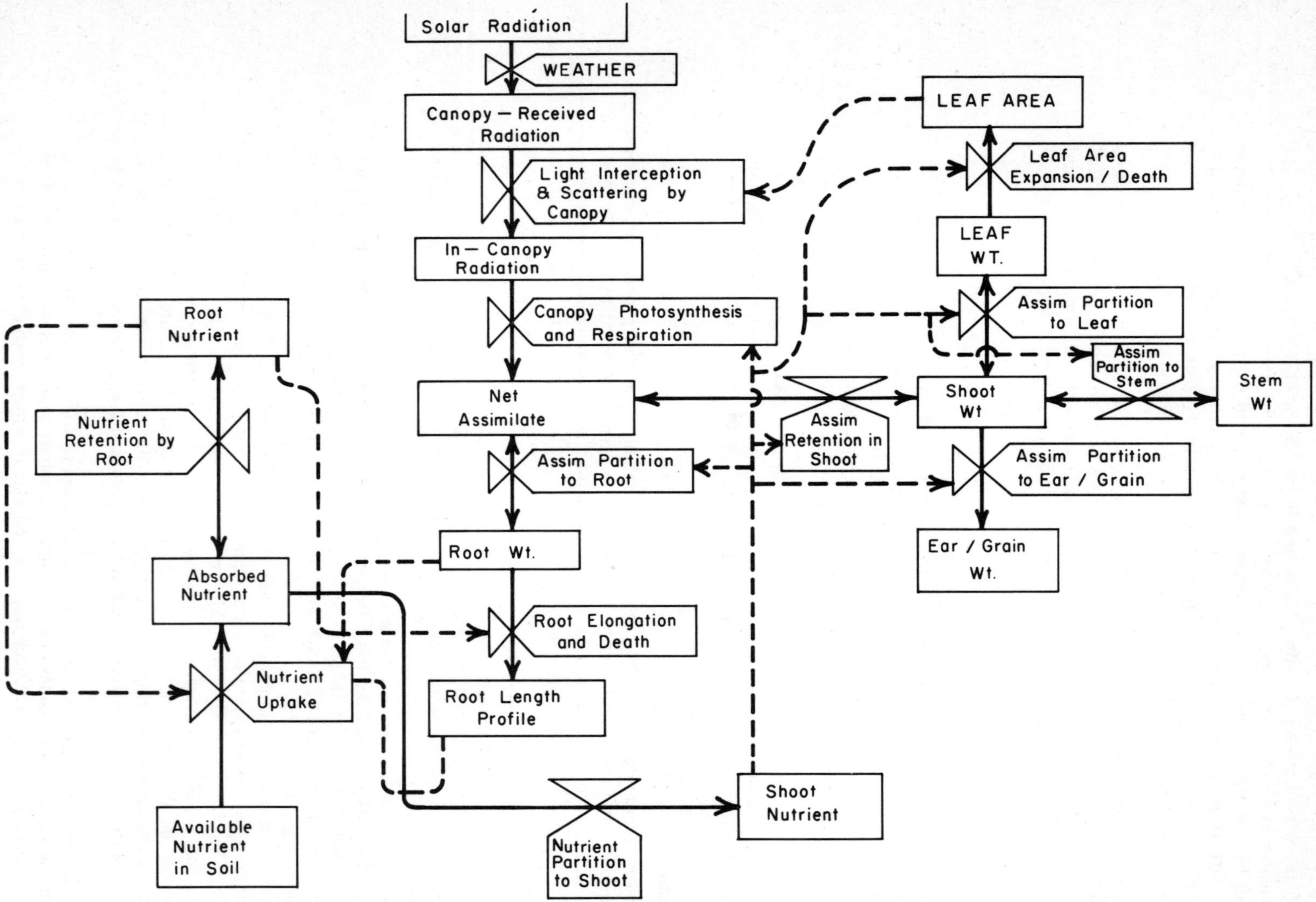

Fig. 1. Assimilation processes in crops. A scheme to represent the main compartments for assimilate and nutrient distribution in a simple model of a cereal crop. The model represents the use of photosynthetically useful radiation in the generation of assimilate, and the partitioning of assimilate for the growth of crop shoot and root. The pattern of continuing use of soil resources is recognized to be dependent upon crop morphology as well as total dry matter production. Thus nutrient uptake is to be calculated in terms of root profile distribution, while the pattern of further growth and assimilation depends upon the nutrient status and structure of the developing canopy. Rectangular boxes are used for measurable variables whose values at any time describe the state of the system at that instant. These boxes are connected by continuous lines to represent material (or energy) flows. The "valve" symbols represent processes which affect the use and distribution of dry

LAI increases to an optimum value, but thereafter declines (e.g. plagiophile canopies such as subterranean clover), and a second in erectophile canopies such as maize and sorghum in which a plateau of productivity is reached as LAI increases. In both cases, maximum productivity occurs at LAI values sufficient to give complete light interception. McCree and Troughton (1966*a*) have presented experimental data which argue against an optimal LAI even in the case of clover-type canopies, because E is itself dependent upon LAI at high levels of LAI, and the light-dependence of photorespiration means that a fixed compensation point cannot be assumed (McCree and Troughton 1966*b*) as has been the case in some models.

The effects of nutrients, particularly nitrogen, on leaf area and tiller production have frequently been demonstrated (Gregory 1937; Watson 1947; Bunting and Drennan 1966; Langer 1966). The total leaf area per plant in gramineous crops is dependent upon the number of tillers and the rate at which leaves are produced (Gott 1961) and senesce on each tiller (Watson 1947; Ryle 1964), together with the maximum area of each leaf. The degree to which nutrients affect all of these parameters depends upon the stage of growth and the conditions of nutrient balance. In wheat plants grown at high density (Puckridge 1963) the maximum area of successive leaves varied according to the supply of nitrogen. In the absence of added nitrogen, leaf 4 was the largest leaf and subsequent leaves were markedly smaller. At an intermediate nitrogen level (50 kg/ha) leaves 4 and 5 were of similar size (21.3 cm^2) and the progressive decline in the size of later leaves was not so severe. At the highest level of nitrogen (235 kg/ha) leaf 5 was the largest leaf but leaves 4-7 were all within the range 25.8-29.6 cm^2.

In most assimilation models the canopy is represented as a series of layers of unit leaf area index. This arrangement allows account to be taken of gradations in leaf aging and photosynthesis rate, angles of leaf display and increases in the amount of non-leaf assimilating surface (sometimes of considerable proportions, Ross and Nilson 1967) during the later stages of growth. Waggoner (1969) simulated the microclimate of a maize canopy, generating profiles of temperature, vapour pressure, insolation per unit leaf area and CO_2 which were generally in fair agreement with observed values. Loomis and Williams (1969) have reviewed the modelling of productivity in relation to canopy architecture and have concluded that while the pattern of leaf display has been shown to be of critical importance, progress in the search for "ideal" canopy structures for different situations requires the interfacing of canopy and microclimate models. Analyses of this type have recently been developed on a theoretical basis by Paltridge (1970, 1972). The results suggest that different canopy architectures may produce greatly different in-canopy profiles of some meteorological variables including leaf and air temperature and specific humidity, particularly under water stress. His models suggested "ideal" types of canopy for optimal growth of pastures in different environments and that the maximum yield may frequently be determined by factors which affect CO_2 transfer within the canopy.

Nutritional stress may act upon canopy structure and function in different ways. In some cases nutrient deficiencies (e.g. N, P, K, S, Mn, Zn, Mo) cause premature senescence of the older, lower leaves and the rate of expansion and final size of young leaves is restricted. Deficiencies of copper and calcium may affect the upper leaves first (Cartwright and Hallsworth 1970; Wallace 1961). Such effects modify the light relations of the canopy and the usual source-sink relationships for assimilate and nutrient distribution. Wallace, Frohlich and Ashcroft (1967) found that several nutrient deficiencies in tobacco produced increases in leaf temperature of 2°-5°C relative to ambient. This response was consistent with effects on stomatal movement and transpiration rates (Wallace and Frohlich 1965).

Ross, Henzell and Ross (1972) simulated a legume-grass mixture with different levels of soil nitrogen and used this to study the relative effects of increasing the values of twenty parameters individually, each by 20 per cent, at one soil nitrogen level. This sug-

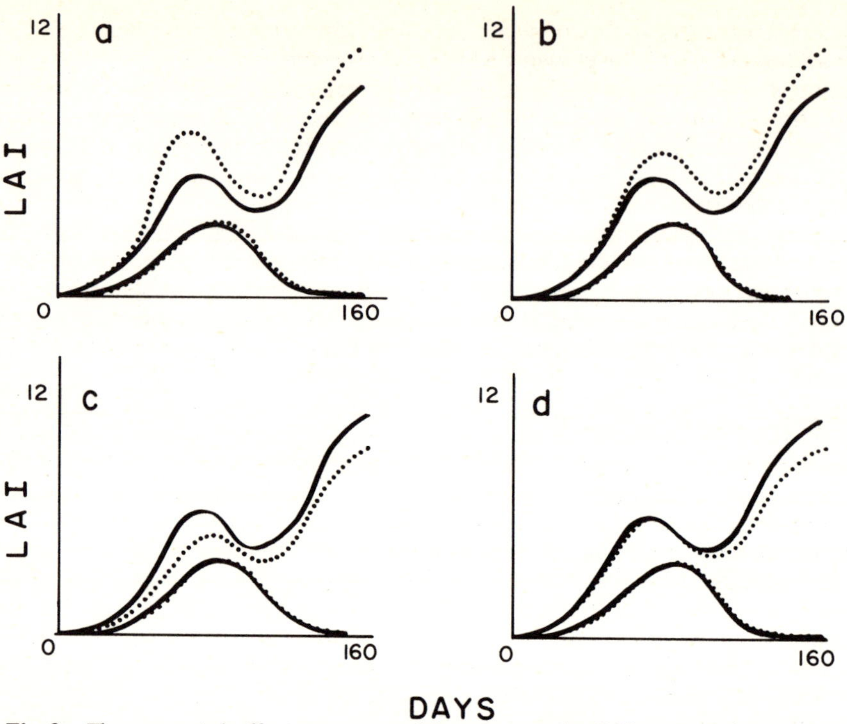

Fig. 2*a*. The computed effect on grass and legume growth of 20 per cent increases in legume values for (*a*) efficiency of conversion of assimilate to leaf area, (*b*) leaf longevity, (*c*) photosynthesis rate, (*d*) extinction coefficient. —— = standard conditions, ... = 20 per cent increase in parameter. Upper lines for clover. (From Ross, Henzell and Ross 1972, reprinted, by permission from *J. Appl. Ecology*.)

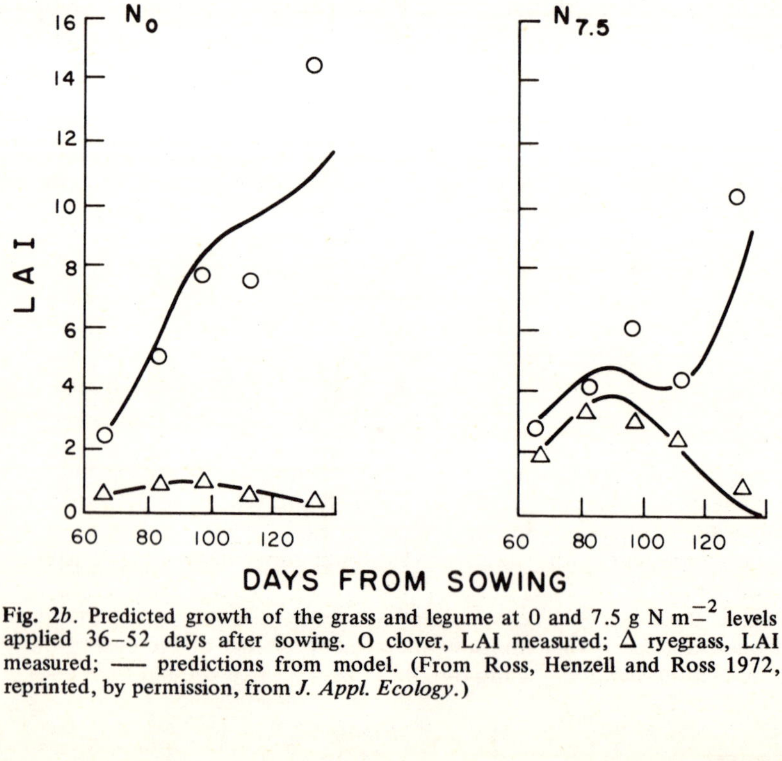

Fig. 2*b*. Predicted growth of the grass and legume at 0 and 7.5 g N m^{-2} levels applied 36–52 days after sowing. O clover, LAI measured; Δ ryegrass, LAI measured; —— predictions from model. (From Ross, Henzell and Ross 1972, reprinted, by permission, from *J. Appl. Ecology*.)

gested that increases in transmission of light through leaves did not increase productivity as much as increasing the efficiency of use of photosynthate for leaf area production, increasing leaf longevity, or increasing photosynthesis rate, in descending order of importance (fig. 2*a*). At 60 days, a 20 per cent increase in efficiency of conversion of assimilates to leaf area in the legume led to approximately 60 per cent increase in LAI of the legume, but an increase in leaf longevity had little effect. At 80 days however, the increased leaf longevity had led to some 20 per cent increase. A 20 per cent increase in leaf resistance to photosynthesis decreased LAI by approximately 20 per cent while the same increase in extinction coefficient of light through the crop (determined by leaf thickness and geometry) was almost without effect at leaf area indices below 10. Figure 2*b* shows the close agreement between measured growth and that predicted by the model at two nitrogen levels in a ryegrass-clover sward.

Assimilation processes

Many studies have reported effects of nutrition on rates of assimilation per se, e.g. deficiencies of nitrogen (Ryle and Hesketh 1969), potassium and magnesium (Peaslee and Moss 1966), sulphur (Herath and Ormrod 1972) have been found to depress photosynthesis. Analysis of these responses in terms of the diffusive resistances controlling gaseous exchange (so far as it is experimentally possible at present) has been reported by Ryle and Hesketh (1969), Nevins and Loomis (1970), Osman and Milthorpe (1971) for nitrogen, Terry and Ulrich (1973), Cartwright et al. (1974) for phosphorus and Fisher and Hsiao (1968) and Graham and Ulrich (1972) for potassium. Increases in both stomatal and the residual diffusive-plus-photochemical resistance are found, the magnitude depending upon the species and nutrient in question. Table 1 (from Cartwright et al. 1974) shows the effects of insufficient phosphorus applications (63 kg/ha), adequate (625 kg/ha) and supra-optimal (1,510 kg/ha) phosphorus applications (as superphosphate) on assimilation parameters of the 4th leaf of wheat grown in a controlled environment cabinet. The deficient and adequate plants differed little in dark or light respiration but differed greatly in photosynthesis rate due to smaller stomata on the deficient plants, lower photochemical efficiency of the deficient leaves and their increased resistance to transport of CO_2 in the leaf.

Table 1. Effects of phosphate nutrition on assimilation parameters

	Phosphate level (kg superphosphate/ha)		
	63	625	1,510
Photosynthesis ($ng\ CO_2\ cm^{-2}s^{-1}$)	21.8	41.1	33.9
Dark respiration ($ng\ CO_2\ cm^{-2}s^{-1}$)	2.50	2.36	1.81
Photorespiration ($ng\ CO_2\ cm^{-2}s^{-1}$)	4.18	4.28	5.20
Compensation point ($\mu l . l^{-1}$)	78	33	40
Resistances ($s.m^{-1}$)			
Total resistance	2,734	1,295	1,671
Stomatal resistance	423	166	233
Residual* (transport + biochemical)	2,301	1,129	1,439

SOURCE. Data from Cartwright et al. (1974). Data on the 4th leaf of wheat, *var.* Halberd, grown in a controlled environment in a lateritic podzolic soil.

* By difference.

Theoretical examination of the factors affecting net photosynthesis in a single leaf has been made using simulation models (Waggoner 1969; Lommen et al. 1971). Among the factors predicted to be major determinants of productivity, respiration (especially photorespiration, Zelitch 1971), sensitivity to CO_2 and minimum stomatal resistance should be capable of some manipulation, e.g. by genetic selection. Although maximum attainable photosynthesis under ideal conditions (for which surveys among species and genotypes have been made, e.g. Hesketh 1963) did not appear to be an effective controller of assimilation in Waggoner's analysis, he suggested the factors controlling the sensitivity of the assimilating system to light and CO_2 would effectively determine productivity under less than ideal conditions. These analyses are directly relevant to the effects of nutrients and water on crop canopy structure and function.

The rate of net assimilation (NAR) in single leaves increases during leaf expansion and declines during leaf senescence. Hopkinson (1964) showed that NAR in cucumber leaves reached a maximum value at about the time of complete leaf expansion, but began to decline almost immediately and by the time the leaves had reached full expansion they were net exporters of phosphorus. Such a rapid decline in metabolic activity is not typical of all plants, e.g. wheat leaves can sustain a relatively high rate of activity for some time after full expansion. A decline of 30 per cent from maximum activity in wheat leaf 3 extended over 28 days in the studies of Doodson, Manners and Myers (1964). The decline in photosynthesis rate with leaf age is further complicated by nutritional status; Moss and Peaslee (1965) observed assimilation rates of 40-36 mg CO_2 dm^{-2} hr^{-1} in leaves 2-11 on well-fertilized maize plants but in the equivalent leaves of potassium-deficient plants the rates were found to decline from 33 to 1 mg CO_2 dm^{-2} hr^{-1} as a function of leaf potassium content.

The maintenance of photosynthetic activity in the flag leaf of wheat has been shown to be regulated by the capacity of sinks to use assimilate (King, Wardlaw and Evans 1967). The retardation of senescence in bean leaves treated with cytokinin (Adedipe and Fletcher 1970) was associated with increased turnover of phosphorylated intermediates (e.g. adenosine tri-phosphate), indicating retention of metabolic control of energy production and photosynthate utilization and, no doubt, of protein synthesis. Although cytokinins appear to exert their effects in conjunction with other factors (Hall 1973), roots are the sites of cytokinin synthesis within the plant (Fox 1969) and this has important implications for control of shoot-root interrelations, leaf longevity, and efficiency in nutrient utilization (Vose 1962; Troughton 1968).

The effects of environmental factors (light, water, nutrients and temperature) on the distribution of photosynthate from leaves has been reviewed by Wardlaw (1968). It is unlikely that the translocation process itself could be rate controlling for assimilate transfer (Milthorpe and Moorby 1969). For the present, models describing dry matter partitioning with relevance to crop growth can be framed in general terms only (Thornley 1972).

NUTRIENT UPTAKE

Root growth, distribution and morphology

Uptake of nutrients and water by a crop is dependent upon continued root growth and activity. The physiological control of root growth and function under field conditions is not well understood and this deficiency is part of a more general problem concerning the control of assimilate partitioning between leaves, fruit, stems and roots. Apart from root elongation, important aspects of root function in ion uptake include the absorbing ability of the root, the production of root hairs and the extension of the root system by fungal associations (mycorrhizas). The data of Ozanne, Keay and Biddiscombe (1969) and figure 3 (from Barrow 1975) show that top:root ratio varies markedly with nutrition and time and thus simple allocation of a constant percentage of assimilates to roots is too simplified a procedure in simulating root growth, with or without nutrient stress.

Byrne and Tognetti (1969) empirically modelled root growth in relation to water use by a *Stylosanthes* pasture by using exponential functions for tap root and lateral growth, with growth of laterals ceasing at 15 bars suction. Another approach comes from detailed studies of root growth by Hackett and Rose (1972) who indicated the tendency for the mean length of barley roots, $\bar{l}$, to be roughly constant with time. When the rate of increase of volume of the highest order of laterals became linear, laterals of the next highest order emerged suggesting that the root developed new growing points to match the exponential growth of the shoots. The constancy of $\bar{l}$ suggested a method for

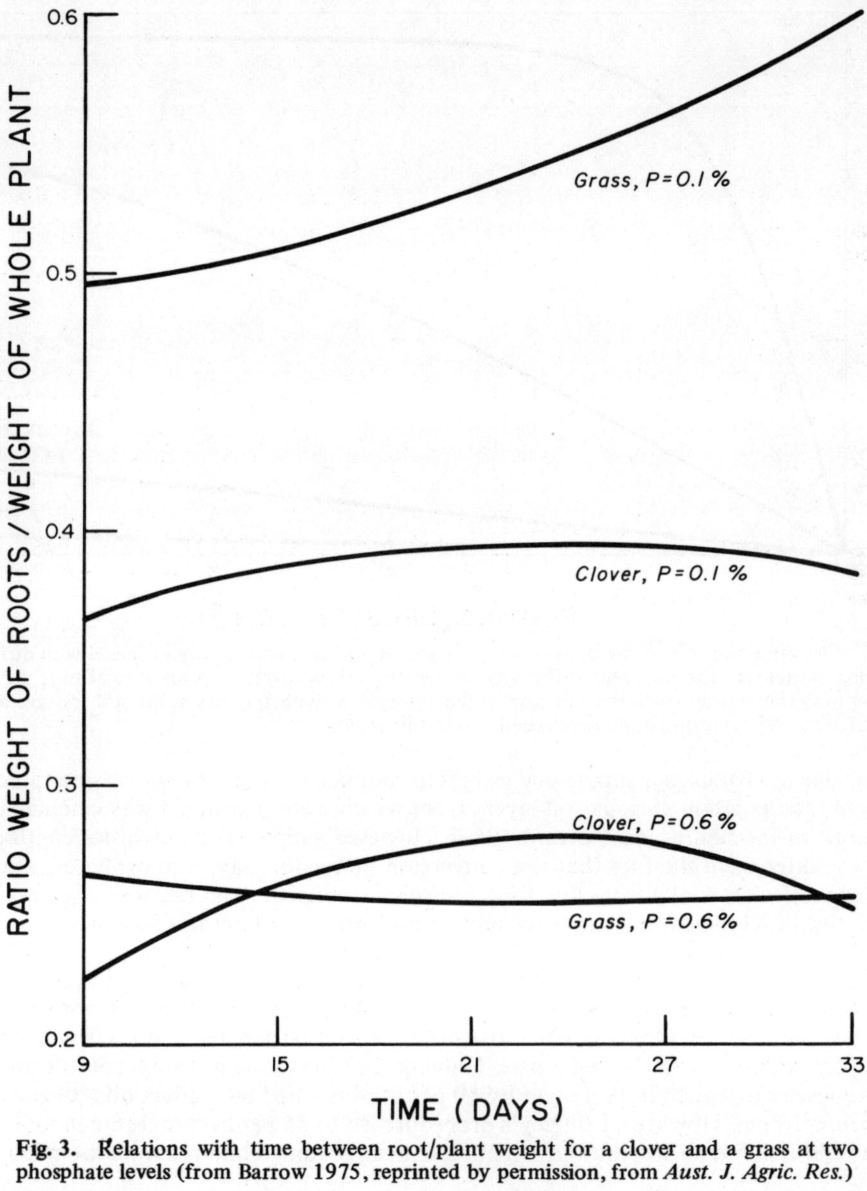

Fig. 3. Relations with time between root/plant weight for a clover and a grass at two phosphate levels (from Barrow 1975, reprinted by permission, from *Aust. J. Agric. Res.*)

generating root morphology in plant models. However $\bar{l}$ can be affected markedly by such factors as nutrition, soil temperature and variety, e.g. data derived from Hackett (1968) with solution-grown barley gives $\bar{l}$ for the variety Proctor at 15.9 mm and 31.6 mm under phosphate deficient and phosphate sufficient conditions respectively, and 18.9 mm and 27.5 mm respectively for Maris Badger. Bowen and Rovira (unpublished) found $\bar{l}$ for the wheat variety Sonora was 21.0 (S.E.$\pm$1.8) mm at 10°C soil temperature and 13.4 ($\pm$0.7) mm at 5°C soil temperature. Nevertheless, the method has possibilities for further development.

The generation of root length (suitably weighted) by such methods is useful in general models, e.g. ion uptake may be calculated approximately by using relationships between uptake, root concentration and diffusivity of particular ions in soil as shown by Barley (1970) (fig. 4).

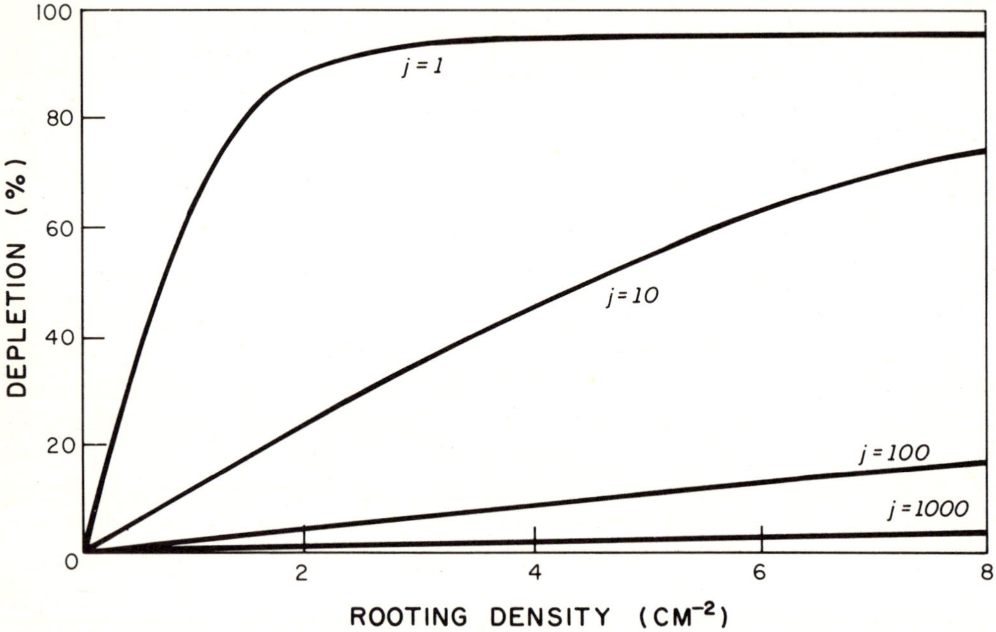

Fig. 4. The influence of rooting density on depletion of a labile part of nutrient in soil with differing buffering capacities. The apparent diffusivity of ions in soil is inversely proportional to j, the differential buffering power which is a function of the absorption characteristics of the soil and soil water. (From Barley 1970, reprinted, by permission, from *Adv. Agron.*)

Paltridge (1970) used a simple dry weight to root density (length) conversion factor to generate root length in various soil layers from which water extracted was calculated by the model of Passioura and Cowan (1968). However simple root weight to length conversions suffer from the fact that the conversion factor for, say, a phosphate sufficient plant can differ markedly from that for a phosphate deficient plant because of changes in the relative development of main axes and various orders of laterals (Bowen, Cartwright and Mooney 1974).

Lungley (1973) developed a root model using rates of elongation and branching frequencies of root axes of each order. Figure 5 illustrates his graphical output using published data on cereal roots. The most sensitive root parameter was the frequency of first order laterals along the main axis. Halving the inter-branch distance in the surface 10 cm increased growth in that layer by 80 per cent, but had little effect on root distribution below 10 cm. However, Lungley's procedure led to 25 per cent increase in total root weight and a different distribution might have been obtained had the total assimilate for

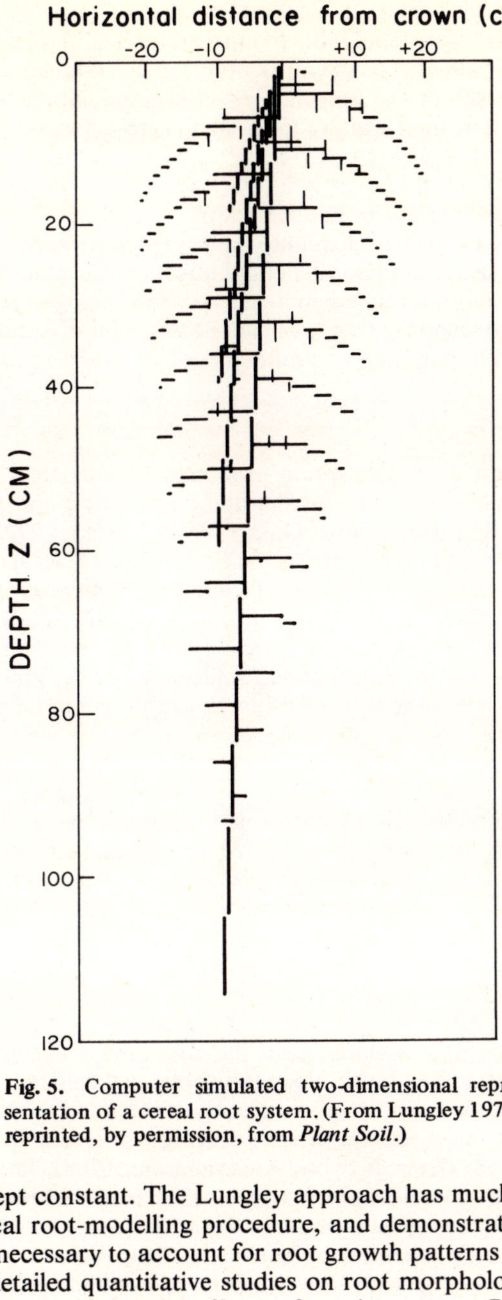

Fig. 5. Computer simulated two-dimensional representation of a cereal root system. (From Lungley 1973, reprinted, by permission, from *Plant Soil*.)

root growth been kept constant. The Lungley approach has much to recommend it as a general and empirical root-modelling procedure, and demonstrates the particular types of data that will be necessary to account for root growth patterns under given soil conditions. While most detailed quantitative studies on root morphology have been on solution grown plants, and little on the effects of nutrient stress, Bowen, Cartwright and Mooney (1974) studied root geometry of wheat plants at three soil phosphate applications. They measured the criteria used in the Lungley model; soil phosphate deficiency had little effect on formation of seminal and adventitious roots or on the frequency of first order laterals but halved the frequency of second order laterals. Phosphate deficiency also decreased elongation rates of all axes except the first and decreased the elongation rates of all first order and second order laterals.

Two factors which need to be distinguished in the development of plant nutrition and root morphology models are: (i) general effects of assimilate distribution, i.e. morphological effects not specifically due to nutrient level, and (ii) effects on root growth and morphology specifically due to the nutrient (or other variable) under study. It is possible that the effects observed by Bowen, Cartwright and Mooney are not specifically due to phosphate and that their data reflects the general allocation of assimilates when they are in short supply. However other nutrients have been shown to have specific effects, e.g. Hackett (1968) found potassium deficiency completely inhibited the formation of secondary laterals by barley in solution culture.

Generally soil nutrients will be localized, due to heterogeneity in soil itself or the localized placement of fertilizers and this could affect root profiles profoundly. Hackett (1972) found that not only did nitrate increase root branching in solution-grown wheat plants, but also that the response was localized at the point of application. Proliferation of roots around fertilizer particles is commonly observed in the field.

Ion uptake by roots

The plant exerts a control on absorption processes as indicated by specificity in ion uptake. Detailed mechanisms of ion uptake and competition and antagonism between ions have been frequently reviewed (e.g. Dainty 1969; Epstein 1972) and will not be discussed here. Curvilinear stepwise uptake-concentration isotherms have been demonstrated in ion uptake studies for a range of cations and anions. One interpretation of such data is that the steps represent the saturation of carrier systems which may be analyzed by Michaelis-Menten kinetics (Epstein 1972), but other interpretations are also possible (Briggs, Hope and Robertson 1961; Anderson 1972). A mechanistic interpretation of these kinetics is not without hazards as they represent uptake by a heterogeneous set of cells and unless special precautions are taken to grow plants aseptically, microorganisms on the root surface can distort the uptake data (Bowen and Rovira 1966; Barber 1968).

There appear to be limits below which plants cannot achieve net uptake and these are of the order of 10^{-7}M for nitrate and 10^{-8}M for phosphate and potassium (Barley 1970). Barrow (personal communication) has suggested that the limit for grasses for phosphate is lower than for legumes, thus leading to a greater responsiveness of many grass species to phosphate applied at sub-optimal levels. Mosse (1973) has suggested one reason for the spectacular increase in phosphate uptake following inoculation of crop plants with mycorrhizal fungi may be that they can absorb phosphate from lower soil solution concentrations than can their host plant.

It is becoming increasingly apparent that the commonly accepted view that absorption ability of roots is restricted to the elongating zone just behind the root apex is incorrect and that the main axis can maintain an absorbing ability for many ions for considerable lengths from the apex, the length varying somewhat for the ion concerned (Rovira and Bowen 1968; Drew, Nye and Vaidyanathan 1969). The pattern of uptake is temperature sensitive: pine seedlings growing at 25°C soil temperature show maximum phosphate absorption in the apical centimetre but when grown at 14°C soil temperature, maximum uptake is several centimetres behind the apex (Bowen 1970). The usefulness of high absorption abilities retained by older parts of the root however depends on transfer processes in soil being adequate to supply the particular ion to such root surfaces. Analyses of ion uptake by roots in soil (see Barley 1970) suggest that where ions diffuse to roots very slowly (e.g. phosphate, molybdenum, zinc) their concentration at the root-soil interface is reduced, and maintained near zero because replenishment from soil is slow. Therefore the persistence of absorption ability is of limited use to the plant in the short term for such ions. However, a fractional increase in the radius of the soil cylinder providing ions to the roots by diffusion may become highly significant (e.g. for phosphate) with specialized organs with a longevity of several months, e.g. mycorrhizas of trees (Bowen 1973). Continued root growth is important for the sustained uptake of ions, especially those which diffuse only slowly.

As well as the chemical and physical factors affecting root growth (see Greacen, this volume) the general soil microflora can retard root growth significantly (Bowen and Rovira 1961) and the extreme case of this is root pathogens and pests. It can be anticipated that decreased growth of plants affected by many root diseases, without death occurring, will be due to water and nutrient stress. Microorganisms are also intimately involved in plant adaptations to low nutrient conditions. Most such associations recorded so far involve infection with microorganisms, e.g. legume and non-legume nitrogen fixation, and mycorrhizas of various types. However under low nutrient conditions non-infecting rhizosphere microorganisms trigger extensive lateral root formation of proteaceous plants (proteoid roots) (Lamont and McComb 1974; Malajczuk and Bowen 1974) which leads to greatly increased exploration of soil, increased absorbing ability and effective interception of nutrients leached from leaves and decaying litter.

Root hairs

By enlarging the effective radius of the root by several hundred microns, root hairs can markedly increase the volume of soil explored for poorly mobile ions. Thus Lewis and Quirk (1963) found with ^{32}P autoradiography that the major zone of depletion corresponded to the cylinder formed by the root hairs. Using quantitative autoradiography of ^{33}P (a much softer β emitter) Bhat and Nye (1973) have shown depletion of phosphate from within the root hair zone of rape (some 60 per cent in 3 days) but substantial supply of P well beyond the root hair zone. For more mobile ions, e.g. potassium, root hairs though still important may play a less significant role, depending on the particular soil condition (Drew, Nye and Vaidyanathan 1959). Some of these relations are shown in figure 6.

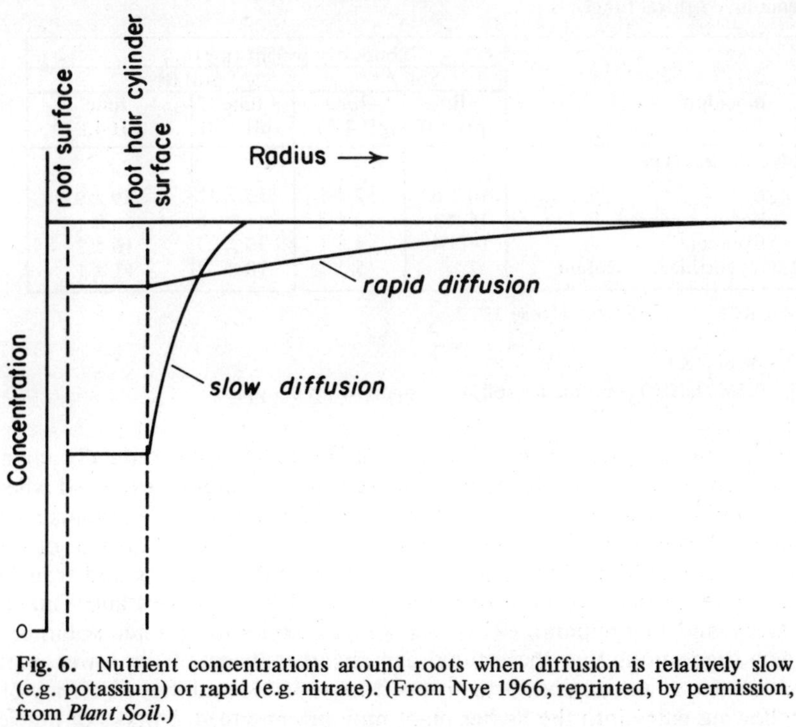

Fig. 6. Nutrient concentrations around roots when diffusion is relatively slow (e.g. potassium) or rapid (e.g. nitrate). (From Nye 1966, reprinted, by permission, from *Plant Soil*.)

The control of root hair production and the effect of soil nutritional factors have received scant attention; calcium is known to be essential for their formation and growth (Cormack 1962). Champion and Barley (1969) found root hairs could not penetrate compact clay with a voids ratio less than 1.0. Some soil microorganisms can cause a marked reduction in root hair production, and microbially induced reductions in root length and root hair growth could reduce the volume of soil contained by the root hair cylinder by up to 60 per cent (Bowen and Rovira 1971). It is possible that elimination of such "sub-clinical" pathogens is a large part of the reason for the large increases in growth of trees and wheat following soil fumigation (Theodorou 1971; Rovira 1972); the neglect of such microbial restrictions on plant growth may have led to serious underestimation of the potential productivity of many soils.

Mycorrhizas

Large increases in uptake of most elements from soil can occur by fungal symbioses with the root, both with ectomycorrhizas of some forest tree species and endomycorrhizal infection of a wide range of species encompassing grasses, legumes, agricultural, forestry and horticultural plants (Bowen 1973; Mosse 1973). Increased uptake is very marked where soils are low in available nutrients, especially poorly diffusible ions. One example of the kinds of responses by endotrophic mycorrhizas is given in table 2 (drawn from Mosse 1972) which shows (i) between fourfold and sevenfold increases in growth of *Paspalum notatum* in two Brazilian soils when inoculated with a highly efficient mycorrhizal fungus (in this case the response was mainly due to increased phosphate uptake), (ii) that some fungal strains are far superior to others, and (iii) an interaction with soil acidity.

Table 2. Response of *Paspalum notatum* seedlings to inoculation with endomycorrhizal fungi

Inoculum	Shoot dry weight (mg) Soil A† + lime (pH 5.0)	Soil A† – lime (pH 4.7)	Soil B† + lime (pH 5.8)	Soil B† – lime (pH 4.8)
Mycorrhizal type				
E3	40 ± 6*	12 ± 1	115 ± 21	59 ± 9
Honey coloured	34 ± 1	10 ± 2	41 ± 8	65 ± 1
Laminate	14 ± 4	4 ± 1	33 ± .7	16 ± 2
No mycorrhizal inoculum	7	5 ± 1	10 ± 2	11 ± 1

SOURCE. Data from Mosse 1972.

* Mean ± S.E.

† 0.5M $NaHCO_3$ soluble P : soil A 1.7 ppm; soil B 2.9 ppm.

Obviously inoculation with selected strains of the fungi could radically change the potential productivity of a soil. Responses to inoculation would be expected where soils are low in available nutrients, and where natural populations of these organisms are low or of a poorly efficient type. Increased uptake results mainly from growth of the fungi into soil for considerable distances, uptake of poorly diffusible ions and translocation and release of the nutrient to the host plant (Bowen 1968; Sanders and Tinker 1971; Hattingh, Gray and Gerdemann 1973). As well as exploring greater volumes of soil there are also suggestions that these fungi can absorb nutrients from lower concentrations in soil solution than can uninfected roots (Mosse 1973) and that efficiency of use of nutrients following entry into the higher plant may be enhanced. Uptake of nutrients by the highly efficient exploration of inter-root soil volumes afforded by mycorrhizal hyphae is not only more conservative of energy than producing more roots, but is also important for plants with low rooting densities competing with plants with high rooting densities. Bevege, Bowen and Skinner (1975) cite an experiment with clover in a nutrient poor soil in which the hyphae external to the root comprised 1 per cent of the dry weight of the plant but led to plant growth two and a half times that of uninoculated plants.

It may be difficult to incorporate root hairs and mycorrhizas into crop models in specific terms. They are likely to be included in a more general way in terms of uptake activity per unit length of root or increase in effective rooting density.

Nutrient distribution within the plant

Initial partitioning. The lack of a soundly based rationale for quantitatively predicting the partitioning and redistribution of dry matter and nutrients between different plant tissues or even simply between shoot and root remains a serious limitation to the prediction of growth in various environments. The problem is widely recognized, and modellers are often forced to use empirical methods such as observed shoot:root ratios (Patefield and Austin 1971; Rose et al. 1972). Thornley (1972) attempted to overcome this problem by developing a model which predicted shoot:root ratios according to the separate activities of the shoot and the root. The shoot and root supplied carbon and nitrogen respectively to substrate pools from which they could be withdrawn according to predetermined carbon:nitrogen ratios for the growth of new tissues. Although the empiricism is seen to be shifted from one ratio to another, the representation of internal competition for available resources is a closer approach to what is needed for a self-generating model of growth.

Data such as that of Keay, Biddiscombe and Ozanne (1970) on eight annual pasture species clearly demonstrate that percentage of phosphate going to the tops of plants is a function of species, the phosphate absorbed and the growth stage of the plant. At low phosphorus concentrations the grasses translocated more of the absorbed phosphate to the tops in the early phases of growth than did clovers. Translocation by each group decreased somewhat between days 29 and 87 but as the plant aged, phosphorus, not only from absorption but also from remobilization in the roots, was translocated to the tops. Species which responded most at low phosphate levels tended to translocate a higher proportion of absorbed phosphorus to the tops. The process of translocation of inorganic solutes in plants has been reviewed by Lauchli (1972).

Remobilization. Remobilization as a significant source of nutrients (e.g. nitrogen, phosphorus, potassium) increases with plant age. Boatwright and Viets (1966) showed that wheat plants adequately supplied with phosphate over the first five weeks could meet their future requirements (if necessary) by retranslocation. It has been suggested (Williams 1955) that during development the mineral nutrients required by a crop to complete its maturation must come increasingly from remobilization. This is because assimilates are diverted towards reproductive growth particularly after flowering so reducing the amount available for the growth of new roots. Since $H_2PO_4^-$, Cu^{2+}, Zn^{2+} and Mn^{2+}, etc. diffuse only very slowly in soil, continued root growth would be necessary to maintain uptake, and redistribution avoids this need. Hopkinson (1964) and Greenway and Gunn (1966) demonstrated the transition exhibited by leaves from net importers of phosphorus when young to net exporters when senescing. Potassium and nitrogen have similar mobility to phosphate in the plant, but calcium is remobilized only with difficulty (Lauchli 1972). In nutrient stress situations there may also be extensive remobilization of nitrogen, phosphorus, potassium and zinc from mature leaves in order to maintain the activity of younger tissues (Williams 1955). By contrast, sulphur, manganese and molybdenum (Williams 1955) and probably copper (Cartwright and Hallsworth 1970) show little mobility from leaves, though rather more from stems. Phosphorus movement in tomato leaves is particularly sensitive to water status (Gates 1964), mobilization from leaves to the stem beginning at an early stage of water stress.

A special aspect of remobilization is the phosphate stored in the fungal component of ectomycorrhizas in times of plentiful nutrient supply, which moves to the higher plant when phosphorus is in short supply (Harley 1969). Similar mechanisms can operate to some extent in higher plants and may involve the use of inorganic polyphosphate reserves (Jeffrey 1964).

Leaching of nutrients from plants. The leaching of nutrients from leaves by rainfall, dew, or irrigation spray (reviewed by Tukey 1970) could be important for some ions. Losses vary between plant species and nutrients, e.g. less than 1 per cent loss of zinc and phosphorus over 24 hours, up to 10 per cent loss for potassium, calcium, magnesium and strontium and up to 25 per cent for sodium and magnesium. Solution adhering to the leaf contains dissolved carbon dioxide so there tends to be a greater extraction of cations than of anions. Loss from older leaves is usually less than from young leaves and lowers the nutrients available for redistribution from the older leaves. Depending on the species, up to 6 per cent of the leaf dry weight can be leached over 24 hours and although most of this is thought to be carbohydrate, losses of other metabolites, hormones, etc. also occur. Organic and inorganic nutrients remaining on the leaf surface form a substrate for growth of phyllosphere microorganisms, some of which have been suggested to fix atmospheric nitrogen non-symbiotically (Ruinen 1965).

Light, continuing rain is a more efficient extractant than heavy falls of short duration. Intermittent rain which permits repeated wetting and drying increases the degree of leaching, so that the occurrence of heavy dews in areas of low rainfall may be an important factor in nutrient loss. Nutrients lost from above-ground parts can be intercepted by lower leaves and roots, but losses resulting from late-season rainfall (after flowering), when root activity is declining, may not be recovered.

NUTRIENT AVAILABILITY

The availability of nutrients in soil to field crops depends upon many chemical, physical and biological factors. The chemical testing methods presently used to assess nutrient availability permit the ranking of soils on a relative basis and may be calibrated to give approximate fertilizer recommendations. The amount of nutrient actually available to a crop over a growing season may depend upon factors independent of the chemical test, such as the particular crop or cultivar, soil water and soil biological activity.

Natural accession

Some accession of nutrients occurs in rainfall but this is usually small except in highly industrialized areas. In most parts of the world nitrogen accession is usually less than 10 kg/ha annually (Moore 1966). Australian figures range from 1.5 kg N/ha at Katherine to some 4.5 kg/ha in Brisbane, Roma and Cairns (Wetselaar and Hutton 1963). Wetselaar and Hutton found no relation between nitrate content of rain water and incidence of lightning and considered most of the salts in rain water on the continent (6-7 kg/ha at Katherine) form part of a terrestrial cycle and are not true accession. In coastal areas salts in rain water come largely from the sea: Hutton and Leslie (1958) estimated 3-8 kg sulphur/ha annually on the coast of Victoria and some 2 kg/ha in most other places in Victoria. However in more highly industrialized countries using fossil fuel, sulphur accession can be ten times this (Stevenson 1968). At twenty-four sites in Victoria, Hutton and Leslie recorded accessions of calcium and potassium of 0.5-2 kg/ha/annum.

Estimates of the nitrogen input by non-symbiotic nitrogen fixing organisms vary widely. Studies by Paul, Myers and Rice (1971) on grasslands in Canada suggest a fixation of approximately 1 kg/ha/growing season but data suggesting a nonsymbiotic fixation of some 50 kg/ha/annum are not uncommon (Moore 1966). In semi-arid environments fixation by blue green algae in surface crusts could be significant; Tchan and Beadle (1955) estimated such fixation in semi-arid areas of western New South Wales would not exceed 2-3 kg N/ha/annum but estimates from other parts of the world have claimed fixation up to ten times this amount (Moore 1966). Nitrogen fixation by microorganisms and lichens in the phyllosphere and on the trunks of trees could be significant over a long period (Dension 1973) but is likely to be small on crops.

Nitrogen fixation by nodulated legumes and symbiotic associations of non-legumes (e.g. species of *Casuarina*) is commonly reported to add 50-200 kg N/ha/annum depending upon the growing conditions. There does not yet seem to have been any attempt to use a systems approach to integrate present knowledge on soil, environmental, and microbial-plant factors affecting nitrogen fixation in legume pastures and grasslands.

Nutrient loss

Models of the leaching and dispersion of materials in soil solution based on laboratory studies have been developed for the movement of pesticides (King and McCarty 1968) and boron (Tanji 1970) in soil columns. A comprehensive model of nutrient movement has been produced by Dutt, Shaffer and Moore (1972). A series of empirical relations were used to account for microbial transformations between organic N, ammonia N, nitrate N and urea N as a function of temperature and soil moisture. Nitrate uptake by the plant was related to root growth and water uptake (given by other sub-models). This then allowed a calculation of nitrate concentration in leachate (of interest in movements of nitrate into ground waters and streams) over thirteen years as a function of fertilizer application. Such approaches are useful provided the appropriate empirical constants can be obtained for each site.

Loss of nutrients by microbial action, e.g. denitrification, does not appear to have been specifically modelled in nitrogen transformation studies but should be capable of inclusion with empirically derived relationships with moisture, temperature and pH.

Nutrients in soil solution

Nutrient accession from applications of mineral fertilizer is a simple input for modelling purposes, but what this represents in terms of nutrient concentrations in soil solution throughout the growing season is not so clear. The principles governing the short-term equilibration of nutrient such as P and K between solid and solution phases are now fairly well established (Sutton and Gunary 1969; Arnold 1969). The system is commonly represented

Nutrient in soil solution	⇄	Labile Nutrient	⇄	Non-labile Nutrient

where the approach to equilibrium between the solution and so-called labile phases is relatively rapid but that between the labile and non-labile phases is extremely slow.

A conceptual representation of the factors affecting nutrient availability and uptake is shown in figure 7. The scheme is possibly too detailed for whole crop simulation, but the principal physico-chemical determinants of availability—the chemical potential of the ion in soil solution, the potential capacity of the soil to rapidly replenish the soil solution and buffer it against ion depletion, the total quantity of nutrient in the soil system and the diffusion of the ion into zones of depletion—are represented.

It is usually necessary to take account of both nutrient potential (an "intensity" parameter) and the immediate potential buffering capacity of the soil (Beckett and White

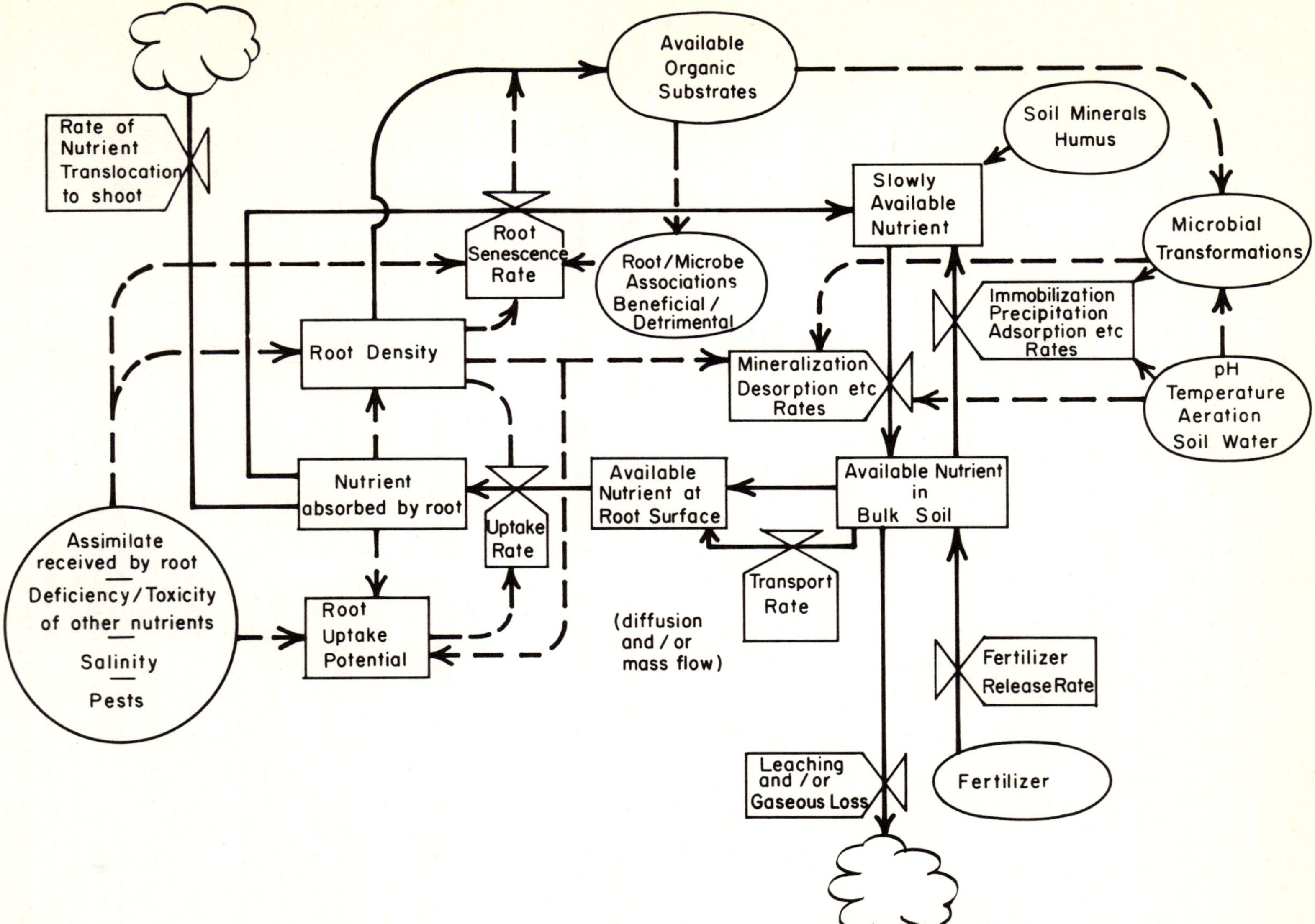

Fig. 7. Nutrient availability and absorption. A scheme to represent the interrelations of some of the soil factors affecting nutrient availability to a crop. The scheme is structured as for figure 1, but in addition the state variables are divided into two subsets: (*a*) those directly involved in considerations of nutrient avail-

1964) for a satisfactory index of nutrient availability, particularly where comparisons are made between different soils. Under controlled conditions, good correlations of these parameters have been obtained with plant uptake of phosphorus (Barrow, Ozanne and Shaw 1965), potassium (Barrow 1966), magnesium (Salmon 1966) and zinc (Tiller, Honeysett and de Vries 1972). Considerable effort has been given to devising extractants and conditions of extraction for simple tests of availability which can take account of both these parameters simultaneously. Amongst similar soils such tests are often reliable, but when applied for predictive purposes in the field to soils of greatly different properties their general utility may be expected to suffer. For individual soils, simple indices of nutrient potential have allowed relative availability throughout the growing season to be monitored (Blakemore 1966; Garbouchev 1966).

Transfer to roots

The processes of diffusion and mass flow affecting the transfer of nutrients to plant roots in soil have been comprehensively reviewed by Olsen and Kemper (1968) and Barley (1970). Earlier theoretical models of uptake by roots dealt principally with idealized situations in which roots were assumed to be uniformly distributed in soil, soil moisture was adequate and the effects of soil microorganisms were ignored. More recently the effects of rooting density and pattern on uptake of nutrients have been examined theoretically by Baldwin, Tinker and Nye (1972), and Baldwin, Nye and Tinker (1973), who calculated that differences in root pattern can decrease the uptake by at least 75 per cent, depending on the value of the diffusion coefficient of the nutrient, root density and time. The relative contributions of diffusion and convection to the uptake of mobile nutrients such as nitrate is likely to depend upon the seasonal conditions and the stage of growth of the crop.

Mineralization

The cycling of nutrients through the soil organic matter pool requires account to be taken of decomposition cycles. The nitrogen mineralization routines of Dutt, Schaffer and Moore (1972) and McGill et al (1973) recognize a balance between the rates of mobilization of carbon and nitrogen which is discussed in detail by Oades and Ladd (this volume). Paul and Domsch (1972) and Dutt, Schaffer and Moore (1972) have simulated conversion from organic nitrogen to NH_4^+ and NO_3^- by using a series of empirical equations for the transfer from one compartment to another, and incorporating functions for the effects of temperature, moisture, and other factors. A simpler representation of nitrogen available to grass from a legume in a mixed sward was achieved by Ross, Henzell and Ross (1972) by relating this nitrogen to rates of senescence of legume material. A cascade of two compartments, legume residues and soil biomass, were each subject to exponential decay. Increasing the exponential decay constant from 25 days to 200 days decreased the inorganic nitrogen pool by some 90 per cent after 140 days of growth, and suggested the need for more detailed examination of the determinants of mineralization rate.

Analysis of pool sizes and transfer rates between pools, where some of these are not accessible to measurement, have been made by deriving unique solutions to a series of equations based on a model and data from compartments which can be measured. This method was used by Till and May (1971) for analyzing sulphur transfer rates in a pasture-sheep-soil ecosystem. It has also been used in a study of caesium-134 transfer in a litter, soil, millipedes-microflora-leachate system. Patten and Witkamp (1967) demonstrated that transfer constants between any two compartments were affected greatly by the combination of other compartments present. The concentration of caesium within various compartments did not necessarily become more stable with increasing complexity of the system. Such models can be used to predict accumulation of material under *similar* conditions where the same transfer rates are likely to occur or to examine potential variation in system behaviour in response to environment change. McGill (1971) and McGill et al. (1973) showed the turnover of organic N was intimately related to carbon metabolism of successions of microbial populations. Nitrogen was mobilized from plant residues more readily than from microbial metabolites into which they became incorporated, and the dynamics of the system indicated adsorption of microbial metabolites onto organic and inorganic colloids increased their resistance to mineralization. McGill et al. (1973) found rates of some processes in the field were quite different from rates measured in the laboratory, e.g. turnover rate of added labelled carbon was fifteen times less in the field—a useful reminder that although laboratory studies have an important function, data collected in the laboratory are not necessarily applicable to the field situation.

SOIL WATER-CROP NUTRITION INTERACTIONS

The supply of water throughout the growing season affects nutrient availability, nutrient absorption and growth processes. In turn, the nutritional status of the crop affects water use per unit of dry matter produced. Williams and Shapter (1955) showed that nutrient uptake decreased under conditions of water stress but it may be difficult to identify whether the effect is primarily due to reduced nutrient availability or to reduced growth and nutrient demand. Incorporation of these concepts into crop simulations requires the interfacing of models of water behaviour in soils and water use by the plant.

The greater resistance to drought of sorghum than maize has been attributed to greater root proliferation and/or root penetration (May and Milthorpe 1962) but these views are in contrast to the proposal of Passioura (1972) that plants with limited root development could conserve soil water for use later in the season. The latter idea requires further exploration of its consequences in the field since apparently excessive root development for water needs may be essential for adequate absorption of nutrients such as phosphate during early growth (Boatwright, Ferguson and Brown 1964) and as an insurance against the ravages of root pathogens. The problem illustrates the dilemma raised by attempts to define optimal conditions for even one component of a complex system without a means of testing the broader implications. This is a difficult job. The development of suitable crop models could assist the conceptual analysis.

Most models of crop growth which include water budgets have divided the root zone into a series of layers through which water is distributed (Byrne and Tognetti 1969; Rose et al. 1972; Paltridge 1972; McCown 1973). Such models usually assume that the soil layers fill to "field capacity" in sequence beginning from the surface layer and when the profile is full, excess water is treated as drainage and run-off. Uptake of water by the crop is calculated as a function of the rooting density in each layer and the total evapotranspiration demand (Rose et al. 1972). More detailed models of the soil water may need to incorporate the hydraulic conductivity of the soil (Ritchie 1972), and of hysteresis in water availability to roots under wetting and drying phases especially in the soil surface. So far, virtually all models of water use by crops have treated the soil layers as uniform, non-swelling and non-cracking. Below we discuss some of the interactions between nutrition and plant water relations which crop models must take into account. We have not considered situations where high salt concentrations lower the water potential itself; such aspects have been discussed by Levitt (1972).

Crop growth and development

Evapotranspiration has been used as a general driving function for models of growth of crops (Rose et al. 1972). Viets (1966) found that whereas the evapòtranspiration/yield function was not affected by soil fertility in wet or normal years, the ratio was increased in dry years for winter wheat fertilized with nitrogen. Viets's (1966) appeal for closer study of the crop density-nutrient-water use complex is certainly pertinent to the Australian situation. Increase in soil fertility in such an environment provides opportunities for greater productivity and "efficiency" of water use in good years, but also introduces the risks occurring with greater variability in total yields from year to year in the uncertain climate of the semi-arid environment.

The response of a crop to water stress depends upon the rapidity with which the stress is imposed, its intensity, duration and upon the level of soil fertility, particularly with respect to nitrogen supply. Water stress reduces the yield of spring cereals at all stages of growth (Slavik 1966). Stress applied during leaf development affected yield by reduction in the number of fertile tillers; during spikelet formation the number of spikelets per spike was decreased; water stress during anthesis decreased the total number of seeds and during grain formation the weight per seed was reduced. The most sensitive stage for yield reduction is when water stress occurs during anthesis (Nix and Fitzpatrick 1969; Wardlaw 1971). When water stress was applied at early heading in spring wheat the number of grains per head was reduced (Dubetz and Bole 1973). Yield responses due to applied nitrogen resulted from increased tillering while individual grain weight was unaffected. The effects of water deficits on various physiological processes have been reviewed by Hsaio (1973).

Assimilation processes

In nutritionally sufficient plants movement of potassium ions (and organic acids) into and out of guard cells lead to changes in turgor pressure associated with opening and closing of stomata. Humble and Hsaio (1969) and Graham and Ulrich (1972) have shown that closure of stomata in potassium deficient sugar beet has two components, one a specific potassium deficiency effect and one due to lowered water uptake caused by a greatly decreased root permeability to water in potassium deficient plants.

Redistribution of nutrients induced by water stress can be expected to influence subsequent photosynthesis by various parts of the plant. In times of stress nitrogen and phosphorus may move prematurely out of mature and ageing leaves to the stem but little loss occurs from the most active and developing leaves and meristematic tissue. On removal of the moisture stress, nutrient transported to the stems is then available to young expanding leaves, thus enhancing their nutrient status and possibly their photosynthesis and growth (Gates 1957), though Loveys and Kriedemann (1973) have shown that there may be a considerable lag in the restoration of photosynthetic rate following the removal of stress.

At a cellular level, water deficits cause an interruption of phosphorylation with consequent decreases in ATP, phosphorylated sugars, RNA and DNA and a suppression of protein synthesis (May and Milthorpe 1962). This may lead to somewhat increased amounts of phosphates and amino acids in the tissues. The documented evidence on tissue fluctuations of nitrogen, phosphorus, potassium, manganese, magnesium and zinc with water level is conflicting (see Crafts 1968). Inhibition of protein synthesis and nitrate reductase activity are relatively sensitive to water stress (Hsaio 1973), e.g. soil moisture suction of 3.0 bars can inhibit nitrate reductase activity of wheat (Plaut 1973).

There is much evidence for reduction of photosynthate translocation in plants under moderate to severe water stress. Stress reduces source strength by reducing photosynthesis and reduces sink strength by inhibiting growth. Restricted translocation is of particular interest in nutrition for its effect on root growth and morphology; no detailed studies of moisture stress on root morphology parameters useful in modelling appear to have been made. Davidson (1969) studied the effects of N, P and soil water status on root/shoot ratios of perennial ryegrass and white clover and found highly significant interactions occurred with the highest ratios, where inbalances also occurred, e.g. for *Lolium perenne* at pF 3.2, root/shoot ratios at P 0 ppm, N 300 ppm and P 300 ppm, N 300 ppm were 3.44 and 0.43 respectively and at pF 2.4 they were 0.82 and 0.46 respectively.

Nutrient uptake processes

Greacen (this volume) has indicated the effects of soil moisture stress on root growth. Nutrient uptake may be reduced under conditions of water stress because of limitation of root growth due to loss of turgor and increase in soil strength (Barley and Greacen 1967). The uptake of poorly diffusing ions such as phosphate is particularly dependent upon continuing root initiation and elongation. In sandy soils the total uptake of phosphate and the efficiency of phosphate uptake per gram of root were reduced by water stress in studies by Watanabe, Olsen and Davidson (1960). In contrast, in a lateritic podzolic soil from Kangaroo Island, South Australia, Fawcett and Quirk (1962) showed that over the suction range 0.2 to 7.9 bars uptake of phosphate by wheat was not greatly affected by water stress if the plants were not wilted. In the latter case it was argued that phosphorus remained available in fine undrained pores even at suctions approaching 15 bars.

As well as root growth inhibition under water stress, root hairs have been reported to die at relatively low water stress levels, followed by marked root suberization (Kramer and Kozlowski 1960); this could be expected to be of consequence to ion uptake and may further increase the importance of mycorrhizas as some of the causal fungi can function at low soil moistures (Bowen 1973).

Very few studies on effects of water stress on uptake with soil grown plants distinguish adequately between effects of water stress on ion transfer to roots and effects on the uptake processes. With solution grown plants there is no evidence of a high sensitivity of ion uptake processes to small turgor changes although water stress or turgor reduction can reduce ion transport to the shoot. However effects on transport of ions in soil are likely to overshadow effects on the nutrient uptake processes. The demand for nutrients in water stressed plants will be reduced because of suppressed growth (Slatyer 1969).

Transfer of ions in soil

In a drying soil diffusion coefficients (D) decrease due to increased tortuosity and discontinuity of liquid films and thus the plants ability to establish and maintain high rooting density becomes more important. In a loam studied by Porter et al. (1960) D of Cl^- decreased from 8.9 x 10^{-7} cm^2s^{-1} at 0.33 bars matric suction to 0.6 x 10^{-7} cm^2s^{-1} at 15.6 bars; D for $H_2PO_4^-$ in a silty clay loam (Olsen, Kemper and Van Schaik 1965) decreased from 6.2 x 10^{-7} cm^2s^{-1} at 0.33 bars matric suction to 0.4 x 10^{-7} cm^2s^{-1} at 6.0 bars. As the water content of the soil declines, the supply of ions such as nitrate which are present almost exclusively in soil solution may be partially offset by the increase in concentration of the ion in solution.

Mathematical models for the uptake of solutes by randomly distributed roots were developed by Baldwin, Nye and Tinker (1973) to examine the relative contributions of diffusion and mass flow to ion uptake. For high values of the term αa/Db (where α is "absorbing power" of roots, a is root radius, D is the diffusion coefficient of the solute and b is the buffering power of the soil) the contribution of mass flow was low or negligible, but where the term had low values a doubling of water flux gave a 2-3-fold increase in nutrient uptake. The model has application in testing the efficiency of different types of root system in a given soil or the performance of a given root system under different soil conditions. A mathematical analysis by Marriott and Nye (1968) indicated that moisture (and convected ion) flux to the root surface at low soil water contents is greatly dependent upon transpiration rate. For a moist sandy loam (pF 2.0) high transpiration rates increased the uptake of Cl^- over diffusive flux by only 12 per cent, but under drier conditions (pF 3.5) an increase of 380 per cent was calculated.

Mineralization

Severe moisture stress decreases the rates of decomposition and mineralization processes. Alternative wetting and drying leads to flushes of decomposition (Birch 1960) which may operate both with seasonal cycles and with periodic wetting and drying of the surface 2-4 cm (usually high in nutrient and organic matter content) within growing seasons.

Wetselaar (1968) studied ammonium and nitrate accumulation in a lateritic red earth (clay loam) and a sand at 25°C under a range of carefully controlled moistures. He found nitrification to decline continually from just above zero suction and to be very low or to cease above 25 to 30 bars. Ammonia production operated with little change or increased up to 50 bars and in these soils, which released approximately 20 ppm NH_4-N plus NO_3-N in three weeks, the cross-over point for NH_4-N dominance over NO_3-N was between 5 and 7 bars. This finding is directly relevant to the nitrogen nutrition of plants in a drying profile under semi-arid conditions. Wetselaar considered the lower sensitivity of ammonification to low soil moisture explained the relatively high NH_4-N values recorded from semi-arid soils at the end of a dry season and concluded that approximately half of the total NO_3-N made available in a growing season is formed in the early part of the wet season. In Wetselaar's soils total NO_3 plus NH_4 was fairly stable (sand) or declined (clay loam) from 23.2 ppm at 50 bars to 7.0 ppm at 2.7 bars after 3 weeks. His data are at variance with those of Robinson (1957) who showed a continual decline of NH_4+NO_3 from field capacity (42 per cent soil moisture) to air-dry soil.

Stefanson (1972) showed denitrification of nitrate-N in two soils at 20 per cent to be between a sixth and a half of that at 24-25 per cent moisture (field capacity) with a very marked stimulative effect of plants. Loss from ammonium-N was little affected by moisture from 21-26 per cent moisture. More detailed studies on mechanisms are necessary before accurate predictions can be made of such losses in relation to soil moisture.

CONCLUSION

Nutritional aspects of soil-plant relations are studied at two levels, either in the field or under controlled conditions, usually by means of pot experiments. Field observations are frequently summarized by regression functions which are valuable for their purposes, but generally tell us little about control mechanisms. In contrast, laboratory studies provide very detailed information from which it may be possible to define potentially causal relationships, but usually under conditions very different from the field. For example, the photosynthetic behaviour of crop canopies cannot be predicted from the considerable data available on the photosynthetic properties of single leaves or isolated plants. Moreover soil water, nutrient and root profiles in pots are markedly different from field profiles.

In the previous discussion we have noted some of the attempts being made in many aspects of crop nutrition to synthesize from the particular to the general by means of simulation studies. Although much development is still required, these studies show considerable potential for testing the implications of what is *thought* to be known, in a context which has practical relevance. The realization of the need to optimize productivity and efficiency within the low energy input systems of Australian agriculture (Harris et al. 1974) imposes greater demands for an integrated understanding of the factors affecting optimal soil utilization in the long term. Multi-disciplinary models represent a valuable contribution to meeting this need.

REFERENCES

Adedipe, N.O., and Fletcher, R.A. (1970). *Plant Physiol.* **46**, 614-7.
Anderson, W.P. (1972). *Annu. Rev. Plant Physiol.* **23**, 51-72.
Arnold, P.W. (1969). In "Ecological Aspects of the Mineral Nutrition of Plants". (Ed. I.H. Rorison). pp. 115-25. British Ecol. Soc. Symp. No. 9. Blackwell, Oxford.
Baldwin, J.P., Nye, P.H., and Tinker, P.B. (1973). *Plant Soil.* **38**, 621-35.
Baldwin, J.P., Tinker, P.B., and Nye, P.H. (1972). *Plant Soil.* **36**, 693-708.
Barber, D.A. (1968). *Annu. Rev. Plant Physiol.* **19**, 71-88.
Barley, K.P. (1970). *Adv. Agron.* **22**, 159-201.
Barley, K.P., and Greacen, E.L. (1967). *Adv. Agron.* **19**, 1-43.
Barrow, N.J. (1966). *Aust. J. Agric. Res.* **17**, 849-61.
———. (1975). *Aust. J. Agric. Res.* **26**, 145-56.
Barrow, N.J., Ozanne, P.G., and Shaw, T.C. (1965). *Aust. J. Agric. Res.* **16**, 61-76.
Beckett, P.H.T., and White, R.E. (1964). *Plant Soil.* **21**, 253-82.
Beek, J., and Frissel, M.J. (1973). "Simulation of Nitrogen Behaviour in Soils". Pudoc, Wageningen.
Bevege, D.I., Bowen, G.D., and Skinner, M.F. (1975). In "Endomycorrhizas". (Eds. F.E. Sanders, B. Mosse and P.B. Tinker). pp. 149-74. Academic Press, London.
Bhat, K.K.S., and Nye, P.H. (1973). *Plant Soil.* **38**, 161-75.
Birch, H.F. (1960). *Plant Soil.* **12**, 81-96.
Blakemore, M. (1966). *J. Agric. Sci.* **66**, 139-46.
Boatwright G.O., Ferguson, H., and Brown, P.L. (1964). *Soil Sci. Soc. Amer. Proc.* **28**, 403-5.
Boatwright, G.O., and Viets, Jr., I.G. (1966). *Agron. J.* **58**, 185-8.
Bowen, G.D. (1968). *Trans. 9th Int. Congr. Soil Sci.* **2**, 219-28.
———. (1970). *Aust. J. Soil Res.* **8**, 31-42.
———. (1973). In "Ecology and Physiology of Ectomycorrhizae". (Eds. G.G. Marks and T.T. Kozlowski). pp. 151-205. Academic Press, New York.
Bowen, G.D., Cartwright, B., and Mooney, J.R. (1974). In "Mechanisms of Regulation of Plant Growth". (Eds. R.L. Bieleski, A.R. Ferguson and M.M. Creswell). pp. 121-5. Bull. No. 12, Roy. Soc., N.Z.
Bowen, G.D., and Rovira, A.D. (1961). *Plant Soil.* **15**, 166-88.
———. (1966). *Nature (Lond.)* **211**, 665-6.
———. (1971). In "Recent Advances in Plant Nutrition". Vol. 1 (Ed. R.M. Samish). pp. 307-16. Gordon and Breach, New York.
Bowling, D.F. (1965). *Nature (Lond.)* **206**, 317-8.
Briggs, G.E., Hope, A.B., and Robertson, R.N. (1961). "Electrolytes and Plant Cells". Blackwell, Oxford.
Brouwer, R., and de Wit, C.T. (1969). In "Root Growth" (Ed. W.J. Whittington). pp. 224-44. Butterworths, London.
Bunting, A.H., and Drennan, D.S.H. (1966). In "The Growth of Cereals and Grasses". (Eds. F.L. Milthorpe and J.D. Ivins). pp. 20-38. Butterworths, London.
Byrne, G.F., and Tognetti, K. (1969). *Agric. Meterol.* **6**, 151-63.
Cartwright, B., and Hallsworth, E.G. (1970). *Plant Soil.* **33**, 685-98.
Cartwright, B., Kriedemann, P.E., Bowen, G.D., Torokfalvy, E., Cunningham, R.B. (1974). In "Mechanisms of Regulation of Plant Growth". (Eds. R.L. Bieleski, A.K. Ferguson, and M.M. Creswell). pp. 142-5. Bull. No. 12, Roy. Soc., N.Z.
Champion, R.A., and Barley, K.P. (1969). *Soil Sci.* **108**, 402-7.
Cormack, R.G.H. (1962). *Bot. Rev.* **15**, 583-612.
Crafts, A.S. (1968). In "Water Deficits and Plant Growth". Vol. 2 (Ed. T.T. Kozlowski). pp. 85-133. Academic Press, New York.
Dainty, J. (1969). In "Physiology of Plant Growth and Development". (Ed. M.B. Wilkins). pp. 455-84. McGraw-Hill, London.
Davidson, R.L. (1969). *Ann. Bot. (Lond.)* **33**, 571-77.
Denison, W.C. (1973). *Scientific American.* **228**, 75-80.
Doodson, J.K., Manners, J.G., and Myers, A. (1964). *J. Exp. Bot.* **15**, 96-103.
Drew, M.C., Nye, P.H., and Vaidyanathan, L.V. (1969). *Plant Soil.* **30**, 252-70.
Dubetz, S., and Bole, J.B. (1973). *Can. J. Plant. Sci.* **53**, 1-5.
Duncan, W.G., Loomis, R.S., Williams, W.A., and Hanau, R. (1967). *Hilgardia.* **38**, 181-205.
Dutt, G.R., Schaffer, M.J., and Moore, W.J. (1972). Ariz. Agric. Exp. Stn. Tech. Bull. No. 196.
Epstein, E. (1972). "Mineral Nutrition of Plants, Principles and Perspectives". John Wiley, New York.

Fawcett, R.G., and Quirk, J.P. (1962). *Aust. J. Agric. Res.* **13**, 194-205.
Fisher, R.A., and Hsaio, T.C. (1968). *Plant Physiol.* **43**, 1953-8.
Fox, J.E. (1969). In "The Physiology of Plant Growth and Development". (Ed. M.B. Wilkins). pp. 85-123. McGraw-Hill, London.
Garbouchev, I.P. (1966). *J. Agric. Sci.* **66**, 399-412.
Gates, C.T. (1957). *Aust. J. Biol. Sci.* **10**, 125-46.
———. (1964). *J. Aust. Inst. Agric. Sci.* **30**, 3-22.
Gott, M.B. (1961). *Aust. J. Agric. Res.* **12**, 547-65.
Graham, R.D., and Ulrich, A. (1972). *Plant Physiol.* **39**, 105-9.
Greenway, H., and Gunn, A. (1966). *Planta (Berl.)* **71**, 43-67.
Gregory, F.G. (1937). *Annu. Rev. Biochem.* **6**, 557-78.
Hackett, C. (1968). *New Phytol.* **67**, 287-300.
———. (1972). *Aust. J. Biol. Sci.* **25**, 1169-80.
Hackett, C., and Rose, D.A. (1972). *Aust. J. Biol. Sci.* **25**, 681-90.
Hall, R.H. (1973). *Annu. Rev. Plant Physiol.* **24**, 415-44.
Harley, J.L. (1969). "The Biology of Mycorrhizas". 2nd Edn. Leonard Hill, London.
Harre, R. (1970). "Principles of Scientific Thinking". Macmillan, London.
Harris, S., Crawford, J.G., Green, T.H., and Honan, N.D. (1974). The Principles of Rural Policy in Australia: A Discussion Report. Aust. Govt. Printing Service, Canberra.
Hattingh, M.J., Gray, L.E., and Gerdemann, J.W. (1973). *Soil Sci.* **116**, 383-7.
Herath, H.M.W., and Ormrod, D.P. (1972). *Plant Physiol.* **49**, 443-4.
Hesketh, J.D. (1963). *Crop Sci.* **3**, 493-6.
Heslop-Harrison, J. (1969). In "Physiological Aspects of Crop Yield". (Ed. J.D. Eastin, F.A. Haskins, C.Y. Sullivan, C.H.M. Van Bavel). pp. 291-316. Amer. Soc. Agron., Madison.
Hopkinson, J.M. (1964). *J. Exp. Bot.* **15**, 125-37.
Hsaio, T.C. (1973). *Annu. Rev. Plant Physiol.* **24**, 519-70.
Humble, G.D., and Hsaio, T.C. (1969). *Plant Physiol.* **44**, 230-4.
Hutton, J.T., and Leslie, T.I. (1958). *Aust. J. Agric. Res.* **9**, 492-507.
Idso, S.B., and de Wit, C.T. (1970). *App. Optic.* **9**, 177-84.
Jeffers, J.N.R. (1971). "Mathematical Models in Ecology". 12th Symp. Brit. Ecol. Soc. Publ. Blackwell, Oxford.
Jeffrey, D.W. (1964). *Aust. J. Biol. Sci.* **17**, 845-54.
Jones, J.G.W. (1970). "The Use of Models in Agricultural and Biological Research". Grassl. Res. Inst., Hurley.
Keay, J., Biddiscombe, E.F., and Ozanne, P.G. (1970). *Aust. J. Agric. Res.* **21**, 33-44.
King, P.H., and McCarty, P.L. (1968). *Soil Sci.* **106**, 248-61.
King, R.W., Wardlaw, I.F., and Evans, L.T. (1967). *Planta (Berl.)* **77**, 261-76.
Kramer, P.J., and Kozlowski, T.T. (1960). "Physiology of Forest Trees". McGraw-Hill, New York.
Lamont, B.B., and McComb, A.J. (1974). *Aust. J. Bot.* **22**, 681-8.
Langer, R.H.M. (1966). In "The Growth of Cereals and Grasses". (Eds. F.L. Milthorpe and J.D. Ivins). pp. 213-55. Butterworths, London.
Lauchli, A. (1972). *Annu. Rev. Plant Physiol.* **23**, 197-218.
Levitt, J. (1972). "Responses of Plants to Environmental Stresses". Academic Press, New York.
Lewis, D.G., and Quirk, J.P. (1963). *Plant Soil.* **26**, 445-53.
Lommen, P.W., Schwintzer, C.R., Yocum, C.S., and Gates, D.M. (1971). *Planta (Berl.)* **98**, 195-220.
Loomis, R.S., and Williams, W.A. (1969). In "Physiological Aspects of Crop Yield". (Ed. J.D. Eastin, F.A. Haskins, C.Y. Sullivan and C.H.M. Van Bavel). pp. 27-47. Amer. Soc. Agron., Madison.
Loomis, R.S., Williams, W.A., and Hall, A.D. (1971). *Annu. Rev. Plant Physiol.* **22**, 431-68.
Loveys, B.R., and Kriedemann, P.E. (1973). *Physiol. Plant.* **28**, 476-9.
Lungley, D.R. (1973). *Plant Soil.* **38**, 145-60.
McCown, R.L. (1973). *Agric. Meteorol.* **11**, 53-63.
McCree, K.J., and Troughton, J.H. (1966*a*). *Plant Physiol.* **41**, 1615-22.
———. (1966*b*). *Plant Physiol.* **41**, 559-61.
McGill, W.B. (1971). Ph.D. Thesis, Univ. of Saskatchewan, Saskatoon.
McGill, W.B., Paul, E.A., Shields, J.A., and Lowe, W.E. (1973). *Bull. Ecol. Res. Comm. (Stockholm).* **17**, 293-301.
Malajczuk, N., and Bowen, G.D. (1974). *Nature (Lond.)* **251**, 316-7.
Marriott, F.H.C., and Nye, P.H. (1968). *Trans. 9th Int. Congr. Soil. Sci.* **1**, 127-34.
May, L.H., and Milthorpe, F.L. (1962). *Field Crop Abstr.* **15**, 171-9.
Milthorpe, F.L., and Moorby, J. (1969). *Annu. Rev. Plant Physiol.* **20**, 117-38.
Monteith, J.L. (1965). *Ann. Bot. (Lond.)* **29**, 17-37.
Moore, A.W. (1966). *Soils Fertil.* **29**, 113-28.
Moss, D.N., and Peaslee, D.E. (1965). *Crop Sci.* **5**, 280-1.
Mosse, B. (1972). *Nature (Lond.)* **239**, 221-3.
———. (1973). *Annu. Rev. Phytopath.* **11**, 171-96.

Nevins,D.J., and Loomis, R.S. (1970). *Crop Sci.* **10**, 21-5.
Nix, H.A. and Fitzpatrick, E.A. (1969). *Agric. Meteorol.* **6**, 321-37.
Nuttonson, M.Y. (1955). "Wheat-Climate Relationships and the Use of Phenology in Ascertaining the Thermal and Photo-Thermal Requirements of Wheat". Amer. Inst. Crop Ecol., Washington.
Nye, P.H. (1966). *Plant Soil.* **25**, 81-105.
Olsen, S.R., and Kemper, W.D. (1968). *Adv. Agron.* **20**, 91-151.
Olsen, S.R., Kemper, W.D., and Van Schaik, J.C. (1965). *Soil Sci. Soc. Amer. Proc.* **29**, 154-8.
Olsen, S.R., Watanabe, F.S., and Danielson, R.E. (1961). *Soil Sci. Soc. Amer. Proc.* **25**, 289-94.
Osman, A.M., and Milthorpe, F.L. (1971). *Photosynthetica (Prague).* **5**, 61-70.
Ozanne, P.G., Keay, J., and Biddiscombe, E.G. (1969). *Aust. J. Agric. Res.* **20**, 809-18.
Paltridge, G.W. (1970). *Agric. Meteorol.* **7**, 93-130.
———. (1972). *Agric. Meteorol.* **10**, 39-54.
Passioura, J.B. (1972). *Aust. J. Agric. Res.* **23**, 745-52.
———. (1973). *J. Aust. Inst. Agric. Sci.* **39**, 181-3.
Passioura, J.B., and Cowan, I.R. (1968). *Agric. Meteorol.* **5**, 121-34.
Patefield, W.M., and Austin, R.B. (1971). *Ann. Bot. (Lond.)* **35**, 1227-50.
Patten, B.C. (1971). "Systems Analysis and Simulation in Ecology". Vol. 1. Academic Press, New York.
———. (1972). "Systems Analysis and Simulation in Ecology". Vol. 2. Academic Press, New York.
Patten, B.C., and Witkamp, M. (1967). *Ecology.* **48**, 813-24.
Paul. E.A., Myers, R.J.K., and Rice, W.A. (1971). In "Biological Nitrogen Fixation". (Eds. T.A. Lie and E.G. Mulder). pp. 495-507. Plant Soil Special Volume. Martinus Nijhoff, The Hague.
Paul, W., and Domsch, K.H. (1972). *Arch, Mikrobiol.* **87**, 77-92.
Peaslee, D.E., and Moss, D.N. (1966). *Soil Sci. Soc. Amer. Proc.* **30**, 120-3.
Peaslee, D.E., Ragland, J.L., and Duncan, W.G. (1971). *Agron. J.* **63**, 561-63.
Plaut, Z. (1973). *Plant Soil.* **38**, 81-94.
Porter, L.K., Kemper, W.D., Jackson, R.D., and Stewart, B.C. (1960). *Soil Sci. Soc. Amer. Proc.* **24**, 460-3.
Puckridge, D.W. (1963). M. Agric. Sci. Thesis, Univ. of Adelaide, Adelaide.
Ritchie, J.T. (1972). *Water Resour. Res.* **8**, 1204-13.
Robinson, J.B.D. (1957). *J. Agric. Sci.* **49**, 100-5.
Rose C.W., Begg, J.E., Byrne, G.F., Torsell, B.W.R., and Goncz, J.H. (1972). *Agric. Meteorol.* **10**, 161-83.
Ross, P.J., Henzell, E.F., and Ross, D.R. (1972). *J. Appl. Ecol.* **9**, 535-56.
Ross, Y.K., and Nilson, T. (1967). In "Photosynthesis in Productive Systems". (Ed. A.A. Nichoporovich). Transl. Edn. Israel Prog. Sci. Trans., Jerusalem.
Rovira, A.D. (1972). *J. Aust. Inst. Agric. Sci.* **38**, 90-4.
Rovira, A.D., and Bowen, G.D. (1968). *Trans. 9th Int. Congr. Soil Sci.* **2**, 209-17.
———. (1973). *Planta.* **114**, 101-7.
Ruinen, J. (1965). *Plant Soil.* **22**, 375-94.
Russell, J.S. (1972). *Soil Sci.* **114**, 387-94.
Ryle, G.J.A. (1964). *J. Br. Grassl. Soc.* **19**, 281-90.
Ryle, G.J.A., and Hesketh, J.D. (1969). *Crop Sci.* **9**, 45-4.
Salmon, R.C. (1966). *Soil Sci.* **101**, 450-4.
Sanders, F.E., and Tinker, P.B. (1971). *Nature (Lond.)* **233**, 278-9.
Scaife, M.A., and Smith, R. (1973). *J. Agric. Sci.* **80**, 353-61.
Slatyer, R.O. (1969). In "Physiological Aspects of Crop Yield". (Ed. J.D. Eastin, F.A. Haskins, C.Y. Sullivan and C.H.M. Van Bavel). pp. 53-83. Amer. Soc. Agron., Madison.
Slavik, B. (1966). In "The Growth of Cereals and Grasses". (Ed. F.L. Milthorpe and J.D. Ivins). pp. 227-40. Butterworths, London.
Stefanson, R.C. (1972). *Plant Soil.* **36**, 129-40.
Stevenson, C.M. (1968). *Q. J. Roy. Met. Soc.* **94**, 56-70.
Sutton, C.D., and Gunary, D. (1969). In "Ecological Aspects of the Mineral Nutrition of Plants". (Ed. I.H. Rorison). pp. 127-34. Blackwell, Oxford.
Tanji, K.K. (1970). *Soil Sci.* **110**, 44-51.
Tchan, Y.T., and Beadle, N.C.W. (1955). *Proc. Linn. Soc. N.S.W.* **80**, 97-104.
Terry, N., and Ulrich, A. (1973). *Plant Physiol.* **51**, 43-7.
Theodorou, C. (1971). *Aust. For.* **35**, 23-6.
Thornley, J.H.M. (1972). *Ann. Bot. (Lond.)* **36**, 419-30.
Till, A.R., and May, P.F. (1971). *Aust. J. Agric. Res.* **22**, 391-400.
Tiller, K.G., Honeysett, J.L., and de Vries, M.P.C. (1972). *Aust. J. Soil Res.* **10**, 151-64.
Troughton, A. (1968). *Ann. Bot. (Lond.)* **32**, 411-23.
Tukey, H.B. (1970). *Annu. Rev. Plant Physiol.* **21**, 305-24.
Viets, F.G. (1966). In "Plant Environment and Efficient Water Use". (Ed. W.H. Pierre, D. Kirkham, J. Pesek and R. Shaw). pp. 259-74. Amer. Soc. Agron., Madison.
Vose, P.B. (1962). *Ann. Bot. (Lond.)* **26**, 425-37.
Waggoner, P.E. (1969). *Crop Sci.* **9**, 315-21.

Wallace, A., and Frohlich, E. (1965). *Nature (Lond.)* 208, 1231.
Wallace, A., Frohlich, E., and Ashcroft, R.T. (1967). *Agron. J.* 59, 386.
Wallace, T. (1961). "The Diagnosis of Mineral Deficiencies in Plants by Visual Symptoms". HMSO, London.
Wang, J'Y (1963). "Agricultural Meteorology". Pacemaker Press, Milwaukee.
Wardlaw, I.F. (1968). *Bot. Rev.* 34, 79-105.
———. (1971). *Aust. J. Biol. Sci.* 24, 1047-55.
Watanabe, F.S., Olsen, S.R., and Danielson, R.E. (1960). *Trans. 7th Int. Congr. Soil Sci.* 3, 450-6.
Watson, D.J. (1947). *Ann. Bot. (Lond.)* 11, 375-407.
Wetselaar, R. (1968). *Plant Soil.* 29, 9-17.
Wetselaar, R., and Hutton, J.T. (1963). *Aust. J. Agric. Res.* 14, 319-29.
Williams, R.F. (1955). *Annu. Rev. Plant Physiol.* 6, 25-42.
———. (1964). In "Grasses and Grasslands". (Ed. C. Barnard). pp. 89-101. Macmillan, London.
Williams, R.F., and Shapter, R.E. (1955). *Aust. J. Biol. Sci.* 8, 435-66.
Wit, C.T. de, and Van Keulen, H. (1972). "Simulation of Transport Processes in Soils". Pudoc, Wageningen.
Zelitch, I. (1971). "Photosynthesis, Photorespiration and Plant Productivity". Academic Press, New York.

CHAPTER 10

Mechanisms and models of plant growth

C.W. Rose and R.J. Millington

INTRODUCTION

The growth of plants both modifies and is modified by environmental parameters. With either natural or man-modified ecosystems, the interpretation of plant-environment interactions is a considerable problem. Approaches to this problem in an agricultural context are evaluated in this paper by considering the advantages and limitations of the type of models employed.

Since early in the development of scientific agriculture, experimentation has been guided and results analyzed using varieties of static statistically based models. Limitations of this type of model are examined. Examples of models are given which allow simulation of certain features of plant-environment interactions. Though currently limited in scope, such models attempt to establish functional, and hopefully causal relations within and between subsystems of the whole.

PLANT-ENVIRONMENT INTERACTIONS: THE PROBLEM OF INTERPRETATION

The methodological problem in agriculture and ecology

Agricultural and ecological scientists are concerned with interpreting the behaviour of complex organisms and communities of organisms, where the performance and even behavioural characteristics of relevant classes of organisms are substantially modified by their continually changing physical and chemical environment. Furthermore, such scientists cannot ignore man's modifications and management for his own purposes, therefore having to deal with a number of different levels of organization or "strata" (in the terminology of Mesarović and Macko 1969). Natural science has commonly progressed by seeking explanation of irregularity or behaviour within one strata in terms of greater order at a lower or prior level of organization. It is thus, for example, that we seek explanation of the variety of chemical reactions in terms of atomic and molecular properties. Agriculture involves the interaction between a relatively wider number of strata, and this poses a considerable methodological problem.

This methodological problem has been partially solved by the application of statistical methods to the design and interpretation of agricultural field experiments. This technique indicates whether or not a particular treatment has any effect on a quantity of interest, (such as yield) and if so what level of significance might be ascribed to the effect when it is in the presence of fluctuations or "errors" due to other uncontrolled variables. Interactions between treatments can also be identified.

This style of approach is of very general application, although the very utility and ubiquity of the technique have perhaps masked its limitations. Accepting the continuing value of traditional statistically based experimentation, what are some of its characteristic deficiencies and limitations?

Limitations to statistically based approaches in agriculture

The observations associated with classical statistical approaches in agricultural research are essentially passive or static in character. No direct appeal to, or dependence on knowledge concerning the casual relationships between growth and plant environment is necessary. Even if measurement of growth and environmental factors through time is made, such information is used rather as an aid to interpretation than in the analysis.

A consequent major limitation is that the results obtained in any particular experiment are essentially specific to the location and time (e.g. season) of the experiment (Nix 1968). Repetition of an experiment through time allows variability in production to be assessed. This implies that in a variable environment an experiment may need to be repeated many times to obtain results which are sufficiently accurate to be of practical value (Odell 1950; Byrne and Tognetti 1969; McCown 1973). Extrapolation of such results to other sites is usually made on the basis of analogy using some form of land classification, using this term to include above-ground environmental factors. Experience and good judgment are necessary to assess the differences in land characteristics between any given location and the experimental site, and thus to establish that an extension of the results would be valid (Nix 1968).

Interpreting plant-environment interaction in an ecological context

In an ecological rather than an agricultural context there are problems of diversity, heterogeneity and a greater lack of data on relevant above- and below-ground environmental factors. Furthermore the principles involved in inter-species competition are only qualitatively understood. Thus, ecology has commonly relied on classification techniques based on species abundance, and numerical techniques of classification. Both have received enhanced emphasis with the advent of computers (e.g. Williams and Lambert 1960).

According to Austin (1971), taxonometric methods have the following three roles: simplification of multivariate data; assistance in hypothesis generation; and in defining a domain with a defined degree of vegetational homogeneity. The use of regression techniques within such a domain can be fruitful.

These approaches, however, need to be complemented by further developments in methods of analysis. Species composition of plant communities and their spatial distribution and dynamics are strongly dependent on plant interactions with a range of primary and resultant microenvironmental factors. Therefore effective progress in analysis in this area would probably be enhanced by a team approach involving collaboration between both biologically and physically based scientists.

The role of systems approaches and simulation

Everyday decisions are made on the basis of mental images and conceptual models. These may have a number of limitations, for example their unstated basic assumptions (Forrester 1971). Mathematics has characteristically provided a suitable language to formalize such models in the physical sciences. In contrast, the description of processes in a mathematical form capable of analysis is not typical in biological sciences. Despite the increasing application of mathematics to this field, less is known quantitatively of mechanisms and principles in the biological than the physical sciences. Furthermore, numerical methods (including simulation) appear to be more suitable than formal mathematics in describing systems which are complex, non-linear, have strong feedback, and are subject to non-analytic or random variation in factors such as inputs. These features are characteristic of many agricultural and biological systems, so that simulation techniques offer particular advantages (Dent and Anderson 1971; Armstrong 1972).

Simulation is a technique in the broad area of dynamic mathematical models (Gordon 1969). It is concerned not only with the building of models, but with the study of their behaviour where a model is regarded as an hypothesis concerning the behaviour of the system being studied. Further, simulation enables the consequence of the hypotheses to be demonstrated in terms of systems behaviour. Comparison of this simulation with the performance of the real system under consideration is needed to test the realism of the simulation, and thus, indirectly, the soundness of the model. Iteration round this loop, of model construction, of simulation, and of validation is a common method whereby the model or hypothesis may be improved (Morley 1972).

The activity of modelling, and the use of systems techniques generally, have proved particularly helpful in organizing research in situations where a team is involved having different disciplinary backgrounds. This is partly because it provides a subject-neutral language and technique through which the hypotheses of subsystem performance can be synthesized, thus clarifying relationships between contributions from various team members.

Under some conditions, interaction between simulation and experimentation can improve the design of experiments, and improve the level of useful observation if the results of simulation are concurrent with the experiment.

Computer models of agricultural or ecological systems also offer a systematic method of extrapolation and prediction in other situations where inputs, such as meteorological data, may be significantly different. Of course, testing is required to determine the success of such predictions.

PLANT-ENVIRONMENT INTERACTION

This very broad and complex topic will be briefly considered from two complementary points of view:

(*a*) The modification of the physical environment by the presence and growth of plants, and

(*b*) The modifications of plant growth by some aspects of the above- and below-ground environment.

The processes in (*a*) are chiefly physical and chemical, and in (*b*) dominantly biological. In both cases there are direct effects and interactive effects, since it is typical that a modification of any one factor will alter other factors. This fact is one reason for controlled environment experimentation where such interaction between factors of the physical environment can be minimized (Evans 1963). Preventing interaction within the plant is much more difficult, or perhaps impossible since truly integrated behaviour is an inherent characteristic of living things.

Modification of the physical environment by plant communities

The characteristics of the physical environment can be described in terms of sets of state attributes (e.g. temperature, vapour concentration) and fluxes (e.g. evaporation rate). In the air and soil, fluxes result from gradients of state attributes via relevant conductivities of soil (Childs 1969) or turbulent diffusivities of the air (Sutton 1953).

A physical basis for understanding the infinite variety of microclimates is given by the conservation principles for heat and water. Assuming a steady state and no advection, energy conservation may be expressed as:

$$R_N = G + H + LE$$

where R_N is net radiation, G heat flux into the soil, H and LE sensible and latent heat fluxes into the atmosphere.

The above equation assumes evaporation at the soil surface, but the site of evaporation soon retreats below the surface of an initially wet soil, particularly under highly evaporative conditions. Development of a dry surface layer results in the phase change of water in evaporation taking place below it. The sensible heat flux in this dry layer is then (G + LE), where G is the sensible soil heat flux remaining after the energy required for the latent heat of evaporation has been provided.

The theory of heat and water fluxes in soil and their interaction has been developed by Philip and de Vries (1957) and tested in the field by Rose (1968*a*, 1968*b*). This flux interaction becomes more important in an environment where temperature gradients in the soil are appreciable (say $>0.5°C\ cm^{-1}$). Analysis applicable to the non-steady diurnal regime is not yet available, so the problem of how the soil dries out is still unresolved, in the analytically predictive sense.

Consideration of the effect of isolated plants or trees in modifying such an environment is more difficult than for an extensive uniform crop. The former is a three-dimensional problem, whereas there is experimental support that exchanges within and between a uniform crop and the atmosphere may be considered using one-dimensional theory (Waggoner and Turner 1972). The degree of uniformity within a crop necessary for transport processes to be adequately represented by one-dimensional theory is difficult to quantify a priori (Philip 1964; Goncz and Rose 1972).

However, when analyzing exchanges between a crop and atmosphere there is support that even with non-steady conditions fluxes can be interpreted with small error in terms of steady state one-dimensional theory (Rose et al. 1972*a*), and similarly for exchanges between row crops and the atmosphere (Graetz 1972). Below ground, the general features of water uptake by a crop can also be described in one-dimensional terms (Rijtema 1965; Rose and Stern 1967).

A vegetation canopy directly modifies the wind profile and turbulent transport characteristics of the atmosphere within and above the vegetation. This modification in turbulent structure and dissipation of momentum in turn modifies the profiles and transport of all other properties of the air (e.g. temperature, water vapour and CO_2 concentration). A vegetation canopy also provides a zone, rather than a single surface for radiation exchange, and spatial distribution of sources and sinks of water vapour and sensible heat.

Different approaches have been developed to improve the analysis of these complex aerial environment interactions. In some cases, the vegetation canopy has been considered to be a continuum and the variables analyzed using calculus (Denmead 1968*a*, 1968*b*, Cowan 1968), while in another method the canopy is represented by strata and algebra is used to solve the sets of equations which describe energy conservation in each stratum (Waggoner and Reifsnyder 1968). The latter authors' model is shown in electrical analogue form in figure 1. The absorbed radiation (S_j) in layer j is partitioned into sensible (H_j) and latent (LE_j) heat fluxes, using analogue resistances which depend on (i) the stomatal and leaf boundary layer resistances (shown horizontally), and (ii) the resistances to turbulent transport in the air (shown vertically). θ_j is the temperature difference between the leaves in the jth stratum and the air at the canopy top. Solution of the equations to give the fluxes requires the following parameters, which must be measured or otherwise inferred:

1. temperature and vapour pressure of
   (*a*) the air above the crop (T_o,e_o)
   (*b*) the air above the ground but below the canopy (T_g, e_g)
2. radiation (*a*) above the crop (S_i)
   (*b*) above the ground but below the canopy (S_g)

and

3. the resistances described above.

Simulation with such a model has been shown to closely match experimentally determined environmental parameters (e.g. evaporative flux, figure 2).

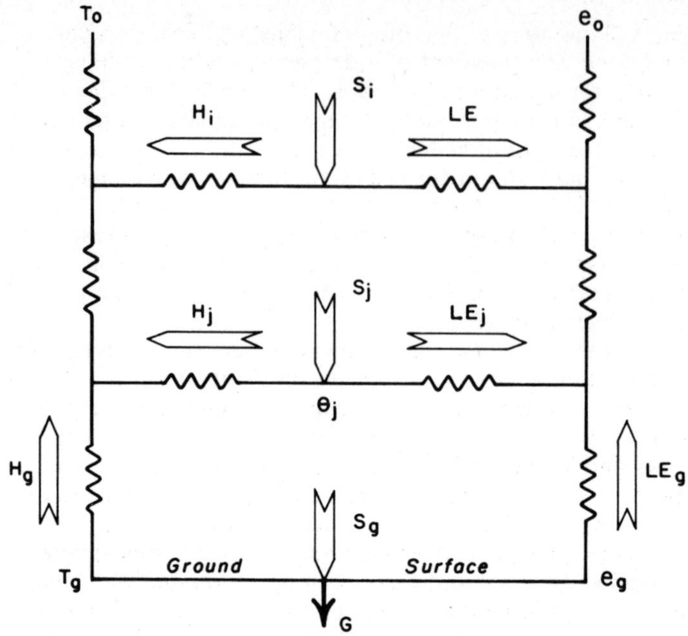

Fig. 1. Outline of model used by Waggoner and Reifsnyder (1968) to partition absorbed radiation S into sensible (H) and latent heat (LE) fluxes. See text for symbols. Horizontal resistances are due to leaf boundary layer and stomata, vertical resistances are associated with transport in the atmosphere.

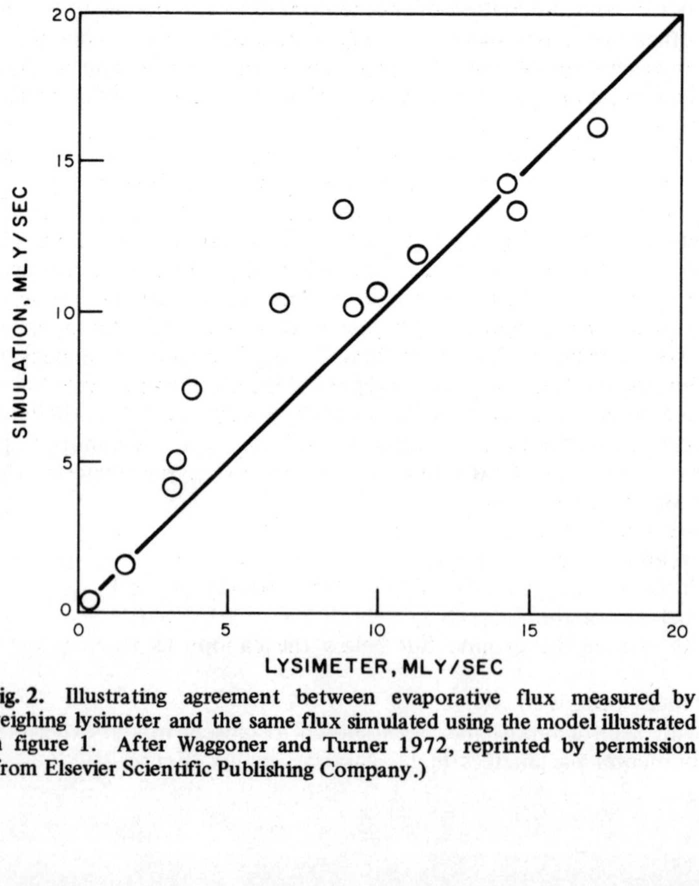

Fig. 2. Illustrating agreement between evaporative flux measured by weighing lysimeter and the same flux simulated using the model illustrated in figure 1. After Waggoner and Turner 1972, reprinted by permission (from Elsevier Scientific Publishing Company.)

Space precludes consideration of the effect of plant communities on the soil environment. However, interactions between the physical environment and plant roots are discussed in chapter 8.

Modification of plant performance by its physical environment

Water availability, temperature, light and nutrients are important physical environmental factors which influence plant performance. The effect of light intensity on photosynthesis in canopies is reviewed below, and the nutritional environment has been discussed in chapter 9.

Water stress and plant performance

More detailed coverage of this subject, and associated literature, can be found in Slatyer (1967), Kozlowski (1968*a*, 1968*b*) and Hsaio (1973). Further, chapter 8 of this book includes the mechanism and models of water uptake by roots.

Plant water content and thus water stress (Weatherley 1950; Barrs 1968) are a result of dynamically varying rates of water supply and loss by the plant, and the characteristic lag of absorption behind transpiration. Consequently, both water content and stress vary considerably between day and night, as well as on longer time scales. It is the effects of such stress, rather than the concepts of how such stresses develop, which will be considered in this section.

Following Kramer (1959) the functions of water in plants may be listed as: a vital constituent of cell protoplasm; an essential reagent in photosynthesis; the solvent by which salts and gases enter and move throughout the plant; and an essential in maintaining sufficient turgidity for growth. Most effects of water stress are attributable either to loss of turgor or dehydration of the protoplasm.

Loss of turgor, the hydrostatic pressure within a cell, results firstly in a drop in rate of cell enlargement, and hence in rate of root and shoot elongation and leaf enlargement. Further loss of turgor can effect rate of cell division, lead to a decrease in stomatal opening, and the latter in turn can reduce transpiration and photosynthetic rate through increased resistance to gas exchange.

The non-aqueous components of protoplasm are proteins, which have considerable hydration shells. Reduction in the water shells on these hydrophilic colloids is associated with changes in their configuration, physical and physiological properties, and associated enzyme activity. Though there is much as yet unknown of the detailed mechanisms involved, protoplasmic dehydration can reduce photosynthetic capacity directly, i.e. not only though reduced stomatal opening (Hsaio 1973); and whilst water stress normally reduces respiration rate, it can also increase it.

Gardner and Ehlig (1963) have shown that a sudden decline in transpiration rate takes place at a leaf water potential, which, at least in the short term, appears characteristic for a particular species (fig. 3).

Transpiration and photosynthesis are often reduced to about the same extent by water stress (Brix 1962), though, using simple theory, a reduction in stomatal aperture should decrease transpiration more than photosynthesis (Waggoner 1966). A more adequate analysis by Cowan and Troughton (1971) indicates a possible range of relationships between transpiration and photosynthesis as leaf resistance varies, with equal effects on both fluxes being one possibility. The relationship between photosynthesis and transpiration is strongly dependent on nutrition and other factors, but nevertheless use has been made of such relationships in interpreting and modelling growth (de Wit 1958; Rose et al 1972*b*).

Water stress during the reproductive phase can have serious and irreversible effect on subsequent development, particularly in determinate species (e.g. grain crops). Grain set

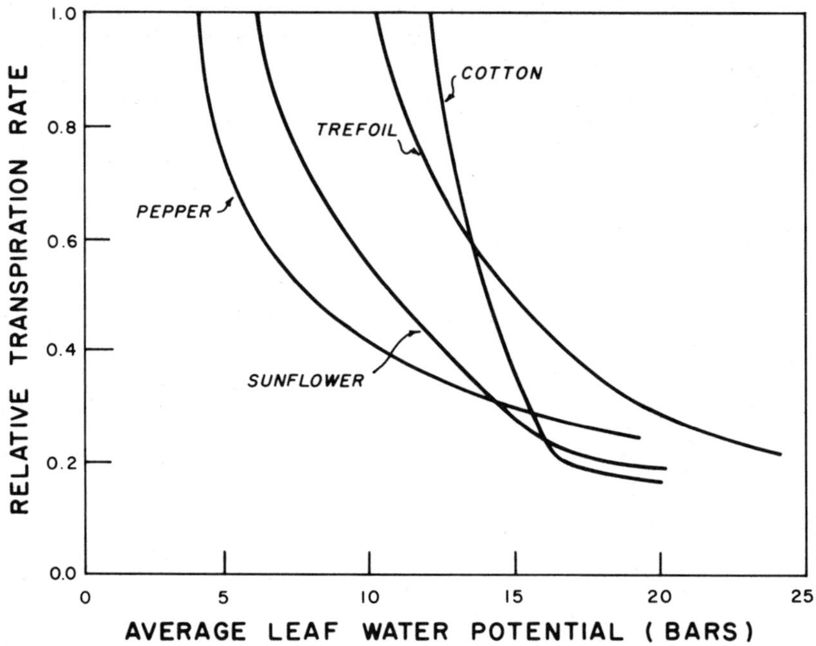

Fig. 3. Relative transpiration as a function of leaf water potential. (After Gardner and Ehlig (1963), reprinted by permission from American Geophysical Union.)

is generally influenced by stress at certain ontogenetic stages—at or shortly after anthesis for wheat (Wardlaw 1971). *Sorghum bicolor*, however, is capable of suspending reproductive development until the water supply is restored (Wilson and Whiteman 1965).

In corn, Denmead and Shaw (1960) and others have shown the tasselling-silking period to be most critical for reduction in subsequent grain yield, though stress during the maturation or grain-filling stage also reduces grain yield appreciably. Analyzing historical records, using a simulation of the soil water regime, Nix and Fitzpatrick (1969) found that the period from ear emergence through anthesis to early grain development appeared to be such a critical ontogenetic period for both wheat and sorghum.

Measurements of grain water status, grain filling, photosynthetic respiration and grain-filling rates in wheat and barley indicate that the grains themselves are relatively resistant to water deficits, except under severe stress (Aspinall 1965). This apparently is a result of assimilate redistribution and an increase in assimilates distributed to the ear from lower plant parts (Wardlaw 1967). However, analyses of depressed final grain weights and earlier grain maturation (than for unstressed plants) indicate that water deficits influence grain development, especially during the cell division-expansion stage, but the effects vary with stress intensity and duration. In crops of a less determinate nature, where reproductive activity may be spread over a longer period, similar critical periods for water stress are less well defined. Although water stress leads to reduced vegetative growth, under some circumstances the desired quality of plant products is improved by moderate water stress (Kramer 1959).

Temperature and plant performance

Although there are strong interactions and correlations between temperature, light intensity and photoperiod, studies of the effects of the individual variables, e.g. temperature, have been possible through experiments in controlled environments (e.g. Vegis 1963).

There is a considerable amount of literature regarding temperature effects on plant growth, and numerous reviews covering particular processes (e.g. photosynthesis, Rabinowitch 1956), particular physiological phases (e.g. Vegis 1963), and particular crops (e.g. rice, Owen 1971*a*), cereals (Friend 1965), and grasses (Alberda 1966). Temperature effects (including frost effects) on the germination, vegetative and reproductive phases of growth in annual plants will be briefly discussed.

Germination. Temperature conditions necessary to break seed dormancy vary among the annual plants. For instance, winter and spring annuals, unlike summer annuals, require low temperatures (Went and Sheps 1969). Since no plant part is directly exposed at this stage of development, frosts generally have little direct effect.

The imbibition of water during germination is subject to temperature effects. The change in average gravimetric water content (W) of corn seeds with time (t) after contact with water, has been investigated over a range of temperatures (Blacklow 1972*a*). An initial exponential phase of water uptake is characteristic of both living and dead seeds. In living seeds this is succeeded by a marked linear phase, prior to germination, and at least up to the time of germination water uptake can be described by:

$$\begin{aligned}\frac{dW}{dt} &= b + ak\ \exp(-kt)\\ &= b + k(f(t)-W)\end{aligned}$$

where b is a measure of the rate of imbition (during the linear phase), k is a measure of the permeability of seeds to water (during the exponential phase), and f(t) is a measure of the changing capacity of the seeds to imbibe water, which depends on the growth of the embryo. It is likely that the above equation will describe imbibition for most species, although it has been derived from data for corn seeds.

Blacklow (1972*a*) also showed that the time from wetting required for the emergence of either radicle or shoot from corn seeds decreased non-linearly with increasing temperature, in such a manner that a temperature can be selected so that the product of duration and temperature above this "threshold" value is very approximately constant. This is a simple constant-temperature example of the more general "day-degree" concept (Langridge and McWilliam 1967) commonly employed in relating stages of physiological development to chronological time and the thermal environment. Blacklow (1972*b*) also found that the rate of elongation of radicle and shoot during early stages of growth increased with temperature from negligible values below 9°C, through a maximum close to 30°C, with a rapid decline at higher temperatures, a constant temperature of 40°C being lethal.

Vegetative phase. During this phase it is the temperature of the apical meristem which is of special importance for rates of plant development (Watts 1971). This will be a soil temperature early in the development of the annual plant, and somewhat closer to air temperature at later stages.

Frosts are common in semi-arid Australia, and both pastures and crops are particularly susceptible to damaging effects. In annual crops where seeds have a vernalization requirement, plant growth is regulated so as usually to avoid the coincidence of frost periods with sensitive stages of development such as flowering. However, where frosts occur during the vegetative phase, the exposed meristematic parts of cereals are most liable to damage (Gott 1961), and cold weather can also reduce the rate of development generally, although plants do not become truly dormant (Olien 1967).

The optimum temperature for maximum growth rates can vary significantly through the life of the crop. Also, some plants need to experience higher temperatures by day than by night—the common field situation (Went 1957). Thus, a considerable programme of controlled environment experimentation is required to fully define the thermal responses of even a single species.

Despite the extensive literature on temperature effects, quantitative comparisons among the variety of species examined is hampered by the non-standardization of other environmental parameters, e.g. light regimes. Despite uncertainties due to such difficulties, distinctly different thermal response curves have been inferred by Fitzpatrick and Nix (1970) for tropical grasses, tropical legumes, and temperate grasses and legumes. Except for tropical grasses the data indicated an optimum temperature allowing maximum dry matter production of the other groups. Fitzpatrick and Nix (1970) used these generalized thermal response curves, together with mean seasonal trends in temperature at a selection of meteorological stations throughout Australia, to infer thermal limitations to the growth of pasture species over the Australian continent. It is interesting that at optimal temperatures for temperate grasses, dry matter produced by tropical grasses is often greater, despite such temperatures being sub-optimal for the latter group.

Reproductive phase. Photo- and thermo-induction requirements may need to be satisfied prior to the commencement of floral initiation. Although a certain sum of degree-days above an appropriate threshold is often an adequate working rule for the completion of either vegetative or reproductive phases, sometimes supplemented by a vernalization requirement (Went and Sheps 1969), there is room for further examination of the fundamental requirements for phasic transition. This whole question is vital for systematic progress in the quest to provide the best practical fit between crop requirements and climatic characteristics. The opportunities in this direction are considerable, especially in a country such as Australia where there is a wide variation in climatic characteristics relevant to plant growth (Fitzpatrick and Nix 1970).

The character of the thermal environment experienced prior to floral initiation can have effects which persist throughout the remainder of the life to the plant, as demonstrated by Owen (1971*b*) for a semi-dwarf wheat. Furthermore, for cereals generally, temperature extremes from floral initiation onward can adversely affect both the mechanisms of fertilization, as has been investigated for rice (Owen 1971*a*) and grain development (Asana and Williams 1965; Owen 1971*c*). Also, low winter temperatures and frosts are particularly damaging at anthesis, and undoubtedly play some part in the retardation of ear development in cereals (Gott 1961).

MODELS OF PHOTOSYNTHESIS AND PASTURE GROWTH

A general case has been argued in earlier sections for continued examination of the role of simulation in agriculture and ecology.

This section is a brief review of current work concerning models of photosynthesis and pasture production, and also of areas where more progress is required. It is designed to complement the water and nutritional models covered earlier in chapters 8 and 9.

Simulation of photosynthetic rate

Photosynthesis, a basic process in plant production, results from a particular type of plant–environment interaction. Due to the extent of relevant information available, there has been considerable activity in developing models to simulate photosynthetic rate. If this is successful on a short period basis, then extension of the approach to cover a more extensive period in the life of the crop requires the adoption of rules for the partition of photosynthate into respiration, photosynthetic and non-photosynthetic tissue (such as roots). There are quantitatively important differences in the character and relative strengths of sinks for photosynthate (at least for row crops) during the different phases of plant growth, and these need to be recognized if the model aims to simulate the essential features of morphological development, which in turn is one requirement for the correct simulation of photosynthesis.

Thus it is necessary to have an ontogenetic routine which predicts the timing of important events in the phasic development of the crop. There is increasing evidence that quantitative rules govern this morphogenesis, as illustrated by the success of some form of heat sum method (Langridge and McWilliam 1967).

Furthermore, an ontogenetic routine is also needed because in some crops there is a possibility that the photosynthetic "capacity" of a plant may exceed that required to satisfy growth and respiration, at least for certain growth phases. In such cases, photosynthetic rate is determined more by the availability of sinks than the availability of carbohydrate (Hackett 1973). The accuracy of simulating photosynthetic rate would then depend on the ability to correctly simulate phasic development, and the size and strength of photosynthetic sinks in each phase.

Answers to such questions are still being sought, and because it is usually assumed that the source is more limiting in (simulated) photosynthesis than the sink, this assumption will be followed.

Both indirect and direct methods of simulating photosynthesis have been reported. Where temperature during the growing season is unlikely to retard growth, rate of increase of crop dry matter M has been obtained by relating it to evaporation rate E via:

$$\partial M/\partial t = mE \tag{1}$$

where m is dependent on nutrition and on leaf area index until light interception by the canopy is complete.

This indirect approach, introduced by de Wit (1958), has been used by Stewart and Hagan (1969), Rose et al. (1972*b*) and McCown (1973), and examples of its use are given below. The utility of this indirect approach depends on there being a relationship between the variables M and E which is not strongly dependent on seasonal characteristics. Figure 4 illustrates such a relationship for the tropical legume

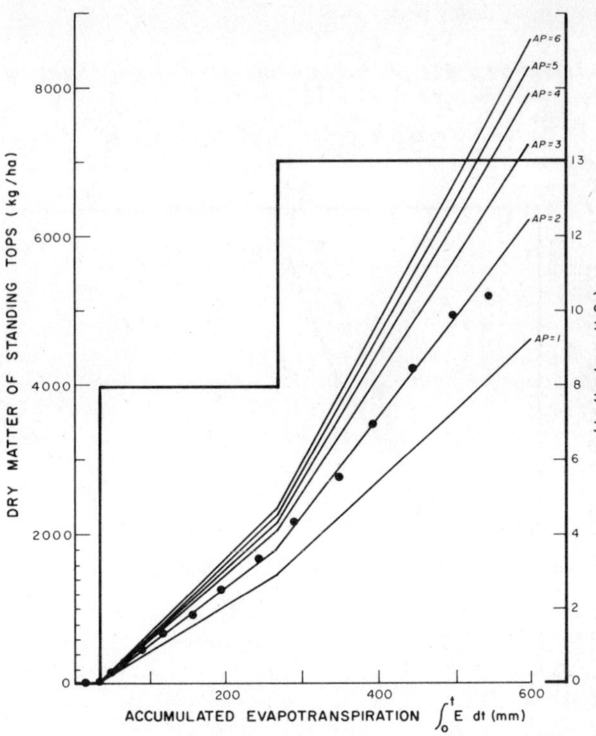

Fig. 4. The family of curves gives the relationship used in the model between crop dry matter M and accumulated evapotranspiration. AP is equivalent rate of superphosphate applied prior to the season. Points are experimental data for AP = 2 (250 kg/ha); Staircase plot gives value of m in equation (1) for AP = 2 (250 kg/ha) (after Rose et al. 1972*b*)

Stylosanthes humilis, where dry matter is related to evapotranspiration accumulated since germination. Though these relationships have a strong dependence on phosphate nutrition, they are reasonably independent of seasonal characteristics, though accumulated evapotranspiration and hence yield are markedly so.

The more commonly used approach in simulating photosynthesis is that which more directly reflects our understanding of the factors affecting photosynthetic rate. In this approach:

(i) The penetration and distribution of light over the leaves of the canopy are first calculated. Given data on crop morphology, this is a physical problem.

(ii) The photosynthetic rate of each leaf or leaf-layer in the canopy is then calculated from the light distribution and a knowledge of the photosynthetic response characteristics to light of individual leaves.

Let us consider these two steps in turn.

Light penetration into plant canopies. This aspect of the problem has received a great deal of attention in almost every major scientific country. Monsi and Saeki (1953), de Wit (1965), Monteith (1965) and Anderson (1966) were early workers in this area.

The penetration of the photosynthetically active wave band (PAR, 400-700 nm approximately) can be estimated in a reasonably accurate yet simple manner as the radiation transmitted and reflected by leaves is only about 10 per cent of the incident radiation. This is not the case for other wavebands, e.g. in the near infra red (700-3,000 nm) where this proportion is about 40 per cent.

Whilst there are a number of problems not coped with by current models of light distribution in crop canopies, good progress has been made in this part of the problem, as illustrated in figure 5, which compares simulated and measured light profiles within a crop canopy. The good agreement between the profiles indicates that success in simulating photosynthesis is more likely to be limited by the second step in the calculation, now described.

Inferring net photosynthetic rate from light distribution in a canopy. The

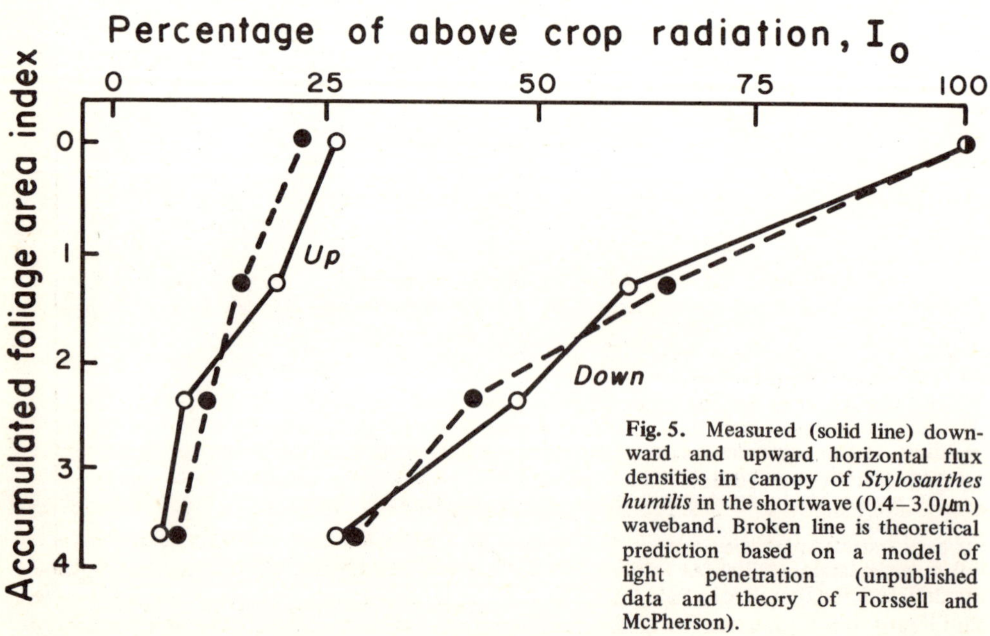

Fig. 5. Measured (solid line) downward and upward horizontal flux densities in canopy of *Stylosanthes humilis* in the shortwave (0.4–3.0μm) waveband. Broken line is theoretical prediction based on a model of light penetration (unpublished data and theory of Torssell and McPherson).

photosynthesis resulting from direct solar radiation is influenced by the angle of leaves presented to it, thus involving the geometrical arrangement of photosynthetic tissue (referred to as crop architecture).

A second factor involved is the relationship between radiation flux density and photosynthetic rate, for the tissue concerned (leaves commonly being the major contributor to photosynthesis). These light-photosynthesis relationships must be obtained experimentally, though assistance in determining the relative significance of these factors, and hence the experimental accuracy required, may be assessed by combining them in a photosynthetic model. By so doing, it becomes apparent that the light flux-photosynthesis relations are more important than the crop architecture. For example figure 6 shows gross photosynthesis rate computed by Ross (1970) in a plant stand for LAI $= 5$ as the total radiation flux density from a clear sky increases with solar elevation. The figure shows that differences due to leaf orientation are quite modest, except at high solar elevations, and that for the range of elevations up to 45° (important to temperate latitudes), crop architecture has very little effect on gross photosynthesis.

In contrast, differences in light flux-photosynthesis relations can have a large effect, not only in temperate but also in tropical latitudes. Differences of 100 per cent and more between the maximum photosynthetic rates of different species are common, and even at a latitude of $\simeq$ 52° (Netherlands) this could lead to 50 per cent differences in daily photosynthesis.

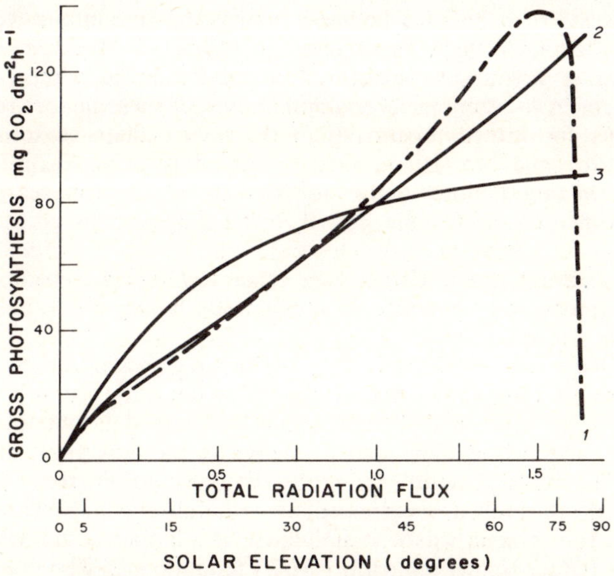

Fig. 6. Computed gross photosynthesis in plant stands as a function of total radiation flux density (corresponding to the solar elevation indicated for various leaf orientations). Leaf area index = 5. 1, vertical leaves. 2, "spherical" orientation (i.e. all orientations equally likely). 3, horizontal leaves (which is very similar to the photosynthesis-light curve for a single leaf). (After Ross 1970.)

Many factors influence the relation between light and photosynthesis. These include species differences (Ludlow and Wilson 1970, 1971), plant nutrition, leaf age and position in the canopy or related factors (e.g. Thorne, 1960), water stress (El-Sharkawy and Hesketh 1964) and other physiological factors. Accuracy in simulating photosynthesis is more likely to be limited by deficiencies in data and understanding of these factors, than by deficiencies in current understanding of light penetration into canopies. However, significant limitations also result from uncertainties concerning photosynthate partition.

Models of pasture growth

Comprehensive studies of natural grassland ecosystems are an important component of the International Biological Program in the U.S.A. Dix and Beidleman (1969) have edited a synthesis of studies on grasslands in an ecosystem framework, and Moore (1970) has given an account of work on Australian grasslands.

In a number of countries, including Australia, there is a rapidly rising interest in developing research on pasture and animal production in which systems analysis techniques play an integrative role. For example this was a major theme in the 1972 Biennial Conference of the Australian Society of Animal Production, reported in vol. 9 of the Proceedings of that Society (1972). The type of approaches being used to relate pasture growth to environmental and management inputs (e.g. water regime and fertilizer application) will be illustrated. Since this output is simply input to other parts of the grazing system, it is understandable that published models of pasture production are commonly simpler than models aimed at assisting plant physiological insight, such as that of de Wit, Brouwer and Penning de Vries (1970).

For reasons discussed above low temperature is an important factor limiting pasture growth in temperate regions. However, water availability is the factor of overriding significance in pastoral areas of Australia. It is important both per se and through strong interactions with nutrition, and thus occupies a major role in pasture production models. The basis and application of methods commonly used to provide simulations of water availability, in terms of a balance between inputs and outputs, is well illustrated by Slatyer (1960), Fleming (1964) and Chapman (1968). In the temperate regions of Australia, the most favourable moisture regime is in the winter, while optimum temperatures occur in the summer. The combination of these factors results in the best overall conditions for growth occurring in the intermediate seasons of spring and autumn (Fitzpatrick and Nix 1970). Over much of tropical Australia, the moisture regime is the dominant limiting factor for growth, as the wet season occurs when temperatures are favourable for the growth of tropical grasses and legumes.

A number of models have been developed in which pasture production is related to water supply (e.g. Brockington 1970; Rose et al. 1972*b*; Hutchinson 1972; McCown 1973). Two examples will be briefly discussed. Both use water balance techniques similar to that of Slatyer (1960).

The model of Rose et al. (1972*b*) uses the indirect method of implying photosynthetic rate from evaporation rate (fig. 4). Such empirical relationships must be obtained for the species and soil type concerned. The abscissa in figure 4 is obtained by integrating evaporation rate, and this is calculated as shown in the flow chart of figure 7, which shows the complete model structure, using the symbolism of Forrester (1969). The input variables required are applied superphosphate, rainfall and pan evaporation.

Growth of the root system was also simulated, and if inadequate water was available to it then E was reduced, thus reducing $\partial M/\partial t$ (equation 1). The growing season was terminated using non-critical assumptions concerning the duration of dry top soil. The model has been examined by comparing predicted and observed growth during six seasons other than the two used to derive the relations used in the model. An example of such a comparison is given in figure 8.

The model of McCown (1973) was also obtained using the tropical legume *Stylosanthes humilis*. It illustrates an important use of models in predicting the frequency characteristic of production in the long term, based on experimental data from a restricted number of years.

From the simulated water balance, the quantity $[(E/PE) - 0.3]$ was calculated, where PE is potential evaporation. No growth occurred when $E/PE = 0.3$, so the above quantity, regarded as the fraction of a day during which water availability did not limit transpiration, was summed to give the total duration of active growth for the season in question. The relative probability (in the long term) of any given seasonal growth duration

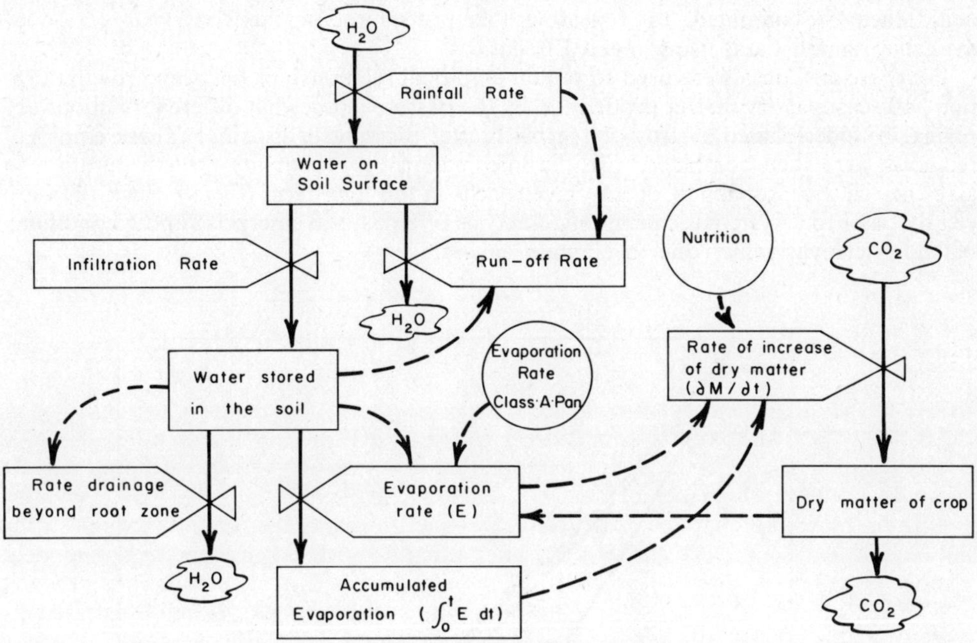

Fig. 7. Generalized continuous system-type presentation of the model of Rose et al. (1972*b*) used to simulate the growth of ungrazed *Stylosanthes humilis* pastures.

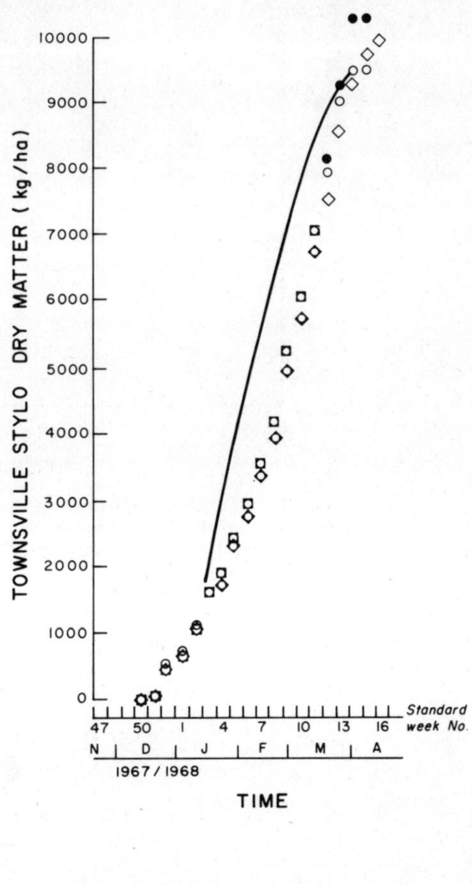

Fig. 8. Line represents experimentally measured growth of *Stylosanthes humilis* pastures as a function of time. Points indicate simulated values using the model outline in figure 7, different symbols referring to different possible assumptions (after Rose et al. 1972*b*).

could then be computed, by repeating this procedure for the sixty-year period of available rainfall data, using mean PE data.

Four seasons' data were used to obtain empirical relationships between growth duration and seasonal dry matter production by the pasture (somewhat different relations applying to underfoliated swards or regrowth after mowing or grazing). These empirical relations were then combined with the information on probability-growth duration to give the relative probability of any given yield in the long term (fig. 9). Such data are clearly basic to a systematic economic analysis of a pastoral enterprise under conditions of high yield variability due to climatic causes.

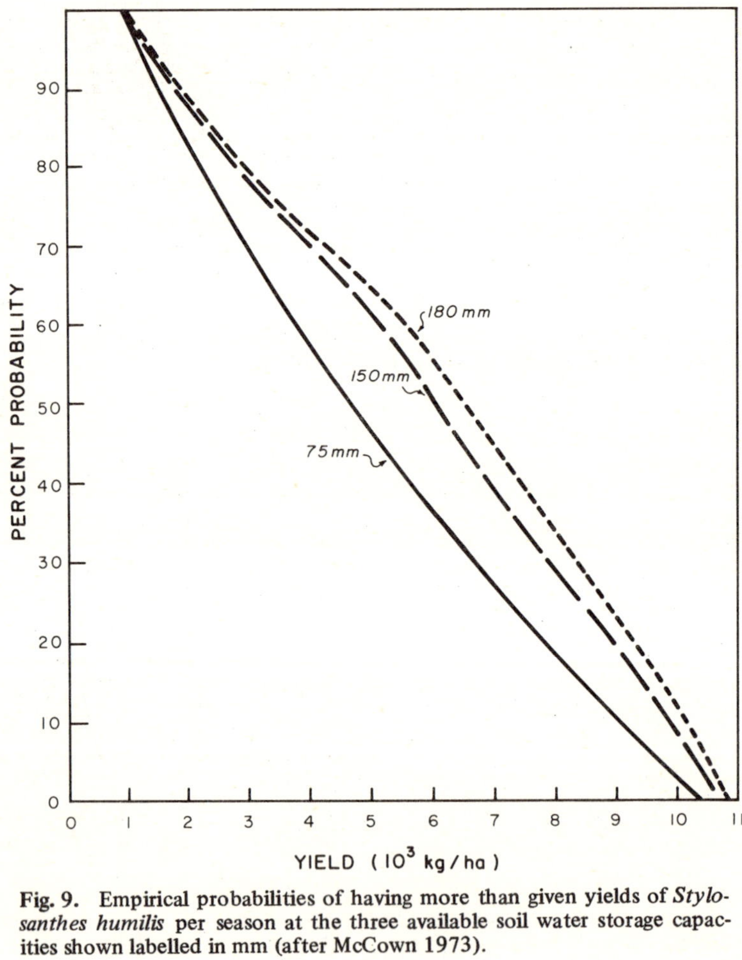

Fig. 9. Empirical probabilities of having more than given yields of *Stylosanthes humilis* per season at the three available soil water storage capacities shown labelled in mm (after McCown 1973).

It is outside the scope of this review to consider the broader animal production and economic setting of such models, and indeed the models described are concerned with but one sub-system. Vickery and Hedges (1972), Arnold and Campbell (1972), and other contributions to vol. 9 of the Proceedings of the Australian Society of Animal Production (1972) provide examples of models including animal production and economic considerations.

ACKNOWLEDGMENTS

The assistance of Ms. T. Komarowski and E.M. Birk, and of numerous colleagues in C.S.I.R.O., is gratefully acknowledged.

REFERENCES

Alberda, T.H. (1966). In "The Growth of Cereals and Grasses". (Eds F.L. Milthorpe and J.D. Ivins). pp. 200-12. Butterworths, London.
Anderson, M.C. (1966). *J. Appl. Ecol.* 3, 41-54.
Armstrong, J.S. (1972). *Proc. Aust. Soc. Anim. Prod.* 9, 179-89.
Arnold, G.W., and Campbell, N.A. (1972). *Proc. Aust. Soc. Anim. Prod.* 9, 23-30.
Asana, R.D., and Williams, R.F. (1965). *Aust. J. Agric. Res.* 16, 1-13.
Aspinall, D. (1965). *Aust. J. Agric. Res.* 16, 265-75.
Austin, M.P. (1971). In "Mathematical Models in Ecology". (Ed. J.N.R. Jeffreys). pp. 61-81. Blackwell, Oxford.
Australian Society of Animal Production (1972). Proceedings of the 9th Biennial Conference, Canberra, 1972.
Barrs, H.D. (1968). In "Water Deficits and Plant Growth". Vol. 1 (Ed. T.T. Kozlowski). pp. 236-347. Academic Press, New York.
Blacklow, W.M. (1972*a*). *Crop Sci.* 12, 643-6.
———. (1972*b*). *Crop Sci.* 12, 647-50.
Brix, H. (1962). *Physiol. Plant.* 15, 10-20.
Brockington, N.R. (1970). In "Ann. Rep. Grassl. Res. Inst." pp. 140-9. Grassl. Res. Inst. Hurley.
Byrne, G.F., and Tognetti, K. (1969). *Agric. Meteorol.* 6, 151-63.
Chapman, T.C. (1968). In "Land Evaluation". (Ed. G.A. Stewart). pp. 312-23. Macmillan, Melbourne.
Childs, E.C. (1969). "An Introduction to the Physical Basis of Soil Water Phenomena". John Wiley, New York.
Cowan, I.R. (1968). *Quart. J. Roy. Met. Soc.* 94, 523-44.
Cowan, I.R., and Troughton, J.H. (1971). *Planta (Berl.)* 97, 325-36.
Dale, M.B. (1970). *Ecology.* 51, 2-16.
Denmead, O.T. (1968*a*). In Proc. WMO Seminar Agric. Meteorol. Melbourne, 1966. Bureau of Meteorol. Australia. pp. 71-105.
———. (1968*b*). In Proc. WMO Seminar Agric. Meteorol. Melbourne, 1966. Bureau of Meteorol. Australia. pp. 445-82.
Denmead, O.T., and Shaw, R.H. (1960). *Agron. J.* 52, 272-4.
Dent, J.B., and Anderson, J.R. (1971). "Systems Analysis in Agricultural Management". John Wiley, Sydney.
Dix, R.L., and Beidleman, R.G. (1969). "The Grassland Ecosystem. A Preliminary Synthesis". Range Science Dept. Science Series No. 2. Colorado State Univ., Fort Collins.
El-Sharkawy, M.A., and Hesketh, J.D. (1964). *Crop Sci.* 4, 51-8.
Evans, L.T. (Ed.) (1963). "Environmental Control of Plant Growth". Academic Press, New York.
Fitzpatrick, E.A., and Nix, H.A. (1970). In "Australian Grasslands". (Ed. R.M. Moore). pp. 3-26. Australian National Univ. Press, Canberra.
Fleming, P.M. (1964). *Proc. 6th Int. Cong. Agric. Engn.* pp. 108-20. Geneva.
Forrester, J.W. (1969). "Principles of Systems". Wright-Allen Press, Cambridge, Mass.
———. (1971). *Technology Review.* 73, 1-16.
Friend, D.J.C. (1965). *Can. J. Bot.* 43, 345-53.
Gardner, W.R., and Ehlig, D.F. (1963). *J. Geophys. Res.* 68, 5719-1733.
Goncz, J.H., and Rose, C.W. (1972). *Agric. Meteorol.* 9, 405-19.
Gordon, G. (1969). "System Simulation". Prentice-Hall, Englewood Cliffs, N.J.
Gott, M. Beth (1961). *Aust. J. Agric. Res.* 12, 547-65.
Graetz, R.D. (1972). Ph.D. Thesis, Australian National Univ., Canberra.
Hackett, C. (1973). *Aust. J. Biol. Sci.* 26, 1057-71.
Hsaio, T.L. (1973). *Annu. Rev. Plant Physiol.* 24, 519-70.
Hutchinson, K.J. (1972). *Proc. Aust. Soc. Anim. Prod.* 9, 79-86.
Kozlowski, T.T. (Ed.) (1968*a*). "Water Deficits and Plant Growth". Vol. 1. Academic Press, New York.
———. (Ed.) (1968*b*). "Water Deficits and Plant Growth". Vol. 2. Academic Press, New York.
Kramer, P.J. (1959). In "Water and Its Relation to Soils and Crops". (Ed. M.B. Russell). pp. 51-70. Academic Press, New York.
Langridge, J., and McWilliam, J.R. (1967). In "Thermobiology". (Ed. A.H. Rose). pp. 231-92. Academic Press, New York.

Ludlow, M.M., and Wilson, G.L. (1970). *Aust. J. Agric. Res.* **21**, 183-94.
———. (1971). *Aust. J. Biol. Sci.* **24**, 449-70.
McCown, R.L. (1973). *Agric. Meteorol.* **11**, 53-63.
Mesarovic, M.D., and Macko, D. (1969). In "Hierarchical Structures". (Eds. LL. Whyte, A.G. Wilson, and D. Wilson). pp. 29-50. American Elsevier, New York.
Monsi, M., and Saeki, T. (1953). *Jap. J. Bot.* **13**, 22-52.
Monteith, J.L. (1965). *Ann. Bot. (Lond.)* **29**, 17-37.
Moore, R.M. (1970). In "Australian Grasslands". (Ed. R.M. Moore). pp. 87-98. Australian National Univ. Press, Canberra.
Morley, F.H.W. (1972). *Proc. Aust. Soc. Anim. Prod.* **9**, 1-8.
Nix, H.A. (1968). In "Land Evaluation". (Ed. G.A. Stewart). pp. 77-87. Macmillan, Melbourne.
Nix, H.A., and Fitzpatrick, E.A. (1969). *Agric. Meteorol.* **6**, 321-37.
Odell, R.J. (1950). *Agron. J.* **42**, 328-35.
Olien, C.R. (1967). *Annu. Rev. Plant Physiol.* **18**, 387-408.
Owen, P.C. (1971*a*). *Field Crop Abstracts* **24**, 1-8.
———. (1971*b*). *Expl. Agric.* **7**, 33-41.
———. (1971*c*). *Expl. Agric.* **7**, 43-7.
Philip, J.R. (1964). *J. Appl. Meteorol.* **3**, 390-5.
Philip, J.R., and de Vries, D.A. (1957). *Trans. Amer. Geophys. Un.* **38**, 222-32.
Rabinowitch, E.I. (1956). "Photosynthesis and Related Processes". Vol. 2, Pt. 2. Interscience, New York.
Rijtema, P.E. (1965). *Agric. Res. Rep.* No. 659. Pudoc, Wageningen.
Rose, C.W. (1968*a*). *Aust. J. Soil Res.* **6**, 31-44.
———. (1968*b*). *Aust. J. Soil Res.* **6**, 45-57.
Rose, C.W., Begg, J.E., Byrne, G.F., Goncz, J.H., and Torssell, B.W.R. (1972*a*). *Agric. Meteorol.* **9**, 385-403.
Rose, C.W., Begg, J.E., Byrne, G.F., Torssell, B.W.R., and Goncz, J.H. (1972*b*). *Agric. Meteorol.* **10**, 161-83.
Rose, C.W., and Stern, W.R. (1967). *Aust. J. Soil Res.* **5**, 11-19.
Ross, J. (1970). In "Prediction and Measurement of Photosynthetic Productivity". (Proc. IBP/PP Technical Meeting, Trebon, 1969). pp. 29-44. Pudoc, Wageningen.
Slatyer, R.O. (1960). CSIRO Aust. Div. Land Res. Reg. Surv. Tech. Paper No. 13.
———. (1967). "Plant Water Relationships". Academic Press, London.
Stewart, J.I., and Hagan, R.M. (1969). *Proc. Congr. Int. Comm. Irrig. Drain.*, 7th, Mexico City, Question 23, R. 32, p. 23. 505-30.
Sutton, O.G. (1953). "Micrometeorology". McGraw-Hill, New York.
Thorne, G.N. (1960). *Ann. Bot. (Lond.)* **24**, 356-71.
Vegis, A. (1963). In "Environmental Control of Plant Growth", (Ed. L.T. Evans). pp. 265-310. Academic Press, New York.
Vickery, P.J., and Hedges, D.A. (1972). *Proc. Aust. Soc. Anim. Prod.* **9**, 16-22.
Waggoner, P.E. (1966). In "Plant Environment and Efficient Water Use". (Eds. W.H. Pierre, D. Kirkham, J. Pesek, and R.H. Shaw). pp. 49-68. Amer. Soc. Agron., Madison.
Waggoner, P.E., and Reifsnyder, W.E. (1968). *J. Appl. Meteorol.* **7**, 400-9.
Waggoner, P.E., and Turner, N.C. (1972). *Agric. Meteorol.* **10**, 113-23.
Wardlaw, I.F. (1967). *Aust. J. Biol. Sci.* **20**, 25-38.
———. (1971). *Aust. J. Biol. Sci.* **24**, 1047-55.
Watts. W.R. (1971). *Nature (Lond.)* **229**, 46-7.
Weatherley, P.E. (1950). *New Phytol.* **49**, 81-97.
Went, F.W. (1957). "The Experimental Control of Plant Growth". Chronica Botanica, Waltham.
Went, F.W., and Sheps, L.O. (1969). In "Plant Physiology". Vol. 5A (Ed. F.C. Steward). pp. 299-406. Academic Press, New York.
Williams, W.T., and Lambert, J.M. (1960). *J. Ecol.* **48**, 689-710.
Wilson, G.L., and Whiteman, P.C. (1965). Univ. of Qld. Papers (Dept. of Botany). **4**, 233-9.
Wit, C.T. de (1958). Transpiration and crop yields. Inst. Biol. Chem. Res. Field Crops Herbage, Mededel, 59.
———. (1965). *Agric. Res. Rep.* **663**, Pudoc, Wageningen.
Wit, C.T. de, Brouwer, R., Penning de Vries, F.W.T. (1970). In "Prediction and Measurement of Photosynthetic Productivity". (Proc. IBP/PP Technical Meeting, Trebon, 1969). pp. 47-70. Pudoc, Wageningen.

Part D

Management and use of semi-arid soils in Australia

CHAPTER 11

Cultivation and fallowing practices

H.J. Sims

Crop production in Australia was developed initially on experience from northern Europe, with new settlers using techniques and knowledge familiar to them. This was relatively suitable in the higher rainfall areas, but, as production expanded into drier areas, the need for some modifications became apparent.

Most of the crop production in the semi-arid parts of Australia has been from winter-grown cereals and crops with a similar growth cycle. Experimental data are now being accumulated in regard to some potential cropping areas in semi-arid tropical parts of the continent.

FALLOWING–DEVELOPMENT AND ADOPTION

The development of fallowing has been of particular importance in crop production in the semi-arid parts of Australia. Professor Lowrie of Roseworthy College indicated that, in 1888, veteran farmers stated that they had not achieved financial success until they adopted bare fallow (Callaghan and Millington 1956, p. 115).

The practice of fallowing in Australia apparently originated as a folk innovation in South Australia. In 1860-61, 27,540 ha were fallowed, and this increased to 359,200 ha by 1900-01 (Dunsdorfs 1956, p. 142). Meinig (1970, p. 113) indicated that there was controversy in South Australia about fallowing methods and seedbed preparation during the 1870s.

Howell (1911) gives a long definition of fallowing:

> By the term "bare fallow" we, of course, understand here land which has been broken up by the plough; but even this preliminary operation might take place in different ways, and at different times of the year. The ground, after ploughing, might be left in a rough condition, and the preparation might take place in the autumn or winter, or even late in the spring or early summer; the preparation might cease with the simple operation of ploughing ... or it might be followed by frequent use of various implements ... Bare fallowing, with all the various acts of tillage included in this practice, has the primary object, under our rainfall conditions, of conserving moisture already in the soil.

He contrasted this objective with that of deep ploughing, the primary purpose of which was to let water into the subsoil, so that it would not be subject to the evaporating forces active at the surface. Meinig (1970) states, however, that South Australian experience between 1869 and 1884 was that yields were greater from shallow cultivation ("scratching") than from deep ploughing.

It appears that Australian fallowing methods suitable for the local soil and moisture conditions were developed independently of the American "dry farming" methods (Callaghan and Millington 1956, chapter 7). The methods which were developed by farmers have been evaluated and further adapted by research workers in various parts of Australia. The evaluation, particularly in the earlier years, was primarily on an agronomic basis, and an appreciation of the soil factors evolved later, and rather slowly.

Cornish (1949), Dunsdorfs (1956) and Meinig (1970), in discussing wheat yield changes prior to 1900, each refer to the role of fallowing in increasing yields. However, the first yield data appear to be those reported by Howell (1904, 1911). In the Wimmera district and its margins, mean wheat yields on fallow from eight experiments were 910 kg/ha compared with 750 kg/ha on non-fallow. Another set of results, mean of five experiments at Winiam (Nhill district), showed a much larger yield advantage, viz. fallow 1,135 kg/ha, stubble 620 kg/ha. In the Northern District (Wycheproof, Katunga, Pine Lodge) in 1903-04, mean yields were nearly 270 kg/ha better on fallow than on stubble land. A point of interest here was that a light dressing of sulphate of ammonia (about 10 kg/ha N) increased the yield on stubble to approximately that on fallow, leading to the comment that, in that year, the principal advantage of fallowing appeared to be that of nitrate accumulation. Other Victorian data, giving district wheat yields in 1908, were presented by Paterson (1916) (table 1). The comment was made that, as the main object is to store up moisture, the benefits are naturally more marked in the drier districts.

Table 1. Effect of fallowing on wheat yields, Victoria, 1908

District	Mean average rainfall	Wheat yields		Gain from fallowing	
		Unfallowed land	Fallowed land		
	mm	kg/ha	kg/ha	kg/ha	per cent
Mallee	330	176	386	210	119
Wimmera	456	385	793	408	106
Northern	475	272	637	365	134
Western	730	903	1,203	300	33

Results of South Australian experiments were reported from time to time in the *Journal of the Department of Agriculture, South Australia.* Fourteen years' data between 1912 and 1928 at Roseworthy (average rainfall 445 mm) are presented by Birks and Cole (1930). These experiments on weakly solonized brown (Mallee)—transitional red-brown earth soils compared time of fallowing (early and late), depth of fallowing and frequency of cultivation. From these results, they recommended, for local conditions, an early ploughing with a cultivation and rolling as soon as possible afterwards (by the end of August at the latest), and subsequent shallow summer cultivations as required to keep the surface open and free from weeds.

Further north in South Australia, at Booborowie (average rainfall 450 mm) on a fine-textured heavy red loam with a tendency to run together after heavy rain, average yields for 13 years (1916-28) were: early fallow 1,850 kg/ha; late fallow, after 1 September, 1,695 kg/ha; and non-fallow, ploughed in autumn, 1,340 kg/ha (Scott and Bristow 1929). Within the early fallows, best returns were obtained by leaving the land in the rough during the winter, giving a good cultivation in early spring and thorough tillage whenever weed growth or surface crusting rendered it necessary.

At Merredin, Western Australia, in the zone of grey and brown calcareous solonized soils, for a six-year period to 1929, early fallow (first week June) yielded 17 per cent higher than late fallow (first week August), the mean yields being: early fallow 1,550 kg/ha, late fallow 1,300 kg/ha. A comparison with non-fallow is quoted for one year only, the yields being: fallow 1,490 kg/ha and non-fallow 1,080 kg/ha (Thomas and Langfield 1930).

In New South Wales, the practice of fallowing appears to have been adopted less quickly than in other States (Anderson 1911), although recommendations for its use go back at least to the turn of the century (Peacock 1901). Considerable advisory information is included in the various editions of the *Farmers' Handbook*, e.g. Synnott 1946.

Up to the 1920s, explanations of the value of fallow were based mainly on American data (e.g. King 1895, 1901; Hardy W. Campbell, cited by Callaghan and Millington 1956). Prior to the mid-1920s the only published data dealing with soil measurements in Australia appear to be:

(i) Richardson (1912, 1925). The data (soil moisture) are summarized below (date of sampling 1 May 1912):

	Longerenong	Rutherglen
Sampling depth, m	0–1.52	0–1.22
Fallow, water %	29.1	15.5
Non-fallow, water %	25.7	11.4
Difference, %	3.4	4.1
Rain equivalent, mm	66	64

(ii) Paterson and Scott (1912*c*). A site near Geelong (outside the semi-arid zone) was used. Soil data for the top 0.45 m as at 28 February 1912 are summarized as follows:

	Worked fallow	Neglected fallow	Oat stubble
Water %	21.7	12.4	10.1
Nitrate nitrogen, kg/ha	166.2	65.7	36.4

The neglected fallow received no further cultivation after the initial cultivation in September 1911. The general conclusions drawn were that fallowing serves the double purpose of storing soil moisture and supplying nitrate, and that a fallow may do little good if neglected. Paterson and Scott (1912*a*, 1912*b*) reported data from laboratory studies on the effect of soil moisture and other factors on soil nitrification.

(iii) Lees 1926. At Wagga, soil moisture content was examined in the first 1.15 m of soil at periods from October 1923 to February 1924, when fallow contained 17.8 per cent moisture and non-fallow 11.6 per cent. The difference of 6.2 per cent was calculated as equivalent to 113 mm of rain.

Richardson (1925, p.44) summarized the then current outlook on fallow. He pointed out that rainfall was the most important factor limiting yields over a considerable portion of the wheat belt in Australia, and that conservation of soil moisture was therefore of fundamental importance, and that systems of cultivation which would conserve as much soil moisture as possible must be used. He also drew attention to some "worn out" soils and stressed the importance of maintaining soil fertility for a permanent system of agriculture.

FALLOWING–MOISTURE OR NITRATE?

A re-examination of the reasons for fallow was found necessary in the late 1920s and early 1930s, especially in South Australia, Western Australia and Victoria.

Work outside the semi-arid area at the Waite Agricultural Research Institute in South Australia led to the questioning of moisture conservation as the main reason for fallowing. The work of Richardson and Fricke (1931) and Richardson and Gurney (1933) found, in general, little, if any, moisture difference between fallow and non-fallow.

Prescott (1933), on the basis of South Australian Department of Agriculture data, suggested that the nitrogen mineralization explanation of the benefits of fallowing might apply more widely than was formerly recognized. He pointed out that there was a high correlation between crop yield and growing season rainfall, but none between fallow period (August-March) rainfall and the yield of the subsequent crop. This situation existed even at the relatively low rainfall conditions of Veitch in the Murray Mallee. On the basis of more complete data, Cornish (1950) considered the effect of available soil moisture at planting to have only a small effect on yield in South Australia.

In Western Australia, Teakle (1928), and Teakle and Burvill (1930*a*, 1930*b*) showed that fallowing was associated with moisture conservation, nitrate accumulation and increased yields. Their general conclusion was that, in dry seasons or districts, the advantage of a fallow was due to moisture conservation. In favourable seasons or localities, the advantage was associated with soil nitrate and soil tilth.

Fallow cultivation experiments initiated by the Victorian Department of Agricuture in the 1920s gave information on grain yields as influenced by time and, sometimes, depth of fallowing, the timing of subsequent cultivations and the substitution of hard grazing by sheep for cultivation at Carwarp (Mullett 1929), Walpeup (Mallee Research Station 1946) and Longerenong (Hore and Sims 1955). Detailed studies on moisture conservation and nitrate accumulation were undertaken in other experiments. Some of the earlier nitrate accumulation data for Werribee, Longerenong and Rutherglen are contained in a report by Skene (1930). Summer fallow had the highest content but, at Rutherglen, winter fallow was fairly close to it. Sampling was done to 0.9 m in four depths.

Leeper (1952, p. 96), making reference to other Victorian work at Longerenong and Rutherglen, stated that plots kept bare by handweeding, i.e. without cultivating the surface, retained as much water in the top 0.9 m and gave as good a crop as one kept under clean cultivation, and both were much superior to non-fallow. In another experiment at Longerenong in 1933, also cited by Leeper (1952), F. Penman found 35.5 per cent water in the top 0.9 m on fallow and 29.2 per cent on stubble areas. Sims (1948) showed that, at Walpeup in 1938, both moisture and nitrate nitrogen contents were increased by fallowing, and that both were decreased, in comparison with normal fallow, by the use of a cover crop for soil erosion control.

In the northeast of Victoria under higher rainfall (560 mm), Morrow and Hayman (1940) found that, following a period of subterranean clover ley, fallowing was not necessary. Analyses of soil from experimental sites in 1937 showed 22.5 ppm of nitrate nitrogen on the clover land unfallowed plots, 13.5 ppm on land fallowed before cropping and 5.0 ppm on old stubble land (sampling depth not indicated).

On brownish-red sandy loam soil at Griffith in New South Wales, West (1930) showed that the maintenance of a surface mulch (100 mm) did not result in any more moisture conservation than where there was no mulch.

Woodroffe (1949), reviewing Waite Institute and other data, concluded that, in years with less than 250 mm growing season rainfall, moisture conservation was the more important factor, but in years when growing season rainfall exceeded 420 mm, the benefit of fallowing was entirely due to nitrate accumulation. Woodroffe (1950) also indicated that, in South Australia, water accumulation by bare fallowing might amount to 75 mm. This could make an important contribution, as most districts received less than 420 mm of seasonal rain.

In a summary on the then current views on the principles of fallowing, Leeper (1952) pointed out that weed control was of prime importance. Fallowing prevented growth of weeds and consequently loss of soil moisture. Cultivation was not necessary if weeds were killed by another means. This conclusion was in contrast to the older philosophy that cultivation, by breaking the capillary channels which carry the water to the surface, prevented the capillary rise of water, and that cultivation was necessary after each rain to break the capillary tubes.

FALLOWING AND SOIL EROSION

During the 1930s and 1940s, the practice of fallowing was seriously criticized in connection with both wind and water erosion. Numerous reports and publications discussed the general aspects of soil erosion, but relatively few dealt specifically with

Australian research data.

Soil erosion in arable areas is not a simple problem. Factors other than fallowing which are related to the occurrence of soil erosion are land use, crop rotation systems, hydrology and grazing management, including rabbit control. Seasonal conditions during the 1930s and 1940s tended to accentuate the problem. In Australian literature on soil erosion, there is considerable overlap between the semi-arid and the better rainfall areas.

Reference was made by Herriot (1944) to the distribution and general causes of the problem in South Australia. Herriot (1951) later gave a brief history of the erosion situation in that state, indicating that Agricultural Bureau members were expressing concern about it in 1916.

In Victoria, committees of enquiry into soil erosion were appointed in 1917, 1933 and 1938 (O'Carroll 1940). The last Committee (McIlroy et al. 1938) considered that the losses caused by erosion were already very great. Further general information on the position in Victoria at that time was given at a symposium arranged by the Victorian Institute of Surveyors (1940).

A Soil Conservation Act was passed in New South Wales in 1938. This provided for a Soil Conservation Service and (together with other things) for experimental and research work, investigations and surveys.

Extension advice regarding soil erosion control was based for a long time on local experience together with overseas technical data, e.g. Herriot (1945*a*, 1945*b*), Beare (1951), Bagnold (1941). Chepil (various, cited in Chepil and Woodruff 1963) showed the various processes—suspension, saltation and surface creep—by which soil particles are moved by the action of wind. Work in North America on wind erosion and its control has been reviewed by Chepil and Woodruff (1963). This review shows the fundamental importance of reducing the erosive force of the wind at the soil surface by surface roughness, by placing barriers in the path of the wind, or by burying the erodible particles by tillage. The value of vegetative material, either alive or dead, was stressed. As McCalla and Army (1961) point out, the basic methods to control wind erosion are to create a soil condition resistant to erosion, and to shelter the soil from wind. The latter method, among other practices, includes stubble mulching.

In some areas, soil surveys provided basic soil data on which to base extension advice. These included surveys of an area in the mid-north of South Australia (Stephens et al. 1945), part of northern Victoria (Downes 1949), and the northern marginal areas of South Australia (Blackburn and Baker 1951). Comparative data on the effect of cultivation (and cropping) in rotations on water stable aggregates in some of these surveys were obtained by the method of Downes and Leeper (1939). The scheme of Taylor and Stephens (1943) was used for mapping land use and erosion incidence.

Various aspects of cultivation practices in relation to soil erosion in semi-arid and the better rainfall areas were discussed by Teakle (1939), Burvill (1945, 1948) and Lightfoot (1952) for Western Australia; by Hore (1940, 1945), Forster (1940) and Sims (1948) for Victoria; by Duggan et al. (1941), Powell, Bowmaker and Dargin (1941) and Condon and Stannard (1956-57) for New South Wales; by Herriot (1944, 1954) for South Australia; and by Ladewig and Skinner (1950) and Rosser (1964) for Queensland. The general principles of cultivation in relation to soil erosion control are clearly set out in Herriot's (1954) article on "Tillage".

The general summation of these data is that the maintenance of a cloddy surface was essential, the use of the mouldboard plough was preferred to disc and scarifying implements which pulverized the soil, that spring and early summer working of fallows should be avoided, and that maximum use should be made of grazing animals in controlling weeds; while contour banks were recommended for water erosion control under certain conditions.

The influence of topography on water erosion was shown in a survey in Western

Australia (Teakle 1939), which indicated that gullying appeared where cultivated slopes exceeded 3-3.5 per cent, and that gully and sheet erosion were severe when slopes ranged from 6 to 10 per cent.

Work at the Mallee Research Station, Walpeup, summarized in an internal report by Sims and Blackburn (1945) showed that dust caught at various heights above fallowed land was higher in plant nutrients than uneroded soil. An erosion index showed that soil loss was most serious where stubbles had been burnt prior to fallowing. Cultivation treatments that left the soil in a rougher condition, e.g. scarified early and ploughed in the spring, and ploughing to 125 mm, considerably reduced the soil loss. The use of cover crops sown on the fallow in the late winter or early spring and grazed to leave some stalks to protect the ground surface was also quite effective (see also Hore, Sims and McCann 1948, p. 475).

Geytenbeek (1951) reported very considerably reduced soil losses at the Waite Institute in 1949 by the use of a straw mulch. Barrow (1955) found better water penetration into a soil where stubble had not been burnt compared with the adjacent burnt area. Kaleski (1956) discussed soil losses on experimental sites on the steepest section (9 per cent slope) of the arable lands at the Wagga Soil Conservation Station. Losses from fallowed plots were much more than where there was vegetative protection to the soil. Logan (1960) found that run-off from cropping land was about double that from good pasture land.

The effect of the loss of topsoil on crop production has been examined in a number of experiments (Hore and Sims 1954; Barr 1957; Daniel 1969; Hamilton 1970; and Marsh 1972). In general, agronomic data are presented. Crop yields and grain nitrogen percentage were lowered by the loss of topsoil. Only a limited amount of soil data is included in these reports.

Once topsoil is lost, how long does it take for the new surface soil to regain satisfactory productivity? Leeper (1953) considered it to be less than was often stated. Sims and Mann (1965), continuing work on the site used by Hore and Sims (1954), found that growing barrel medic pasture for four years greatly improved subsequent wheat yields. Mean yields after this medic ley were not significantly different from those on the plots that had lost no topsoil, while grain protein was only slightly lower.

Soil erosion in Australia is now much less extensive than in the 1940s. Rowan and Downes (1963) discussed this situation in respect of wind erosion in Victoria, and the reasons for the improvement. In Australia generally, the situation is due largely to improved crop rotation systems, including the use of legume-based pastures. However, there is now a better appreciation of the effects of unsuitable cultivation practices.

In summary, a fundamental approach to combating soil erosion on cultivated lands includes the following principles:

(*a*) The reduction of the force of the wind or the water on the soil.
Surface roughness is of prime importance in reducing wind or water speed at the soil surface. Land cultivated after a pasture ley retains its cloddiness better than cultivated land. The incorporation of crop residues or other vegetative material maintains surface roughness longer than is possible on bare land. Farming practices, therefore, should make maximum use of all vegetative residues. The choice of implements for such a system is quite important. Contour cultivation and structures such as contour banks reduce water speed. Fallows well ridged at right angles to prevailing winds are more resistant to wind erosion.

(*b*) The increase in the size and the stability of the soil aggregates.
Normally this is achieved by crop rotations which include a period of pasture. Cultivation should be done in a manner which will pulverize the soil as little as possible. Therefore, implements of a type that do a minimum of damage to soil aggregates should be chosen, and such implements should be worked at lower speeds to lessen the damage.

INCORPORATION OF STUBBLE RESIDUES

This aspect of erosion control has received particular attention in North America, and good summaries are given by Zingg and Whitfield (1957), McCalla and Army (1961), and Hanway (1970). The objective of stubble mulching is the maintenance of crop residues continuously on the soil surface to give protection against wind and water erosion. The residues must be anchored in the soil so that the wind will not blow them away. Residues that are covered with soil decay rapidly, and thus the management must be such that the residues give protection throughout the fallow period.

The value of the incorporation of stubbles in erosion control in Australia has been stressed in many general publications. Some difficulties have been experienced, however, in regard to implements for handling residues. General comments on various aspects of these problems have been made by, e.g., Hore (1940), Herriot (1948, 1954), Lang (1948), Powell (1948), Sims (1948), Geytenbeek (1953), Beare (1962) and Waggamba Shire Extension Committee (1972). Some implements which have been widely used in North America, such as rod weeders and subsurface sweeps, have not become popular in Australia.

On the other hand, where soil erosion is not a problem, there appear to be no Australian data that indicate that the incorporation of stubbles increases cereal yields, while decreases have been shown in a number of instances, e.g. Synnott (1946, p. 229), Hore and Sims (1955), McKeown and McCulloch (1962), Rooney, Sims and Tuohey (1966), Macadam and Southwood (1968), and Mason, Toms and Gartrell (1969). Victorian experiments were summarized more completely by Sims (1965), as in table 2.

Table 2.

Centre	Soil	No. of years in test	Wheat yields after stubble burnt	Wheat yields after stubble not burnt
			kg/ha	kg/ha
Walpeup	Solonized brown soil	17	1,975	1,995
Longerenong	Grey clay	16	2,800	2,660
Kaniva	Grey clay	4	2,115	1,895
Diggora	Red-brown earth	4	2,345	1,880
Dookie	Red-brown earth	8	1,800	1,490
Rutherglen	Podzolic Red-brown earth intergrade	11	2,310	1,940

The results in table 2 refer to wheat grown on fallow at all centres except Rutherglen, where the wheat was grown on stubble after an earlier crop on ley. In addition to lower yield, the protein content of the grain was lower where the stubble was not burnt at all locations except Walpeup. Earlier data on the effect of cultivation practices on grain protein were reported by Hore (1955).

Rooney, Sims and Tuohey (1966) showed that soil nitrate nitrogen in the surface 15 cm at sowing was higher on plots where stubbles were burnt than where stubbles were retained. However, after twelve years (i.e. six crops), the total nitrogen on summer fallow, stubble burnt, was 0.094 per cent compared with 0.098 per cent where the stubbles were incorporated. Teakle (1956), in discussing the protective value of stubble, also indicated that it was no value as a fertilizer and that it immobilizes soil nitrogen.

Kimber (1967), in examining a possible reason for yield depressions following the incorporation of wheat stubble, showed that straw rotted for periods of up to six weeks inhibited the growth of wheat and oat seedlings grown under aseptic conditions. Prolonged storage of the straw, or the weathering of wheat stubble in the field, reduced the inhibitory effect.

FALLOWING—A BETTER BASIC UNDERSTANDING

During the past twenty-five years, there has been a closer look at the effects of fallowing, particularly in relation to the wider use of pasture legumes in cropping rotations and the consequent increase in soil nitrogen.

Work prior to the 1939-45 war had clearly demonstrated that, in some circumstances in the wetter margins of the cereal belt, moisture conservation was not a major reason for fallowing. During the 1940s and 1950s, species and cultivars of subterranean clover and medics that will grow satisfactorily in the lower rainfall environments became available to, and were adopted by, farmers. Thus the areas with increased soil nitrogen were extended further into the lower rainfall parts of the cereal-growing areas.

These changed conditions occur particularly in Western Australia. Large areas of sandy soils have been brought into cropping, and subterranean clover pastures have been established widely in areas receiving 355 to 510 mm annual rainfall. Wild (1959) suggested that fallowing might be supplanted under such conditions, but considered the control of certain diseases and pests affecting cereal crops could become more difficult on non-fallowed land. The results of Fisher (1962*a*) showed that fallowing was not required in these areas, and also (Fisher 1962*b*) that a succession of cereal crops without fallow, after the pasture phase, increased overall production. One hazard in adopting non-fallow was the web worm. Button (1963) found that delaying seeding for at least three weeks after the initial working would reduce the incidence of this pest. On heavy land, however, e.g. Merredin, the benefits of fallowing were still apparent (Millington 1954), presumably due to moisture conservation under such conditions.

In New South Wales, Kohn, Storrier and Cuthbertson (1966) showed that at Wagga, on red-brown earth sites (Gombalin clay loam) which were under subterranean clover for four or five years, the length of fallow had little influence on the conservation of rainfall and that mineral nitrogen concentration in weedfree, uncultivated conditions was as high as on fallow. They concluded that, in their environment, weed control was more important than moisture conservation or mineral nitrogen accumulation. Lipsett and Simpson (1965) also showed, on subterranean clover ley on red-brown earth sites near Junee in southern New South Wales, that fallowing was not necessary to release nitrogen for a wheat crop.

In South Australia, French (1963) showed little moisture was stored on lighter soils without clay subsoils, but that appreciable gains were obtained on the heavier soils. Fallowing gave higher levels of available nitrogen at seeding, but this aspect became less important as soil fertility was increased by legume pastures, and fallowing was of benefit only on heavier soils, where up to 124 mm of additional water was stored. As a general guide for land with a good growth of legumes and without a special weed problem, he indicated that fallowing should be carried out only if

(*a*) a clay subsoil was present
(*b*) July-August rainfall exceeded 100 mm
(*c*) seasonal rainfall was less than 460 mm.

The data on which the above conclusions and recommendations were based are detailed in French (1966).

Holmes, Greacen and Gurr (1960), by observations on a bare fallowed soil, showed that a fine tilth containing soil crumbs predominantly of 2.5 mm diameter size was more effective in reducing water loss than either the untilled state or coarse clods at the surface.

Schultz (1971) further developed the South Australian work by examining in more detail, and at more frequent intervals, the effect of fallow on water conservation and on crop growth. Working on three soils—a black earth soil, a red-brown earth soil and a deep sandy soil—he concluded that, in the winter rainfall cereal areas of southern Australia, fallow was unlikely to conserve water other than from the previous winter and spring. Further work (Schultz 1972), on a hard-setting red-brown earth, showed that fallows modified with gypsum, straw or hexadecanol conserved more water than traditional fallow. Crops on land with greatest water conservation produced highest wheat yields with greatest water use efficiency.

Kijne (1968) discussed the effect of treatment with cetyl alcohol on evaporative losses from soils. Both evaporation and infiltration were affected. Techniques for testing were examined also (Kijne 1972).

In Victoria, on the solonized brown soils, Walpeup sandy loam, Sims and Mann (1957) found that, while the general yield level was greatly influenced by rainfall while the crop was growing, yield differences between fallowing treatments could be explained mainly by differences in soil moisture at sowing. Conserved moisture equivalent to 25 mm of rainfall was associated with an increase in yield of approximately 289 kg/ha. Wells (1971*a*), also at Walpeup, on sandy soils (Kattyoong sand) and in the presence of skeleton weed, showed that, under these conditions, fallowing was worthwhile in most years. Nitrogen accumulation was the key factor on these sandy soils, which emphasized the importance of raising their nitrogen status by medic pasture.

In the Wimmera, on grey clays that had been under grazed medic pasture for some years, Rooney, Sims and Tuohey (1966) obtained highest yields by fallowing in July or August, which was the same time as had been found best prior to the introduction of medics. Tuohey, Robson and Rooney (1972) showed that the yield variations which occurred were associated mainly with moisture in the 300-600 mm soil layer. The role of nitrate accumulation in determining yield responses to fallow was only of minor importance. The experimental crops were generally weedfree, and thus weeds were not a factor in producing yield differences. It was concluded also that, although the Wimmera would appear to meet the criteria suggested by French (1963), the aspects of climate suggested by him were not the important ones in this area. Here, the most promising predictors of yield response to fallowing were the April to August rainfall before the commencement of winter fallow, and the September rainfall before the commencement of spring fallow.

Another aspect of the increased yields from fallow is that of control of soil-borne diseases, e.g. root diseases (Sims, Meagher and Millikan 1961).

In discussing the situation in Queensland, Miles (1957) pointed out that, in most districts where the term "dry farming" was used, the average rainfall was not particularly low—between 500 and 750 mm. Although this is much higher than in the semi-arid or dry farming areas of southern Australia, evaporation is much higher, particularly in the winter. He mentioned four factors influencing the Queensland position, namely, uneven rainfall distribution during the growing period, the very large variability in rain per year or per season, high rainfall intensity with considerable run-off, and high evaporation and transpiration rates during the greater part of the year.

University of Queensland workers (Waring, Fox and Teakle 1958*a*, 1958*b*; Waring and Teakle 1960) found a very different pattern of moisture accumulation from that which occurs in southern Australia. In a short fallow, they found the mean value for available water at planting was 122 mm. Long fallow resulted in much higher moisture accumulations, reaching 250 mm in one instance. Wheat yields were found to be correlated sufficiently strongly with available soil moisture at planting to suggest that this could be used successfully for yield prediction. On the other hand, correlations between mineral nitrogen at planting time and grain yield were usually non-significant. Martin and Cox (1956) studied moisture contents and fluctuations, and also mineral nitrogen levels and nitrifying capacity. Mineral nitrogen levels under fallow did not suggest nitrogen deficiency.

Wells (1951) and Allen and George (1956) examined water storage and its utilization in central Queensland, at Biloela. The alluvial clay soils there could hold approximately 200 mm of available water to 1.22 m, and this was capable of producing 2,000-2,700 kg/ha of wheat.

The advantage of stored soil moisture for summer crops was emphasized in 1946-47 (Miles 1957). At Biloela, Wells (1951) showed the importance of water availability up to the flowering period of sorghum, it being immaterial whether this water was wholly held in the soil at planting or whether it was partly made up by post-planting additions.

Information kindly supplied by Oxenham (1973) indicates that farmers in the major grain-growing areas of Queensland have the choice of growing winter or summer crops. The general rule is to use a short fallow for successive winter crops, or for successive summer crops. There is now, however, a concerted effort by many farmers to switch from winter to summer cropping and vice versa. This results in a fallow of intermediate length. Only about 5-8 per cent of the farmers on the Darling Downs consciously persist with the once popular long fallow for wheat. The main reasons for the decline in popularity of these long fallows is that they are often associated with a nutritional disorder (Hart 1962, and other research by the Queensland Department of Primary Industries) and that they are often inefficient in terms of moisture storage.

Research work into moisture conservation, fallowing and aspects of soil nitrogen has been encouraged by grants under the Wheat Research Act. Work supported is listed in the annual reports made under the Act, e.g. Wheat Research Act (1972).

Although there is little cropping in the semi-arid areas of tropical Australia, experiments on various aspects of cultivation treatments have been undertaken at both Katherine, Northern Territory, and Kununurra, Western Australia. On some soils at Katherine, their inability to conserve moisture over a period is a difficulty (Slatyer 1954). The influence of land preparation on crop yield on Tippera clay loam soil is discussed by Katherine Research Station (1959), Phillips (1959) and Phillips and Norman (1966).

In the Ord River Valley, Beech and Norman (1964, 1968), Wetselaar and Beech (1968), and Beech, Norman and McIntyre (1968) showed that wet season cultivation and clean fallow resulted in yield benefits which were associated with greater soil nitrogen mineralization and, in some instances, reduced weed infestation. Beech, Norman and McIntyre (1968) showed also that the nitrate nitrogen which was accumulated during the summer season showed to advantage over fertilizer applied at sowing time—it appeared about three times as effective as ammonium sulphate and almost five times as effective as urea at sowing.

The research work undertaken over the past twenty-five years to give a better basic understanding of fallowing, moisture conservation, nitrogen mineralization and accumulation and aspects of soil structure has resulted in many improved techniques, e.g. Quirk (1950), McIntyre and Stirk (1954), Holmes (1956), Keraitis and Ferguson (1957), Fox and Page-Hanify (1959), Fox, Page-Hanify and Cullinane (1964), Malcolm (1964), and Freney and Wetselaar (1967).

CULTIVATION—MINIMUM TILLAGE

It has been shown in many experiments that one of the main benefits of cultivation is that of killing weeds. Therefore, the possibility of substituting other means of killing weeds, e.g. by the use of herbicides, has been examined under quite a few conditions. From this angle, cultivation and other treatments can be compared on the basis of efficiency and cost of killing weeds. Also, chemicals more suited for preseeding weedkilling, either because of better selectivity properties or because they are non-persistent, have been developed.

American and European work on minimum tillage has stimulated interest in Australia in the possibility of reducing cultivation in cereal cropping. Most of this work has been done under conditions of rainfall higher than is received in the semi-arid cultivated areas of Australia.

An early report of the usage of a herbicide to replace part of the normal cultivation programme is that of Myers and Lipsett (1958). On red-brown earth soils near Temora in southern New South Wales, they obtained better results from spraying skeleton weed than from cultivation when both were carried out 50 days prior to sowing. The yield increases were due to improved nitrogen availability to the crop. McCann and Wells (1966) found this method ineffective at Walpeup on solonized brown soils (Kattyoong sand and Winkie sand), partly because of the relatively small accumulation of available nitrogen and partly because of the persistence of spray residues.

In regard to the preparation of standard fallows, exploratory work at Walpeup (Mallee Research Station 1964), mainly on Walpeup sandy loam, indicated that the use of 2,4-D could be an aid to but not a complete replacement for cultivation, due mainly to the lack of control of grassy weeds. Further work there by Wells (1971*b*) on Kattyoong sand soils infested with skeleton weed showed that, compared with cultivated fallows, chemical fallowing increased yields. Increased soil nitrate and soil moisture at sowing occurred, reflecting the better weed control given by the herbicides.

In southern New South Wales, also in the presence of skeleton weed, Kohn, Storrier and Cuthbertson (1966) did not have success on a red-brown earth (Gombalin clay loam), in fallow preparation, from spraying during the summer; but spraying after the "break" rains in early autumn gave a reasonably good control of weeds and grain yields as good as those of other treatments. The need to wait for germinating weeds to present a reasonable surface could delay sowing and reduce yields.

Unpromising results were obtained also by Greenwood et al. (1970) on a shallow red-brown earth of the Merredin series in a low rainfall area of Western Australia. Where cultivation was replaced by herbicides, and the grain sown in a narrow slit, nitrogen stress was greater, and growth rate, nitrogen uptake and grain yield were lower. The application of nitrogen at 75 kg/ha at sowing brought yield during the vegetative stage up to that of the cultivated sites, but did not fully do so for grain yield.

Schultz (1971) on three sites—a black earth soil, a red-brown earth soil and a deep sandy soil—in South Australia, reported moisture storage efficiency and nitrate accumulation from conventional and chemical fallows. In a wet year, yields from the two treatments were similar, but in a dry year the conventional fallow significantly outyielded the chemical fallow.

The use of chemicals in the preparation of land for sowing on non-fallow in Western Australia, especially in regard to earlier planting which gives increased yields, was discussed by Pearce (1969). Experiments undertaken in 1966 at Chapman, Wongan Hills and Merredin Experiment Stations showed that a herbicidal spray, used in lieu of a cultivation, generally gave good weed control and wheat yields generally not significantly different from those on plots cultivated at the same time as the herbicidal application. Pearce considered an important disadvantage could be the slower release of soil nitrogen on uncultivated land. Paraquat was considered the most satisfactory herbicide for this chemical preparation.

Sowing without cultivation has also been examined by Reeves and Smith (1973) at Rutherglen—outside the semi-arid zone. Overall, sowing without cultivation on Rutherglen loam, a transitional soil between a podzol and a red-brown earth, gave satisfactory results. Paraquat was a successful herbicide, but direct cash outlay for the herbicide was estimated as being greater than for cultivation. Cook (1972) discussed the economics of a spray-sow technique on Eyre Peninsula and calculated, on equal yields, the cost break-even point for the use of chemicals was when the preseeding time (under local conditions) was two hours per ha. At Canberra, also outside the semi-arid zone, Hutchings and Hicks (1971) found that, on an alluvial flat with a history of intermittent cropping interspersed with subterranean clover leys over twenty-five years, minimal tillage techniques were less satisfactory than conventional tillage for wheat establishment and grain production.

CULTIVATION–IMPLEMENTS

The choice of implements is of importance in fallow preparation chiefly in respect of the effectiveness of the implement in weed control, in the handling of trash, and in leaving the soil in a condition more likely to resist erosion. Soil type can be of importance in all these aspects. General reference to implements has been made in many of the articles on fallow preparation. Historical aspects of the development of implements for Australian conditions are given by Wheelhouse (1966).

Willis (1959) looked at the tillage problem on a red-brown earth at Wagga to give engineering information, but found unexpected crop responses to tillage methods, which he considered suggested that secondary tillage is more than a weedkilling operation. Further work, however (Willis and Willatt 1965), showed that, in a more "normal" season, there was no marked response to tillage treatment, only the mouldboard plough plots yielding lower, possibly because of a greater infestation of skeleton weed. Allen (1961) disagreed with some of Willis's earlier conclusions and pointed out that, on clay loams at Biloela, there was no significant difference between the disc and the scarifier in regard to soil moisture conservation, weed growth and crop growth and yield.

When cropping without fallow, the choice of implements could be of greater importance. Fisher (1962*c*) in Western Australia, on sandy loam or sandy surfaced soils, found the use of the mouldboard or the disc plough produced higher yields than did the use of tyned implements–rigid tyne cultivator and chisel plough. This difference was related to weed control. Bath (1951), on podzol—red-brown earth intergrade soils at Rutherglen, also found the use of the mouldboard plough a surer method of seedbed preparation (without fallow) than the disc plough, disc harrows or rotary hoe.

In cultivating land liable to wind erosion, experience in South Australia and Victoria has shown that the use of implements at fast speeds leaves the soil in a condition more prone to wind erosion. At Wongan Hills in Western Australia, on a sandy loam overlying clay, which had not been cropped for five years, Fisher (1966) found that higher cultivating speeds tended to increase wheat yields, the effect being greater with the scarifier than with the disc. Fisher's work was under unfallowed ley land conditions, and soil erosion risks, although not mentioned, were probably low.

The need to prepare a level seedbed, as well as the choice of implements to achieve this, is mentioned often, but supporting data appear to be lacking. On the other hand, data in Victorian Department of Agriculture records show that a cultivation of a fallow in autumn, to 125 and 175 mm respectively, to destroy the seedbed, did not have any serious effect on the yield of the subsequent wheat crop on a grey clay soil at Longerenong or a podzol-red-brown earth intergrade at Rutherglen (non-clover land).

Operating costs can be an important factor influencing the choice of implements. Ryan (1972) found that, in the Victorian Wimmera, the variable costs of breaking up ground were halved where a scarifier was used instead of a mouldboard plough. Also, the cost of cultivating when a light cultivator (cultivating bar) was used instead of the scarifier in fallow preparation was approximately halved. His general conclusion was that the larger the implement pulled by the tractor, the lower the variable costs. The machinery costs of fallowing and cropping operations in the Wimmera, based on information obtained in a survey done at the same time, are discussed further by England, Hunt and Lloyd (1973).

At Katherine, Basinski, Phillips and Norman (1964) recorded rates of work and fuel consumption to estimate cultivation cost differences and break-even points for peanuts. Under their conditions, increasing the depth of ploughing from 50 mm to 200 mm, and increasing inter-row cultivations from one to three were profitable.

Another aspect of implement usage is that of timeliness of operation, in respect of operational efficiency, possible earlier sowing, and soil effects.

CULTIVATION—DEPTH OF CULTIVATION

Depth of cultivation often causes controversy, and has been mentioned or investigated from time to time in fallow preparation and other cultivation experiments already discussed. Experiments on deep placement of fertilizer do not appear to have been undertaken in relation to cropping in semi-arid areas of Australia. Compaction or plough sole layers are more likely in the wetter areas. It is well known that cereal (and other) roots explore the soil to considerable depths. Harbison (1959) recorded wheat roots to a depth of 2.44 m on an alluvial clay loam soil at Biloela.

Early workers, such as Howell (1911), considered deep ploughing was necessary to allow deep penetration of water for effective moisture conservation, but his results do not show any marked effects. While some increases, generally small, have been recorded, the overall Australian position is that little is gained by deeper cultivation of normal soils. This is in agreement with the conclusion of Khan (1963)—that, on a world basis, it is only exceptionally that deep tillage has increased crop yields in semi-arid areas.

CULTIVATION—SOME SOIL STRUCTURE PROBLEMS

It has generally been considered that one of the objectives of cultivation is the preparation of a suitable seedbed. Smaller seeds require a finer seedbed than larger seeds. In cereals, it is customary to prepare a finer seedbed for wheat than for barley or oats.

Continued cultivation in Australia has led to a decline in structure of some soils, especially of the red-brown earth types. Howell (1911) reported poor germination on red-brown earths in the Nathalia-Cobram area of Victoria in 1901-2—due to heavy rains after seeding, and surface crusting. Early Australian studies showing this decline of structure were made by Downes and Leeper (1939) and Clarke and Marshall (1947). The latter showed that, on two red-brown earth soils at the Waite Institute, structure deterioration as measured by water stable aggregates was greatest in the initial years of cultivation and that there was a parallel decline in organic matter.

Greacen and Perkman (1953), working at Griffith, New South Wales, on a site ranging from Hanwood loam to Hanwood sandy loam and on which cereals had not been grown, showed that cultivation reduced aggregate stability. Skene (1955) discussed the decline, with cultivation, of soil physical condition generally. Stoneman (1962, 1964), at Merredin, Western Australia, in the zone of grey and brown calcareous solonized soils, showed a decline in structure of six wheat belt soils, and an improvement in wheat yield, by 25 per cent, following structure improvement by krilium. Clarke (1963) discussed generally the balancing of periods of cultivation, which destroyed structure, and of periods of vigorous grass, which restored it, in the maintenance of soil fertility.

The studies of Rovira and Greacen (1957) using six diverse soils showed that disrupting soil aggregates in a shearing machine resulted in increased microbial activity. Thus, shearing action (cultivation) exposed organic matter previously protected from soil microorganisms. This gives a better understanding of how and why soil structure deteriorates with cultivation.

Craswell, Saffigna and Waring (1970), working with mainly clay soils of two soil series from the Darling Downs, the Waco and Cecilvale, showed that cultivated soils were homogeneous with regard to organic matter content of aggregates of different sizes, while the virgin soils were relatively heterogeneous. Under fallow without protection by vegetation, macroaggregates break down to their constituent microaggregates. The effect of aggregate size on mineralization is being studied by Bell and Thitipoca (1972).

Particular attention has been given to seedling emergence problems on a red-brown earth soil at the Waite Institute in South Australia by McIntyre (1955) and Millington (1959, 1961). Poor soil aeration in wet years caused a reduction in wheat establishment and growth. McIntyre (1958*a*, 1958*b*) studied the surface seal or crust and its formation in relation to raindrop action. Other aspects of seedling emergence on this soil type, and especially soil strength, have been examined by Barley (1963, 1964) and Barley, Farrell and Greacen (1965).

Implements, especially harrows and rollers, have been widely used at critical times to ameliorate soils liable to setting in southern Australia. In the more northerly areas, Arndt (1962) and Gunton and Kerr (1972) found rollers effective.

Arndt (1965*a*, 1965*b*) has studied the mechanism of seal formation on the Tippera sandy clay soil at Katherine. In this soil type, a natural tendency to form surface seals is accentuated by cultivation. He found the mechanisms of seal formation quite complex, and that the development of a method for measuring impedance over a range of soil and crop conditions was not possible. He developed a model that provided a quantitative assessment for the impedance met with by seedlings.

The use of suitable soil amendments is another approach to the improvement of soil structure, cultivation characteristics and seedling emergence in difficult soils. Gypsum has been the main amendment used commercially (Sims and Rooney 1965; Scott, Evans and Noble 1968; and Matheson 1969); but krilium has also been tested (Stoneman 1962, 1964). The use of suitable pastures in the rotation has also been widely and successfully practised.

CONCLUSION

The practice of fallowing is widely used in crop production in semi-arid areas of Australia. From data published in Commonwealth Year Books, it is estimated that, for the Victorian, South Australian, New South Wales and Western Australian wheat crops, approximately 80 per cent, 40 per cent, 40 per cent and 30 per cent, respectively, are sown on fallowed land.

This review of cultivation-soil-plant factors in crop production in the semi-arid areas of Australia reveals many contradictory results and statements, especially in relation to the practice of fallowing. Many of these apparent inconsistencies are related to rainfall amount and incidence, and soil factors. A clearer definition and understanding of these factors would facilitate the drawing of general conclusions. A re-examination, including an overall analysis of Australian data, could help in such a clarification.

It is clear also that soil type and the range of "available" moisture-holding capacity are of vital importance. In Australia, it is conventional to use a soil water potential of −15 bars in estimating wilting point. Work in the U.S.A. (Papendick, Cochran and Woody 1971) and Canada (Yang and de Jong 1972) indicates that the lower limit of water extraction by the wheat plant may reach −40 bars.

The work done in Queensland clearly shows the high capacity to store water in the black earth clay soils there. The limited capacity of sandy soils to store water, together with a concentration of rains in the winter and spring, gives an explanation for the low proportion of the crop being grown on fallow in Western Australia.

While fallowing results in marked yield improvements in suitable environments, only a relatively small proportion of the rainfall is conserved. Examples are Waring, Fox and Teakle (1958*a*), 17 per cent; Sims and Mann (1957), 18.5 per cent; South Australia generally, as estimated from French (1966), less than 20 per cent, and at the Waite Institute in the 1949-50 season (Butler and Prescott 1955), 30 per cent. Waring, Fox and Teakle (1958*b*) indicated that U.S.A. and Canadian data showed moisture conservation efficiencies ranging from 4 to 50 per cent of the rainfall with most of the results in the range of 20-30 per cent. Doughty (1958) gives 34 per cent in a review of Canadian work. From these data, one can ask what are the reasons for the low conservation achieved, and is there any other possible means of increasing the efficiency of water conservation? Certainly moisture in the surface mulch can be lost fairly quickly and, thus, moisture received in light falls of rain is usually lost quickly. A better assessment of the efficiency of water conservation would be obtained from data on moisture accretion and retention in the soil below the mulch. Howell (1911) realized the importance of accretion when he suggested that deep ploughing might let water into the subsoil. However, experimental results have not shown a yield advantage from deeper ploughing. It could be queried whether soil moisture data are fully recorded on balance sheets, as sampling was only done to a limited depth, and some moisture could be conserved below the sampling depth—but this amount would be relatively small. In respect of lessening evaporation, the work done on evaporation suppressants so far has not been particularly encouraging.

An aspect of plant growth investigations under controlled conditions is that such data are normally obtained under conditions where moisture is not limiting. Such data are not of use in a systems approach to growth under semi-arid conditions. Techniques for collecting data for use in models for semi-arid conditions need to be developed so that the mechanisms applying where moisture is limiting can be better understood.

The preparation of a seedbed by cultivation has been traditional practice in Australia, as elsewhere. The importance of weed control is well established. A better understanding of the function of the seedbed in relation to seed germination, soil aeration, soil strength, moisture entry and penetration, and nitrogen transformations would help in the development of minimum cultivation practices.

Australian experience shows that the efficiency of conserved moisture is lower than that in the Pacific northwest of the United States of America. Research at Washington State University (Guettinger 1966) has shown that wheat needs 100 mm of water to grow the plant before it produces seed. After 100 mm are subtracted from the total, each remaining 25 mm of moisture produces 470 kg/ha of grain—a figure very much in excess of the 289 kg/ha found by Sims and Mann (1957). In addition, moisture conservation in fallows in the Pacific northwest may be greater than is the case in Australia, and the dust and stubble mulch used there might be more efficient in moisture conservation than the coarser mulch used in Australia (R.J. Cook, Washington State University, personal communication 1974). An understanding of the reasons for the differences in these results might improve moisture conservation and utilization efficiency in Australia.

There are, however, varying degrees of efficiency with which water is used in different parts of the U.S.A. (Viets 1971), the data he cites being 350 kg/ha in eastern Washington and 150 kg/ha of grain per 25mm of soil water in the Northern Great Plains.

It has been indicated throughout this review that tillage is a means of manipulating water and nitrogen management and seedling environment zones within the soil surface. In this regard, Larson (1963) considers the emphasis of tillage should change towards devising systems which maximize water storage in the neighbourhood of the plant. This is extremely important for those semi-arid areas with predominantly summer rainfall. In such areas, moisture inputs and moisture stress coincide. Under these conditions, traditional European cultural methods can be unsatisfactory, and are often at variance with the broad aim of conserving moisture (Arndt 1966).

Another possibility of water manipulation is that of the development of micro water sheds to concentrate water around the base of the plant in row cropping. This method, with wheat sown in deep furrows in 356 mm rows, was noted during a personal visit to eastern Washington, U.S.A. There appear to be no Australian data on this practice, but some experiments, comparing sowing wheat in the furrows and between them were undertaken at Longerenong Agricultural College, Victoria, in the early 1930s, but no yield benefit was recorded (Victorian Department of Agriculture unpublished data).

The spectacular yield increases following soil treatment with chloropicrin by officers of the Victorian Plant Research Institute, Burnley, and the Division of Soils, Adelaide (J.W. Meagher, Victorian Plant Research Institute, Burnley, personal communication) pose the question, are there biological reasons for fallowing in addition to that of nitrogen mineralization? Beneficial effects of fallowing in regard to the control of root diseases and cereal cyst nematode are well established, e.g. Sims, Meagher and Millikan (1961); Meagher and Rooney (1966). Other biological aspects, including that of soil antagonists, are receiving some attention.

It is apparent also that, despite its many advantages, cultivation has a number of adverse effects on the soil and its capacity to produce high-yielding crops. While much work has been done on various aspects of these problems, it is clear that further work is required.

Current developments in Queensland tend towards maintenance of rough surfaces and use of press wheel planters to compensate for departure from traditional fine seed beds and deeper planting. There is also, in summer rainfall areas, a considerable movement towards opportunist cropping with a departure from field schedules (G.B. Stirk, Division of Soils, C.S.I.R.O., Queensland, personal communication).

Costs of cultivation in relation to its efficiency, and to crop production, have seldom been examined. Cost-benefit analyses are necessary in this regard, and also for the evaluation of possible new practices, including herbicidal use and minimum tillage.

In addition to economic costs, more emphasis on the energy used in cultivation and in the production of agricultural chemicals used as herbicides can be expected in the future. Further examination is needed of the factors responsible for the benefits obtained from fallowing and the energy costs, whether for cultivation or herbicides, that are required to obtain these benefits.

Currently, systems analysis and crop modelling are rapidly moving fields of research and will no doubt play an important part in various aspects of investigations on soils, soil water movement and water balance. Work is being done in various divisions of the C.S.I.R.O. (e.g. Ross 1971), and at Macquarie University (Sydney) and the University of Queensland, and more should be heard of this in the not so distant future.

REFERENCES

Allen, G.H. (1961). *Qld. Agric. J.* **87**, 303.
Allen, G.H., and George, R.W. (1956). *Qld. J. Agric. Sci.* 13, 19-46.
Anderson, H.C.L. (1911). *Agric. Gaz. N.S.W.* 22, 577-9.
Arndt, W. (1962). Personal communication (3rd Aust. Conf. Soil Sci., Canberra).
———. (1965*a*). *Aust. J. Soil Res.* 3, 45-54.
———. (1965*b*). *Aust. J. Soil Res.* 3, 55-68.
———. (1966). Personal communication (4th Aust. Conf. Soil Sci., Brisbane).
Bagnold, R.A. (1941). "The Physics of Blown Sands and Desert Dunes". Methuen, London.
Barley, K.P. (1963). *Soil Sci.* **96**, 175-80.
———. (1964). *Aust. J. Sci.* **26**, 224-5.
Barley, K.P., Farrell, D.A., and Greacen, E.L. (1965). *Aust. J. Soil Res.* 3, 69-79.
Barr, D.A. (1957). *J. Soil Conserv. Serv., N.S.W.* 31, 27-32.
Barrow, P. McK. (1955). *J. Dep. Agric. S. Aust.* **58**, 421.
Basinski, J.J., Phillips, L.J., and Norman, M.J.T. (1964). *Aust. J. Exp. Agric. Anim. Husb.* **4**, 295-99.
Bath, J.G. (1951). *J. Dep. Agric. Vict.* **49**, 601-7.
Beare, J.A. (1951). *J. Dep. Agric. S. Aust.* **54**, 315-8.
———. (1962). *J. Dep. Agric. S. Aust.* **66**, 134-41.
Beech, D.F., and Norman, M.J.T. (1964). *Aust. J. Exp. Agric. Anim. Husb.* **4**, 197-205.
———. (1968). *Aust. J. Exp. Agric. Anim. Husb.* **8**, 66-71.
Beech, D.F., Norman, M.J.T., and McIntyre, G.A. (1968). *Aust. J. Exp. Agric. Anim. Husb.* **8**, 72-90.
Bell, L.C. and Thitipoca, H. (1972). Research Projects, Dep. Agric. Univ. Queensland. 1972:25.

Birks, W.R., and Cole, T.A. (1930). *J. Dep. Agric., S. Aust.* 3, 801-7.
Blackburn, G., and Baker, R.M. (1951). *J. Dep. Agric., S. Aust.* 55, 78-88, 155-62.
Burvill, G.H. (1945). *J. Dep. Agric., West. Aust.* 22, 3-10.
———. (1948). *J. Dep. Agric., West. Aust.* 25, 293-300.
Butler, P.F., and Prescott, J.A. (1955). *Aust. J. Agric. Res.* 6, 52-61.
Button, J.A. (1963). *J. Dep. Agric. West. Aust.* 4, 625-30.
Callaghan, A.R., and Millington, A.J. (1956). "The Wheat Industry in Australia". Angus and Robertson, Sydney.
Carrol, Dorothy (1939). *Geol. Mag.* 76, 6-23.
Chepil, W.S., and Woodruff, N.P. (1963). *Adv. Agron.* 15, 211-302.
Clarke, A.L. (1963). *J. Dep. Agric. S. Aust.* 66, 394-400.
Clarke, G.B., and Marshall, T.J. (1947). *J. Coun. Sci. Ind. Res. Aust.* 20, 162-75.
Condon, R.W., and Stannard, M.E. (1956-57). *J. Soil Conserv. Serv. N.S.W.* 12, 192-202; 13, 17-26, 68-74 and 1111-28.
Cook, A.V. (1972). *J. Dep. Agric. S. Aust.* 75, 174-7.
Cornish, E.A. (1949). *Aust. J. Sci. Res. Ser. B.* 2, 83-137.
———. (1950). *Aust. J. Sci. Res. Ser. B.* 3, 178-218.
Craswell, E.T., Saffigna, P.G., and Waring, A.S. (1970). *Plant Soil.* 33, 383-92.
Daniel, P.R. (1969). *J. Soil Conserv. Serv. N.S.W.* 25, 327-9.
Doughty, J.L. (1958). *Emp. J. Exp. Agric.* 26, 85-91.
Downes, R.G. (1949). Coun. Sci. Ind. Res. Aust. Bull. No. 243.
Downes, R.G., and Leeper, G.W. (1939). *Proc. Roy. Soc. Vic.* 52, 1-20.
Duggan, W.C., Hodgson, S.C., Nicholson, G.D., Pearson, A., and Carn, K.G. (1941). *Agric. Gaz. N.S.W.* 52, 506-16, 529.
Dunsdorfs, E. (1956). "The Australian Wheat Growing Industry 1788-1948". Melbourne Univ. Press, Melbourne.
England, P.J., Hunt, C.A.G., and Lloyd, A.G. (1973). University of Melbourne School Agric. Forest. Bull. No. 1.
Fisher, H.M. (1962*a*). *J. Dep. Agric. West. Aust.* 3, 172-84.
———. (1962*b*). *J. Dep. Agric. West. Aust.* 3, 282-97.
———. (1962*c*). *J. Dep. Agric. West. Aust.* 3, 524-37.
———. (1966). *J. Dep. Agric. West. Aust.* 7, 190-4.
Forster, H.C. (1940). *J. Dep. Agric. Vict.* 38, 257-9, 296-302.
Fox, W.E., and Page-Hanify, D.S. (1959). *Soil Sci.* **88**, 168-71.
Fox, W.E., Page-Hanify, D.S., and Cullinane, R.E. (1964). *Aust. J. Exp. Agric. Anim. Husb.* 4, 191-2.
French, R.J. (1963). *J. Dep. Agric. S. Aust.* 67, 42-8, 76-9.
———. (1966). M. Agric. Sci. Thesis, Univ. of Adelaide, Adelaide.
Freney, J.R., and Wetselaar, R. (1967). CSIRO, Aust. Div. Plant Ind. Tech. Paper No. 23.
George, R.W. (1949). *Qld. Agric. J.* **68**, 69-70.
Geytenbeek, P.E. (1951). *J. Dep. Agric. S. Aust.* 54, 391 and 400.
———. (1953). *J. Dep. Agric. S. Aust.* 56, 299-304.
Greacen, E.L., and Perkman, O. (1953). *Aust. J. Agric. Res.* 4, 193-203.
Greenwood, E.A.N., Boyd, W.J.R., Whitehead, J.A., and Titmanis, Z.V. (1970). *Aust. J. Exp. Agric. Anim. Husb.* **10**, 763-7.
Guettinger, D.L. (1966). *Crops Soils Mag.* **18**(9), 14-5.
Gunton, J.L., and Kerr, J.C. (1972). *Qld. Agric. J.* **98**, 609-11.
Hamilton, G.J. (1970). *J. Soil Conserv. Serv., N.S.W.* **26**, 118-23.
Hanway, D.G. (1970). Proc. Inter. Conf. Mech. Dryland Farming. FAO, Deere and Co., 102-22.
Harbison, J. (1959). *Qld. Agric. J.* **85**, 659.
Hart, J. (1962). *Qld. Agric. J.* **88**, 42-4.
Herriot, R.I. (1944). *J. Dep. Agric. S. Aust.* **48**, 112-20.
———. (1945*a*). *J. Dep. Agric. S. Aust.* **48**, 349-51.
———. (1945*b*). *J. Dep. Agric. S. Aust.* **48**, 443-5.
———. (1948). *J. Dep. Agric. S. Aust.* 52, 26.
———. (1951). *J. Dep. Agric. S. Aust.* 55, 59-63.
———. (1954). *J. Dep. Agric. S. Aust.* **58**, 91-100, 140-9.
Holmes, J.W. (1956). *Aust. J. Appl. Sci.* 7, 45-58.
Holmes, J.W., Greacen, E.L., and Gurr, C.G. (1960). *Trans. 7th Int. Cong. Soil Sci.* 1, 188-94.
Hore, H.L. (1940). *J. Dep. Agric. Vict.* 38, 219-28.
———. (1945). *J. Dep. Agric. Vict.* 43, 233-40, 263-6.
———. (1955). *J. Dep. Agric. Vict.* 53, 305-10.
Hore, H.L., and Sims, H.J. (1954). *J. Dep. Agric. Vict.* 52, 241-50.
———. (1955). *J. Dep. Agric. Vict.* 53, 263-4.
Hore, H.L., Sims, H.J., and McCann, J.M. (1948). *J. Dep. Agric. Vict.* 46, 449-56, 468-78, 480.
Howell, F.J. (1904). *J. Dep. Agric. Vict.* 2, 772-80.

———. (1911). "Soil Problems in Wheat Growing". George Robertson, Melbourne.
Hutchings, R.J., and Hicks, J.S. (1971). Preprints of Papers. 1971 Nat. Agric. Mach. Conf., Melbourne, 4, 1-9.
Kaleski, L.G. (1956). *J. Aust. Inst. Agric. Sci.* 22, 120-8.
Katherine Research Station. Progress Report 1946-56. CSIRO Aust., 1959.
Keraitis, K., and Ferguson, A.W. (1957). *J. Aust. Inst. Agric. Sci.* 23, 248-9.
Khan, A.R. (1963). *World Crops.* 15, 68-71.
Kijne, J.W. (1968). *Aust. J. Soil Res.* 6, 67-73.
———. (1972). *Neth. J. Agric. Sci.* 20, 145-53.
Kimber, R.W.L. (1967). *Aust. J. Agric. Res.* 18, 361-74.
King, F.H. (1895). "The Soil". (Ed. L.H. Bailey). Macmillan, New York.
———. (1901). "A Textbook of the Physics of Agriculture". 2nd Edn. Madison.
Kohn, G.D., Storrier, R.R., and Cuthbertson, E.G. (1966). *Aust. J. Exp. Agric. Anim. Husb.* 6, 233-41.
Ladewig, J.E., and Skinner, A.F. (1950). *Qld. Agric. J.* 70, 14-25, 63-72, 187-93.
Lang, S.B. (1948). *J. Dep. Agric. S. Aust.* 52, 144.
Larson, W.E. (1963). *Neth. J. Agric. Sci.* 11, 100-9.
Leeper, G.W. (1952). "Introduction to Soil Science". 2nd Edn. Melbourne Univ. Press, Melbourne.
———. (1953). *J. Aust. Inst. Agric. Sci.* 19, 86-8.
Lees, R.D. (1926). *Agric. Gaz. N.S.W.* 15, 109-10.
Lightfoot, L.C. (1952). *J. Dep. Agric. West. Aust.* 1, 249-56.
Lipsett, J., and Simpson, J.R. (1965). *CSIRO Aust. Div. Plant Ind. Field Sta. Rec.* 4, 9-12.
Littler, J.W. (1964). *Qld. J. Agric. Sci.* 21, 137-42.
Logan, J.M. (1960). *J. Soil Conserv. Serv. N.S.W.* 16, 95-111.
Macadam, R.D., and Southwood, O.R. (1968). *Agric. Gaz. N.S.W.* 79, 490-2.
McCalla, T.M., and Army, T.J. (1961). *Adv. Agron.* 13, 125-96.
McCann, J. McC., and Wells, G.J. (1966). *J. Dep. Agric. Vict.* 64, 315-9.
McIlroy, W., Brake, J., Strom, H.G., Lakeland, W.J., and O'Carroll, J.D. (1938). Report of Committee Appointed to Investigate Erosion in Victoria. Government Printer, Melbourne.
McIntyre, D.S. (1955). *Aust. J. Agric. Res.* 6, 797-803.
———. (1958*a*). *Soil Sci.* 85, 185-9.
———. (1958*b*). *Soil Sci.* 85, 261-6.
McIntyre, D.S., and Stirk, G.B. (1954). *Aust. J. Agric. Res.* 5, 291-6.
McKeown, N.R., and McCulloch, R.N. (1962). *Aust. J. Exp. Agric. Anim. Husb.* 2, 197-203.
Malcolm, C.V. (1964). *Aust. J. Exp. Agric. Anim. Husb.* 4, 189-90.
Mallee Research Station. Experimental Work and Results. (1946) *J. Dep. Agric. Vict.* 44 (1946): 465-76.
Mallee Research Station, Walpeup. Field Day Notes. Dep. Agric. Vict. 1964.
Marsh, B. a'B. (1972). *J. Dep. Agric. West. Aust.* 13, 121.
Martin, A.E., and Cox, J.E. (1956). *Aust. J. Agric. Res.* 7, 169-83, 184-93.
Mason, M.G., Toms, W.J., and Gartrell, J.W. (1969). *J. Dep. Agric. West. Aust.* 9, 582-3.
Matheson, W.E. (1969). *J. Dep. Agric. S. Aust.* 72, 402-6.
Meagher, J.W., and Rooney, D.R. (1966). *Aust. J. Exp. Agric. Anim. Husb.* 6, 425-31.
Meinig, D.W. (1970). "On the Margins of the Good Earth". Rigby, Adelaide.
Miles, L.G. (1957). *J. Aust. Inst. Agric. Sci.* 23, 190-5.
Millington, A.J. (1954). D.Sc. (Agric.) Thesis, Univ. of Western Australia, Perth.
Millington, R.J. (1959). *Aust. J. Agric. Res.* 10, 487-94.
———. (1961). *Aust. J. Agric. Res.* 12, 397-408.
Morrow, J.A., and Hayman, R.H. (1940). *J. Dep. Agric. Vict.* 38, 205-10.
Mullett, H.A. (1926). *J. Dep. Agric. Vict.* 24, 641-45.
———. (1929). *J. Dep. Agric. Vict.* 27, 1-8.
Myers, L.F., and Lipsett, J. (1958). *Aust. J. Agric. Res.* 9, 1-12.
O'Carroll, J.D. (1940). In "Soil Erosion in Victoria". Dep. Lands and Survey, Melbourne.
Oxenham, B.L. (1973). Personal communication. Qld. Dep. Prim. Ind.
Papendick, R.I., Cochran, V.L., and Woody, W.M. (1971). *Agron. J.* 63, 731-4.
Paterson, J.W. (1916). "Nature in Farming". Lothian, Melbourne.
Paterson, J.W., and Scott, P.R. (1912*a*). *J. Dep. Agric. Vict.* 10, 275-82.
———. (1912*b*). *J. Dep. Agric. Vict.* 10, 393-400.
———. (1912*c*). *J. Dep. Agric. Vict.* 10, 521-8.
Peacock, R.W. (1901). *Agric. Gaz. N.S.W.* 12, 787-90.
Pearce, G.A. (1969). *J. Dep. Agric. West. Aust.* 10, 134-7.
Phillips, L.J. (1959). CSIRO Aust. Div. Land Res. Reg. Surv. Tech. Paper No. 1.
Phillips, L.J., and Norman, M.J.T. (1966). *Aust. J. Exp. Agric. Anim. Husb.* 6, 48-55.
Powell, E.C. (1948). *Agric. Gaz. N.S.W.* 59, 505-9.
Powell, E.C., Bowmaker, E.W., and Dargin, H.J. (1941). *Agric. Gaz. N.S.W.* 52, 613-20.
Prescott, J.A. (1933). Proc. 5th Pacific Sci. Cong., 1933: 2657-67. Univ. Toronto Press, Toronto.
Prescott, J.A., and Piper, G.R. (1930). *J. Agric. Sci.* 20, 517-31.

Quirk, J.P. (1950). *Aust. J. Agric. Res.* 1, 276-84.
Reeves, T.G., and Smith, I.S. (1973). *J. Dep. Agric. Vict.* 71, 74-5.
Richardson, A.E.V. (1912). *J. Dep. Agric. Vict.* 10, 329-38.
———. (1925). Dep. Agric. Vict. Bull. No. 55 (New Series).
Richardson, A.E.V., and Fricke, E.F. (1931). *J. Dep. Agric. S. Aust.* 35, 57-86.
Richardson, A.E.V., and Gurney, H.C. (1933). *Emp. J. Exp. Agric.* 1, 193-205.
Rooney, D.R., Sims, H.J., and Tuohey, C.L. (1966). *J. Dep. Agric. Vict.* 64, 403-10.
Ross, P.J. (1971). CSIRO Aust. Div. Soils. Tech. Mem 22/71.
Rosser, J. (1964). *Power Farm. Aust. N.Z.* 73 (6), 39.
Rovira, A.D., and Greacen, E.L. (1957). *Aust. J. Agric. Res.* 8, 659-73.
Rowan, J.N., and Downes, R.G. (1963). *Soil Conserv. Auth. Vict. Bull.* T.C.2.
Ryan, T.J. (1972). *J. Dep. Agric. Vict.* 70, 362-70.
Schultz, J.E. (1971). *Aust. J. Exp. Agric. Anim. Husb.* 11, 236-42.
———. (1972). *Aust. J. Exp. Agric. Anim. Husb.* 12, 299-305.
Scott, B.J., Evans, L.H., and Noble, J.C. (1968). *Agric. Gaz. N.S.W.* 79, 493-8.
Scott, R.C., and Bristow, A.E. (1929). *J. Dep. Agric. S. Aust.* 32, 838-47.
Sims, H.J. (1948). *J. Dep. Agric. Vict.* 46, 241-8.
———. (1965). Personal communication. (Aust. Cereal Agron. Conf., Horsham.)
Sims, H.J., and Blackburn, G. (1945). Dep. Agric. Vict., Internal Report.
Sims, H.J., and Mann, A.P. (1957). 2nd Aust. Soil Sci. Conf., Melbourne. 2: Paper No. 49, 1-8.
———. (1965). Personal communication (Aust. Cereal Agron. Conf., Horsham.)
Sims, H.J., Meagher, J.W., and Millikan, C.R. (1961). *Aust. J. Exp. Agric. Anim. Husb.* 1, 99-108.
Sims, H.J., and Rooney, D.R. (1965). *J. Dep. Agric. Vict.* 63, 401-9.
Skene, J.K.M. (1930). M. Agric. Sci. Thesis, Univ. Melbourne, Melbourne.
———. (1955). *Power Farm. Aust. N.Z.* 64 (3), 7-13, 113-5.
Slatyer, R.O. (1954). *J. Aust. Inst. Agric. Sci.* 20, 46-7.
Stephens, G.C., Herriot, R.I., Downes, R.G., Langford-Smith, T., and Acock, A.M. (1945). Coun. Sci. Ind. Res. Aust. Bull. No. 188.
Stoneman, T.C. (1962). *J. Dep. Agric. West. Aust.* 3, 493-7.
———. (1964). *Aust. J. Exp. Agric. Anim. Husb.* 4, 363-6.
Synnott, K. (1946). "The Farmers' Handbook". 5th Edn. N.S.W. Dep. Agric., Sydney.
Taylor, J.K., and Stephens, C.G. (1943). *J. Coun. Sci. Ind. Res. Aust.* 16, 33-6.
Teakle, L.J.H. (1928). *J. Dep. Agric. West. Aust.* 5, 444-8.
———. (1939). *J. Dep. Agric. West. Aust.* 16, 297-304.
———. (1956). *Power Farm. Aust. N.Z.* 65 (6), 13.
Teakle, L.J.H., and Burvill, G.H. (1930*a*). *J. Dep. Agric. West. Aust.* 7, 119-34.
———. (1930*b*). *J. Dep. Agric. West. Aust.* 7, 215-20.
Thomas, I., and Langfield, J.H. (1930). *J. Dep. Agric. West. Aust.* 7, 279-83.
Tuohey, C.L., Robson, A.D., and Rooney, D.R. (1972). *Aust. J. Exp. Agric. Anim. Husb.* 12, 414-9.
Victorian Institute of Surveyors (1940). "Soil Erosion in Victoria". Dep. Lands and Survey, Melbourne.
Viets, Frank G., Jr. (1971). In "Drought Injury and Resistance in Crops". (Eds. K.L. Larson and J.D. Eastin). pp. 57-76. Crop Science Society of America, Spec. Pub. No. 2. Madison.
Waggamba Shire Extension Committee. (1972). Handling crop stubbles in the Waggamba Shire. Qld. Dep. Prim. Ind.
Walker, B.S. (1950). *J. Soil Conserv. Serv. N.S.W.* 6, 209-10.
Waring, S.A., Fox, W.E., and Teakle, L.J.H. (1958*a*). *Aust. J. Agric. Res.* 9, 205-16.
———. (1958*b*). *Aust. J. Agric. Res.* 9, 717-29.
Waring, S.A., and Teakle, L.J.H. (1960). *Aust. J. Agric. Res.* 11, 27-41.
Wells, G.J. (1971*a*). *Aust. J. Exp. Agric. Anim. Husb.* 11, 313-9.
———. (1971*b*). *Aust. J. Exp. Agric. Anim. Husb.* 12, 320-7.
Wells, W.G. (1951). *Qld. Agric. J.* 73, 109-15.
West, E.S. (1930). *J. Coun. Sci. Ind. Res. Aust.* 3, 97-105.
Wetselaar, T., and Beech, D.F. (1968). *Aust. J. Exp. Agric. Anim. Husb.* 8, 59-65.
Wheat Research Act 1957. (1972) 14th Annu. Rep. 1971, Dep. Prim. Ind., Canberra.
Wheelhouse, F. (1966). "Digging Stick to Rotary Hoe: Men and Machines in Rural Australia". Cassell, Melbourne.
Wild, A.S. (1959). *J. Dep. Agric. West. Aust.* 8, 249-50.
Willis, A.H. (1959). Univ. of New South Wales School of Mechanical Engineering–Report.
Willis, A.H., and Willatt, S.T. (1965). Univ. of New South Wales School of Mechanical Engineering. Report No. 1965/AE.6.
Woodroffe, K. (1949). 27th Meeting ANZAAS, Hobart, Section K.
———. (1950). Br. Commonw. Scient. Off. Conf., Rep. Spec. Conf. Agric., Aust., 1949, pp. 425-36.
Yang, S.J., and de Jong, E. (1972). *Agron. J.* 64, 574-8.
Zingg, A.W., and Whitfield, C.J. (1957). U.S. Dep. Agric. Tech. Bull. No. 1166.

CHAPTER 12

Fertilization

A.N. Smith

SOILS OF THE SEMI-ARID REGIONS, THEIR NUTRIENT STATUS AND SUITABILITY FOR CROPPING

In the main, seven soil groups are used for cropping in the semi-arid regions of Australia. They are: red-brown earths, mallee soils, non-calcic brown soils, yellow earths, solodized solonetz and solodic soils, grey clays and black earths. The names of the soil groups are those of Stace et al. (1968).

More wheat is grown on the red-brown earths than on any other soil group, and wheat is by far the main crop in the semi-arid areas. Typically, cropping is taken in rotation with a pasture containing a legume such as subterranean clover or a medic, or in rotation with lucerne. Only the black earths, and possibly some of the grey clays, are fertile by world standards; all the rest are inherently infertile in that they all require added nitrogen and phosphorus to produce good crops, some are also deficient in trace elements, and many are susceptible to erosion by wind and water.

THE EFFECT OF FERTILIZERS ON PLANT GROWTH. CEREALS

Phosphatic fertilizers

The work of Custance and Lowrie at Roseworthy Agricultural College in South Australia during the 1880s showed the value of superphosphate for wheat growing. Since that time the results of many experiments have been published, and the results of many more remain unpublished; but they show that on the above soils, responses can usually be obtained to applications of 100-150 kg, and occasionally up to 400 kg/ha of single superphosphate (9.7 per cent P). This experience is not surprising considering the generally low status of phosphate in Australian soils (Wild 1958). Recent research has been directed towards the investigation of the residual values of previous applications and the efficiency of alternative and cheaper forms of phosphorus. It is generally accepted by farmers in southern Australia that superphosphate will produce a yield increase; this acceptance is exemplified in the observation that 99 per cent of the wheat grown in Victoria in 1968-69 received a phosphate dressing (McClelland and Price 1970).

Acceptance of the usefulness of superphosphate is not yet general amongst farmers in northern New South Wales or in Queensland; but wheat has been shown to respond (Esdaile and Colwell 1963; Bradley and Fitzsimmons 1964; Hibberd, Whitehouse and Duncan 1969), and so has maize (Marley 1969).

Residual value

Wheat crops in South Australia (mainly grown at the Waite Institute) which had received from 0 to 3,750 kg/ha of single superphosphate in the past showed little, if any, response in the yield of grain when current rates were applied, if previous applications had exceeded 1,900 kg/ha (Woodroffe and Williams 1953). Grain yield could, in many cases, have been limited by the nitrogen status of the soil, and the authors recommended the use of superphosphate during the legume phase but not during the cereal phase of the rotation. Wheat in a wheat:fallow:barley rotation, on land which had received 3,000 kg/ha of superphosphate over twenty-three years, could be grown for five years before a clearly defined response to current dressings developed (Piper and de Vries 1964); but stubble-sown barley showed a response after the first year.

After eight cycles of a wheat:oats:fallow rotation in a permanent fertilizer trial at Walpeup in the Victorian Mallee, the current effect of superphosphate application was decreasing at an increasing rate (McClelland 1969). This trend did not continue when the rotation was changed to wheat:medic pasture:medic pasture:fallow for a further three cycles. Despite high levels of previously applied superphosphate the current effect was frequently greater than that on a virgin soil. Experiments nearby showed little or no response to nitrogen, and it was concluded that the increase in current effect resulted from a change in the availability of phosphorus in the soil after medic pasture had been grown. When ten cycles of a wheat:oats:fallow rotation had been completed, an application of 31 kg/ha of superphosphate produced a response of 300 kg/ha of wheat, of which 56 per cent was attributable to residual phosphate (McClelland 1968).

On a lateritic red earth, superphosphate had a residual value for sorghum of 50 per cent after one year, and 8 per cent after seven years (Arndt and McIntyre 1963).

It is clear that, even though Australian soils are low in phosphorus and have high sorption capacities in many cases, appreciable residual effectiveness can be built up after about 2,500 kg/ha of single superphosphate has been applied.

Alternative forms

The universal need for increased profitability has led to a renewal of interest in the possibility of using cheaper forms of phosphorus. Ground rock phosphate and calcined Christmas Island C Grade phosphate have been tested. Rock phosphate was inferior to superphosphate for wheat (Sims 1959; Rudd 1972*b*), and sorghum (Arndt and McIntyre 1963); calcined rock was much inferior for wheat (Mason and Cox 1969; McClelland and Price 1970), with yields not significantly greater than those with no fertilizer (Rudd 1972*b*), or even lower (Lipsett and Williams 1970).

Nitrogenous fertilizers

Interest in nitrogen as a fertilizer has increased in recent years, for three possible reasons:

(*a*) The gradual decline in the price of wool following the boom years of the early fifties led to an increase in the number of crops taken during a rotation.

(*b*) Cereal production has expanded into some areas in which legumes will not grow satisfactorily.

(*c*) Fertilizers have become progressively cheaper per unit of nitrogen. In 1957 nitrogen in fertilizers cost 38 cents per kg N, in 1970 the price was 17 cents (Landy 1970).

Callaghan and Breakwell (1937) reviewed earlier work on nitrogen fertilizers for cereals and concluded that there was no need for nitrogen in South Australia (if the crop was grown in a reasonably broad rotation on a well-prepared fallow). This same conclusion was reached in Queensland where the mineral nitrogen at planting and the subsequent grain yield of crops on the Darling Downs were not significantly correlated; the nitrogen recovered in the crop was only half that available in the soil at planting (Waring and Teakle 1960). There was no increase in yield when nitrogenous fertilizers were applied to wheat in a wheat:pasture:pasture:fallow rotation in the Victorian Mallee, but there was a response on a wheat:fallow rotation (McClelland 1970). Responses to nitrogen have been recorded for wheat (Littler 1961, 1963; Sims 1961, 1964; Colwell

1963*a*, 1963*b*; Esdaile and Colwell 1963; Russell 1963; Bradley and Fitzsimmons 1964; Mason 1968*a*, 1968*b*; Smith and Spence 1969*a*; Mason and Rowley 1969; Corbin 1972), barley (Toms 1960; Sims 1964; J. Russell 1967; Littler, Kelso and Marley 1969), oats (Toms 1959; Sims 1964), sorghum (Littler 1967), and maize (Marley 1969). In some cases there was no difference in the yield increases produced by different types of fertilizer, e.g. between sulphate of ammonia and anhydrous ammonia (Sims 1961), or between urea and anhydrous ammonia (Mason 1968*b*; Mason and Rowley 1969), but urea was inferior to sulphate of ammonia on Mallee soils (McClelland and Griffiths 1972).

The response of cereals to nitrogen is extremely sensitive to environmental factors, particularly climate; in favourable seasons responses can be obtained to as much as 60 or 70 kg/ha. Mean responses in grain yield for crops in South Australia were lower than those in more humid wheat-growing areas of Australia, but some crops were as good (J.S. Russell 1967). The difficulty was to predict the sites. Yield response was positively correlated to winter rainfall, and negatively related to high maximum temperatures in the latter part of the growing season (Russell 1968). In a black earth, with a summer rainfall pattern, there was no response to nitrogen by dryland plots of wheat; but when water was previously stored in the soil there were responses of ca. 370 kg/ha per 125 kg of urea (Swartz and White 1966). An improved supply of nitrogen to wheat results in an increased size of the plants which may subsequently suffer increased moisture stress in the late spring (Storrier 1965*a*). Application of nitrogen fertilizers leads to an earlier depletion of soil water (Barley and Naidu 1964), but the lower grain yield with increased nitrogen supply may be due, not only to the competition amongst the plants in the denser crop for the limited water resources, but also to other special effects independent of that on the dry matter, namely, the effect of the leaf area ratio and mean shoot size, and the delay in the time of flowering (Fischer and Kohn 1966). Application of nitrogenous fertilizers delayed flower initiation fourteen days, and ear emergence eight days (Halse et al. 1969).

Other nutrients

As cereal growing has extended into new areas, as sources of fertilizer have changed, as productivity has increased, and as old land has become deficient the need has been demonstrated for nutrients other than nitrogen and phosphorus.

Lack of sulphur limited the dry matter production of oats when high rates (90 kg N/ha) of fertilizer were applied (Crofts and Blair 1966; Blair and Crofts 1970), however no response has been demonstrated in the yields of cereal grain (Corbin 1969; Mason and Cox 1969; Rudd 1970; Lipsett and Williams 1971). Wheat and oats were more sensitive than lucerne and white clover to sulphur deficiency and they required more sulphur to achieve maximum dry matter production; however, cereals recovered a higher proportion of the applied sulphur from the soil and they used applied sulphur more efficiently for dry matter production (Holford 1971).

In contrast with crops grown on semi-arid soils in other countries, the recorded responses to potassium in Australia are extremely few. Where there is a greater plant production on a poorer soil, as for example when grazing oats are grown on granitic soils with high rates of nitrogen application (Crofts 1966) there may be a response to potassium; however for normal cropping the semi-arid soils of Australia are well supplied with potassium as, for example, Rasmussen (1969) showed for the Darling Downs in Queensland.

Some soils of the semi-arid areas are deficient in trace elements. Responses have been recorded to copper, zinc, molybdenum, manganese, iron, and possibly boron. Copper and zinc deficiency occurs in cereals in Western Australia, and in South Australia. Wheat responds to applications of 5.5 kg/ha $CuSO_4$ and 0.84 kg/ha ZnO (Gartrell

1969; Gilbey, Greathead and Gartrell 1970); however, some soils sorb copper strongly and increased rates must be applied (Fitzpatrick 1962). There have been responses to zinc on black earths by wheat (Duncan 1967*b*), maize (Duncan 1967*a*; Hibberd 1970), and linseed (Duncan 1967*c*). Superphosphate may contain in excess of 1000 ppm Zn; that made from rock phosphate from Nauru has the highest zinc content. This impurity may be sufficiently high to correct deficiencies on some soils (Fitzpatrick 1962). Responses have been demonstrated to iron (Fitzpatrick and Sprivulis 1963), manganese (Smith and Toms 1958), and molybdenum (Doyle et al. 1965; Lipsett and Simpson 1971).

THE EFFECT OF FERTILIZERS ON PLANT GROWTH. PASTURES

In southern Australia pastures are usually grown in rotation with crops. In northern New South Wales and in Queensland other species such as lucerne are being introduced into what has been, in many cases, continuous cropping to wheat and sorghum or maize. The response of pasture, or alternative crop, to fertilizer, and the way in which the pasture is utilized will have a direct effect on the response of subsequent cereal crops to fertilizer.

Phosphatic fertilizers

The growth of legume pasture increases the organic matter content of the soil. This increase may be unrelated to amounts of superphosphate applied (Hingston 1959; Russell 1960*a*), but this finding may be explained by the observation that increasing the rate of superphosphate application above 70-100 kg/ha/year does not result in equivalent increases in the amount of nitrogen accumulated (Russell 1960*a*). In New South Wales each 125 kg of superphosphate produced an increase of 95 kg N/ha (Donald and Williams 1954). In Western Australia there was a progressive increase in the yield of herbage dry matter and nitrogen with increased levels of superphosphate up to 35 kg/ha, and of root dry matter yield up to 42 kg (Watson and Lapins 1969). During the first four years after establishment of subterranean clover pastures on a coarse sand the increase in organic matter content was largely independent of fertilizer application, but the rate of increase was lower in subsequent years on the plots with lower superphosphate applications (Watson 1969).

Rate of organic matter accumulation may be limited by insufficient sulphur (Donald and Williams 1954; Williams and Donald 1957), in which case a partial substitution of gypsum for superphosphate might be advisable, or the use of a fertilizer with a lower P/S ratio. On sandy soils leaching may be the reason for low sulphur levels (Barrow 1966; Glencross and Cox 1969; Watson 1969).

There is no doubt that sulphur deficiency would be widespread were it not for the fortunate occurrence of sulphur in single superphosphate; the omission of one year's dressing can bring about deficiency in a pasture (Cox 1968*b*). Sulphur is necessary for lucerne on some soils (Jones 1970; Holford 1971).

Whilst an inadequate supply of sulphur may limit pasture production on some soils, on many others it is the phosphorus content of the fertilizer to which the pasture plants respond (McLachlan and Norman 1962); in such circumstances an increase in the P/S ratio might be advantageous.

Residual value

Superphosphate applied in previous years has a residual effectiveness which may be high (Barrow 1966; Cameron and McGowan 1969), or low (Hughes and Searle 1964), depending on soil type. The residual effectiveness of phosphorus and sulphur was found to be related to the number of years since last application rather than to the total amount of superphosphate applied (McLachlan and Norman 1962).

Alternative forms

Rock phosphate has been compared frequently with superphosphate as a fertilizer for pastures. Initial yields from equivalent weights of phosphorus as rock phosphate are generally lower than those from superphosphate (Bath and Cameron 1960; McLachlan 1960; Spencer 1960), but total yields during the seven years following an application may be similar (McLachlan 1960).

Other comparisons have been made between superphosphate and calcined Christmas Island C Grade phosphate (Smith 1968; Mason and Cox 1969; Lipsett and Williams 1970), with the calcined rock always giving lower yields.

Calcium magnesium phosphate was less effective than superphosphate on a podzol, but more effective on a krasnozem (McLachlan and Norman 1969). The increase in effectiveness was related to the slower release of the phosphate from the fertilizer rather than to any specific reaction with the soil. Basic slag also had a better residual value for both clover and wheat grown in pots with a red-brown earth (Smith 1968); this was ascribed to a higher solubility of the residual fertilizer over the solubility of the reaction products between monocalcium phosphate and soil.

Phosphate rock from a deposit in Queensland when calcined at 1450°C for 12 hr was as effective as superphosphate in promoting yield increase in stylo (Fergus, Gilkes and White 1970); both soils used in the experiment were acid (pH 4.6 and 4.4 in 0.01M $CaCl_2$).

Other nutrients

If a pasture does not contain a legume component then nitrogen fertilizers may be necessary, but for pastures with subterranean clover or a medic, or for lucerne, the nitrogen-fixing capabilities of the symbiotic microflora ensure an adequate supply.

Potassium is necessary for pastures in some areas (Toms and Fitzpatrick 1961; Lines and Marrett 1962), and its usage is increasing (*Commonw. Agricst.* 1966).

In certain situations other nutrients such as calcium, magnesium, iron, manganese, copper, zinc, molybdenum, cobalt and boron may be lacking, but the nutrients chiefly needed are undoubtedly phosphorus and sulphur.

Mineralization of organic matter

Soils in semi-arid areas have periods of dryness but these are followed by periods in which they are moist and mineralization of organic matter can occur. It has been shown that the longer a soil remains in the dry state (Birch 1959), and the greater its carbon content (Birch 1960), the greater are the amounts of nitrogen which are mineralized when the soil is moistened. It is impossible to generalize as to the effect of these periods of drying and wetting on the release of plant nutrients from organic matter since rainfall patterns and temperatures during the dry season vary over the semi-arid areas, but it is clear that the supply of such nutrients as are constituents of the organic matter (such as nitrogen, sulphur, and to a smaller extent, phosphorus) is affected to a great extent by the rate of mineralization. Thus the nutritional value (as distinct from the physical and structural value) of organic matter built up under pasture is dependent on the extent of its breakdown.

PASTURES AND THE GROWTH OF SUBSEQUENT CROPS

Our interest in pastures and their response to fertilizers is in the way in which they change soil fertility and the effect that this has on the response of crops to fertilizer.

Average yields of wheat in South Australia have increased where farmers have introduced subterranean clover leys into their rotations (Goode 1963; Bicknell 1964; Webber 1964), and in Western Australia wheat yields on light land have increased 50 per cent since the introduction of clover ley farming (Shier 1962). Grain protein was higher in wheat in first crops after superior clover ley than in first crops on average clover land; a superior ley was defined as one that had been down for three years or longer, and which had received at least 200 kg/ha of superphosphate during the previous three years (Taylor 1966). In Western Australia soil nitrogen under clover increased at a steady rate of 47 kg N/ha/yr over a five-year period, and each additional year of clover led to the production of an additional 135 kg/ha of wheat grain in the first crop (Watson 1963). Lucerne in a rotation on the Darling Downs in Queensland improved the yield of a following wheat crop when moisture was not limiting (Harty et al. 1966).

Whilst there is much evidence to show that the yield of a legume pasture will increase, in deficient situations, when a fertilizer is applied, there has been relatively little attempt, until recently, to relate this increased production of herbage to animal production. It may appear obvious that there is no economic justification for the application of high levels of superphosphate to pasture unless additional production is utilized through an increased stocking rate (Carter and Day 1970), but to many farmers the application of superphosphate to pastures is an article of faith rather than an agronomic necessity. Stocking rate × superphosphate trials are being conducted in many areas. The conclusions drawn from them are not uniform, some experiments showing an increased phosphorus requirement under grazing (Ozanne and Howes 1971), whilst others show that less superphosphate is needed at high stocking rates (McLachlan and Norman 1966; McLachlan 1968) or that there is no significant response in animal production to phosphate application (Kohn 1974); these latter conclusions will obviously have a great influence on the rates of fertilizer applied to pastures in the future. If much less phosphate is put on the pastures more may need to be applied to crops, for there will be a lower residual from the pasture phase, and in fact the clovers and grasses will be existing on residual phosphate from the cropping phase.

When pastures are ploughed and cultivated in preparation for a cereal crop, the top 8 or 10 cm of soil are mixed; this can lead to a reduction of phosphate availability on soils with high sorption capacity (Williams and Simpson 1965); this may, in extreme cases, lead to a sharply reduced residual effect of the phosphate previously applied to the pasture.

REACTIONS BETWEEN FERTILIZERS AND SOIL

Phosphates

Apart from the extremely small quantities of nutrients applied to soils as organic fertilizers or as inorganic mineral amendments, all the fertilizers with which the farmer is concerned are soluble inorganic chemical compounds. They are applied at fairly low rates in terms of the total weight of soil in a field, and whilst they are apparently evenly spread on a broad scale, close examination will reveal that there is some distance separating the crystals or the granules of the fertilizer. This means that the soil in immediate contact with the fertilizer particle will be greatly enriched with nutrient when the compound dissolves. If the fertilizer has been spread on the surface, then only the soil under the particle will be in contact with it. If the fertilizer is sown with the seed, as superphosphate is, then the granules will be completely surrounded by soil.

When a water-soluble phosphate compound such as monocalcium phosphate (MCP) is buried in moist soil, water vapour moves into the particle from the soil and forms an almost saturated solution of the compound. The composition of the saturated solutions of four common fertilizer phosphates is shown in table 1. It is concentrated and highly acidic solutions such as these which come into contact with the soil.

Table 1. Composition of saturated solutions of phosphate salts at 25°C (Data from Huffman and Taylor 1963)

Salt	pH	Saturated solution composition (M)		
		Ca	NH_4	P
$Ca(H_2PO_4)_2.H_2O + CaHPO_4$ (stable triple point)	1.01	1.34		4.50
$Ca(H_2PO_4)_2.H_2O + CaHPO_4.2H_2O$ (metastable triple point)	1.48	1.44		3.98
$(NH_4)\ H_2PO_4$	4.00		2.90	2.90
$(NH_4)_2HPO_4$	8.08		7.62	3.86

When MCP pellets were allowed to dissolve in a soil a wet zone formed at the pellet site, and this zone advanced into the soil as dissolution progressed (Huffman 1962). Vapour phase transfer accounted for much of the moisture in the pellet, and the movement of the solution into the soil was under the capillary action of the surrounding soil (Lehr, Brown and Brown 1959). The solution leaving the MCP granule dissolves large amounts of calcium, iron, aluminium, and manganese. For example, after a reaction time of twenty-one days in a fine sandy loam at 0.4 moisture equivalent (i.e. in a moderately dry condition) the concentration of the solution 10 mm above the fertilizer band was Ca 1.08 M, P 4.03 M, Fe 187 mM, Al 395 mM, and Mn 12.7 mM (Lindsay and Stephenson 1959*a*, 1959*b*). The concentration of the iron is several thousand times greater than that normally encountered in soil solution.

The movement of the solution into the soil continues until the concentration of the ions is reduced by dilution or precipitation, and no concentration gradient remains. Phosphate is precipitated as compounds of calcium, iron, or aluminium, and a new range of compounds is formed; the composition of these reaction products depends both on the type of fertilizer and the type of soil. The occurrence and properties of iron and aluminium phosphates in soil have frequently been the subject of research, and this has been reviewed (Smith 1965*a*). The application of phosphate fertilizer to soil results then not in the simple increase in the phosphate concentration of the soil solution, it is merely the first step in a process involving dissolution, creation of zones of high acidity, movement of concentrated solutions, and precipitation of new compounds which themselves undergo dissolution. Condensed phosphates (polyphosphates) are being increasingly used as fertilizers in North America, and they have been introduced into Australia. Whilst the reactions between orthophosphates and soil, which have just been described, take place within a matter of hours, those with condensed phosphates require some days for initiation and then continue for weeks or months (Philen and Lehr 1967).

In Australia by far the greater part of the superphosphate used is spread on the surface of the soil. In this situation the concentration of nutrients in the fertilizer solution was found to be Ca 1.5 M, P 4.0 M, S 0.023 M; the pH of the solution ranged from 1.60 to 1.92 (Williams 1971*a*). These fertilizer solutions entered the soil mainly by a process of diffusion; the sulphur content was generally 1/50—1/100 of the phosphorus content, and the solution was found to displace sulphate and organic sulphur compounds from acid soils.

Ammonium compounds

The research on fertilizer:soil reaction products described above was done on a micro scale; on the macro scale superphosphate does not produce any overall losses in plant nutrients, or any changes in pH. This is not the case with nitrogenous fertilizers containing the ammonium (NH_4^+) ion. It has been known for a long time that these compounds reduce the base status and pH of soils. For example, it requires the equivalent of 100-120 kg of calcium carbonate to replace the bases (i.e. calcium, magnesium, potassium) displaced by the application of 100 kg of sulphate of ammonia. Urea is an ammonium fertilizer since it is changed to ammonium carbonate in soil before being nitrified by bacteria. The bases displaced are subject to the leaching action of percolating water, but this does not necessarily mean that they are washed below the plant roots, for this depends on the depth of rooting and the intensity of leaching: but the increased acidity does mean that the solubility of such toxic elements as aluminium and manganese is increased, and the availability of essential nutrients such as sulphur, copper, zinc and boron may be decreased.

Ammonium ions added from fertilizers may be fixed by the soil in a non-available form; montmorillonite soils are particularly noteworthy in this regard (Black and Waring 1972).

LOSSES OF FERTILIZERS FROM SOIL

Leaching

Whilst an environment may be categorized as semi-arid there may yet be periods during the year when the rainfall is sufficient to cause a considerable movement of water down through the soil profile, i.e. leaching can occur.

The ammonium ion, being positively charged, is electrically attracted to the negatively charged soil colloids, but when it is nitrified and changed into the negatively charged nitrate ion there is repulsion and the nitrate is free to move with the soil water. Whether leached nitrate is completely lost to the plants again depends on the depth of rooting and the rate of water movement. In a red-brown earth in southern New South Wales the surface 45 cm of soil supplied the greater part of the nitrogen taken up by the wheat crop (Storrier 1962), but nitrate leached during the autumn and winter was retained in the surface 75 cm and was subsequently taken up by the crop (Storrier 1965*b*).

Phosphate is an anion which is strongly sorbed in acid soils, for example all the phosphate applied up to twenty-six years previously was accounted for in the top 10 cm (Donald and Williams 1954). However, on lighter soils phosphate may move down the profile and some instance of this are shown in table 2.

Table 2. Losses of phosphate from soils by leaching

Fertilizer	Rainfall	Losses	Depth	Reference
	mm	%	cm	
Superphosphate	965	70	0–15	Hingston 1959
Superphosphate	559	17–81	0–10	Ozanne, Kirton, and Shaw 1961
Superphosphate	549	>50	0–20	Russell 1960*b*
Rock phosphate	549	75–89	0–5	Russell 1960*b*
Rock phosphate	559	56–85	0–10	Ozanne, Kirton, and Shaw 1961
Superphosphate Rock phosphate	>2,500	Only when >225 kg/ha P_2O_5 applied	0–10	Bruce 1968

It was suggested (Russell 1960*b*) that application of lime or gypsum should help to retain phosphate, and indeed an application of lime enhanced the growth of subterranean clover (Paton and Loneragan 1960; Ozanne, Kirton and Shaw 1961). Less phosphorus was leached, but the apparent reduction in leaching was in whole (Ozanne, Kirton and Shaw 1961) or in part (Paton and Loneragan 1960) due to the fact that increased plant uptake reduced the amount of phosphorus that was available for leaching.

Leaching appears to be greatest in soils with low organic matter (Ozanne, Kirton and Shaw 1961); the relationship is not a direct one for in many cases soils which do not sorb phosphorus very strongly have low organic matter contents; however, in laboratory studies when granules of superphosphate were leached in the presence of plant material, the dissolution of the phosphate was sometimes retarded (Jones and Bromfield 1969).

The lack of reaction between sandy soils and phosphate ions, which gives the opportunity for leaching, also brings about a high concentration of phosphate in the soil solution if water is adequate and drainage is restricted; toxicity has been demonstrated in young clover and oat plants under such conditions (Loneragan, Carroll and Snowball 1966).

Volatilization and denitrification

Nitrogen may be lost by volatilization as ammonia (NH_3) from fertilizers. Losses from surface-applied urea can occur on both acid and alkaline soils, but losses from solid ammonium fertilizers do not usually occur in soils with pH below 7.0. Volatilization loss can be reduced by incorporating the fertilizer into the soil.

When ammonium fertilizers are nitrified they change through the intermediate nitrite (NO_2^-) to the nitrate (NO_3^-) ion, which is of course a constituent of nitrate fertilizers. If conditions in a soil are largely anaerobic, and there is adequate organic matter, gaseous nitrogen and oxides of nitrogen can be produced from the nitrate by biological denitrification. For four wheat soils in growth chambers the rate of loss of nitrogen was 1-15 mg N/kg soil/week when nitrate nitrogen was applied, and 1-4 mg when ammonium nitrogen was applied as fertilizer (Stefanson 1972).

Uptake in crops and animals

The nitrogen and phosphorus in the grain of an average wheat crop can amount to 20 and 2.5 kg/ha respectively. At higher rates of superphosphate application the proportion of added phosphorus which is removed becomes progressively lower (Smith 1965*c*). The nitrogen removed in the crop is more than the amount commonly applied in fertilizers, whilst the phosphorus removed is almost always less.

PARTICLE SIZE

Generally, where sorption capacity is high the size of phosphate fertilizer granules should be greater than for soils where sorption is low. This has been confirmed for pastures, which responded better to coarse particles (0.56 cm diameter, 0.28 cm height, 130 mg weight) on soils which had a high sorption capacity and which were subject to leaching (Williams and Lipsett 1969); but pasture gave a better response to fine particles (<60 mesh) applied in autumn and winter (Muir 1966). When superphosphate was applied to the surface of a moist soil, a hemi-spherical zone was formed beneath each granule. The size of the zones was governed by the sorption capacity of the soil, the size of the particle, and the soil moisture (Williams 1971*b*). The optimum diameter of particle, considering both phosphorus and sulphur, was 2.5 mm; this same size was found to be optimum for gypsum in a theoretical study (Millington and Powrie 1968). Leaching of sulphate from gypsum and superphosphate was markedly reduced when the granule size was 2 mm or greater (Aylmore, Karim and Quirk 1971). In contrast, yields of wheat grown with fine, medium, or coarse superphosphate were not significantly different from yields with the commercial fertilizer (Rudd 1972*a*).

Because of the known tendency of sulphate to leach, sources of sulphur other than gypsum have been investigated. Elemental sulphur of 100 mesh size was only slightly inferior to gypsum in producing yield increases in pastures in the year of application, but it was superior in the second year (Weir 1967); particles finer than 200 mesh oxidized quickly and were a poor source of sulphur, those between 40 and 100 mesh were the best compromise between too fast and too slow (Barrow 1971*a*). Elemental sulphur (Weir, Barkus and Atkinson 1963), pyrrhotite, Fe_7S_8, (Barrow 1971*b*), and pyrites (Banath 1969) are effective sources for plants if fine enough.

FERTILIZER PRACTICES

Placement

Germinating seeds can be easily damaged. Urea drilled with the seed delayed wheat germination and reduced wheat yields (Mason, Parkin and Patrick 1969; Mason et al. 1970); urea is worse in this regard than ammonium nitrate (Mason 1971) or sulphate of ammonia (Price 1971), and generally it is safer to broadcast the fertilizer rather than drill it with the seed. If the fertilizer is an ammonium phosphate compound then it must be drilled with the seed for best response from the phosphate. Seeds vary in their susceptibility to damage in the order rape>linseed≫oats>wheat>barley (Mason 1971). Germination damage decreases as moisture levels increase, but if dry weather or topography limit moisture supply, then symptoms of damage soon appear (Smith and Woodward 1968).

Time of application

When superphosphate is placed on the surface of the soil capillary uptake of soil water is more important than hygroscopic uptake in the wetting of the granules. For hygroscopic uptake a relative humidity of 90 per cent or more is necessary (Williams 1969). Particles on a dry soil are exposed to an atmospheric humidity of well below 90 per cent, thus superphosphate resting on a soil will remain unchanged unless it rains.

There are frequently cost advantages in having superphosphate spread by a contractor, and for the cereal farmer anything which reduces the amount of material he has to transport to the paddock and put through the drill at the busy seeding time is attractive. Where crops respond to the application of superphosphate drilling with the seed gives the greatest yield increase (Sims and Andrew 1958; McDowell et al. 1961; Smith 1965*b*; Loutit, Stallwood and Cox 1968). Where superphosphate was top-dressed before (Rudd 1972*c*), or after (Smith 1967) sowing, the nearer the application was made to the time of sowing the more closely the yield increase approached that obtained with combine drilling.

In experiments such as these it is difficult to separate the time and the placement effects since, whilst top-dressing can be carried out at any time, placement of the fertilizer with the seed can only occur at sowing. In pot experiments in which MCP was applied seven, thirty-three, and fifty-four days after sowing to plants of three wheat varieties growing under different day lengths to give a range of times of phasic development, there was no significant increase in grain yield when phosphate was applied later than 14 days after the initiation of spikelet primordia (Smith 1969*a*). The response of different varieties to phosphatic fertilizer may be influenced by their sensitivity to day-length.

In most cases best responses from crops are obtained when nitrogenous fertilizers are applied at seeding, but other findings are not exceptional. Response to urea seems to be largely affected by season. Sometimes application as close as possible to seeding gives the best results (Littler 1963; Mason 1968*b*), at others later applications after seeding give higher yields (Mason, Rowley and Quayle 1972); any reduction in yield due to post-seeding application may be small (McClelland and Griffiths 1972). For anhydrous ammonia the time of application is less important if the soil is moist and has some clay content (Mason 1968*b*).

RECOMMENDED RATES

Notwithstanding the responses which occur from time to time to large applications of phosphorus and nitrogen fertilizers, the rates recommended (tables 3 and 4) are conservative, ranging up to 125 kg of superphosphate for pastures and 250 kg for crops, with nitrogen rates up to 180 kg urea/ha. These application rates are not high by world standards, for example for the whole of the continental United States in 1970 there was an average application of 123 kg/ha of *plant nutrients* ($N + P_2O_5 + K_2O$) (Hargett 1973).

The rates quoted are based on various criteria and they are district or regional recommendations. Attempts have been made to bring the regional recommendation down to a paddock basis (e.g. Leslie and Hart 1967); but since response to added nutrient is largely affected by the level of available nutrient in the soil, analysis is an obvious aid in the increase in the accuracy of recommendations. Soil testing services available in Australia were surveyed in 1966 (Skene et al.) and later reviewed (Smith 1971).

Whilst soil testing may give a more accurate indication of the rate of fertilizer to apply than does a flat recommendation, it still has to be shown that a soil test can do better than an adviser who, with local and personal knowledge, can integrate all the factors likely to affect yield.

MODIFICATION OF PRACTICE BY ENVIRONMENT

Whatever way the farmer arrives at the rate of fertilizer he puts on his paddock, it is plain that the resultant yield will be dependent strongly on climatic and other environmental factors. Fertilization is rarely, if ever, carried to the limit of attempting to produce maximum yield per ha, but rather the aim is to obtain maximum economic return per ha, or per dollar invested in fertilizer. The danger of the production of excessive dry matter in wheat is exemplified by the number of papers which mention the problem of "haying-off" (Colwell 1963*a*; Taylor 1965; Dann 1969).

Climate may lower the yield ceiling, but weeds compete for nutrients, and pests and diseases consume nutrients already incorporated into plant material. As an example, skeleton weed (*Chondrilla juncea L.*) is a notorious competitor with wheat in New South Wales and Victoria, especially on land of low fertility. It can be controlled by spraying with herbicide; the response of a wheat crop to the spraying of skeleton weed was analogous to the effect of a direct application of nitrogenous fertilizer (Myers and Lipsett 1958). Strong growth of subterranean clover will choke the weed, but other legumes may not be as effective (Wells 1969).

If rainfall is excessive some nutrients will be leached; even in semi-arid areas precipitation may greatly exceed evapo-transpiration at certain times during the year. Impermeability in the soil will reduce leaching, but the resulting waterlogging may reduce the availability of soil phosphorus (Williams and Simpson 1965; Smith 1969*b*).

Table 3. Recommended rates of application of nitrogenous and phosphatic fertilizers to pastures

State	Phosphorus kg/ha	Nitrogen kg/ha	Conditions	Reference
N.S.W.	12		Responsive basaltic soils	McLachlan and Norman 1964
S.A.	19 4.8		Establishment phase After 12 years	Powrie 1963
S.A.	12–24 12–18		Mallee Red-brown earth	French and Rudd 1967
S.A.	22 11	22 0	Sandy loams Clay loams	Seelinger 1968
W.A.	19 3		Light land, no history Heavy land, more than 2,250 kg/ha superphosphate in the past	Cox 1969

Table 4. Recommended rates of application of nitrogenous and phosphatic fertilizers to cereals

State	Phosphorus kg/ha	Nitrogen kg/ha	Conditions	Reference
N.S.W.	22 11		510–560 mm annual rainfall 410–460 mm annual rainfall	Smith and Spence 1969*b*
N.S.W.		45	Where high rates of superphosphate applied to third and fourth crops	Smith and Spence 1969*a*
N.S.W.		22–45 0	Second and third crops after nil clover pasture First crop after good clover	Storrier 1969
N.S.W.	11–22 0	28–56 0	Low fertility, high rainfall High fertility, low rainfall	Corbin 1972
Queensland		26–52		Littler 1961
Queensland		0–67	Depends on key	Leslie and Hart 1967
S.A.	12–24 6–9		Mallee Red-brown earth	French and Rudd 1967
S.A.	12 12	12 24	Light land Heavy land	Reuter 1967
S.A.	9–13	22		Seelinger 1968
Tasmania	40		Krasnozems	Russell and Stackhouse 1972
Victoria	10 10	21 28	Mallee Mallee with skeleton weed	McCann 1968
W.A.		26 13	Soils low in N Drier districts	Toms and Burvill 1961
W.A.		26	New land Second and third crops after legume	Roberts et al. 1966
W.A.	4–23		Based on soil type and previous history	Cox 1968*a*
W.A.		38 0	Old light land High rainfall Heavy land Low rainfall	Mason 1968*a*
W.A.		38 0	Old light land Over 460 mm annual rainfall Heavy land Under 330 mm annual rainfall	Halse 1969

FERTILIZING A CEREAL CROP

If the crop is to be sown after pasture it is less likely to need nitrogen than if the crop is to be sown on stubble land. The factors affecting the amount of nitrogen in the soil after pasture have been discussed. If nitrogen is applied, the rate will be decided more on the basis of cropping history than on any other factor. Fertilizers applied with the seed frequently reduce germination and delay emergence, but soil moisture is an important determinant in this; application before sowing, or top-dressing after sowing does not cause contact damage and may give only slightly reduced yields. Climate plays a large part in affecting the response of the crop to nitrogen; in summer rainfall areas where the crop is grown on water stored in the soil, and since this can be estimated, some of the uncertainty is taken out of the determination of fertilizer rates.

In southern Australia the crop is virtually always sown with superphosphate (single or concentrated). For best results the fertilizer will be drilled with the seed. The rate of application will be decided upon after considering some, or all, of such factors as past experience of crop responses, length of previous cropping, length of previous ley, phosphate applied to pasture or crop and its likely residual value, management of previous pasture, weed burden, and soil analysis.

THE FUTURE

Fertilizers

The trend in new fertilizers is towards an increase in the concentration of plant nutrients brought about by the use of new "high analysis" products and by compounding. The most concentrated nitrogenous fertilizer, anhydrous ammonia, is already an article of commerce; but it has not gained wide acceptance, except in Queensland, possibly because of the need to handle it under pressure and because of the cost of the special equipment necessary for its application. The most concentrated phosphatic fertilizer is elemental phosphorus, but trials with it have shown no attractive features. The production of phosphoric acid by the electric furnace process has enabled the manufacture of such concentrated products as superphosphoric acid (33 per cent P). This product is the starting point for the manufacture of polyphosphates, and these can be readily ammoniated to give highly concentrated NP fertilizers. Their slow reaction with the soil has been mentioned, and they have an additional advantage in that they are powerful sequestering agents for metals and thus can serve as efficient carriers for trace elements. The general drawback to compound fertilizers is their fixed ratio of nutrients, but this is not a great disadvantage if a range of products is marketed; because of their economic attractiveness they are the materials of the future.

Alternative crops

Subterranean clover for all its success as a pasture plant has a number of undesirable features; its seed is difficult to harvest, it may be toxic, it is shaded out by grasses, and it does not respond to summer rain (Donald 1962). It is true that its main undesirable feature, the high oestrogen content, can be markedly reduced by selection, but nevertheless, if there were to be a large-scale replacement of subterranean clover by a more agronomically suitable but less efficient nitrogen-fixing legume, then the overall fertility of soils under rotation would decrease.

The growth of alternative crops such as sunflower, rape, linseed, or safflower will not alter the general fertility very much, but if legumes such as lupins, field peas, lentils, or gram can be grown, and sold, it would mean that soil fertility was no longer dependent on the profitability of sheep or cattle.

Pollution

The pollution problem with fertilizers is that of their leaching into groundwaters and streams. There is little problem with the phosphate in phosphatic fertilizers except for some areas in Western Australia; however, superphosphate may contain as much as 2 per cent of flourine (Caro 1964), and from 18 to 91 ppm of cadmium (Williams and David 1973). With long continued use of superphosphate these toxic elements may build up to undesirably high concentrations. There is a problem with nitrates, and indeed high nitrate levels have been measured in water supplies after periods of high rainfall; the major part of this nitrate comes from the mineralization of organic matter under pastures and is thus part of the general agricultural pollution problem, but nevertheless the use of slowly-available nitrogen sources such as triazines, oxamide, IBDU, or metal ammonium phosphates would reduce the pollution and, of course, improve the recovery of added nitrogen by plants.

Increased use of fertilizers

Soil testing is on the increase, but that does not necessarily mean that there will be an overall increase in the use of fertilizer; there may be a more efficient use of fertilizer.

Tissue testing for nitrogen in crops would probably lead to an increased use of nitrogen. If the post-seeding application of nitrogenous fertilizer is almost as efficient as drilling with the seed, then it might be better to wait to see how the plants grow. Nitrogen deficiency can occur in soils of good status if there is continued heavy rain; the later the application is made the shorter will be the growing period left for which climatic guesses have to be made.

The climate cannot be altered, and it sets the yield ceiling. Plants can however be altered and if breeders and physiologists can decide what type of plant to breed, and if the question is resolved as to whether the activities of breeders or the increase in soil fertility is responsible for recent increases in yield (Warren 1971), then we may get new varieties which are better adapted to the environment.

Conservation of fertility

The increase in acidity which is concomitant with the use of ammonium fertilizers is probably not important in the short term, but it must deplete fertility in the long term, and there will be an acceleration in this process if the usage of nitrogenous fertilizers increases markedly.

Crop rotations involving a legume pasture maintain a stable level of soil fertility. If crops become more important because of pressure on world food supplies then, unless we can grow pulses for export, we shall have to rely on fertilizer nitrogen to increase the number of crops in a rotation. Both nitrogen and phosphorus have been applied in the past with no thought of maintenance of fertility, but only for immediate profit. The practice has led to the build up of phosphorus in the soil but not of nitrogen. If continuous cropping with cereals becomes an economic necessity it will be essential to apply nitrogen with a view to the maintenance of fertility; there is no reason why this cannot be done, but continuous cropping is an unstable system which requires a high degree of farming skill.

In the immediate future the best insurance of the conservation of the fertility of the soil in the semi-arid regions is the continued profitability of the animals consuming the pasture in the rotation.

REFERENCES

Arndt, W., and McIntyre, G.A. (1963). *Aust. J. Agric. Res.* **14**, 785-95.
Aylmore, L.A.G., Karim, M., and Quirk, J.P. (1971). *Aust. J. Soil Res.* **9**, 21-32.
Banath, C.L. (1969). *Aust. J. Agric. Res.* **20**, 697-708.
Barley, K.P., and Naidu, N.A. (1964). *Aust. J. Exp. Agric. Anim. Husb.* **4**, 39-48.
Barrow, N.J. (1966). *Aust. J. Exp. Agric. Anim. Husb.* **6**, 9-16.
———. (1971*a*). *Aust. J. Exp. Agric. Anim. Husb.* **11**, 211-6.
———. (1971*b*). *Aust. J. Exp. Agric. Anim. Husb.* **11**, 217-22.
Bath, J.G., and Cameron, I.H. (1960). *J. Dep. Agric. Vict.* **58**, 459-67.
Bicknell, K.G. (1964). *J. Dep. Agric. S. Aust.* **67**, 385-7.
Birch, H.F. (1959). *Plant Soil.* **11**, 262-86.
———. (1960). *Plant Soil.* **12**, 81-96.
Black, A.S., and Waring, S.A. (1972). *Aust. J. Soil Res.* **10**, 197-207.
Blair, G.J., and Crofts, F.C. (1970). *Aust. J. Exp. Agric. Anim. Husb.* **10**, 768-73.
Bradley, J., and Fitzsimmons, R.W. (1964). *Agric. Gaz. N.S.W.* **75**, 1487-97.
Bromfield, S.M. (1960). *Aust. J. Agric. Res.* **11**, 304-16.
Bruce, R.C. (1968). *Qld. J. Agric. Anim. Sci.* **25**, 77-9.
Callaghan, A.R., and Breakwell, E.J. (1937). *J. Dep. Agric. S. Aust.* **41**, 326-44.
Callaghan, A.R., and Millington, A.J. (1956). "The Wheat Industry in Australia". Angus and Robertson, Sydney.
Cameron, I.H., and McGowan, A.A. (1969). *Aust. J. Exp. Agric. Anim. Husb.* **9**, 617-21.
Caro, J.H. (1964). In "Superphosphate: Its History, Chemistry, and Manufacture". U.S. Dep. Agric., Washington.
Carter, E.D., and Day, H.R. (1970). *Aust. J. Agric. Res.* **21**, 473-91.
Colwell, J.D. (1963*a*). *Aust. J. Exp. Agric. Anim. Husb.* **3**, 51-61.
———. (1963*b*). *Aust. J. Exp. Agric. Anim. Husb.* **3**, 190-7.
Commonw. Agricst. 36 (1966), 9-12.
Corbin, E.J. (1969). *Agric. Gaz. N.S.W.* **80**, 428.
———. (1972). *Agric. Gaz. N.S.W.* **83**, 39-44.
Cox, W.J. (1968*a*). *J. Dep. Agric. West. Aust.* **9**, 97-100.
———. (1968*b*). *J. Dep. Agric. West. Aust.* **9**, 440-2.
———. (1969). *J. Dep. Agric. West. Aust.* **10**, 218-21.
Crofts, F.C. (1966). *Agric. Gaz. N.S.W.* **77**, 258-62.
Crofts, F.C., and Blair, G.J. (1966). *J. Aust. Inst. Agric. Sci.* **32**, 142.
Dann, P.R. (1969). *Aust. J. Exp. Agric. Anim. Husb.* **9**, 625-9.
Donald, C.M. (1962). *J. Aust. Inst. Agric. Sci.* **28**, 171-8.
Donald, C.M., and Williams, C.H. (1954). *Aust. J. Agric. Res.* **5**, 664-87.
Doyle, R.J., Parkin, R.J., Smith, J.A.C., and Gartrell, J.W. (1965). *J. Dep. Agric. West. Aust.* **6**, 699-703.
Duncan, O.W. (1967*a*). *Qld. J. Agric. Anim. Sci.* **24**, 293-300.
———. (1967*b*). *Qld. J. Agric. Anim. Sci.* **24**, 287-92.
———. (1967*c*). *Qld. J. Agric. Anim. Sci.* **24**, 301-7.
Esdaile, R.J. (1965). *Agric. Gaz. N.S.W.* **76**, 155-9.
Esdaile, R.J., and Colwell, J.D. (1963). *Agric. Gaz. N.S.W.* **74**, 282-4.
Fergus, I.F., Gilkes, R.J., and White, R.E. (1970). *Aust. J. Sci.* **32**, 405-7.
Fischer, R.A., and Kohn, G.D. (1966). *Aust. J. Agric. Res.* **17**, 281-95.
Fitzpatrick, E.N. (1962). *J. Dep. Agric. West. Aust.* **3**, 323-7.
Fitzpatrick, E.N., and Sprivulis, R. (1963). *J. Dep. Agric. West. Aust.* **4**, 735-7.
French, R.J., and Rudd, C.L. (1967). *J. Dep. Agric. S. Aust.* **70**, 336-40.
Gartrell, J.W. (1969). West. Aust. Dep. Agric. Tech. Bull. No. 3.
Gilbey, D.J., Greathead, K.D., and Gartrell, J.W. (1970). *J. Dep. Agric. West. Aust.* **11**, 70-2.
Glencross, R.N., and Cox, W.J. (1968). *J. Dep. Agric. West. Aust.* **9**, 492-5.
———. (1969). *J. Dep. Agric. West. Aust.* **10**, 202-6.
Goode, J.R. (1963). *J. Dep. Agric. S. Aust.* **66**, 358-65.
Halse, N.J. (1969). *J. Dep. Agric. West. Aust.* **10**, 79-81.
Halse, N.J., Greenwood, E.A.N., Lapins, P., and Boundy, C.A.P. (1969). *Aust. J. Agric. Res.* **20**, 987-98.
Hargett, N.L. (1973). "1972 Fertilizer Summary Data". Tennesee Valley Auth., Muscle Shoals.
Harre, E.A. (1971). "Fertilizer Trends 1971". Tennesee Valley Auth., Muscle Shoals.
Harty, R.L., Kelso, W.T., Bygott, R.B., and Seton, D.H.C. (1966). *Qld. J. Agric. Anim. Sci.* **23**, 261-70.
Hibberd, D.E. (1970). *Qld. J. Agric. Anim. Sci.* **27**, 89-94.
Hibberd, D.E., Whitehouse, M.J., and Duncan, O.W. (1969). *Qld. J. Agric. Anim. Sci.* **26**, 457-63.
Hingston, F.J. (1959). *J. Aust. Inst. Agric. Sci.* **25**, 209-13.
Holford, I.C.R. (1971). *Aust. J. Agric. Res.* **22**, 879-84.
Huffman, E.O. (1962). *Proc. Fertil. Soc.* No. 71.

Huffman, E.O., and Taylor, A.W. (1963). *Agric. Food Chem.* **11**, 182-7.
Hughes, J.D., and Searle, P.G.E. (1964). *Aust. J. Agric. Res.* **15**, 377-83.
Jones, O.L., and Bromfield, S.M. (1969). *Aust. J. Agric. Res.* **20**, 653-63.
Jones, R.M. (1970). *Aust. J. Exp. Agric. Anim. Husb.* **10**, 749-54.
Kohn, G.D. (1974). *Aust. J. Agric. Res.* **25**, 525-35.
Landy, J.M. (1970). *J. Aust. Inst. Agric. Sci.* **36**, 214-23.
Lehr, J.R., Brown, W.E., and Brown, E.H. (1959). *Soil Sci. Soc. Amer. Proc.* **23**, 3-7.
Leslie, J.K., and Hart, J. (1967). *Qld. Agric. J.* **93**, 331-4.
Lindsay, W.L., and Stephenson, H.F. (1959*a*). *Soil Sci. Soc. Amer. Proc.* **23**, 12-8.
———. (1959*b*). *Soil Sci. Soc. Amer. Proc.* **23**, 18-22.
Lines, G.E., and Marrett, P. (1962). *J. Dep. Agric. S. Aust.* **65**, 506-9, 546.
Lipsett, J., and Simpson, J.R. (1971). *J. Aust. Inst. Agric. Sci.* **37**, 348-51.
Lipsett, J., and Williams, C.H. (1970). *Aust. J. Exp. Agric. Anim. Husb.* **10**, 783-9.
———. (1971). *Aust. J. Exp. Agric. Anim. Husb.* **11**, 59-63.
Littler, J.W. (1961). *Qld. Agric. J.* **87**, 721-4.
———. (1963). *Qld. J. Agric. Sci.* **20**, 323-44.
———. (1967). *Qld. Agric. J.* **93**, 193-6.
Littler, J.W., Kelso, W.T., and Marley, J.M.T. (1969). *Qld. J. Agric. Anim. Sci.* **26**, 309-17.
Loneragan, J.F., Carroll, M.D., and Snowball, K. (1966). *J. Aust. Inst. Agric. Sci.* **32**, 221-3.
Loutit, A., Stallwood, P., and Cox, W.J. (1968). *J. Dep. Agric. West. Aust.* **9**, 418-21.
McCann, J. McC. (1968). *J. Dep. Agric. Vict.* **66**, 131-3.
McClelland, V.F. (1968). *Aust. J. Agric. Res.* **19**, 1-8.
———. (1969). *Aust. J. Exp. Agric. Anim. Husb.* **9**, 622-4.
———. (1970). *Aust. J. Exp. Agric. Anim. Husb.* **10**, 450-4.
McClelland, V.F., and Griffiths, J.B. (1972). *J. Aust. Inst. Agric. Sci.* **38**, 219-21.
McClelland, V.F., and Price, G.H. (1970). *J. Aust. Inst. Agric. Sci.* **36**, 141-2.
McDowell, T.E., Smith, J.A.C., Toms, W.J., and Carroll, M.D. (1961). *J. Dep. Agric. West. Aust.* **2**, 869-70.
McLachlan, K.D. (1960). *Aust. J. Agric. Res.* **11**, 513-23.
———. (1963). *Aust. J. Exp. Agric. Anim. Husb.* **3**, 184-9.
———. (1968). *Aust. J. Exp. Agric. Anim. Husb.* **8**, 33-9.
McLachlan, K.D., and Norman, B.W. (1962). *Aust. J. Agric. Res.* **13**, 836-52.
———. (1964). *Aust. J. Exp. Agric. Anim. Husb.* **4**, 152-7.
———. (1966). *Aust. J. Exp. Agric. Anim. Husb.* **6**, 22-4.
———. (1969). *Aust. J. Exp. Agric. Anim. Husb.* **9**, 341-9.
Marley, J.M.T. (1969). *Qld. J. Agric. Anim. Sci.* **26**, 303-8.
Mason, M.G. (1968*a*). *J. Dep. Agric. West. Aust.* **9**, 133-40.
———. (1968*b*). *J. Dep. Agric. West. Aust.* **9**, 198-208.
———. (1971). *Aust. J. Exp. Agric. Anim. Husb.* **11**, 662-9.
Mason, M.G., and Cox, W.J. (1969). *Aust. J. Exp. Agric. Anim. Husb.* **9**, 99-104.
Mason, M.G., Loutit, A., Smith, J.A.C., Highman, D., and Stallwood, P. (1970). *J. Dep. Agric. West. Aust.* **11**, 63-4.
Mason, M.G., Parkin, R.J., and Patrick, H. (1969). *J. Dep. Agric. West. Aust.* **10**, 56-7.
Mason. M.G., and Rowley, A.M. (1969). *Aust. J. Exp. Agric. Anim. Husb.* **9**, 630-5.
Mason, M.G., Rowley, A.M., and Quale, D.J. (1972). *Aust. J. Exp. Agric. Anim. Husb.* **12**, 171-5.
Millington, R.J., and Powrie, J.K. (1968). *Trans. 9th Int. Congr. Soil Sci.* **3**, 385-93.
Muir, B.C.M. (1964). *J. Dep. Agric. Vict.* **64**, 348-50.
Myers, L.F., and Lipsett, J. (1958). *Aust. J. Agric. Res.* **9**, 1-12.
Ozanne, P.G., and Howes, K.M.W. (1971). *Aust. J. Agric. Res.* **22**, 81-92.
Ozanne, P.G., Kirton, D.J., and Shaw, T.C. (1961). *Aust. J. Agric. Res.* **12**, 409-23.
Paton, D.F., and Loneragan, J.F. (1960). *Aust. J. Agric. Res.* **11**, 524-9.
Philen, O.T. Jr., and Lehr, J.R. (1967). *Soil Sci. Soc. Amer. Proc.* **31**, 196-9.
Piper, C.S., and de Vries, M.P.C. (1964). *Aust. J. Agric. Res.* **15**, 234-72.
Powrie, J.K. (1963). *J. Dep. Agric. S. Aust.* **66**, 294-7.
Price, G.H. (1971). *J. Aust. Inst. Agric. Sci.* **37**, 343-4.
Rasmussen, T.S. (1969). *Qld. J. Agric. Anim. Sci.* **26**, 475-82.
Reuter, D.J. (1967). *J. Dep. Agric. S. Aust.* **70**, 252-9.
Roberts, F.J., Fitzpatrick, E.N., Toms, W.J., Greenwood, E.A.N., and Oliver, G. (1966). *J. Aust. Inst. Agric. Sci.* **32**, 72-92.
Rudd, C.L. (1970). Dep. Agric. S. Aust. Exp. Rec. No. 5, 28-31.
———. (1972*a*). *J. Aust. Inst. Agric. Sci.* **38**, 63-4.
———. (1972*b*). *Aust. J. Exp. Agric. Anim. Husb.* **12**, 65-9.
———. (1972*c*). Dep. Agric. S. Aust. Exp. Rec. No. 7, 4-9.
Russell, J. (1967). *Tasm. J. Agric.* **38**, 305-8.
Russell, J., and Stackhouse, K.M. (1972). *Tasm. J. Agric.* **43**, 227-31.

Russell, J.S. (1960*a*). *Aust. J. Agric. Res.* **11**, 902-26.
———. (1960*b*). *Aust. J. Agric. Res.* **11**, 927-47.
———. (1963). *Aust. J. Exp. Agric. Anim. Husb.* 3, 319-25.
———. (1967). *Aust. J. Exp. Agric. Anim. Husb.* 7, 453-62.
———. (1968). *Aust. J. Exp. Agric. Anim. Husb.* **8**, 223-31.
Sauchelli, V. (1965). "Phosphates in Agriculture". Reinhold, New York.
Seelinger, M.T. (1968). *J. Dep. Agric. S. Aust.* 72, 90-3.
Shier, F.L. (1962). *J. Dep. Agric. West. Aust.* 3, 698-710.
Sims, H.J. (1959). *J. Dep. Agric. Vict.* 57, 759-65.
———. (1961). *Aust. J. Exp. Agric. Anim. Husb.* 1, 140-3.
———. (1964). *J. Dep. Agric. Vict.* **62**, 1-7.
Sims, H.J., and Andrew, R.A. (1958). *J. Dep. Agric. Vict.* 56, 9-10.
Skene, J.K.M., Hosking, W.J., Leverington, K.C., Russell, J.S., and Williams, C.H. (1966). *J. Aust. Inst. Agric. Sci.* 32, 159-72.
Smith, A.N. (1965*a*). *J. Aust. Inst. Agric. Sci.* 31, 110-26.
———. (1965*b*). *Agric. Gaz. N.S.W.* **76**, 200-3.
———. (1965*c*). *Aust. J. Exp. Agric. Anim. Husb.* 5, 152-7.
———. (1967). *Agric. Gaz. N.S.W.* **78**, 539-40.
———. (1968). *Plant Soil.* 29, 144-55.
———. (1969*a*). *Physiol. Plant.* 22, 371-8.
———. (1969*b*). *Agrochimica.* **13**, 235-42.
———. (1971). *J. Aust. Inst. Agric. Sci.* **37**, 296-302.
Smith, A.N., and Spence, T.B. (1969*a*). *Agric. Gaz. N.S.W.* **80**, 291-2.
———. (1969*b*). *Agric. Gaz. N.S.W.* **80**, 293-4.
Smith, A.N., and Woodward, H.A. (1968). *Agric. Gaz. N.S.W.* **79**, 150-1.
Smith, S.T., and Toms, W.J. (1958). *J. Dep. Agric. West. Aust.* 7, 65-70.
Spencer, K. (1960). *J. Aust. Inst. Agric. Sci.* **26**, 63-7.
———. (1963). *Aust. J. Exp. Agric. Anim. Husb.* 3, 180-3.
Stace, H.C.T., Hubble, G.D., Brewer, R., Northcote, K.H., Sleeman, J.R., Mulcahy, M.J., and Hallsworth, E.G. (1968). "A Handbook of Australian Soils". Rellim, Adelaide.
Stefanson, R.C. (1972). *Aust. J. Soil Res.* **10**, 183-95.
Storrier, R.R. (1962). *Aust. J. Exp. Agric. Anim. Husb.* 2, 185-92.
———. (1965*a*). *Aust. J. Exp. Agric. Anim. Husb.* 5, 317-22.
———. (1965*b*). *Aust. J. Exp. Agric. Anim. Husb.* 5, 323-8.
———. (1969). *Agric. Gaz. N.S.W.* **80**, 76-8.
Swartz, G.L., and White, B.J. (1966). *Qld. J. Agric. Anim. Sci.* 23, 397-406.
Taylor, A.C. (1965). *Aust. J. Exp. Agric. Anim. Husb.* 5, 491-4.
———. (1966). *Agric. Gaz. N.S.W.* 77, 34-7.
Toms, W.J. (1959). *J. Dep. Agric. West. Aust.* **8**, 619-20.
———. (1960). *J. Dep. Agric. West. Aust.* 1, 943-4.
Toms, W.J., and Burvill, G.H. (1961). *J. Dep. Agric. West. Aust.* 2, 321-8.
Toms, W.J., and Fitzpatrick, E.N. (1961). *J. Dep. Agric. West. Aust.* 2, 763-73.
Waring, S.A., and Teakle, L.J.H. (1960). *Aust. J. Agric. Res.* **11**, 27-41.
Warren, J.F. (1971). *J. Aust. Inst. Agric. Sci.* **37**, 303-13.
Watson, E.R. (1963). *Aust. J. Agric. Res.* **14**, 796-807.
———. (1969). *Aust. J. Agric. Res.* **20**, 447-56.
Watson, E.R., and Lapins, P. (1969). *CSIRO Aust. Div. Plant Ind. Field Stn. Rec.* **8**, 17-27.
Webber, G.D. (1964). *J. Dep. Agric. S. Aust.* 67, 354-60.
Weir, R.G. (1967). *Agric. Gaz. N.S.W.* **78**, 486-91.
Weir, R.G., Barkus, B., and Atkinson, W.T. (1963). *Aust. J. Exp. Agric. Anim. Husb.* 3, 314-8.
Wells, G.J. (1969). *Aust. J. Exp. Agric. Anim. Husb.* 9, 521-7.
Wild, A. (1958). *Aust. J. Agric. Res.* 9, 193-204.
Williams, C.H. (1969). *Aust. J. Soil Res.* 7, 307-16.
———. (1971*a*). *Aust. J. Soil Res.* 9, 83-94.
———. (1971*b*). *Aust. J. Soil Res.* 9, 95-106.
Williams, C.H., and David, D.J. (1973). *Aust. J. Soil Res.* **11**, 43-56.
Williams, C.H., and Donald, C.M. (1957). *Aust. J. Agric. Res.* **8**, 179-89.
Williams, C.H., and Lipsett, J. (1969). *Aust. J. Agric. Res.* **20**, 265-78.
Williams, C.H., and Simpson, J.R. (1965). *Aust. J. Agric. Res.* **16**, 413-27.
Woodroffe, K., and Williams, C.H. (1953). *Aust. J. Agric. Res.* **4**, 127-50.

CHAPTER 13

Crop sequential practices

A.L. Clarke and J.S. Russell

INTRODUCTION

Permanent crop production in any environment requires a stable ecosystem. In particular soil fertility needs to be maintained. Under semi-arid conditions special problems arise, for aridity results in low and variable yields, in reduced soil cover and hence increased risks of wind and water erosion, and in difficulties in growing soil-restoring species such as legumes.

One critical factor affecting the stability of ecosystems is the feed-back relationship—shown in figure 1—between crop yield and soil organic matter. Soil organic matter levels initially affect plant yield; plant residues in turn affect the level of soil organic matter. Eventually equilibrium is reached between the amount of residues returned and soil organic matter level. Long-term trends in soil organic matter and plant yield are related and indicate the direction of change towards an ultimate equilibrium.

One practice that has evolved in semi-arid areas is fallowing—that is, leaving the soil cultivated to conserve water for use by a following crop. Although this practice increases crop yield, reduces yield variability, and allows cropping to extend into drier areas, it also promotes more rapid decline in soil organic matter, which in the long term has to be balanced against the immediate yield advantages.

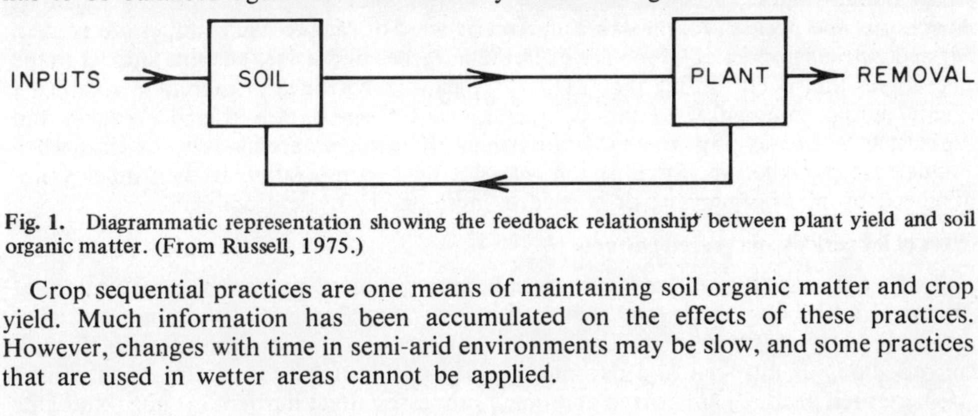

Fig. 1. Diagrammatic representation showing the feedback relationship between plant yield and soil organic matter. (From Russell, 1975.)

Crop sequential practices are one means of maintaining soil organic matter and crop yield. Much information has been accumulated on the effects of these practices. However, changes with time in semi-arid environments may be slow, and some practices that are used in wetter areas cannot be applied.

THE EFFECT OF SEQUENTIAL PRACTICES ON SOIL CHEMICAL PROPERTIES AND NUTRIENT CYCLING

Different crop systems may affect soil chemical properties in a number of ways. Firstly specific plants may affect nutrient cycles by removing, adding or distributing nutrients. Secondly the tillage practices used in the culture of certain crop plants may modify the rate of loss of organic matter and nutrients. Thirdly the inclusion of ley periods and grazed pastures affects both nutrient cycling and organic matter levels.

Effect of nutrient removal by crop plants

The removal of nutrients is related to the yield and nutrient concentration. Plant yield in semi-arid areas is generally less than in wetter areas. Concentration of nutrients may be higher than in more humid areas but generalization on a climatic basis is not possible because soil factors are also important. The yield and nitrogen, phosphorus and potassium concentration of Gabo wheat grain sampled at a number of centres in South Australia with differing annual rainfall are shown in table 1. Low yield does not necessarily imply high nutrient concentration.

Table 1. Nutrient removal in Gabo wheat grown at six centres in South Australia in 1957

Site	Annual rainfall mm	Yield of wheat kg/ha	Concentration of nutrients in grain (%)			Nutrient removal in grain (kg/ha)			Soil levels (0–15 cm) kg/ha*		
			N	P	K	N	P	K	N	P	K
Wanbi	310	868	2.21	0.20	0.27	19.2	1.74	2.34	450	87	248
Minnipa	350	800	2.26	0.18	0.27	18.1	1.44	2.16	2,010	220	1,040
Roseworthy	430	1,076	2.03	0.18	0.28	21.8	1.94	3.01	2,740	410	840
Jamestown	450	796	1.65	0.19	0.27	13.1	1.51	2.15	1,160	420	460
Turretfield	460	716	1.61	0.29	0.33	11.5	2.08	2.36	1,610	514	1.150
Kybybolite	510	2,454	2.53	0.27	0.36	62.1	6.63	8.83	1,810	180	214

SOURCE. From J. S. Russell, unpublished.

* N, P as total, K as exchangeable K.

Although the amounts of nutrients removed are low in relation to soil levels, in the case of phosphorus they can be significant. The gradual decline in wheat yields over the period 1850-1900 in Australia has been attributed partly to the removal of phosphorus in grain (Donald 1963). The large increases in yield following the subsequent use of applied phosphorus support this hypothesis.

Effect of tillage practices

Tillage practices can influence nutrient loss in two ways. Firstly the frequency and intensity of cultivation can affect the rate of breakdown of organic matter. This aspect is covered more fully in chapter 11.

Secondly the extent of nutrient depletion depends to some extent on how the crop residues are treated. When residues are burnt to facilitate cultivation for the next crop, nitrogen and sulphur are lost as gases. Minerals that remain in the ash, such as phosphorus and potassium, may be removed by wind or rain. When residues are grazed, ingested nutrients, most of which are excreted in faeces and urine, become subject to the same losses that occur during the grazing of pastures. Nitrogen in excreta is volatilized rapidly during summer when the stubbles of winter cereals are grazed (Watson and Lapins 1969). Losses of nutrients, and particularly nitrogen, are likely to be least when residues are ploughed in, for nitrogen released by decomposition is then quickly immobilized by microorganisms or sorbed as ammonium by soil colloids.

Effect of ley periods and grazed pastures

In a soil left uncultivated and with a volunteer or sown pasture several important processes occur. Organic matter accumulates at the soil surface. Nutrients including nitrogen, phosphorus and sulphur are chemically incorporated into this material. Significant additions of nitrogen and carbon can also occur under favourable conditions.

Where such pastures are grazed additional processes affect nutrient cycling. Nutrients absorbed by plants are continually being returned to the pasture as faeces and urine, in forms which can be re-absorbed. Such recycling has been shown for sulphur by Till and co-workers (for example, Till and May 1971), by labelling soil sulphur with ^{35}S. Although relatively few nutrients are removed in livestock products, large quantities of some elements, notably nitrogen, are lost from excreta during the recycling process.

Of the elements likely to be significant in semi-arid areas, the most important are nitrogen, phosphorus, sulphur and potassium.

Nitrogen

Nitrogen supply is much affected by mechanisms effecting both accretion and loss. Williams (1962) listed a number of factors—environmental, nutritional and managerial—that are likely to influence nitrogen accretion.

Accretion. Of the environmental factors climate and soil determine the pasture species that survive and fix nitrogen effectively. For example, soil characteristics largely determine whether annual *Medicago* species grow satisfactorily on cereal land in northern South Australia (French, Matheson and Clarke 1968). In semi-arid winter rainfall areas in Australia the most satisfactory species are self-regenerating winter annuals, such as *Trifolium subterranean*, *T. hirtum*, *Medicago truncatula* and *M. littoralis*. As conditions become more arid many cultivars of these species cannot persist, and mechanisms such as hard-seededness become important in survival. In the semi-arid summer rainfall areas most crops are grown on fertile black earths, grey clays and red earths. Lucerne (*M. sativa*) can grow and persist for short periods on some of these soils but no self-regenerating tropical legumes that will persist on the heavy clay soils, such as the "brigalow" (Isbell 1962), have yet been found. On the other hand perennial tropical legumes such as *Macroptilium atropurpureum* have persisted in areas of light soil with rainfall as low as 625 mm, and self-regenerating annuals such as *Stylosanthes humilis* can persist on the poorer soils in the lower latitudes.

Nutritional factors in part determine the yield of legume in a sward, and so influence accretion, for total legume yield determines the amount of nitrogen added to the soil-plant system through *Rhizobium* fixation. Watson (1963) found nitrogen accumulation in the soil increased linearly with the amount of nitrogen present in clover herbage grown on the soil. The incorporation of fixed nitrogen into stable organic compounds, which have more or less constant C:N:S:P ratios (Williams and Steinbergs 1958) requires an adequate supply of inorganic sulphur and phosphorus, and hence lack of these nutrients may limit the process. Williams and Donald (1957) concluded that phosphate supply limited clover growth, and hence gross nitrogen fixation, but nitrogen incorporation was limited by the availability of inorganic sulphur, since the sulphur source, superphosphate, has a lower ratio of sulphur to phosphorus than soil organic matter. Watson (1969) also considered that the lack of sulphur, caused by the leaching of sulphur excreted by grazing animals, may have restricted nitrogen build up in a sandy soil. Russell (1960) found that nitrogen accumulated in proportion to "sheep grazing days" and so presumably to pasture growth.

In newly sown leys early legume growth is often sparse. Penman (1949) considered this caused slow nitrogen accumulation during the early years of a ley at Rutherglen, Victoria, and suggested long leys (eight-ten years) alternating with several years of intensive cropping. Goode (1963) established clover pastures rapidly with high seeding rates and achieved very high rates of nitrogen accumulation (see table 5). Watson (1963) recommended short leys (two years) alternating with one crop sown without fallow, and found that under this system subterranean clover volunteered profusely in the year following the crop. On "mallee" soils in South Australia a similar sequence, involving alternate years of cereal crop and medic pasture, is practised (French, Matheson and Clarke 1968).

Of the managerial factors grazing intensity markedly affects the botanical composition of a pasture. Low grazing pressure allows tall-growing grasses to shade legumes, whereas heavy grazing suppresses annuals such as subterranean clover and medics by reducing seed production. Best subterranean clover growth occurs at moderate grazing intensity. However, there is no clear evidence that nitrogen accumulation is affected similarly. McLachlan (1968) found that the total nitrogen content of soil under a pasture that had been heavily grazed for fifteen years (0.164 per cent) significantly exceeded that where grazing had been light (0.101 per cent). But Simpson, Bromfield and Jones (1974) observed the opposite: annual nitrogen accretion, during four years, at high and low stocking rates was 45 and 58 kg/ha respectively.

Losses. Most of the nitrogen ingested by livestock is returned to the soil surface as urea and ammonium nitrogen in urine (Whitehead 1970). Urine raises soil pH and this can lead to rapid volatilization of ammonia (Doak 1952; Watson and Lapins 1969) especially during warm, dry weather. Watson and Lapins found that up to 54 per cent of nitrogen applied as urine to ryegrass in pots was volatilized in fifty-four days; losses were most rapid during summer and were stimulated by periodic wetting which promoted urea hydrolysis. Under milder conditions that permit continuous pasture growth, less volatilization may occur. Thus Walker, Orchiston and Adams (1954) considered losses from urine in New Zealand do not exceed 20 per cent. Losses through leaching and denitrification may also be greater under urine patches than elsewhere, and are increased by heavy rain.

Nitrogen in faeces is likely to be lost through volatilization unless it is incorporated into the soil. Gillard (1967) measured 80 per cent nitrogen loss from faeces dried in the sun. The rate of incorporation in the soil depends largely on the activity of soil fauna (Barley 1959*b*; Gillard 1967).

Nitrogen may also be "lost" during grazing because excreta are distributed unevenly within the paddock. Nutrients accumulate in stock camps at the expense of the rest of the grazed area (Hilder 1964). Also in any one year excretions affect only a fraction—usually less than 10 per cent with cattle—of the paddock area (Petersen, Woodhouse and Lucas 1956).

Because of such losses the accumulation of soil nitrogen may be no faster when a pasture is grazed than when the herbage is cut and removed. Watson and Lapins (1964) found the difference in accumulation over two years—0.017 per cent under grazing and 0.009 per cent under cutting—was not statistically significant, and concluded that clover butts, roots and burrs are the main source of soil nitrogen. They considered, in fact, that grazing animals tend to deplete soil organic matter by excreting nutrients in forms that are readily lost. A more likely general situation is that some of the returned nitrogen is retained in the soil, and accumulation is thereby enhanced.

Other nutrients

Much less consideration has been given to changes occurring under pasture in the status of nutrients other than nitrogen. On many soils in Australia the application of superphosphate ensures an adequate supply of phosphorus and sulphur for crops. During a pasture period, these nutrients accumulate in organic matter approximately in proportion to the accumulation of nitrogen. Values of the ratio of C:N:S:P in soil organic matter, determined by Williams and Steinbergs (1958) for some Australian soils used to grow cereals, are given in table 2. Taking the values for "all soils", it can be calculated that accompanying an annual nitrogen increment of 75 kg/ha* there are increments also of 1125 kg/ha of organic matter, 9.5 kg/ha of organic sulphur and 5.0

Table 2. Mean ratios of C:N:S:P in the organic matter of various Australian surface soils (Data from Williams and Steinbergs 1958)

Soil group	No. of soils analyzed	C	N	S	P
Podzolic	56	157	10	1.24	0.58
Red-brown earth	34	151	10	1.26	0.70
Black earth	9	156	10	1.23	0.80
"Mallee"	7	148	10	1.72	0.36
Grey and brown soil of heavy texture	4	114	10	1.89	0.54
All soils	155*	150	10	1.26	0.66

* Including 45 soils not listed above.

* Approximately the mean for legume-grass pastures quoted in table 5.

kg/ha of organic phosphorus. During a pasture period soil organic matter is therefore a major "sink" for both sulphur and phosphorus. As both nutrients are released again during the cultivation phase–sulphur at a rate matching that of nitrogen, and phosphorus somewhat slower–organic matter then serves as a reserve of these nutrients.

Potassium can become deficient depending on the initial concentration in the soil, the rate of plant removal and the rate of replenishment of exchangeable potassium by non-exchangeable forms. With pastures two factors can influence potassium deficiency. Firstly, the removal of hay or green fodder can rapidly reduce exchangeable potassium levels. Secondly, as the work of Hilder (1964) shows, grazing can result in the concentration of potassium in camping areas.

The accumulation of organic matter under pasture increases soil cation-exchange-capacity (Williams and Donald 1957; Russell 1961) and the reverse occurs under cultivation (Williams and Lipsett 1961). The exchange sites concerned may be occupied by cations supplied with fertilizers or brought to the surface by plant roots, or by H^+ ions which decrease soil pH. However, under most circumstances the changes so induced are not likely to be quantitatively significant in land that is sequentially cropped.

THE EFFECT OF SEQUENTIAL PRACTICES ON SOIL PHYSICAL PROPERTIES

Many top-soils show marked deterioration in physical properties as their organic matter is depleted by cultivation, but plant communities, and in particular dense pastures, are known to improve these properties. In addition, plant cover helps to preserve existing favourable structure by protecting the soil against rain and wind.

Processes of structure improvement

Greenland (1971) recognized two processes in structure improvement under pasture. Firstly, individual particles of sand and silt, and clusters or "domains" of clay particles, are moulded into aggregates, and, secondly, these aggregates are stabilized by bonding agents. Russell (1971) noted a third, namely, the improvement of the soil's tilth, that is its tendency to remain friable rather than become sticky when wet. As this process is very slow, taking decades for completion (Russell 1971), we can neglect it when considering the effect of short-term leys grown in sequential crop systems.

Moulding of aggregates

The arrangement of soil particles into aggregates, with pore spaces between, results from forces exerted during swelling and shrinking as the soil wets and dries, from compression caused by roots penetrating or expanding in fine pores, and from moulding as the particles pass through the alimentary tracts of earth worms and other soil fauna. The relative importance of these aggregating agents undoubtedly varies with soil texture, depth in the soil, and many other factors, but little quantitative information is available. Greenland (1971) considered that the growth of roots alone cannot cause the increases in aggregate porosity that have been observed under pasture, and ascribed a major aggregating role to soil fauna. However, Barley (1959*b*) found that under productive pasture at Waite Institute only 2 per cent of the soil's surface was ingested by earth worms annually. Unless other soil fauna are much more active, faunal action will be too slow to have large aggregating effects in short-term leys.

As roots are important agents in drying surface soil, and faunal population and activity are related to the supply of plant material (Barley 1959*a*; Hutchinson and King 1970), it is clear that the growth and distribution of roots have a major influence on aggregation under pasture.

Another possibility, as yet unexplored, is that treading by grazing animals may influence the rate of formation or the physical properties of aggregates in surface soil. The investigations of Edmond (1963) and Gradwell (1968) in New Zealand, in which heavy treading by a confined flock of sheep was used to simulate much longer periods of treading in the field, are not relevant to semi-arid conditions. Here compression of wet soil by treading is commonly followed by drying and localized cracking. The physical effects of treading may in some respects resemble those of tillage or freezing and thawing.

Stabilization of aggregates

The stabilization of soil aggregates has been discussed by Harris, Chesters and Allen (1966), Greenland (1971) and Russell (1971). It is agreed that polysaccharides of high molecular weight are the most important stabilizers of arable soils and those under short-term leys. These substances are produced in pores, both within and between the aggregates, by bacteria that are small enough to live in the micro-pores concerned. By virtue of their long chain molecular structure and their strong adsorption to soil particles, especially clay, the polysaccharides bind the aggregates into water-stable entities. However, the substances, which are produced mainly by the decomposition of roots, themselves undergo further microbial breakdown, and so new polysaccharide must continually be produced for aggregates to be kept stable. Hence for maximum stabilizing effect there needs to be an adequate supply of roots that are readily accessible to bacteria living in fine soil pores. This perhaps explains why Clarke, Greenland and Quirk (1967) found that soil was stabilized more effectively under both rye grass (*Lolium rigidum*) and *Phalaris tuberosa* when their growth was increased by nitrogen fertilizer, and rye grass, with its more finely branched roots was more effective than *Phalaris* (fig. 2).

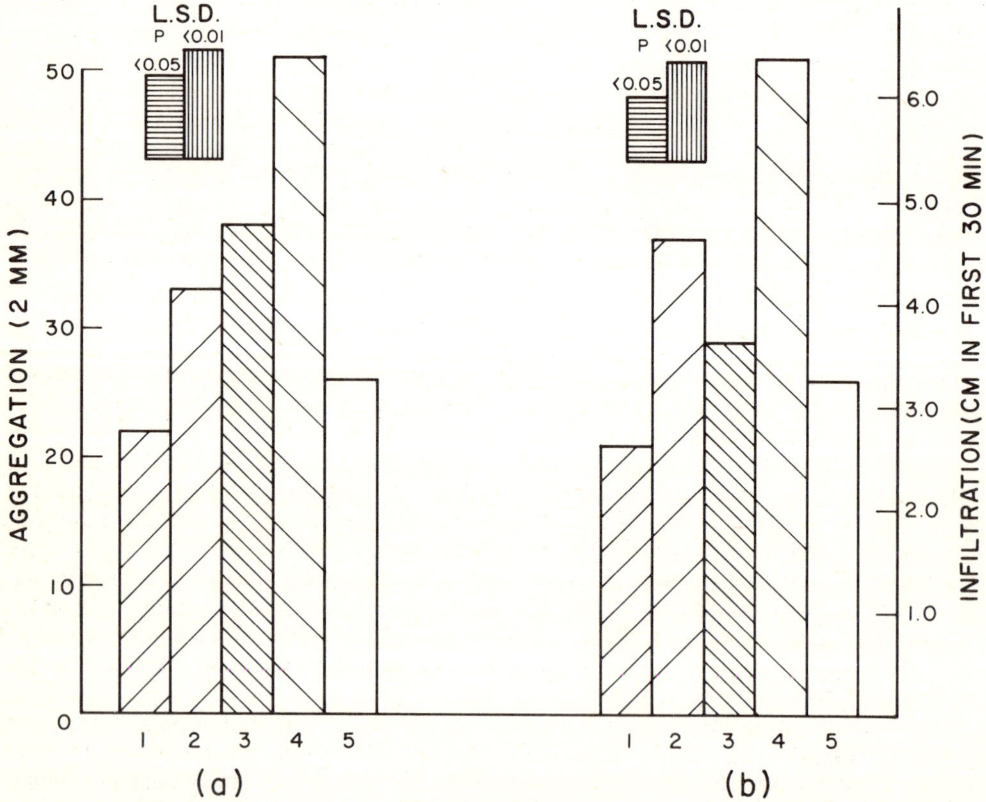

Fig. 2. (*a*) Aggregation (2 mm) and (*b*) infiltration of water measured with a ring infiltrometer, after three years of pasture. 1, *Phalaris*, no nitrogen; 2, *Phalaris* + 87 kg/ha nitrogen annually; 3, Wimmera rye grass, no nitrogen; 4, rye grass + nitrogen; 5, subterranean clover. (Reprinted , by permission, from Clarke, Greenland and Quirk 1967.)

Other organic compounds besides polysaccharides also stabilize aggregates. Greenland, Lindstrom and Quirk (1962) demonstrated that such materials accumulate under long-term pastures, for the removal of polysaccharides from old pasture soils by treatment with sodium periodate caused relatively little reduction in aggregate stability. The hyphae of some fungi also bind aggregates together by enmeshing them (Swaby 1949) or by stabilizing them with mucilage (Bond and Harris 1964; Harris, Chesters and Allen 1966). The presence of fine calcium carbonate, as in some solonized brown soils, and high concentrations of aluminium and iron hydroxides in krasnozems, also have marked effects on aggregate stability. Although there may be important interactions between these components and organic compounds, the structure of these soils is less sensitive to changes in organic matter content than that of other soils.

Soil texture also influences aggregate stability. Soils with high silt and fine sand contents, such as the surface layer of many Australian red-brown earths, are particularly prone to structural breakdown when intensively cultivated (Greenland 1971).

Observed changes in structure due to sequential practices

Effect of pasture

Changes in the structure of a soil under pasture are reflected in changes in its porosity or density. The bulk density under high-yielding pasture is generally lower than under sparse pasture. The difference is clearly related to the accumulation of organic matter, as Russell (1960) showed for old pasture plots at Kybybolite, South Australia. Comparing treatments in this experiment, bulk density (0-6 cm) decreased by 0.02 g/cm^3 for each 100 kg/ha accumulation of nitrogen (0-15 cm).

A period of pasture also makes individual aggregates more porous (Clarke, Greenland and Quirk 1967). Aggregates from virgin soils or old pasture have a higher proportion of large pores which are drained at moderate suctions; such soils are less sticky when wet (Russell 1971). Pores left between aggregates when roots decay may be important in transmitting water to the subsoil. Figure 2, taken from Clarke, Greenland and Quirk (1967), shows the effect of three years of various pasture species and nitrogen treatments on the infiltration rate of a loamy fine sand. For both grasses infiltration rate increased linearly ($P < 0.001$) with increasing yield of dry matter during the preceding three years, suggesting that old root holes served as channels for water flow. The effect of improved pasture in increasing infiltration under field conditions has been clearly shown by run-off experiments in New South Wales at Wellington (Logan 1960*a*) and Wagga Wagga (Jones 1961).

As organic matter accumulates in the soil the quantity of water held at a given suction usually increases. The "available water capacity" is thereby increased if the water content at field capacity increases more than that at wilting point. Russell and Shearer (1964) concluded that half of the additional water held by a sandy soil under an improved pasture, compared with unimproved, was unavailable to plants; the increase in available water capacity, after forty years, was only 3.8 mm. A larger increase may have been recorded had a lower suction–say 0.10 bar instead of 0.33 bar–been used to assess field capacity. Nevertheless, it seems unlikely that available water retention can be significantly increased in arable soils by short pasture leys.

Effect of tillage

In practice tillage is used to prepare a satisfactory soil structure for seedling growth. Certainly this is much easier if initially the soil is composed of clods which crumble readily into aggregates of a desirable size (mostly 1-5 mm), but with skilful use of implements an adequate seed-bed can be prepared even from poorly structured soil. The main concern, however, is whether the seed-bed will retain its structure through the life of the crop, and particularly during the germination and emergence of the seedlings. If the aggregates present after tillage are unstable, they slake when beaten by rain-drops to produce a dense surface crust. As the permeability of this crust may only by $\frac{1}{2,000}$ of that of the underlying soil (McIntyre 1958), the entry of water and air and the emergence of seedlings are impaired (Millington 1959). If the structure is particularly weak and rainfall is heavy, most or all of the seed-bed eventually collapses, producing a dense mass which is too coherent to permit satisfactory penetration by roots (Barley and Greacen 1967).

Soil protection

A dense cover of crop or pasture protects soil from the impact of rain drops and the erosive action of wind. Traditional cultivation removes this cover and exposes bare, loose soil. The period of exposure varies from a few weeks, when a quick-growing crop is sown immediately after the soil is first cultivated, to nearly a year when the crop is preceded by bare fallow. During this period the soil is highly vulnerable to erosion—see, for example, Logan (1960*a*, 1960*b*) and Jones (1961).

Practices relating to plant residues

Farming techniques, and particularly methods of tillage, have been developed to reduce the period of soil exposure by leaving plant residues as a cover on the soil's cultivated surface. The practice is known as "stubble mulching" or "trash farming". Tillage implements have been developed to cultivate and sow without burying residues. Trash farming has received greatest attention in North America, where the practice has been extensively reviewed by McCalla and Army (1961).

In Australia, too, trash farming has been widely recommended (see, for example, Herriot 1954). However, the practice has not been critically examined under Australian conditions, and although it has been adopted successfully by some farmers it has not gained general acceptance. Its disadvantages are, firstly, the practical difficulty of cultivating and sowing through the straw and the cost of special equipment required to do so; and secondly, the lower grain yield that may result when straw is present, due to the physically poorer seed-bed, induced nitrogen deficiency (Littler 1964), or toxic effects of rotting straw (Kimber 1967). Also, when a second crop is grown, without fallow, the period of soil exposure is so brief that the erosion risk is not high, even if the stubble is destroyed. Furthermore, as this crop may be under-sown with legume pasture, removal of the straw to permit seeding into a firm, fine seed-bed may be justified, as this leads to better pasture establishment and hence a faster increase in soil fertility.

Stubble mulching is likely to prove most successful where land still carrying stubble is fallowed for another crop. This happens when a succession of wheat crops is grown on fallow after an extended period of pasture. Here the stubble will persist as a protective mulch for several months, but will be dispersed in time for seeding. Stubble mulching is also of value in areas such as the Darling Downs where summer and winter crops are grown in succession.

Stubble mulching has been shown in many North American experiments to reduce soil losses by wind and water erosion. Further investigations of the practice in Australia on soils where maintenance of a desirable structure is difficult may well be justified.

Use of particular species

A special example of the use of plant communities to protect soil is the practice of growing cereal rye to stabilize sand dunes in the "mallee" lands of southern Australia. Of the cereals, rye is by far the most resistant to sand-blast. The practice is to sow rye, with light dressings of superphosphate and ammonium sulphate, on the upper dune slopes and wheat or barley on lower slopes and inter-dune flats (e.g. Blencowe 1957). With fertilizer the rye soon grows high enough to stabilize the sand. It produces a large bulk of tough and unpalatable straw which persists for two or even three years. Not only the dunes but the flats between are protected from the effects of erosion.

THE LONG-TERM EFFECT OF CROP SEQUENCES ON CROP YIELDS AND SOIL PROPERTIES

The overall effect of a crop sequence on soil fertility is the sum of the effects of individual sequential practices. For a sequence to be sound in the long term, practices that exploit fertility, such as cultivation and nutrient removal, need to be balanced by restorative practices, such as the growth of pastures and the addition of fertilizer. Nye and Greenland (1960) have shown that in the tropics the apparently wasteful system of shifting agriculture, in which arable cropping alternates with periods under scrub, can be balanced and moderately productive. In temperate regions, such as northern Europe, balanced rotations have evolved in which arable and livestock farming are integrated and soil fertility is maintained.

In semi-arid regions practices that are satisfactory in humid temperate regions can not be readily adopted. This is particularly so in subtropical areas where the soils, crops and climate are far removed from temperate agriculture. In these areas sequential crop systems are still evolving and there is still much to be learned about the maintenance of soil fertility and the development of economic practices.

Both trends in yield and in soil properties need to be considered in assessing crop sequences. It is possible, for example, for crop yields to be maintained for a period using exploitative practices, such as fallowing and frequent cultivation, but such systems are unstable in the long term because they cause rapid decline of soil organic matter.

As agricultural ecosystems are complex and crop yields and soil properties change only slowly, it is often difficult to isolate specific factors affecting trends. These factors are most easily identified where yields increase rapidly, or where large and specific inputs, such as nitrogen, are added to the system, and these circumstances rarely apply in semi-arid environments.

Information on the effect of cropping sequences on crop yield and soil properties is best obtained by analyzing long-term trends within agricultural ecosystems and the results of long-term sequential experiments.

Analysis of trends within agricultural ecosystems

Yield trends

Various mathematical procedures are available for the time-series analysis of long-term yield trends on a local or national basis. Although it is possible to identify major trends it is frequently difficult to isolate causal factors, due to the possibility of changes in the area of crops in the statistical divisions under consideration, and to changes in cultural practices.

Cornish (1949) calculated yield trends for commercial wheat crops growing in "hundreds" in South Australia. Considering the period 1896-1941, he showed that when land receiving superphosphate was cropped intensively, wheat yields initially increased (phase 1), then became relatively constant (phase 2), and in some instances finally declined (phase 3). He considered that phase 1 reflected the cumulative benefits of phosphate application and improvements in cultivars and cultural techniques, and phases 2 and 3 the increasing effects of a more severely limiting nitrogen supply. The length of phases 1 and 2, that is the time taken for wheat yields to begin to decline, depends on how long it takes to deplete initial soil nitrogen to a critical level. Cornish found this varied from twenty years, for sandy and initially poor soils, to over fifty years for heavier and inherently richer red-brown earths. He prepared 10 type curves to describe yield trends evident in the 296 hundreds surveyed, and related these curves to major soil groups.

Donald (1963, 1965) reviewed trends in wheat yields on an Australia-wide basis. He produced a graph (fig. 3) depicting national yield trends for the period 1870-1960, which

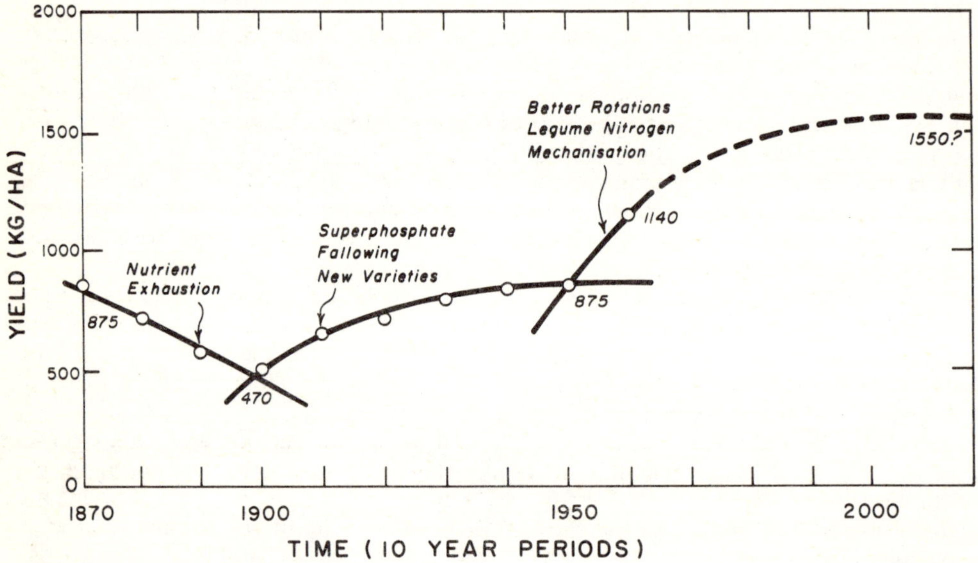

Fig. 3. Trends in Australia wheat yields (decennial means) during the past century. (Reprinted, by permission, from Donald, 1963.)

agreed closely with those given for South Australia by Cornish (1949), and, like Cornish, he considered the processes of exhaustion and later replenishment of soil phosphate and nitrogen to be dominating factors in determining these trends. On the other hand, on the basis of New South Wales data, Warren (1969) contended that changes in the distribution and quality of the land being cropped, caused by economic factors, have much larger effects on regional, state or national wheat yields than nutrient or genetic factors. Although the analytical methods used by Warren were insensitive, it is clear that when and where rapid changes in the area being sown occur marked trends in yield may result.

There is evidence (Donald 1963, 1965), that widespread increases in yield resulted during the 1950s following the adoption of longer ley periods and the increased growth of legumes. Donald considered that a new positive trend had thus been established which he speculated could lead to a further substantial rise in national wheat yields (fig. 3). However it is also true that the 1950s were a period of above average rainfall. Certainly within restricted areas marked increases in yield have been recorded since legume leys have been successfully established. French, Matheson and Clarke (1968) observed that wheat yields were 540-860 kg/ha higher in the 1950s than in the 1930s where annual *Medicago* spp. had become well established ("medic" areas), but where legume growth was generally weak ("non-medic" areas) increases were mostly less than 270 kg/ha. They attributed the contrast primarily to soil fertility effects. Goode (1963) quoted a farm-scale example showing very large effects of legume leys on cereal yield.

Russell (1973) examined crop yield trends in several environments. National yield trends in dryland Australian crops in recent years have been much smaller than those in the United States. For example, linear trend values as high as 100 kg/ha/year were shown for sorghum in the United States for 1944-68, whereas the value for Australia for the same period was only 12 kg/ha/year. The low value for Australia is partly related to the large expansion in area of this crop, especially in drier areas; in more favourable areas, such as the eastern Darling Downs of Queensland, linear trends as high as 50 kg/ha/year were recorded. The year to year variability of Australian crops was considerably greater than that observed in the United States.

This study also showed that significant yield trends in wheat and barley yields had occurred in some areas in Australia during 1935-67 but rates of increase were generally less than 20 kg/ha/year. On Yorke Peninsula, South Australia, in fifty-one years (1917-67) the yield of wheat increased only slightly faster than that of barley. As there were marked genetic changes in wheat during this period, but not in barley, the results suggest that the increased yields were related to soil, nutrient and management factors rather than genetic changes.

J.S. Russell (1973) emphasized the importance of the genotype-nutrient interaction as a factor increasing yields of non-leguminous crops in humid environments. However, this interaction is less significant in semi-arid environments because water limitation and year to year variability reduce the potential maximum yield.

Soil trends

It is difficult to obtain precise quantitative data on soil trends in large agricultural systems. Treated and virgin soils can be compared, but the difficulty of obtaining enough comparisons for a comprehensive analysis, and uncertainties regarding supposedly virgin samples, restrict the use of this technique. Data from long-term experiments are more reliable: comparative treatments of sampled areas are more accurately known, and sampling and analytical procedures are more readily standardized.

Analysis of sequential experiments

Most of the long-term sequential experiments in Australia have involved areas of winter rainfall and wheat has been the major crop of these systems. In the semi-arid summer rainfall areas, where sorghum is the most important crop, long-term experiments with this crop are in progress, but few data are yet available.

Yield trends

Long-term sequential experiments in Australia have attempted to answer four main questions:

(1) The effects of fallow on crop yield and soil properties.

(2) The effects of pasture periods of varying length and nature on yields of crop and pasture and on soil properties. The pastures have included annual legumes of the *Trifolium* and *Medicago* genera, biennial legumes such as *Lablab purpureum*, and perennial legumes such as *Medicago sativa*. They have also included short-term grass species such as Wimmera rye grass (*Lolium rigidum*) and *Sorghum almum*.

(3) The effects of fertilizer practice, principally the amount of superphosphate.

(4) The effects of different crop sequences, including other cereals such as oats and barley but also non-cereals such as field peas.

Surprisingly there has been much less interest in long-term comparisons of cultivation practices or of the effect of handling residues by mulching or burning. This is in contrast to North America where many experiments on these aspects have been carried out (Hume 1940; Oveson and Hall 1957).

There have been other pioneering experiments at Katherine in the Northern Territory on some of the effects of continuous cropping of peanuts and cotton (Arndt 1961*a*, 1961*b*; Phillips and Norman 1962).

A summary of the yield effects of sequential crop experiments in southern Australia is shown in table 3. Mean yields of wheat after fallow are invariably higher than under continuous wheat, and in long-term experiments often more than three times as high. Generally wheat in fallow-wheat-pasture systems yields more than in fallow-wheat systems.

Table 3. The effect of cropping sequence on the mean yield of wheat

Location*	Mean yield of wheat (kg/ha)†						Reference‡
	WW	FW	FWO	FWp	FW pp	PW	
Werribee, Vic. (35)	400	1,100	–	1,210	–	–	1
Rutherglen, Vic. (29)	460	1,540	–	1,450	–	1,300	1
Longerenong, Vic. (31)	640	2,110	2,320	2,310	–	–	1
Booborowie, S.A. (13)	–	1,800	1,920	1,850	–	–	2
Waite Institute, S.A. (46)	770	1,870	2,100	2,400	–	1,520	3
Chapman, W.A. (16)	360	540	–	770	1,070	–	4
Wongan Hills, W.A. (16)	310	870	–	1,160	1,500	–	4
Walpeup, Vic. (27)	–	1,510	1,700	1,710	1,770	–	5

* Period of trial in years in brackets.
† W = wheat, F = fallow, O = oats, p = pasture, P = peas.
‡ 1, Sims and Jardine (1949); 2, Wark (1942); 3, Waite Agricultural Research Institute (1972*a*); 4, Drover (1956); 5, Elliott and Jardine (1972).

Experiments have shown that continuous cropping of wheat leads to a rapid decline in grain yield with time. Decreases in yield under continuous cropping can result from increasing weed and disease infestations as well as from nutrient depletion.

Yield trends under fallow-wheat were of considerable interest in Australia in the 1930s because this sequence was used very commonly in wheat growing. Analyses of various long-term experiments (Forster 1939; Wark 1942; Woodroffe 1949; and Sims and Jardine 1949) showed little yield trend for wheat grown on fallow-wheat rotations. Yields of stubble-sown crops of oats and barley, however, tended to decline, indicating a decreasing capacity of the soil to provide mineral nitrogen without a period of fallow.

Elliott and Jardine (1972) found a positive, near linear increase in yield with all sequences in experiments at Walpeup in the Victorian Mallee during the first fifteen years of the 1940-68 period. But after 1955, yields under fallow-wheat declined (fig. 4). It is possible that the increase in yield was associated with residual effects of superphosphate applications (McLelland 1968). Although yields for sequences diverged slowly, approximately in the order that could be explained by the effects of rotations on soil nitrogen, the limited soil data were inconclusive and the reasons for the decline of yield under fallow-wheat are unexplained. Nitrogen availability is unlikely to have been a factor in the early years of the experiment, and later yield differences between the fallow-wheat and the fallow-wheat-wheat systems seem much too large for this explanation.

One of the oldest long-term sequential experiments in Australia is at the Waite Institute. Orthogonal polynomials have been fitted to yield trends of wheat over the period 1925-71 (table 4). Significant declines in the yield of wheat have occurred in continous wheat and fallow-wheat sequences (fig. 5). Although wheat yields have also decreased in the fallow-wheat-pasture and fallow-wheat-oats-pasture sequences the decreases are less and the coefficients are not significant. In the fallow-wheat sequence yields over the first twenty years were maintained at a high level but have since decreased markedly; on the other hand yields have been substantially maintained in the fallow-wheat-pasture and fallow-wheat-oats-pasture sequences.

In recent years the question of the optimum balance between periods of pasture ley

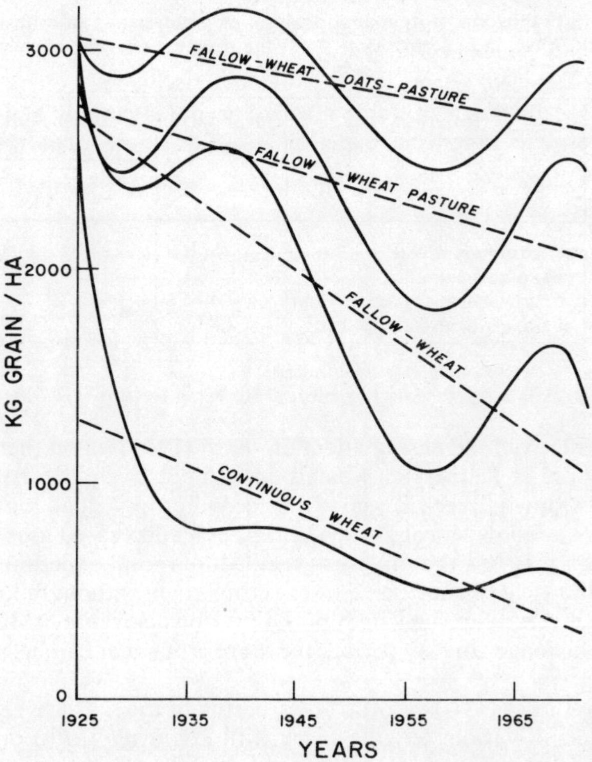

Fig. 4. Yield trends of wheat from several rotations in the permanent rotation experiment at the Waite Institute.

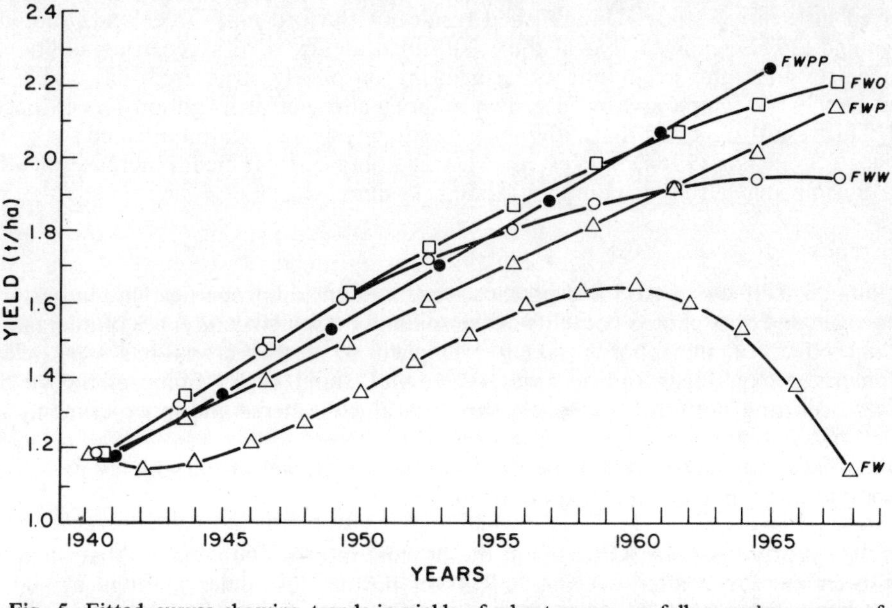

Fig. 5. Fitted curves showing trends in yields of wheat grown on fallow under a range of cropping sequence; the site has previously grown only three crops F = fallow, W = wheat, O = oats, P = pasture. (Reprinted, by permission, from Elliott and Jardine 1972.)

Table 4. Orthogonal polynomial analysis of yield trends of wheat in the long-term rotation experiment at the Waite Institute over the 47-year period 1925–71

Crop	Rotation	Mean yield kg/ha	Linear trend kg/ha/yr	
			Value	Standard deviation
Wheat	Continuous wheat	808	−21.4**	5.9
Wheat	Fallow-wheat	1,882	−35.2***	7.3
Wheat	Fallow-wheat-pasture	2,402	−14.9	9.5
Wheat	Fallow-wheat-oats-pasture	2,852	−9.3	9.3

Significance ** $P < 0.01$, *** $P < 0.001$.
† Based on data from Waite Agricultural Research Institute (1972*b*).

and cropping has received increasing attention. Bath (1949) found that when old subterranean clover pasture at Rutherglen was cropped successively to wheat without fallow, yield declined only slowly over six years; the second crop yielded most. When the land was cropped under a fallow-wheat system, yield per ha decreased more rapidly and total yield over six years was less than half as great. More recent experiments at Rutherglen have shown that the yield of successive wheat crops sown without fallow declined linearly at rates related inversely to the length of the previous clover ley (Mullaly, McPherson and Bath 1965); the longer the ley period, the more crops that can be grown satisfactorily after ploughing.

Experiments by Littler (1967) on the black earths of the Darling Downs have shown the beneficial effects of various lengths of ley with *Medicago sativa* on both crop yields and soil properties. Periods under lucerne had beneficial effects on subsequent crops of wheat even up to the sixth crop. Whitehouse (1967) showed that the beneficial effects on yield were associated with increases in soil nitrogen.

The effect on crop yields of improving soil structure by growing pasture leys has received little direct study. Unpublished results of Burford and Greenland, quoted by Greenland (1971), showed that at the Waite Institute, even with generous additions of nitrogen fertilizer and irrigation, wheat growing on poorly structured, old arable land failed to yield as much as a crop grown without nitrogen or irrigation on old pasture land. This result suggested that differences in soil physical condition affected the growth of wheat. Stoneman (1964) in Western Australia obtained significant increases in wheat yield by improving soil structure by adding krilium.

Soil trends

The long-term effects of crop and pasture sequences on soil properties, and in particular on nitrogen and phosphorus contents of the soil and on soil structure, are of interest. The residual effect of phosphorus fertilizer applied to cereal crops has been clearly demonstrated (e.g. Piper and de Vries 1964; McLelland 1968). Some of the yield increases occurring both in broad scale agricultural ecosystems and in experiments may be due to accumulation of residual phosphorus.

Data on the changes in soil properties have been obtained for varying ley periods and for long-term crop sequential experiments.

Effects of pasture periods. Changes in the organic nitrogen content of Australian soils have been measured after varying lengths of pasture ley under a range of soil and climatic conditions (table 5). Most increases can be associated with legume growth, but increases in soil nitrogen have also been measured under substantially grass pastures. The rates of accumulation in the absence of legume are appreciable, especially when

Table 5. Mean annual increments of soil nitrogen in Australian soils under pasture

Location	Soil description	Period of measurement (years)	Pasture composition*	Depth (cm)	Mean increment (kg/ha/yr)	Reference†
Kojunup, W.A.	Lateritic sand	2	G–C	0–10	81	1
Merridin, W.A.	Calcareous	2	M	0–25	39	2
	solonized soil	2	G	0–25	39	2
Turretfield, S.A.	Red-brown earth	2	G–C	0–10	182	3
Truro, S.A.	Red-brown earth	1	G	0–12	155	4
		3	G	0–12	22	4
Walpeup, Vic.	Solonized brown	6	G–M	0–15	27	5
	soil	6	G	0–15	18	5
Longerenong, Vic.	Grey soil of	4	G–M	0–15	117	6
	heavy texture	4	G	0–15	88	6
		8	G–M	0–15	93	5
		8	G	0–15	57	5
Rutherglen, Vic.	Transitional podzol	8	G–C	0–15	78	5
Dookie, Vic.	Red-brown earth	10	G–C	0–15	93	5
Crookwell, N.S.W.	Podzolic	26	G–C	0–10	47	7
Kybybolite, S.A.	Solodized solonetz	35	G–C	0–15	52	8
Waite Inst., S.A.	Red-brown earth	36	G–C	0–20	67	9

* G–C = grass + clover, G–M = grass + medic, C,M or G = clover, medic or grass alone.

† 1, Watson and Lapins (1964); 2, Parker (1957); 3, Goode (1963); 4, Clarke (1970); 5, Mullaly et al. (1967); 6. Meagher and Rooney (1966); 7, Donald and Williams (1954); 8, Russell (1960); 9, Greenland (1971).

quantities of nitrogen that are present in the grasses at soil sampling, and which must also have come from the soil, are added. The evidence points clearly to non-symbiotic fixation.

The amount of readily mineralized nitrogen that accumulates in the soil as a result of the period of pasture is of more significance to future crop growth than the total nitrogen build-up. Greenland (1971) discussed the significance of partially humified, "light fraction" organic matter, which builds up rapidly under pasture and mineralizes readily following cultivation. However, the effects of management and other factors on "light fraction" accumulation have not yet been elucidated.

It is known, though, that nitrogen mineralizes more rapidly in materials of low rather than high C:N ratio. The likely practical advantage of maintaining low C:N ratios in plant residues before the land is cultivated for cropping is therefore evident. Simpson and Bromfield (1974) observed higher concentrations of mineral nitrogen in the soil, and higher yield in the following barley crop, at high compared with low stocking rate. (Note that *total* soil nitrogen was affected conversely by stocking rate—see page 281.) Under heavy grazing more clover was maintained in the pasture and residues with high C:N ratios were removed. Rapid mineralization of nitrogen during cultivation is particularly important where crops are grown without fallow (Penman 1949).

Effects of crop sequence. Data showing concentrations of soil nitrogen in several long-term experiments are given in table 6. Clearly, higher concentrations of soil nitrogen are maintained when fallow is omitted; less nitrogen is removed and more accumulates through fixation. Even with continuous cropping to wheat, concentrations of soil nitrogen were greater than under the fallow-wheat sequence. The effects of adding one or two years of pasture to the sequence are also evident.

Cereal yield and soil nitrogen data provide an opportunity to prepare nitrogen balance sheets for the simple sequence experiments. In table 7, quantities of nitrogen removed in produce are compared with measured decreases in soil nitrogen, for several long-term experiments. In general, where cereals were grown on fallow, these two quantities were

Table 6. Gravimetric nitrogen concentrations (percentage) of surface soil from long-term (23–36 years) rotation experiments

Location	Soil texture	Annual rainfall (mm)	Rotation*						Reference†
			FW	FWp	FWpp	WW	PW(1) PWO(2) PWR(3)	Wpp	
Wongan Hills, W.A. ‡	Loamy sand	360	0.021	0.028	0.037	0.025	–	–	1
Chapman, W.A. ‡	Sandy loam	460	0.021	0.027	0.040	0.042	–	–	1
Waite Institute, S.A. ‡	Fine sandy loam	620	0.091	0.116	–	0.129	0.173(1)	0.138**	2
Rutherglen, Vic. ‖	Silty loam	570	0.045	0.050	–	0.065	0.061(1)	–	3
Longerenong, Vic. ‖	Loam	420	0.079	0.096	–	0.081	0.106(2)	–	3
Werribee, Vic. ‖	Loam – clay loam	500	0.061	0.080	–	0.072	0.073(3)	–	3

* F = fallow, W = wheat, p = pasture, P = peas, O = oats (grain), R = rape.
† 1 = Drover (1956); 2 = Greenland (1962); 3 = Penman (1949).
‡ 0–10 cm depth.
‖ 0–15 cm depth.
** 23 years WOp, 12 years Wpp

Table 7. Removal of nitrogen in hay and grain and measured loss of soil nitrogen, under narrow cereal rotations, in some long-term experiments.
Some values are approximate because it has been necessary to estimate, from limited information, nitrogen concentrations of hay and grain and bulk densities of soil.*

Site	Rotation†	Period (years)	N removed in produce (kg/ha)	Measured loss of soil N (kg/ha)	Reference Ø
Werribee, Vic. ‡	F,O_H	35	700	510	1,2
	F,W		375	290	
	O_H,O_H		775	290	
	W,W		265	24	
Rutherglen, Vic.	F,W	36	435	435	1,2
Vic ‡	W,W		260	–50	
Waite Inst.	F,W	36	685	3,140	3,4
S.A. ‖	W,W		585	2,265	
Chapman, W.	F,W	23	115	415	5,6
W.A. *	W,W		150	140	
Wongan Hills,	F,W	23	165	80	5,6
W. A.	W,W		125	25	

* Estimations used in calculations.

	N%	Bulk density (g/cm^3)
‡	Hay, 0.8	–
	grain, 1.8	–
‖	grain, 1.8	–
**	grain, 1.5	1.3

† F = fallow, O_H = oats for hay, W = wheat for grain:

1, Sims and Jardine (1949); 2, Penman (1949); 3, Greenland (1962); 4, Waite Agricultural Research Institute (1962); 5, Shier and Cullinane (1948); 6, Drover (1956).

similar. Under continuous cropping, however, soil depletion was substantially less than product removal in four of six instances. Most likely some nitrogen fixation by non-symbiotic microorganisms occurred under the cereals.

At the Waite Institute, under both fallow-wheat and continuous wheat sequences, and at Chapman under fallow-wheat, substantial losses of nitrogen occurred from the soil other than through product removal. Such losses could have resulted from leaching, but Greenland (1971) concluded from the work of McGarity and Myers (1968), Burford and Greenland (1970) and others that denitrification generally causes more serious nitrogen losses in Australian wheat soils. However, it should be pointed out that rainfall at the Waite Institute (620 mm annually) exceeds that of most of the Australian wheat-belt (mostly < 500 mm), and is particularly heavy during winter (April-November). Losses by both processes at the Waite Institute are therefore likely to be higher than elsewhere.

High levels of soil nitrogen produced by clover leys frequently increase the vegetative growth of the following crop more than grain yield; thousand grain weights are commonly depressed, and sometimes the crop matures prematurely, a phenomenon known as "haying off". Grain yield may even be lower than where nitrogen is less abundant. The effect has been attributed to the premature exhaustion of soil water by the nitrogen-stimulated crop (for example, Storrier 1962), but observations by Storrier (1965) and Dann (1969) indicate that other factors related to the physiology of the crop may be involved.

The important question regarding soil physical properties is whether short periods of pasture (one-four years) in the crop sequence can increase the stability of the cultivated soil sufficiently to prevent serious failure and crusting of the seed bed. Millington (1959) showed that the apparent density of a loam sown to wheat was much affected by the cropping sequence, being highest for fallow-wheat. Two years' pasture between successive fallow-wheat phases prevented structural failure during two drier seeding periods (rainfall in one month < 70 mm), but not in a wet period (rainfall 130 mm). With three years' pasture and no fallow between crops, failure did not occur even during the wet seeding period. The effect measured by Millington of apparent density on wheat establishment is shown in figure 6.

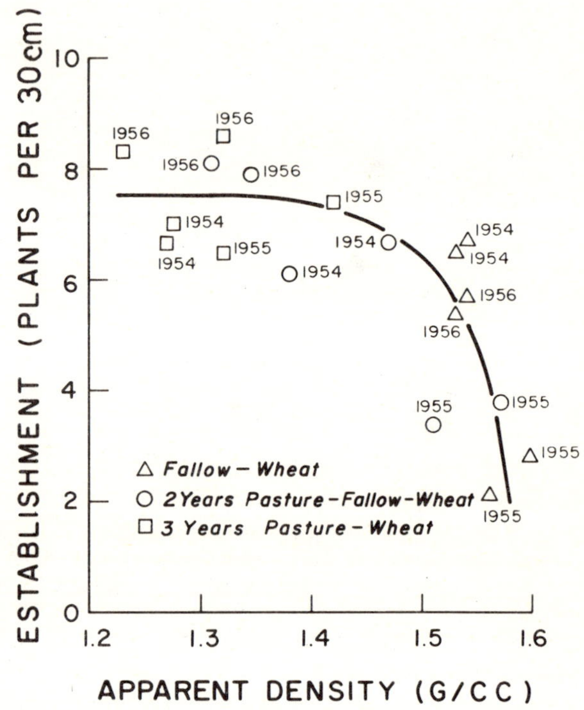

Fig. 6. Relationship between the establishment of wheat seedlings and the apparent density of cultivated soil, as influenced by the previous cropping sequence; rainfall during one month after seeding was: 1954, 70 mm; 1955, 130 mm; 1956, 65 mm. (Reprinted, by permission, from Millington 1959.)

Using wet sieving to measure structure stability, Stephens et al. (1945) found that after long periods of fallow-wheat-pasture and fallow-wheat sequences aggregation (0.2 mm) of Belalie loam was, respectively, 84 per cent and 33 per cent of that of uncultivated soil. A decade of pasture, following fallow-wheat, restored aggregation to 60 per cent of the virgin level. Similar beneficial effects of including pasture in cropping sequences have been measured, using wet sieving techniques, by Newton and Drover (1956) and Smith (1969).

Greacen (1958) has concluded that wet sieving techniques commonly exaggerate structural improvements because agitation of the soil during sieving is too gentle. Figure 7, from his paper, shows large effects of pastures on the stability of a silty loam as measured by wet sieving. However, when soil from successive 2.5 cm layers was subjected to intense artificial rain, the "stable" surface soil (0-2.5 cm) dispersed more readily than "unstable" soil from the deepest layer (12.5-15 cm); with soil from the three-year pasture field, infiltration rates after twenty minutes of rain were, respectively, 16 and 41 mm/hr for these layers. In the field, infiltration was much faster when the soil surface (0-10 cm) was inverted with a mould board plough than when it was disturbed without inversion with a chisel plough. Clearly, assessments of structure based on a single measurement of aggregate stability should be treated cautiously. The effects of structure-restoring treatments need to be measured through the following cropping phase, by a range of methods related to the requirements of the growing crop.

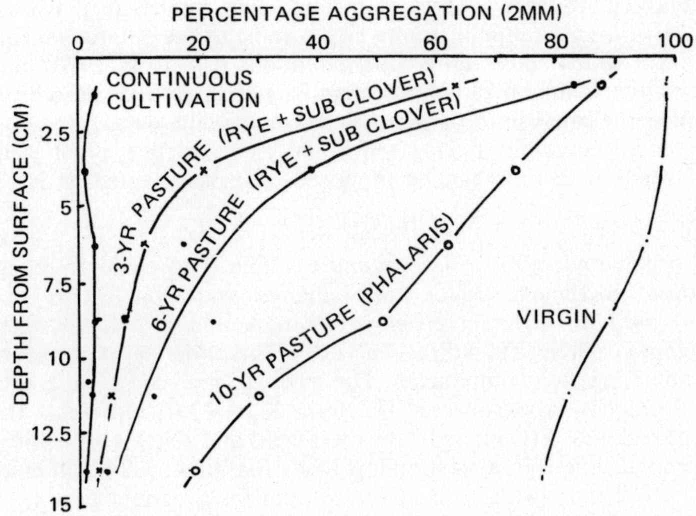

Fig. 7. Profiles of aggregate stability, measured by wet sieving, in a loam following several cultural treatments. (Reprinted, by permission, from Greacen 1958.)

To achieve rapid structural improvement during the pasture phase of a cropping sequence, pastures that have the maximum effect in forming and stabilizing soil aggregates are needed. In general aggregation increases proportionally with pasture yield, but differences between species, related to the production and distribution of their roots, may occur (Troughton 1961; Andrew 1965; Clarke, Greenland and Quirk 1967). For short-term leys in areas of winter rainfall, vigorous Wimmera rye grass (*Lolium rigidum*), either accompanied by a legume or fertilized with nitrogen, is likely to be the most effective structure builder.

Continuous cereal cropping without fallow may not be harmful to the soil structure (Newton and Drover 1956). Under this system depletion of organic matter is less rapid than under fallow-cereal (see above), and presumably the cereal roots are able to promote a moderate degree of aggregation. Provided removed nutrients are replaced by fertilizers, and diseases and weeds are satisfactorily controlled, such a mono-cultural system can probably be maintained permanently. In most Australian cereal districts, however, some fallowing is necessary to conserve water, and hence a period of pasture to restore soil structure is desirable.

THEORETICAL MODELS OF LONG-TERM SOIL AND YIELD TRENDS

Analysis of long-term sequential experiments

Much quantitative information has accumulated in various parts of the world on trends in plant yield and in soil properties with different sequential crop practices. Although trends have frequently been observed in long-term experiments the integration of plant and soil data has not been possible. Conclusions that have been drawn have frequently been qualitative only.

The development of procedures for fitting mathematical models to data has allowed these relationships to be examined more fully, and coefficients to be estimated, provided a history of cultural treatments and plant yields, and soil data for different treatments at different times, are available.

These mathematical models may be simple algebraic equations, in which case programs which calculate least squares from non-linear models (e.g. Ross 1971) are suitable. However in some situations it may be desirable to use differential equations. In this case a program which solves differential equations (e.g. Ross 1973) may be combined with a least squares minimization program. Russell (1975) has described the use of such a program in the analysis of sequential soil and plant data.

Various differential equations linking sequential crop practice, plant yield and soil organic matter trends with time can be proposed. A general equation is

$$\frac{dN}{dt} = -K_1(t) * N + K_2 + K_3(t) * Y(t) \qquad (1)$$

where N is soil organic nitrogen or carbon and t is time. This equation recognizes that the decomposition coefficient, $K_1(t)$, may change with time. This occurs with agricultural land used for different crops in rotation, and in a fallow-wheat-pasture system $K_1(t)$ changes each year in a three-year cycle. Equation 1 also considers the addition term as comprising two components. The term K_2 represents the addition to soil organic matter from non-crop sources. The term $K_3(t) * Y(t)$ represents the addition term due to plant residues. Y(t) refers to the crop yield and $K_3(t)$ is a coefficient related to the specific crop at time t. In a rotation it is likely that there will be different K_3 coefficients and $K_3(t)$ will vary with crop in the rotation at a certain time t.

The main advantage of such an approach is that it allows the effects of crop yield within a rotation on soil organic matter levels to be examined. In a wheat-oats-pasture rotation it can be postulated that each of these crops has specific decomposition coefficients, i.e. K_{1_w}, K_{1_o}, K_{1_p} and that $Y_w(t)$, $Y_o(t)$ and $Y_p(t)$ represent yields of these crops at specific times. Thus in a wheat-oats-pasture rotation the sequential change of $K_1(t)$ with time is K_{1_w}, K_{1_o}, K_{1_p}, K_{1_w}, etc., and of $K_3(t) * Y(t)$ is $K_{3_w} * Y_w(t_1)$, $K_{3_o} * Y_o(t_2)$, $K_{3_p} * Y_p(t_3)$, $K_{3_w} * Y_w(t_4)$ etc. where $Y_w(t_1)$, for example, represents the yield of wheat in the first year of the rotation and $Y_o(t_2)$ represents the yield of oats in the second year.

It is possible to fit equation (1) to plant and soil data from long-term field experiments and this has been done using annual crop yields and soil data from several experiments (Russell 1975). The parameters K_1, K_2 and K_3 were calculated for specific crops in these experiments.

The main limitations in the estimation of coefficients for crop sequences in semi-arid areas are the lack of long-term experiments in these areas and of associated soil data.

Relationship between plant yields and soil organic matter

One of the advantages of the approach discussed above is that the feedback relation between plant yield and soil organic matter level can be quantified. At a given equilibrium level of soil nitrogen content and plant yield, and where $K_1(t)$ and $K_3(t)$ do not vary, i.e with continuous treatments, it can be assumed that

$$\frac{dN}{dt} = 0 \qquad (2)$$

At this point the equilibrium soil level is N_E and the plant yield level is Y_E; thus

$$-K_1 * N_E + K_2 + K_3 * Y_E = 0 \qquad (3)$$

and

$$N_E = \frac{K_2 + K_3 * Y_E}{K_1} \qquad (4)$$

From the estimated coefficients it is possible to plot the effect of crop yield on equilibrium soil level. The intercept on the Y axis when Y_E is zero is K_2/K_1 and the slope of the line relating plant yield and soil equilibrium level is K_3/K_1.

This indicates the extent to which changes in crop yield affect soil organic nitrogen content. Such studies are a first step towards quantifying the effects of sequential crop practices and of understanding the relationship between crop yield and soil properties.

SUMMARY

There are special difficulties in maintaining the stability of ecosystems when crops are grown in semi-arid environments. This is related to the need for fallowing and cultivation to store soil water, to the low and variable yields which restrict economic inputs, and to the low rainfall which limits the pasture species that will grow and their production.

Sequential cropping practices involving periods of legume-based pastures have enabled crop yields and soil organic matter levels to be maintained on a wide range of soils in regions of Australia receiving winter rainfall. In semi-arid areas with summer rainfall there is a lack of suitable legumes for the heavy soils used for cropping and sequential practices are still evolving.

Further development of mathematical procedures for the integration of crop yields and soil trends with time and the collection of suitable soil and plant data should enable crop relationships in semi-arid ecosystems to be quantified, and optimum systems to be simulated.

REFERENCES

Andrew, W.D. (1965). *Aust. J. Exp. Agric. Anim. Husb.* 5, 133-6.
Arndt, W. (1961*a*). CSIRO Div. Land Use Res. and Reg. Surv. Tech. Paper No. 16.
———. (1961*b*). CSIRO Div. Land Use Res. and Reg. Surv. Tech. Paper No. 17.
Barley, K.P. (1959*a*). *Aust. J. Agric. Res.* 10, 171-8.
———. (1959*b*). *Aust. J. Agric. Res.* 10, 179-85.
Barley, K.P., and Greacen, E.L. (1967). *Adv. Agron.* 19, 1-43.
Bath, J.G. (1949). Br. Commonw. Scient. Off. Conf., Rep. Spec. Conf. Agric., Aust., 1949, pp. 448-57.
Blencowe, J.P. (1957). *J. Dep. Agric. S. Aust.* 60, 233-7.
Bond, R.D., and Harris, J.R. (1964). *Aust. J. Soil Res.* 2, 111-22.
Burford, J.R., and Greenland, D.J. (1970). Proc. 11th Int. Grassl. Congr., pp. 458-61.
Clarke, A.L. (1970). Proc. 11th Int. Grassl. Congr., pp. 461-5.
Clarke, A.L., Greenland, D.J., and Quirk, J.P. (1967). *Aust. J. Soil Res.* 5, 59-68.
Cornish, E.A. (1949). *Aust. J. Sci. Res. Ser. B*, 2, 83-137.
Dann, P.R. (1969). *Aust. J. Exp. Agric. Anim. Husb.* 9, 625-9.
Doak, B.W. (1952). *J. Agric. Sci.* 42, 162-71.
Donald, C.M. (1963). *Aust. J. Sci.* 25, 386-95.
———. (1965). *Aust. J. Sci.* 27, 187-98.
Donald, C.M., and Williams, C.H. (1954). *Aust. J. Agric. Res.* 5, 664-87.
Drover, D.P. (1956). *J. Soil Sci.* 7, 219-25.
Edmond, D.B. (1963). *N.Z. Jl. Agric. Res.* 6, 265-76.
Elliott, B.R., and Jardine, R. (1972). *Aust. J. Agric. Res.* 23, 935-44.
Forster, H.C. (1939). *J. Dep. Agric. Vict.* 37, 130-3.
French, R.J., Matheson, W.E., and Clarke, A.L. (1968). "Soils and Agriculture of the Northern and Yorke Peninsula Regions of South Australia". Dep. Agric. S. Aust., Adelaide.
Gillard, P. (1967). *J. Aust. Inst. Agric. Sci.* 33, 30-4.
Goode, J.R. (1963). *J. Dep. Agric. S. Aust.* 66, 358-65.
Gradwell, M.W. (1968). *N.Z. Jl. Agric. Res.* 9, 127-36.
Greacen, E.L. (1958). *Aust. J. Agric. Res.* 9, 129-37.
Greenland, D.J. (1962). 3rd Aust. Conf. Soil Sci., Canberra. Paper No. 88.
———. (1971). *Soils Fertil.* 34, 237-51.
Greenland, D.J., Lindstrom, G.R., and Quirk, J.P. (1962). *Soil Sci. Soc. Amer. Proc.* 26, 366-71.
Harris, R.F., Chesters, G., and Allen, O.N. (1966). *Adv. Agron.* 18, 107-69.
Herriot, R.I. (1954). *J. Dep. Agric. S. Aust.* 58, 140-9.
Hilder, E.J. (1964). *Proc. Aust. Soc. Anim. Prod.* 5, 241-8.
Hume, A.N. (1940). S.D. Agric. Exp. Stn. Bull. No. 344.
Hutchinson, K.J., and King, K.L. (1970). *Search.* (Syd.) 1, 42-3.
Isbell, R.F. (1962). CSIRO Aust. Div. Soils, Soils and Land Use Ser. No. 43.
Jones, H.R. (1961). *J. Soil Conserv. Serv. N.S.W.* 17, 156-69.
Kimber, R.W.L. (1967). *Aust. J. Agric. Res.* 18, 361-74.
Littler, J.W. (1964). *Qld. J. Agric. Sci.* 21, 137-42.
———. (1967). Qld. Dep. Prim. Ind. Pasture Wheat Rotation Trial Field Day Publ., pp. 5-11.

Logan, J.M. (1960*a*). *J. Soil Conserv. Serv. N.S.W.* **16**, 95-111.
———. (1960*b*). *J. Soil Conserv. Serv. N.S.W.* **16**, 214-27.
McCalla, T.M., and Army, T.J. (1961). *Adv. Agron.* **13**, 126-96.
McGarity, J.W., and Myers, R.J.K. (1968). *Soil Sci. Soc. Amer. Proc.* **32**, 812-7.
McIntyre, D.S. (1958). *Soil Sci.* **85**, 185-9.
McLachlan, K.D. (1968). *Aust. J. Exp. Agric. Anim. Husb.* **8**, 33-9.
McLelland, V.F. (1968). *Aust. J. Agric. Res.* **19**, 1-8.
Meagher, J.W., and Rooney, D.R. (1966). *Aust. J. Exp. Agric. Anim. Husb.* **6**, 425-31.
Millington, R.J. (1959). *Aust. J. Agric. Res.* **10**, 487-94.
Mullaly, J.V., McPherson, J.B., and Bath, J.G. (1965). Aust. Cereal Agron. Conf., Horsham, Paper D3.
Mullaly, J.V., McPherson, J.B., Mann, A.P., and Rooney, D.R. (1967). *Aust. J. Exp. Agric. Anim. Husb.* **7**, 568-71.
Newton, K., and Drover, D.P. (1956). *J. Soil Sci.* **7**, 226-34.
Nye, P.H., and Greenland, D.J. (1960). Commonw. Bur. Soils, Tech. Commun. No. 51, CAB, Harpenden.
Oveson, M.M., and Hall, W.E. (1957). Oreg. Agric. Exp. Stn. Tech. Bull. No. 39.
Parker, C.A. (1957). *J. Soil Sci.* **8**, 48-59.
Penman, F. (1949). Br. Commonw. Scient. Off. Conf., Rep. Spec. Conf. Agric., Aust., 1949. pp. 457-65.
Petersen, R.G., Woodhouse, W.W., and Lucas, H.L. (1956). *Agron. J.* **48**, 444-9.
Phillips, L.J., and Norman, M.J.T. (1962). CSIRO Div. Land Use Res. and Reg. Surv. Tech. Paper No. 20.
Piper, C.S., and de Vries, M.P.C. (1964). *Aust. J. Agric. Res.* **15**, 234-72.
Ross, P.J. (1971). CSIRO Aust. Div. Soils Tech. Paper No. 6.
———. (1973). CSIRO Aust. Div. Soils Tech. Paper No. 20.
Russell, E.W. (1971). *J. Soil Sci.* **22**, 137-51.
Russell, J.S. (1960). *Aust. J. Agric. Res.* **11**, 902-26.
———. (1961). *Aust. J. Agric. Res.* **12**, 273-85.
———. (1973). *J. Aust. Inst. Agric. Sci.* **39**, 156-66.
———. (1975). *Soil Sci.* **120**, 37-44.
Russell, J.S., and Shearer, R.C. (1964). *Aust. J. Agric. Res.* **15**, 91-99.
Shier, F.L., and Cullinane, W.P. (1948). *J. Dep. Agric. West. Aust.* **25**, 351-61.
Simpson, J.R., and Bromfield, S.M. (1974). *Aust. J. Exp. Agric. Anim. Husb.* **14**, 479-86.
Simpson, J.R., Bromfield, S.M., and Jones, O.L. (1974). *Aust. J. Exp. Agric. Anim. Husb.* **14**, 487-94.
Sims, H.J., and Jardine, R. (1949). Br. Commonw. Scient. Off. Conf., Rep. Spec. Conf. Agric., Aust., 1949. pp. 438-47.
Smith, S.T. (1969). *J. Dep. Agric. West. Aust.* **10**, 530-3.
Stephens, C.G., Herriot, R.I., Downes, R.G., Langford-Smith, T., and Acock, A.M. (1945). Coun. Sci. Ind. Res., Aust. Bull. No. 188.
Stoneman, T.C. (1964). *Aust. J. Exp. Agric. Anim. Husb.* **4**, 363-6.
Storrier, R.R. (1962). *Aust. J. Exp. Agric. Anim. Husb.* **2**, 185-92.
———. (1965). *Aust. J. Exp. Agric. Anim. Husb.* **5**, 317-22.
Swaby, R.J. (1949). *J. Gen. Microbiol.* **3**, 236-54.
Till, A.R., and May, P.F. (1971). *Aust. J. Agric. Res.* **22**, 391-400.
Troughton, A. (1961). *J. Br. Grassl. Soc.* **16**, 1-5.
Waite Agricultural Research Institute (1962). Biennial Rept. 1960-61, Univ. of Adelaide, Adelaide.
———. (1972*a*). Biennial Rept. 1970-71, Univ. of Adelaide, Adelaide.
———. (1972*b*). Annual and Biennial Repts. 1925-33 to 1970-71, Univ. of Adelaide, Adelaide.
Walker, T.W., Orchiston; H.D., and Adams, A.F.R. (1954). *J. Br. Grassl. Soc.* **9**, 249-73.
Wark, D.C. (1942). *Trans. Roy. Soc. S. Aust.* **66**, 133-41.
Warren, J.F. (1960). *J. Aust. Inst. Agric. Sci.* **35**, 240-52.
Watson, E.R. (1963). *Aust. J. Agric. Res.* **14**, 796-807.
———. (1969). *Aust. J. Agric. Res.* **20**, 447-56.
Watson, E.R., and Lapins, P. (1964). *Aust. J. Agric. Res.* **15**, 885-94.
———. (1969). *Aust. J. Exp. Agric. Anim. Husb.* **9**, 85-91.
Whitehead, D.C. (1970). Commonw. Bur. Pastures and Field Crops, Bull. No. 48, CAB, Hurley.
Whitehouse, M.J. (1967). Qld. Dep. Prim. Ind. Pasture Wheat Rotation Trial Field Day Publ. pp. 12-14.
Williams, C.H. (1962). *J. Aust. Inst. Agric. Sci.* **28**, 196-205.
Williams, C.H., and Donald, C.M. (1957). *Aust. J. Agric. Res.* **8**, 179-89.
Williams, C.H., and Lipsett, J. (1961). *Aust. J. Agric. Res.* **12**, 612-29.
Williams, C.H., and Steinbergs, A. (1958). *Aust. J. Agric. Res.* **9**, 483-91.
Woodroffe, K. (1949). Br. Commonw. Scient. Off. Conf., Rep. Spec. Conf. Agric., Aust., 1949. pp. 425-36.

CHAPTER 14

Development and reclamation of secondary salinity

A.J. Peck

INTRODUCTION

Throughout the world saline and sodic soils are relatively more common in arid and semi-arid zones. In the broadest sense this is a consequence of the excess of potential evaporation over rainfall, resulting in little discharge of the liquid water in which soluble materials are exported from an area. In this respect Australia is no exception. Average yearly potential evaporation exceeds median rainfall over most of the continent (Austalian Water Resources Council 1968). A very large area of land does not drain to the ocean at all, and a marginal area is drained by surface water flow only in years of exceptional rainfall. A broad survey (Northcote and Skene 1972) indicates that there are $3{\cdot}9 \times 10^7$ ha of saline soils, and 2×10^8 ha of sodic soils in Australia; in all, 33 per cent of the area of the continent is affected by excessive salinity or sodicity.

Not all of the saline and sodic soils of Australia are agriculturally unproductive. Large areas of sodic soils are irrigated, and others are cropped or grazed. Saline soils support only halophytic vegetation, but this is often valuable for low intensity grazing.

In addition to the naturally saline and sodic soils, there are areas of secondary salinization throughout the agricultural regions of Australia. Secondary salinization refers to an increase of salinity and/or sodicity of a soil, caused by land use changes of which agriculture is the most important. This phenomenon is well known in irrigation areas throughout the world, and its control is a recognized part of irrigation management. Methods for maintenance of desirable salinity conditions of irrigated soils are well established (Richards 1954; Reeve and Fireman 1967), although their large-scale implementation may pose great problems such as those of the Murray River in south-eastern Australia (River Murray Commission 1970).

Secondary salinity problems under natural rainfall conditions (often referred to as dryland salinity) are rare, or poorly documented, on a global scale. They usually develop from a shallow, saline water table with or without seepage of liquid water from the soil. There is no essential difference between these forms of the problem, and often an affected area will yield liquid water during that part of the year when rainfall is highest relative to evaporation.

In Canada (Ballantyne 1963; Greenlee, Pawluk and Bowser 1968) and the Northern United States (Paul L. Brown, Carl W. Carlson, and W.O. Willis, personal communications, 1971) saline seepages have developed on agricultural land in recent years. The affected area in Montana has been estimated as 32,000 ha with a yearly increase of about 10 per cent (Clark 1971). Secondary salinity problems in unirrigated soils are reported from New South Wales (Wagner 1957; Logan 1958) but no estimate of the affected area is known. Cope (1958) estimated more than 4,000 ha of affected land in parts of Victoria, and Northcote and Skene (1972) quote an estimate of a further area of 5,000 ha elsewhere in that state. A total area of secondary salinization of 13,000 ha in South Australia was estimated by Matheson (1968). However, the most extensive areas of affected soils appear to be located in Western Australia. A survey in 1962 (Lightfoot, Smith and Malcolm 1964) revealed that, of land previously cropped or sown to pasture, a total 124,000 ha had become visually salt affected, and the area was increasing at a rate of 3,400 ha yr^{-1}. In the region of this survey 32 per cent of farms exhibited the problem, and on 2·8 per cent it affected more than 200 ha of soil.

In a quite different category is the development of scald patches. Scalds occur where a sodic subsurface soil is exposed to the dispersive effects of rainfall following overgrazing and erosion by wind or surface water. A shallow water table does not appear to be necessary for the development of a scald (Jones, 1969). It is not clear whether any change of soil sodicity plays a part in the formation of a scalded soil, and they are not considered in any detail here.

Since all secondary salinity problems have their origins in man's disturbance of the natural environment, it is reasonable to suspect that these areas may be amenable to reclamation by suitable management procedures. Appropriate methods have been developed and applied extensively in irrigation agriculture. They involve drainage to lower the water table, application of excess water to leach salts, and, when necessary, chemical amendments to maintain or improve physical properties of the soil. In the economic context of dryland farming, however, these well-known solutions to salinity problems are not ordinarily possible, so that, if reclamation of affected areas is to be achieved, alternative methods are generally required.

This is not intended to be an encyclopaedic review of investigations of secondary salinity and the reclamation of affected soils. On the contrary, the objective is restricted to a suitably detailed consideration of the processes involved in the development of dryland salinity in farmed areas. By such means the deficiencies of our present understanding will become apparent, and methods for reclamation of affected areas may be viewed in a proper perspective.

EFFECTS OF SALTS ON SOILS AND PLANTS

Many detailed reviews are available of the effects of salts on soils and plants. These include Richards (1954), Bernstein (1962), Allison (1964), Reeve and Fireman (1967), Quirk (1971), Robinson (1971), and Emerson's chapter in this volume (chapter 5). In the context of the present article, only a brief summary of these effects is warranted.

The ions found in appreciable quantities in the soil solution are Na^+, K^+, Ca^{2+}, Mg^{2+}, Cl^-, HCO_3^-, SO_4^{2-}, CO_3^{2-}. Some specific solutes, such as boron, are toxic to plants even at very low concentration, but the purely osmotic role of solutes is usually the dominant effect, even when salts are partially absorbed by plant roots. Other effects of salts on plants can be indirect, for example via the pH of the root environment, or by interference with nutrient uptake.

Soil salinity is often expressed by the electrical conductivity of the solution extracted from a water-saturated sample of the soil. When this is less than 0·4 S m^{-1} at 25°C it has been found empirically that there is little reduction in growth of most crop species, and the soil is considered non-saline.

Another common measure of salinity is the mass of soluble salts in unit mass of dry soil. Northcote and Skene (1972) defined as non-saline sandy loams and coarser soils with less than 0·1 per cent of soluble salts in the surface 0·2 m. For clay loams and clays the limit was taken as 0·2 per cent since these finer textured soils hold more water, even at the lower limit for plant growth.

The concentration of specific ions, or all dissolved salts, in the soil solution is another expression of salinity used in certain circumstances, particularly when considering solute transport in soil.

The ratio of Na^+ to total cations on the exchange surfaces of a soil is expressed as the exchangeable sodium percentage (ESP). A soil with an ESP greater than 15 per cent is generally referred to as sodic, and may display structural deterioration when the electrolyte concentration in the soil solution is low. An important consequence of structural deterioration is a reduction in the hydraulic conductivity of the soil. Northcote and

Skene (1972) refer to soils with ESP in the range 6-15 per cent as sodic, and those with higher ESP as strongly sodic.

As equilibrium, the ESP of a soil has an isotherm-like functional relationship with the relative concentraton of Na^+ in the soil solution expressed by the sodium adsorption ratio (SAR), and defined by

$$SAR = Na/[(Ca + Mg)/2]^{1/2}. \quad (1)$$

In this equation Na, Ca, Mg represent ionic concentrations. Since the SAR is not dimensionless, an increase in concentration of the soil solution which has no effect on the relative abundance of solute ions, will increase the SAR and finally the ESP.

SOURCES OF SALTS IN SOILS

The development of secondary salinity is usually attributed to a redistribution of the salts within a mass of soil. An alternative possibility is that man's activities affect the rate of input of solutes to the soil solution, and so it is necessary to briefly consider the origin of salts in soils.

Perhaps the most obvious source of the salts in a soil is deposition with sedimentary parent material in a marine or otherwise saline environment. Examples are mentioned by Richards (1954), and Collis-George and Evans (1964). Clearly, man's activities can have no effect on the quantity of such connate salt, although they may affect its redistribution, or its release by weathering reactions.

Weathering of mineral particles is usually considered to be the primary source of all solutes in earth's waters. But on the basis of the chemical composition of rocks and river water, it appears that only 1-2 per cent of the Cl discharged by Scandinavian rivers is contributed by contemporary weathering (Eriksson 1955, 1960). Cleaves, Godfrey and Bricker (1970) determined the geochemical balance of a small forested watershed in the eastern United States in detail. They concluded that dissolved solids released by contemporaneous weathering totalled 60 kg $ha^{-1}yr^{-1}$. The major weathering products exported in streamflow were HCO_3 and SiO_2.

Undoubtedly rates of weathering vary greatly in space and time with different minerals and climates, and it is conceivable that man's disturbance of natural ecosystems and hydrological regimes has affected the current rates at certain localities. This is a factor which should not be ignored when considering the source of salts in a salinity problem, but it has not been shown to be important in an agricultural context.

Atmospheric deposition was considered by Anderson (1945) to be the major source of salts found in the rivers and lakes of southern Australia. Detailed studies of the chemical composition of rainfall have been published for southeastern and southwestern Australia by Hutton and Leslie (1958) and Hingston (1958) respectively, but there is no such information for the rest of the continent. The available data show that close to the coast there is a high concentration of salts in rainfall, and the ionic composition resembles that of sea water. Further inland the concentration decreases, and the ionic composition changes due to an increasing contribution of terrestial material.

In addition to deposition of salts in rain, dry fallout is sometimes considered to be of similar magnitude (Gorham 1961). This process might be expected to be most important in arid (dusty) areas, or near industrial centres or active volcanoes, and to be least important in a forest situation. No reliable estimates of the net contribution of dry fallout to the total deposition of salts from the atmosphere are known to the author.

An atmospheric source of direct and indirect importance is carbon dioxide which dissolves in soil water forming HCO_3. Solution may take place directly from the atmosphere, or there may be intermediate biological activity such as photosynthesis, translocation, and ultimately respiration. The indirect importance of CO_2 solution is its effect on acidity of the soil water, and consequently on weathering reactions. It is reasonable to expect that any drastic vegetation change, such as clearing land and planting exotic crops, will affect CO_2 solution in the soil water, and consequently the rates of weathering reactions. However, there does not appear to be any information which would indicate quantitatively the importance of this process.

Eriksson (1955) estimates that in agricultural districts of Sweden the use of potassium fertilizers contributes about 5 kg $ha^{-1}yr^{-1}$ of Cl to the soil. Similarly, applications of insecticides, herbicides and fungicides make small contributions to soil salts. The use of salt to de-ice highways may provide an appreciable local source. A large addition of soluble material usually occurs when animal, domestic or industrial wastes are disposed of on an area of land.

Other sources of salts are associated with biological activity (e.g. NO_3), but these have never been shown to reach a concentration which could be considered as a salinity problem.

The mechanisms of material transport in soils, by which salts are redistributed and may be concentrated in critical areas, are discussed in a later section.

Evidently some detailed work will be necessary to determine the relative importance of the various sources of salt in any particular saline area.

SALT BALANCE

A concept which has proved useful in studies of salinity problems is the salt balance. This is a quantitative statement of the amounts of salt coming into and leaving a chosen area.

Salt balance studies in irrigated areas originated with Scofield (1940) who compared the input of dissolved material in irrigation water with the loss in drainage water. When input exceeds loss there is an accumulation of salt in the area, and continuation of this situation must eventually lead to the development of saline soil.

Studies of salt balances in unirrigated areas appear to originate with Weller (1926) who concluded that the rate of atmospheric deposition of salt on a forested catchment in Western Australia was essentially equal to the rate of loss in streamflow from the area. Subsequently, more detailed observations elsewhere (Juang and Johnson 1967; Douglas 1968; Claridge 1970; Baldwin 1971) have often led to the conclusion that there is a small excess of loss over gain for forested areas, but this is usually attributed to inadequate measurement of the total salt input.

Smith (1962) estimated the salt balance of a large catchment area, much of it farmed, in Western Australia and concluded that there was a considerable net loss of salt in the streamflow.

A more detailed study of the chloride balances of eight forested and seven partly farmed catchments in Western Australia has been reported by Peck and Hurle (1973). Their data are reproduced in table 1. As in other studies, input of Cl is slightly less than loss for forest catchments, and in agreement with Smith (1926) there is a large net loss of Cl from the farmed areas.

Losses of Cl from the farmed catchments studied by Peck and Hurle (1973) are too large to be attributed to error in input estimation, or to release by contemporary weathering. In this situation the net loss of salt indicates decreasing average soil salinity, but the movement of saline water into streams and rivers results in secondary salinity of soils in groundwater discharge areas.

Given sufficient time the rate of loss of salt from these farmed areas must decay, and the areas of secondary salinity can be expected to decline. There are reports that this is occurring already (Smith 1962), but Peck and Hurle (1973) estimate characteristic times of the recovery process in the range 30-400 years. The larger values are more typical of the semi-arid areas where secondary salinity problems are widespread, which agrees with a suggestion by Holmes (1971), and with experience that the recovery rate of saline areas under existing conditions is generally imperceptible.

Using certain simplifying assumptions, Peck and Hurle estimate rates of groundwater

Table 1. Salt (chloride) balances of some catchments in Western Australia as reported by Peck and Hurle (1973).

Catchment name	Deposition of Cl in rain (kg/ha/yr)	Loss of Cl in streamflow (kg/ha/yr)
Julimar	53	78
Seldom Seen	120	160
More Seldom Seen	120	140
Waterfall	110	180
N. Dandalup	130	180
Davies	130	140
Yarragil	100	130
Harris	84	130
- - -	- - -	- - -
Brockman	80	340
Wooroloo	78	420
Dale	24	460
Hotham	48	370
Williams	31	650
Collie East	50	740
Brunswick	110	350

NOTE. Catchments listed above the broken line are forested, but managed for timber production, while those below the broken line include substantial areas of cleared farmland.

discharge under present farming conditions, and under the pristine forest vegetation. Their data are shown as table 2. They support earlier hypotheses (Burvill 1947; Smith 1962) that following agricultural development of naturally vegetated land, there is an increased flow of saline groundwater which leads to secondary salinity in discharge areas. Moreover the estimated changes in groundwater flow are consistent with those reported from many other areas where a comparison has been made between recharge under forest and grassland vegetation (Dagg and Blackie 1965; Holmes and Colville 1968; Boughton 1970; Urie 1971).

Table 2. Estimated discharges of groundwater in pristine conditions and under present farming practice for some Western Australian catchments
(After Peck and Hurle 1973)

Catchment name	Pristine groundwater discharge (mm/yr)	Present groundwater discharge from farmland (mm/yr)
Brockman	8.0	73
Wooroloo	3.9	61
Dale	0.8	24
Hotham	1.9	26
Williams	1.2	37
Collie East	1.7	60
Brunswick	70	500

In Western Australian catchments the continuing deposition of marine salts in rainwater appear to be the source of solutes in groundwaters, since estimated characteristic accumulation times are of order 10,000 years or less (Peck and Hurle 1973). The magnitudes of the accumulation (and pristine equilibration) characteristic times suggest the possiblity that catchment salt balances may not have been in equilibrium before agricultural development, due to climatic change, but this hypothesis has not been properly tested.

Elsewhere in Australia, and overseas, the factors responsible for secondary salinity development may differ from those determined by detailed studies in Western Australia. For example, in grain-producing areas of the United States and Canada the increased groundwater flow is attributed to a fallow-crop or fallow-fallow-crop rotation to conserve soil moisture, and the ionic composition of groundwater differs greatly from that of the ocean (Miller 1971). A reasonable generalization is that secondary salinity of unirrigated soil is associated with an increased flow of groundwater containing appreciable dissolved solids, and this is reflected by a net loss of salts from the area.

TRANSPORT OF SALTS IN SOILS

The mechanisms of salt transport in soils have been reviewed in several recent articles (Gardner 1965; Philip 1969, 1970; Peck 1971) and the reader is referred to these for detailed discussions. Even very small rates of movement of soil water result in convection being the dominant mechanism of solute transport. However, molecular diffusion may remain as the main process of mixing, and may make an important contribution to transport, where there are significant concentration gradients of the dissolved salt. At higher solution flow rates, diffusion and the microscopic convection pattern in a soil interact to disperse solutes, and at still higher velocities the role of molecular diffusion becomes quite negligible.

In the simplest case of a completely inert solute, and a medium with a narrow range of pore sizes, the mixing coefficient (dispersion coefficient) is essentially identical to the coefficient of molecular diffusion (D) of the solute in the soil so long as the dimensionless Peclet number (P) is less than unity (Saffman 1960; Wooding and Peck, unpublished work). The Peclet number is defined by

$$P = |u|l/D \quad (2)$$

where $|u|$ is the magnitude of the average solution velocity, and l is a characteristic pore dimension. Thus diffusion is the dominant mixing mechanism provided that

$$|u|l < D \quad (3)$$

Now we may use Darcy's law to substitute for $|u|$ so that condition (3) becomes

$$(Kl|\nabla\Phi|)/\theta < D \quad (4)$$

where K is the hydraulic conductivity of the soil, Θ its volumetric moisture content, and $|\nabla\Phi|$ is the magnitude of the gradient of hydraulic potential.

Furthermore, for inert solutions the hydraulic conductivity of a porous medium is proportional to the square of its characteristic pore size. That is

$$K = \alpha l^2 \quad (5)$$

where α is a proportionality constant. Substituting now for l in condition (4), and rearranging we have

$$K < \alpha^{1/3}(\Theta D/|\nabla\Phi|)^{2/3} \quad (6)$$

An estimate of l can be made from the matric potential ψ_{pd} at which significant drainage of a soil occurs, using

$$\psi_{pd} = -2\sigma/\rho g l \quad (7)$$

where σ is the surface tension of water, ρ is the density of liquid water, and g is the acceleration due to gravity. Using tabulated values for the constants, and moisture characteristics and conductivity data for a number of soils (such as the data of Gardner 1960) in equations (5) and (7), $\alpha^{1/3}$ is found to differ little from the value 20 $m^{-1/3}$. Typically Θ is about 10^{-1}, and D is of order 10^{-9} m^2s^{-1} for salts in aqueous solution. $|\nabla\Phi|$ usually exceeds unity in unsaturated soil near roots, or near a wetting front during infiltration, but above a water table observations support a characteristic order-of-magnitude of 1. Below a water table, smaller values of $|\nabla\Phi|$ are normal, and the order-of-magnitude value 10^{-2} is reasonable.

Substituting these numerical values in condition (6) indicates that predominantly diffusive mixing of salts can be expected in unsaturated media for $K < 5 \times 10^{-6}$ ms^{-1}, or in saturated media for $K < 10^{-4}$ ms^{-1}. These are only order-of-magnitude estimates, but they suggest that molecular diffusion is the dominant mixing process for inert (non-adsorbed) solutes during water movement other than infiltration and the early stages of redistribution in all but highly permeable soils. (Note that $K = 5 \times 10^{-6}$ ms^{-1} is a very high value for the hydraulic conductivity of an unsaturated soil.)

Based on experimental data for a number of soils, Todd and Kemper (1972) have concluded that diffusion is dominant in solute dispersion for average solution velocities less than about 10^{-7} ms^{-1} which provides general support to the above estimates.

The range of solute velocities in non-uniform or highly structured soils presents particular problems which have received little attention. Passioura (1971), and Passioura and Rose (1971) have considered the relevant problem of solute dispersion in aggregated media where viscous flow is assumed confined to the macropore space, and transport within aggregates occurs by molecular diffusion. The structure of these materials enhances the role of convection in solute dispersion.

Ions which are adsorbed or involved in exchange reactions within a soil are transported at less than the average velocity of the soil solution. Biggar and Nielsen (1967) show examples from their experiments. Gardner (1965) indicates how the ratio of the velocities depends on the adsorption characteristics in a simplified model. It should be noted that although the magnitudes of solute and solution velocity may differ, their directions remain identical.

In summary, (i) transport of salts in soils occurs by convection in the soil water so that a knowledge of the water movement is essential to understanding or predicting salt movement; and (ii) in many practical situations the decay of salt concentration gradients is likely to be a result of molecular diffusion alone.

DISTRIBUTION OF SALTS IN SOILS

At any given time the spatial distribution of salt in a mass of soil reflects the pattern of water movement from sources to sinks and the distribution of sources and sinks of the dissolved material. Bresler (1972) has reviewed methods which may be applied to predict transient salt profiles in simple situations. However certain patterns of salt distribution occur frequently, and their interpretation is discussed in this section.

Salt distribution in bare soil with a shallow water table

Soil salinity is very often associated with a permanent saline water table within about 2 m of the surface. The distribution of salts in soil above the water table may vary considerably with season (Jackson, Blackburn and Clarke 1956), but after an appreciable period without rain it is usual to find an accumulation of salt on the surface with concentration decaying over a distance of order 0·1 m to that of the underlying water.

A classic problem in soil-water movement is the steady evaporation from a shallow water table which remains at a sensibly constant depth by groundwater movement. The case of a non-uniform soil profile has been analyzed (Hadas and Hillel 1972), but for simplicity the uniform soil problem first investigated by Philip (1957) will be considered here. See Hillel (1971) for a general review of this problem.

If z is defined as positive downwards, and $\psi_p = o$ at $z = z_0$ (the water table), then rearrangement and integration of Darcy's law for one-dimensional vertical flow gives

$$z_o - z_* = \int_{\psi_{p*}}^{o} \frac{d\psi_p}{1 + (E/K)} \qquad (8)$$

E is the steady evaporation, and $K(\psi_p)$ the hydraulic conductivity-matric potential characteristic of the soil. Equation (8) allows the calculation of the height $z_0 - z_*$ above the water table at which a chosen capillary potential ψ_{p*} will be found for any chosen value of $E > 0$.

Since K decreases rapidly with increasing magnitude of ψ_p, there is a maximum rate of evaporation which can be sustained from a water table at any given depth. Gardner (1958) used an empirical equation to approximate the $K(\psi_p)$ characteristic, and obtained an expression of the form

$$E_{max} = C/Z^n \tag{9}$$

where Z is the depth of the permanent water table and C and n are constants. Soil characteristics affect the empirical factor n which ranges upwards from about 2. In some uniform sands values of more than 10 have been reported (Anat, Duke and Corey 1965).

Although it is only an approximation, equation (9) indicates that the maximum rate of sustained evaporation from a water table increases quite rapidly as the water table is brought closer to the soil surface. Finally, with very shallow water table, atmospheric conditions rather than transport in the soil will limit the evaporation rate.

The upward movement of water which is subsequently evaporated results in an accumulation of salts at the soil surface at the rate cE where c is the concentration of solutes in the soil water. This accumulation may be subsequently leached downwards by rain, but it may firstly reach a level sufficient to affect plant growth. Jackson, Blackburn and Clarke (1956) report observations from South Australia where the salinity of the top 25 mm of soil varied from 0.019 per cent in winter to 3.2 per cent in summer.

Talsma (1963) has considered the effect of water table depth on surface accumulation of salts in the Murrumbidgee Irrigation Areas. In the context of irrigation, he considered a critical depth to avoid salinity problems was that which permitted an E_{max} of no more than 1 mm day^{-1}. He determined critical water table depths for ten soils, and found values ranging from 0·8 to 6.6 m. The latter is exceptional as the next largest depth was 2·01 m.

Over a very long period of time, much smaller evaporation rates, corresponding to deeper water tables, may be important in the development of a natural distribution of salts in the soil profile.

In their study of non-uniform soil profiles Hadas and Hillel (1972) showed that layering can greatly reduce the steady evaporation rate, especially when the order is coarse textured soil at the surface and finer material below. Their analysis could be applied to investigate upward movement of confined groundwater through a layer of low hydraulic conductivity, which is the situation found by Smith (1962, 1966) and Bettenay, Blackmore and Hingston (1964) in studies of secondary salinity of valleys in Western Australia.

Once there is an enhanced concentration of salts at the soil surface due to evaporation, a downward diffusive transport of salt opposes the upward convective flux. In the simplest such situation an equilibrium results, which has been investigated by Gardner (1965) who showed that the salt profile is described by

$$c = c_o \exp(-Ez/D). \tag{10}$$

c is salt concentration in the soil water at depth z (measured positive downwards from the soil surface), c_o that at the soil surface, and D is the diffusion coefficient, or dispersion coefficient if the influence of convection is significant.

Gardner (1965) found that equation (10) matched experimental data for NO_3 distribution in the top 0·1-0·2 m of soil after a rainless period with no irrigation. Very similar near surface Cl distributions have been reported by Smith (1962), Talsma (1963), and Bettenay, Blackmore and Hingston (1964).

Salt distribution in bare soil with a deep water table

In some situations secondary salinity has been observed in areas where there is no evidence of a shallow water table (Teakle and Burvill 1938; Smith 1962). Then upward movement, if any, of saline groundwater will be extremely slow, and other characteristics are responsible for the accumulation of salt.

This problem has not been studied in detail, but it appears that salt is merely redistributed within the top 1 m of soil during the year, with little downward loss (Smith 1962). The chloride ion at least is known to move with soil water at relatively low moisture contents (Marshall and Gurr 1954). The explanation of this salinity problem lies in local factors (the rainfall and evaporation patterns, and water transporting properties of the soil profile) which limit penetration of the annual wetting front so that essentially all rain is subsequently evaporated.

As shown by Smith (1962), after a rain-free period salt is concentrated at the soil surface, and the salt profile is of the form predicted by equation (10).

Salt distribution in the plant root zone

The uptake of soil water by plant roots which exclude solutes must result in an enhanced solute concentration within the soil profile. This problem has been considered by Gardner (1967) who assumed a steady state situation with salts becoming more concentrated as the soil solution moved down through the soil profile. If water with average salinity c_o is supplied to the soil surface at the average rate P (for instance by rain containing appreciable dissolved salts), then the equilibrium salt concentration profile is c(z) given by

$$c(z) = Pc_o/(P - \int_0^z w dz). \qquad (11)$$

w(z) is the rate of water uptake by roots per unit volume of soil at depth z.

Salt profiles of the general form predicted by equation (11) have been reported by Bettenay, Blackmore and Hingston (1964), but they have not been analyzed in detail. The author's unpublished data from forested sites with deep water tables in Western Australia show a peak of salt concentration in the soil water about 5 m below the surface. At greater depths there is a marked reduction of concentration as shown (fig. 1). This decay of concentration cannot be accounted for by equation (11), which suggests that further developments are needed in the analysis of salt profiles.

Just as evaporation from bare soil can be maintained at a steady rate by transport of water upwards from a fixed water table, so can a steady state of evapotranspiration be maintained. Whisler, Klute and Millington (1968) have analyzed this problem in which water is removed from the soil at different depths due to the functioning of plant roots. If the soil water contains solutes which are rejected by the roots, then there will be enhancement of concentration, but this problem has not been analyzed to determine the form of the resultant salt profile.

Areal distribution of salinity

As a general rule, groundwater investigations show that water table topography is similar to that of the overlying land surface. An inevitable consequence is that groundwater flows from upland areas towards valley floors, and downslope in valleys. This movement of water carries solutes with it, as discussed above. They become concentrated in the soil water by evaporation from the soil surface, and by exclusion during water uptake by plant roots.

When the salts are continuously deposited on the soil surface at a known rate, and some other simplifying conditions are met, the areal distribution of groundwater salinity can be related to the distribution of the groundwater recharge (or discharge) rate (Eriksson and Khunakasem 1969).

Observations of groundwaters in farmed hillsides and valleys have been reported by Smith (1962) and Bettenay, Blackmore and Hingston (1964). These data are

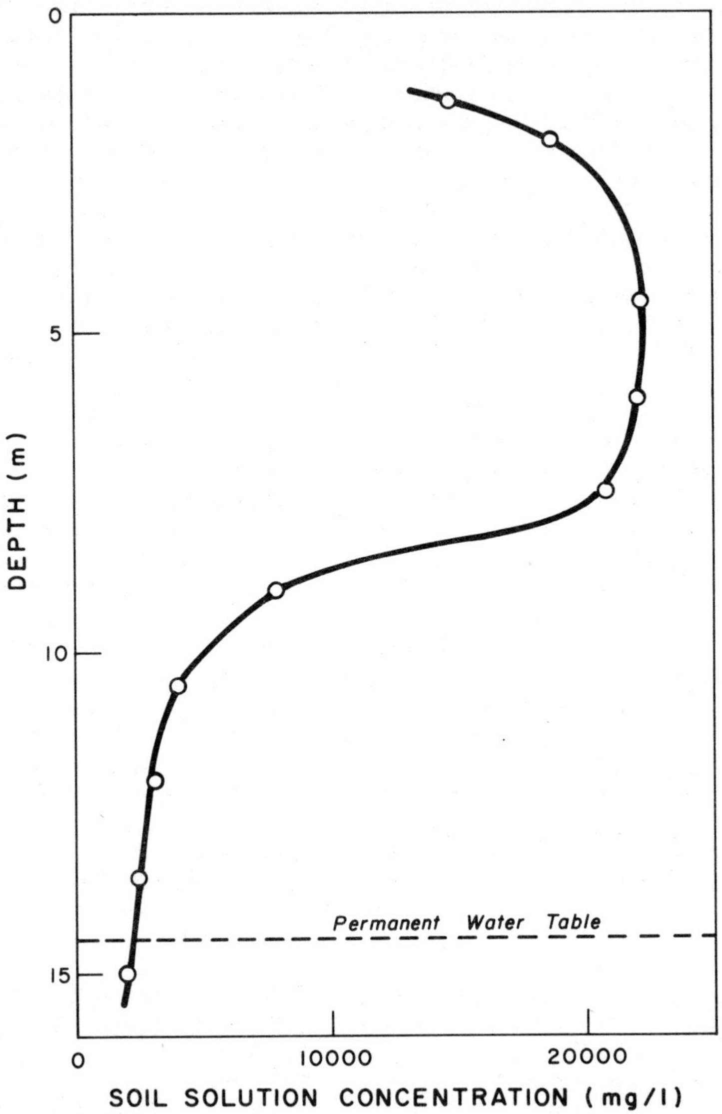

Fig. 1. Salt profile above a deep water table at a forest site east of Collie, Western Australia.

characterized by an increase of salinity downslope which suggests that recharge may be greatest in the uplands, and the valley floors may be areas of groundwater discharge. It should be noted, however, that discharge areas are not restricted to valleys. Freeze and Witherspoon (1967, 1968) relate several water table configurations and geologic settings to groundwater discharge, and Toth (1971) has provided a general discussion of discharge phenomena.

Clearly the potential for secondary salinity will be greatest in areas of discharge where there is the greatest natural salinity of groundwater.

DEVELOPMENT OF SECONDARY SALINITY

Descriptions of the development of dryland secondary salinity have been provided by

Wood (1924), Teakle and Burvill (1938, 1945), Burvill (1947), Cope (1958), Smith (1962), Bettenay, Blackmore and Hingston (1964), and Clark (1971). Some hydrologic aspects of the problem have been discussed by Holmes (1971). However it is appropriate to consider a synthesis of the mechanisms discussed in previous sections. This provides a contemporary hypothesis for secondary salinity in Australia. Catchment-wide observations to test this hypothesis are being made by C.S.I.R.O. staff, but results will not be available for several years.

Before native vegetation is replaced by agricultural crops and pastures, processes of salt and water transport in the landscape are near equilibrum. Over many years salts deposited in rain have accumulated within the soil-water system until the rate of loss of salts in streamflow and subsurface seepage from an area closely matches the average input rate. The greater time taken to accumulate or lose salt in more arid regions makes an equilibrium less likely due to the lagged effect of past environmental disturbance, such as climatic change.

In upland areas, under native vegetation salinity increases with depth in a near-surface region as a consequence of the uptake of water by roots. There is some small average yearly recharge, and groundwater of reasonable quality may be found. Seasonal changes in the salinity profile near the soil surface reflect the propagation into the soil of fluctuating surface moisture conditions caused by the local rainfall and evaporation pattern.

Similarly the salinity profile near the surface fluctuates during the year at lowland sites. Under native vegetation the water table is usually quite deep, but it probably provides water to plant roots in excess of local rainfall, and much higher salinities are found here than in upland groundwaters.

Figure 2(*a*) is a schematic valley cross-section illustrating equilibrium distributions of water and solutes. Liquid water may be lost down the valley by groundwater seepage, or in an ephemeral flow over the soil surface. Either route will export soluble salts. Even when the water table is several metres down, some salts reach the soil surface by uptake in the transpiration stream of deep rooted plants, followed by leaching of their aerial parts and plant debris.

Where the valley drains a sufficiently large area, even though the (space-time) average recharge rate under the native vegetation is quite small, the groundwater table is often so close to the soil surface that salts accumulate there. These are the permanently saline valleys and playa lakes found in some areas. Elsewhere saline groundwater discharges directly into streams with little evidence of soil salinity in the undisturbed environment.

Replacement of the natural perennial vegetation by agricultural crops and pastures, which are usually annuals and relatively shallow rooted, significantly reduces the loss of water by evapotranspiration. A transient situation ensues as the environment moves towards a new hydrologic equilibrium. The reduced loss of water increases groundwater recharge, and modifies salt profiles in recharge areas.

Valley sites which had discharged groundwater through deep-rooted vegetation may become temporary recharge areas with appropriately modified salinity profiles.

Following clearing of the native vegetation, the water table begins to rise. Depending on the initial depth, and the patterns of recharge and groundwater movement, the rising water table phase may continue for many years. An increased seepage flow through the greater area of water-saturated material may be detectable.

The rising water table may intersect the soil surface resulting in the development of a seepage area or spring. In some situations where the increased recharge is separated from saline groundwater by soil profile characteristics, or simple density stratification, these seeps may be initially of relatively low salinity. However their salinity generally increases due to mixing processes within the soil. Evaporation increases the salinity of seepage water to a value which can be much greater than that of the underlying groundwater.

Near seepage areas, and wherever the water table is less than about 2 m deep, there can be an appreciable upward solution transport to the soil surface where water is evaporated and salts are deposited in rainless months. Figure 2(*b*) illustrates the situation. Depending on local factors, these areas may have the salt accumulations leached by rain, or they may be wet-weather seepage areas.

The increased concentration of salts at the soil surface will inhibit plant growth. Furthermore, the increased SAR will increase sodicity in seepage and nearby areas, so that the soil will tend to disperse when leached by rain.

Due to the combined effects of a high water table, reduced vegetation cover, and dispersion of bare sodic soils, secondary salinity patches are serious erosion hazards when

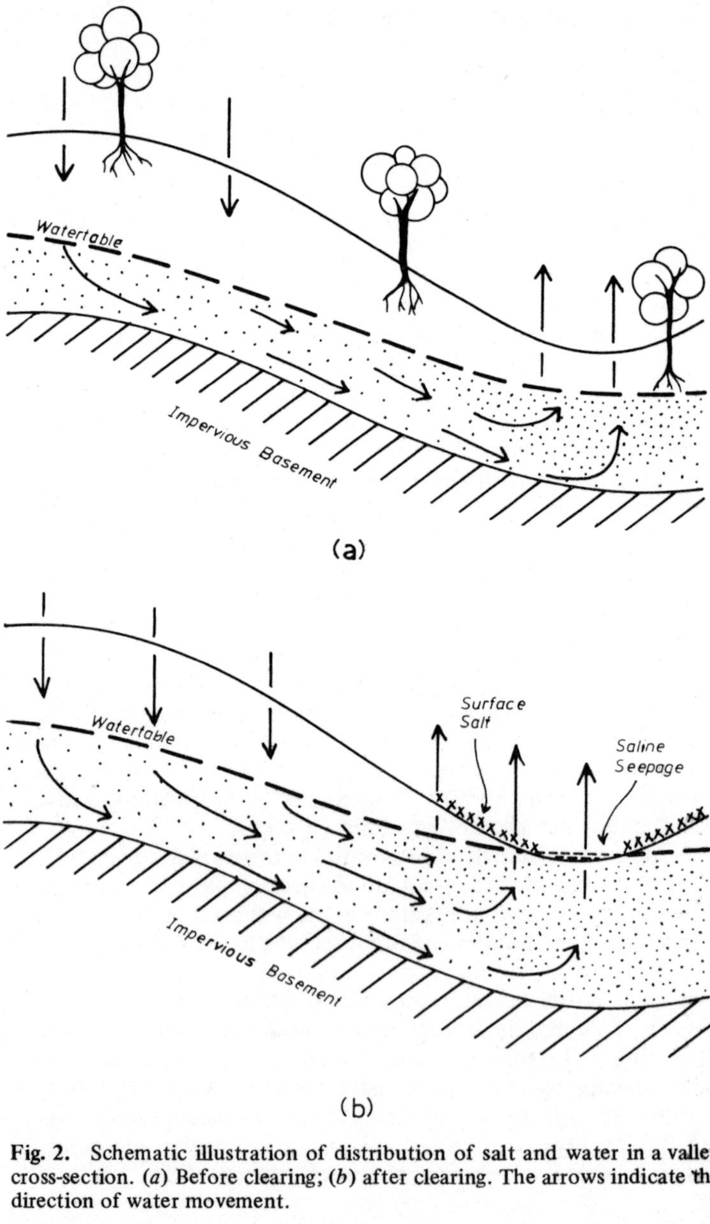

Fig. 2. Schematic illustration of distribution of salt and water in a valley cross-section. (*a*) Before clearing; (*b*) after clearing. The arrows indicate the direction of water movement.

rain falls. These same factors increase the probability of flooding, which is in any case usually enhanced by the changed land use in agricultural development.

The output of salt in seepage areas ultimately finds its way to streams or rivers, and the development of secondary salinity is associated with a net loss of salt from the landscape. Although streamflow may be increased by the removal of native vegetation, there is evidence of an even greater removal of salt in the streams since stream salinity increases (fig. 3).

The extension of areas of secondary salinity may continue for many years due to further clearing in hydrologically related parts of the landscape. Even without continued clearing, the water table may rise somewhat intermittently towards a new equilibrium, due to variations of rainfall from year to year. Abnormally high rainfall will lead to a marked increase in water table level, and extension of areas of secondary salinity, particularly in land with little relief. Water tables will fall due to reduced recharge in years of subnormal rainfall, and this will reduce the rate of accumulation of salt at the soil surface. However there may be little reduction of the saline area since this requires a leaching excess of rain.

In time, the area affected by secondary salinity will reach a maximum, and subsequently the average rate of loss of salt from the landscape will decrease until, with continuation of the present land use and climate, an equilibrium will be re-established with the average rate of atmospheric salt deposition. As groundwater salinity decreases, there may be increased plant water use, particularly around discharge sites, and some local decrease of water table levels may occur. However the undisturbed original levels are unlikely to be re-established under these circumstances.

At the new equilibrium the greater discharge of groundwater from well-drained farmed areas will result in stream salinities slightly lower than those in the undisturbed landscape. In the poorly drained landscape characteristic of semi-arid regions in Australia, however, the elevated water table after clearing will cause an extension of the playas and saline valleys which may continue for hundreds or thousands of years, or be permanent.

RECLAMATION OF SECONDARY SALINITY

With the objective of renewed production, there are broadly two approaches to the reclamation of land affected by secondary salinity. The first is to introduce productive halophytic plants on salt patches, and the second is to modify the local hydrological regime in such a way that soil salinity is reduced to a tolerable level for crops or pastures.

In some circumstances, however, renewed agricultural production may not be the primary object of reclamation. For example, the value of high quality streamflow may be such that economic return from vegetation is only a secondary consideration. Such conflicts of multiple land use are arising in Western Australia. It has been estimated that in the South-West Coast drainage division, which includes much of the state's agricultural land, not less than 47 per cent of the total streamflow is saline (Australian Water Resources Council 1965). There is evidence too that the salinity of many streams in this region has been increased by traditional farming practices (fig. 3, and Peck and Hurle 1973). Accordingly this discussion is not restricted to reclamation methods which are economic in agriculture.

Use of salt tolerant plants

The growth of halophytes on salt patches is valuable for several reasons: cosmetic, erosion control, amelioration and economic production. In many cases this will be the

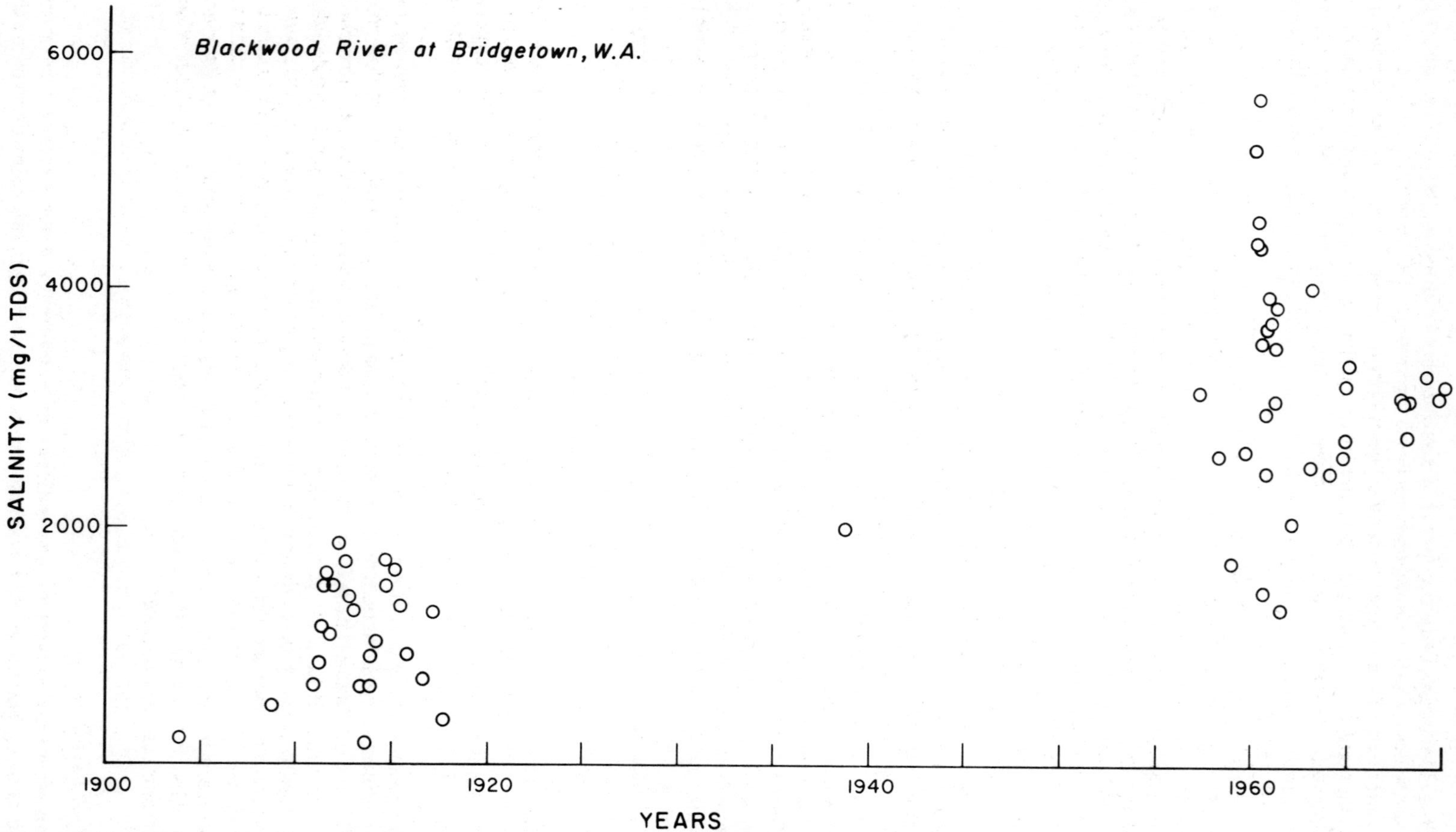

Fig. 3. Historic variation of salinity of the Blackwood River at Bridgetown, Western Australia, attributed to agricultural development in the catchment area. Data are from Wood (1924), Western Australian Government Railways Department, and Western Australian Department of Agriculture.

simplest and most generally successful method for handling a salinity problem, and it is recommended by Australian Departments of Agriculture (Wagner 1957; Cope 1958; Smith 1961; Matheson 1968; Malcolm 1969). These articles stress the importance of cultivation, fertilization and grazing control in establishing successful plant cover.

Halophytes bring about significant changes to the salinity profile relative to that in bare soil (Smith 1962). This is due to the extraction of water from the whole root zone rather than just at the soil suface. Once the surface salinity is reduced, less tolerant species may become established.

Where the accumulation of salts at the soil surface is not associated with a shallow water table, but reflects a redistribution of salt following removal of native vegetation, the initial use of halophytes may be the only practical method for reclamation. These plants may be subsequently replaced by crop or pasture species which are able to tolerate the lower concentrations resulting when salts are distributed through a rooting zone.

Cultivation and mulching

Cultivation is a simple treatment which is known to affect significantly the salinity of at least some surface soils (Smith 1962). The physical disruption of perhaps the top 0.1 m of soil may have two important effects on the local water balance: increased infiltration where this is otherwise restricted by a surface crust of soil with low hydraulic conductivity, and a reduced steady rate of evaporation where a shallow water table exists. The latter effect is in keeping with theory for a non-uniform soil profile as discussed above.

Smith (1962) also reports experiments with a sand mulch 0.05 m deep. This appears to be effective both in increasing leaching of salt downwards during winter and in reducing evaporation and upward movement of salt in summer. Since its hydraulic properties are relatively stable, a sand mulch could be expected to provide a more permanent effect than does cultivation.

It will be apparent that cultivation or a sand mulch affects neither infiltration nor evaporation where saline water seeps from the soil surface.

Since most occurrences of secondary salinity are associated with a shallow water table, a treatment which increases infiltration and reduces evaporation may, therefore, merely redistribute salt in a local area. It is conceivable that a net increase in the area of saline land could result from cultivation or mulching, so the use of these treatments alone should be considered carefully.

Use of drainage

Surface drainage to prevent run-off water from other areas contributing to infiltration on a salt patch may be effective in reducing local water tables in some cases. However, studies in several saline valleys in Western Australia (Smith 1962, 1966; Bettenay, Blackmore and Hingston 1964) show that the groundwater is often confined and under some pressure. In such cases the confining layer will restrict penetration of surface water just as it restricts the rate of upward movement of groundwater, and so surface drainage will be relatively ineffective. It is recommended by the Western Australia Department of Agriculture, however, because prolonged flooding may kill plants and thereby increase surface salinity by a redistribution within the soil.

A conflict arises over handling of surface waters on areas other than salt patches. Malcolm (1961) and the Western Australian Department of Agriculture (1962) have emphasized the importance of contour banks and pasture improvement to retain water in the soil where it falls, and so to minimize run-off and flooding. However enhanced infiltration must tend to increase groundwater recharge, particularly where surface water accumulates, as it may above a contour bank. Any increased recharge will raise water tables and accentuate secondary salinity where the dominant flux of water is not downwards.

The established technique for controlling salinity of irrigated land is to provide some form of underground drainage which lowers the water table, and so reduces the rate of accumulation of salts at the surface in hazardous areas. Details of the technology are described in Luthin (1957). Subsurface drainage for reclamation of secondary salinity in dryland agriculture is not recommended by Australian Departments of Agriculture because of installation costs relative to the value of increased production. In general this is undoubtedly a valid argument but situations are conceivable where subsurface drainage could be economic, and it should not be excluded from consideration as a possible method for the reclamation of a given saline area.

Hydrological management

The reclamation methods mentioned above can have little, if any, effect on the fundamental cause of secondary salinity in dryland agriculture, which is the reduced water usage of the introduced vegetation. Indeed, treatment such as subsurface drainage may alleviate soil salinity at the expense of increased salinity of streams and rivers fed by the drainage waters, and this may be quite unacceptable.

Clearly the most generally satisfactory method for prevention or reclamation of secondary salinity is to develop land management practices which result in a catchment water balance as near as possible to that of the pristine environment. This appears to be feasible since the increased recharge responsible for the problem in semi-arid areas is small relative to the local rates of potential evaporation. (See table 2, and Australian Water Resources Council 1968).

In some areas of Western Australia native vegetation has been left on uplands, and elsewhere a strip along the valley has remained uncleared. These are attempts, based on intuition, to maintain recharge and discharge rates respectively. However there are no data on suitable control areas, or their optimal location, and the development of saline areas does not appear to have been affected. More attention could be given to the use of native vegetation for groundwater control, but Smith (1966) considers that at least 40 per cent of the land would need to be left uncleared, and the loss of agricultural production would then exceed that from the much smaller areas which are now saline.

A management procedure to minimize secondary salinity, which is often emphasized in Australia, is to maintain a plant cover whenever possible (Teakle and Burvill 1945; Smith 1962). Apart from the direct effect of plants on salinity profiles in saline areas, plant water consumption generally exceeds evaporation from bare soil so that recharge over the catchment will be minimized by transpiring plants. Secondary salinity problems in the northern United States are attributed to the increased groundwater recharge under land kept in fallow for one or two summers (Clark 1971).

No serious attention appears to have been given in Australia to selection of plant varieties, and development of management procedures for increased water use, although it is generally acknowledged that this is a desirable objective. There is today a fairly detailed basic understanding of factors affecting plant water use (Pierre et al. 1966) which should be applicable. It is known, for example, that nutrient supply affects evapotranspiration (Viets 1962). Brown (1971) found that nitrogen application at the rate 268 kg/ha increased total water use of a wheat crop in the United States by 94 mm. This is more than the increased recharge estimated for semi-arid farmed catchments in Western Australia (see table 2), and suggests that suitable fertilizer applications may make a significant contribution to salinity reclamation.

Investigations of prevention and reclamation of secondary salinity in the United States are centred on the use of continuous cropping and alternative crops to increase evapotranspiration.

There would seem to be a very good possibility that crop and pasture varieties could be selected, and farm management procedures developed which would increase plant water use and thereby reduce groundwater recharge in the Australian environment. Natural drainage would then lead to a general decay of water table levels, and leaching of salt-affected areas would begin. The time scale for reclamation would depend on local factors, such as the rate of decay of the water table, and the excess of winter rain available for leaching, and may be several years. Chemical amendment such as gypsum may be needed to aid the reclamation of soils which have become sodic.

SUMMARY

Areas of secondary salinity associated with dryland agriculture have developed in all mainland states of southern Australia, particularly in Western Australia. Similar, but less extensive, problems have arisen in some other countries.

In Western Australia, and probably in all southern states, deposition of salt from the atmosphere is the major source. Salt has accumulated in the soil-water system, until the rate of deposition of salt on undisturbed land is very nearly equal to the rate of loss in solution in streamflow. However, replacement of the native vegetation by agricultural crops or pastures disturbs this system, and accelerates the rate of salt loss.

Available data show that the spatial distribution of salt in the landscape conforms to classic patterns. Groundwater salinity increases from upland (recharge) to lowland (discharge) areas and down valleys. Above a shallow water table the salt profile reflects upward movement of water, and evaporation from the soil surface. However, current theory does not account for the salt profile in forested sites.

Salt-tolerant plants can provide some control of, and production from, saline areas, but reclamation by natural leaching under the present hydrologic regime may take several hundred years. The key to rapid and effective reclamation of secondary salinity in most areas is to lower the water table in discharge areas. This can be achieved by reducing groundwater recharge over a catchment, or by artificial drainage.

Drainage is often feasible in irrigation areas, but it is generally uneconomic in dryland agriculture. While drainage can be technically effective in alleviating problems of soil salinity, it fails to improve stream salinity, and this will be important where good quality water is sparse.

Farm management practices can reduce groundwater recharge, and thereby affect secondary salinity. Fertilizer applications to increase crop water use, alternate rotations, and alternate crops are methods being investigated for salinity reclamation in the United States. The amounts of water responsible for secondary salinity in Western Australia appear to be within the range where fertilizers, or different varieties of crop and pasture species, may have important effects on the development and reclamation of secondary salinity.

ACKNOWLEDGMENTS

I am grateful to the Western Australia Department of Agriculture and the Western Australian Government Railways Department who made available data included in figure 3, and to colleagues in C.S.I.R.O. for their critical comments.

REFERENCES

Allison, L.E. (1964). *Adv. Agron.* **16**, 139-80.

Anat, A., Duke, H.R., and Corey, A.T. (1965). Hydrology Paper No. 7, Colorado State Univ., Fort Collins.
Anderson, V.G. (1945) *Aust. Chem. Inst. J. and Proc.* 12, 41-68 and 83-98.
Australian Water Resources Council (1965). Review of Australia's water resources (stream flow and underground resources) 1963. Department of National Development, Canberra.
———. (1968). Review of Australia's water resources: Monthly rainfall and evaporation. Bureau of Meteorology, Melbourne.
Baldwin, A.D. (1971). *Water Resour. Res.* 7, 1007-12.
Ballantyne, A.K. (1963). *Can. J. Soil Sci.* 43, 52-8.
Bernstein, L. (1962). *UNESCO Arid Zone Research* 18, 139-74.
Bettenay, E., Blackmore, A.V., and Hingston, F.J. (1964). *Aust. J. Soil Res* 2, 187-210.
Biggar, J.W., and Nielsen, D.R. (1967). In "Irrigation of Agricultural Lands". (Eds. R.M. Hagan, H.R. Haise and T.W. Edminster). pp. 254-74. Amer. Soc. Agron., Madison.
Boughton, W.C. (1970). Report No. 120, Water Research Laboratory. Univ. N.S.W., Manly Vale.
Bresler, E. (1972). In "Optimizing the Soil Physical Environment towards Greater Crop Yields". (Ed. D. Hillel). pp. 101-32. Academic Press, New York.
Brown, P. (1971). *Agron. J.* 63, 43-6.
Burvill, G.H. (1947). *J. Aust. Inst. Agric. Sci.* 13, 9-19.
Claridge, G.G.C. (1970). Studies in Hydrol. No. 12 (UNESCO). Results of Res. on Rep. and Exp. Basins: Proc. of Wellington Symp., p. 523.
Clark, C. (1971). In "Proceedings Saline Seep-Fallow Workshop". Cooperative Extension Service, Bozeman.
Cleaves, E.T., Godfrey, A.E., and Bricker, O.P. (1970). *Geol. Soc. Amer. Bull.* 81, 3015-32.
Collis-George, N., and Evans, G.N. (1964). *Aust. J. Soil Res.* 2, 20-8.
Cope, F. (1958). Catchment Salting in Victoria. Soil Conserv. Auth. Vict., Melbourne.
Dagg, M., and Blackie, J.R. (1965) *Bull. Int. Assn. Sci. Hydrol.* 10, 63-75.
Douglas, I. (1968). *Earth Sci. J.* 2, 126-44.
Eriksson, E. (1955). *Tellus.* 7, 243-50.
———. (1960). *Tellus.* 12, 63-109.
Eriksson, E., and Khunakasem, V. (1969). *J. Hydrol.* 7, 178-97.
Freeze, R.A., and Witherspoon, P.A. (1967). *Water Resour. Res.* 3, 623-34.
———. (1968). *Water Resour. Res.* 4, 581-90.
Gardner, W.R. (1958). *Soil Sci.* 85, 228-32.
———. (1960). *Soil Sci.* 89, 63-73.
———. (1965). In "Soil Nitrogen". (Eds. W.V. Bartholomew and F.E. Clark). pp. 550-72. Amer. Soc. Agron., Madison.
———. (1967). Proc. Symp. "Isotope and Radiation Techniques in Soil Physics and Irrigation Studies", IAEA, Vienna, pp. 335-41.
Gorham, E. (1961). *Geol. Soc. Amer. Bull.* 72, 795-840.
Greenlee, G.M., Pawluk, S. and Bowser, W.E. (1968). *Can. J. Soil Sci.* 48, 65-75.
Hadas, A., and Hillel, D. (1972). *Soil Sci.* 113, 65-73.
Hillel, D. (1971). "Soil and Water: Physical Principles and Processes". Academic Press, New York.
Hingston, F.J. (1958). CSIRO Aust. Div. Soils Tech. Report 1/58.
Holmes, J.W. (1971). In "Salinity and Water Use". (Eds. T. Talsma and J.R. Philip). pp. 25-40. Macmillan, London.
Holmes, J.W., and Colville, J.S. (1968). *Trans 9th Int. Congr. Soil Sci.* 1, 39-46.
Hutton, J.T., and Leslie, T.I. (1958). *Aust. J. Agric. Res.* 9, 492-507.
Jackson, E.A., Blackburn, G., and Clarke, A.R.P. (1956). *Aust. J. Agric. Res.* 7, 20-44.
Jones, R.M. (1969). *J. Soil Conserv. Serv. N.S.W.* 25, 105-19.
Juang, F.H.T., and Johnson, N.M. (1967). *J. Geophys. Res.* 72(22), 5641-47.
Lightfoot, L.C., Smith, S.T., and Malcolm, C.V. (1964). *J. Dep. Agric. West. Aust.* 5, 396-410.
Logan, J.M. (1958). *J. Soil. Conserv. Serv. N.S.W.* 14, 229-42.
Luthin, J.N. (Ed.) (1957). "Drainage of Agricultural Lands". Amer. Soc. Agron., Madison.
Malcolm, C.V. (1961). *J. Dep. Agric. West. Aust.* 2, 515-8.
———. (1969). *J. Aust. Inst. Agric. Sci.* 35, 38-49.
Marshall, T.J., and Gurr, C.G. (1954). *Soil Sci.* 77, 147-52.
Matheson, W.E. (1968). *J. Dep. Agric. S. Aust.* 71, 266-72.
Miller, M.R. (1971). In "Proceedings Saline Seep-Fallow Workshop". Cooperative Extension Service, Bozeman.
Northcote, K.H., and Skene, J.K.M. (1972). CSIRO Aust. Soil Publ. No. 27.
Passioura, J.B. (1971). *Soil Sci.* 111, 339-44.
Passioura, J.B., and Rose, D.A. (1971). *Soil Sci.* 111, 345-51.
Peck, A.J. (1971). In "Salinity and Water Use". (Eds. T. Talsma and J.R. Philip). pp. 109-23. Macmillan, London.
Peck, A.J., and Hurle, D.H. (1973). *Water Resour. Res.* 9, 648-57.

Philip, J.R. (1957). *Congr. Irrig. Drain. 3* 8, 125-54.
———. (1969). In "Ciba Foundation Symposium on Circulatory and Respiratory Mass Transport". (Eds. G.E.W. Wolstenholme and J. Knight). Churchill, London.
———. (1970). *Ann. Rev. Fluid Mech.* 2, 177-204.
Pierre, W.H., Kirkham, D., Pesek, J., and Shaw, R. (Eds.) (1966). "Plant Environment and Efficient Water Use". Amer. Soc. Agron., Madison.
Quirk, J.P. (1971). In "Salinity and Water Use". (Eds. T. Talsma and J.R. Philip). pp. 79-91. Macmillan, London.
Reeve, R.C., and Fireman, M. (1967). In "Irrigation of Agricultural Lands". (Eds. R.M. Hagan, H.R. Haise and T.W. Edminster). pp. 988-1008. Amer. Soc. Agron., Madison.
Richards, L.A. (Ed.) (1954). U.S. Dep. Agric. Handb. No. 60.
River Murray Commission (1970). Murray Valley Salinity Investigation. Vol. 1.
Robinson, J.B. (1971). In "Salinity and Water Use" (Eds. T. Talsma and J.R. Philip). pp. 193-206. Macmillan, London.
Saffman, P.G. (1960). *J. Fluid Mech.* 7, 194-208.
Scofield, C.S. (1940). *J. Agric. Res. (Washington)* 61, 17-39.
Smith, S.T. (1961). *J. Dep. Agric. West. Aust.* 2, 757-60.
———. (1962). M. Sc. Thesis, Univ. of Western Australia, Perth.
———. (1966). *J. Dep. Agric. West. Aust.* 7, 334-40.
Talsma, T. (1963). *Meded. Landbouwhoge-sch., Wageningen*, 63, 1-68.
Teakle, L.J.H., and Burvill, G.H. (1938). *J. Dep. Agric. West. Aust.* 15, 218-45.
———. (1945). *J. Dep. Agric. West. Aust.* 22, 87-93.
Todd, R.M., and Kemper, W.D. (1972). *Soil Sci. Soc. Amer. Proc.* 36, 539-43.
Toth, J. (1971). *Bull. Int. Assoc. Sci. Hydrol.* 16, 7-24.
Urie, D.H. (1971). *Water Resour. Res.* 7, 1497-510.
Viets, F.G. (1962). *Adv. Agron.* 14, 223-64.
Wagner, R. (1957). *J. Soil Conserv. Serv. N.S.W.* 13, 33-9.
Weller, W.K. (1926). *Aust. Assoc. Adv. Sci.* 18, 633.
Western Australia Department of Agriculture (1962). *J. Dep. Agric. West. Aust.* 3, 773-8.
Whisler, F.D., Klute, A., and Millington, R.J. (1968). *Soil Sci. Soc. Amer. Proc.* 32, 167-74.
Wood, W.E. (1924). *J. Roy. Soc. West. Aust.* 10, 35-47.

Index